PHARMACEUTICAL BIOASSAYS

PHARMACEUTICAL BIOASSAYS

Methods and Applications

Shiqi Peng

Ming Zhao

WILEY

A JOHN WILEY & SONS, INC., PUBLICATION

Published by John Wiley & Sons, Inc., Hoboken, New Jersey.
Published simultaneously in Canada.

Library of Congress Cataloging-in-Publication Data:

Pharmaceutical bioassays : methods and applications / Shiqi Peng . . . [et al.].
p. ; cm.
Includes bibliographical references.
ISBN 978-0-470-22760-2 (cloth)
1. Drugs--Testing. 2. Biological assay. I. Peng, Shiqi.
[DNLM: 1. Biological Assay--methods. 2. Pharmaceutical Preparations--analysis.
QV 771 P5357 2009]
RM301.27.P475 2009
615′.19--dc22

2009007433

Printed in the United States of America.

10 9 8 7 6 5 4 3 2 1

CONTENTS

PREFACE

Twenty years ago when I did my research work as a Humboldt Fellow in Prof. Dr. Dr. h.c. Winterfeldt's laboratory at Hannover University, I clearly perceived a trend that chemists were focusing attention on bioassays corresponding with states of disease free, disease onset, and disease progression, which trend pushed our synthetic chemistry to combine with pharmaceutical bioassay. Ten years ago when I was engaged as the dean of the College of Pharmaceutical Sciences in Peking University, pharmaceutical bioassay was considered as one of the platforms in our postgraduate education. Five years ago when I was engaged as the dean of the College of Pharmaceutical Sciences at Capital Medical University, pharmaceutical bioassay became a platform of chemistry, pharmacy, biology, and medicine and was extensively reflected in our undergraduate and postgraduate education.

The trend of chemistry, pharmacy, biology, and medicine to focus attention on bioassays corresponding with states of disease free, disease onset, and disease progression continues today and leads the discovery of efficacious new human and animal therapeutic agents, one of humanity's most vital tasks. The development of pharmaceutical bioassays is an enormous, demanding activity that requires creativity, a vast range of chemical, pharmaceutical, biological, and medical knowledge, as well as great persistence. Nowadays, no education and research would be complete without some exposure to the ways in which bioactive agents and new medicines are defined, discovered, and developed with the support of pharmaceutical bioassays. For those people interested in chemistry, pharmacy, biology, and medical sciences, such knowledge is arguably mandatory.

Pharmaceutical Bioassay: Methods and Applications represents a unique attempt to describe assays to evaluate the bioactivities of therapeutic moieties resulting from chemical, biological, and natural processes and to explore the mechanism of action of therapeutic moieties in the cells, tissues, and organs of living beings, as well as in living beings themselves. Such assays are highly important in the pharmaceutical industry and are the prime engine for the advancement of chemistry, pharmacy, biology, and medicine.

It would be impossible to be comprehensive in a book of this size, although hopefully there are no important omissions. I apologize that due to the rapid increase of new knowledge, I have had to omit some assays, and due to the limit of my appreciation or ability, some topics have not been treated with the depth that they may deserve. The emphasis, in general, has been on the procedures and applications of pharmaceutical bioassays. I hope that the selection and the arrangement of the material may be found satisfactory and useful to all those working in or entering this field.

I am grateful to the Beijing Municipal Commission of Education, which established the Beijing Area Major Laboratory of Peptide and Small Molecular Drugs for us, and also am grateful to the leaders of Capital Medical University, who provided me with excellent equipment, sufficient funds, and had high confidence in my research related to pharmaceutical bioassays; therefore, I hope that this book may be dedicated to the fiftieth anniversary of the founding of Capital Medical University. I am also indebted to my current co-workers and students who help to keep pharmaceutical bioassay alive for me with their enthusiasm for our research. Finally, I am especially grateful to my wife, who has not complained about the extra time I have needed to work during the past decades.

Shiqi Peng

Capital Medical University

CONTRIBUTORS

Shiqi Peng, Capital Medical University, Beijing, China

Ming Zhao, Capital Medical University, Beijing, China

Chunying Cui, Capital Medical University, Beijing, China

Guohui Cui, Capital Medical University, Beijing, China

1

METHODS AND APPLICATIONS OF ANTICANCER BIOASSAYS

Shiqi Peng

Cancer is a major human disease and causes great suffering and financial loss worldwide. Although a great deal of progress has been made in cancer chemotherapy, the incidence and mortality rate of most forms of cancer remain very high. Continuous efforts have been made in establishing efficient diagnostic methods and developing safe drugs or drug candidates to prevent and treat all human cancers. Extensive interest has been attracted to exploring pathologic mechanism-based assays as *in vitro* and *in vivo* evaluations for anticancer agents. *In vitro* assays usually consist of cell culture systems with neoplastic cell lines from human or other animal tumors as targets. The capacity of test compounds inhibiting the growth or reducing the survival of cancer cells in culture media and the potency of test compounds inducing structural change of cancer cells in culture media are generally correlated with the *in vivo* potency of a cancer therapeutic agent. To directly observe anticancer capacity, test compounds are administered to animals. Thus, great achievements in chemotherapy in recent years have been due to development of anticancer assays. In this chapter, 22 models used in anticancer research are described: 3-(4,5-dimethylthiazol-2-yl)-2,5-diphenyltetrazolium bromide (MTT) assay for six carcinoma cells,[1–12] flow cytometric assay for cell apoptosis,[11] DNA fragmentation assay,[11] Bcl-XL/BH3 interaction assay,[13] dissociation enhanced lanthanide fluoro-immunoassay (DELFIA),[14] Ishikawa cell and rat assay for detecting antiestrogens,[15–23] adenosine triphosphate (ATP) assay for eight cells,[24] alkaline phosphatase (AP) activity assay,[25,26] tumor

Pharmaceutical Bioassays: Methods and Applications. By Shiqi Peng and Ming Zhao

endothelial cell tube formation assay,[27] antiangiogenic assay,[28] *in vivo* hollow fiber assay,[29] VX2 rabbit lung assay,[30] insulin-like growth factor-I (IGF-I)-induced kinase receptor activation assay,[31] insulin-like gD· trkA-induced kinase receptor activation assay,[32] UV spectra-based calf thymus DNA intercalation assay,[33,34] fluorescence spectra-based calf thymus DNA intercalation assay,[33,34] P-glycoprotein pump (P-gp) in MCF-7R cells assay,[35] P-gp-related efflux carrier assay,[36] [^{3}H]substrate transport inhibition assay,[36] lactate dehydrogenase release assay,[36] functional assay of mitochondrial P-glycoprotein (P-gp),[37] and resistance index value assay.[38]

1.1 MTT ASSAY FOR SIX CARCINOMA CELLS[1–12]

Human gastric cancer cells (SGC-7901, SC-M1, and BGC-823), breast adenocarcinoma (MCF-7 and MDA-MB-231), stomach adenocarcinoma (AGS), colon carcinoma cells (HCT-116), and lung (A549 and NCI-H460) and central nervous system carcinoma cells (SF-268) were plated at 37°C for 24 h on 96-well plates at a density of 10^3 to 10^4 cells per well with Dulbecco's Modified Eagle's Medium (DMEM) : F-12 (1 : 1) with phenol red (Sigma, St. Louis, MO, USA) in a humidified atmosphere (5% CO_2). To this medium, L-Gln (4 mM), penicillin (200 units/mL), streptomycin

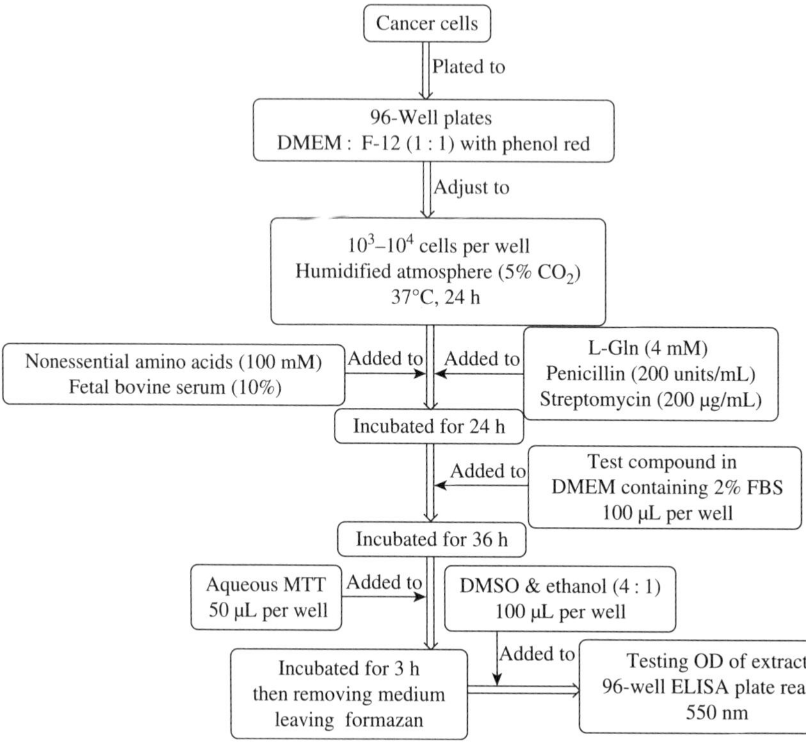

Figure 1.1 Flowchart of MTT assay for cancer cells.

(200 μg/mL), DMEM nonessential amino acids (100 mM), and 10% fetal bovine serum (FBS) were added and incubated for 24 h. Then the cells were treated with the test compound in a medium containing 2% FBS (100 μL per well) and incubated for an additional 36 h. To each well, 50 μL aqueous 3-[4,5-dimethylthiazol-2-yl]-2,5-diphenyl-tetrazolium bromide (MTT; 5 mg/mL) solution was added, and the mixture was incubated at 37°C for another 3 h. After removing the MTT solution, formazan was extracted from the cells in each well with 100 μL of a 4/1 dimethyl sulfoxide (DMSO)/ethanol mixture. Optical density (OD) was measured around 550 nm with a 96-well enzyme-linked immunosorbent assay (ELISA) plate reader. All assays were repeated three times. Viability of the cells treated with test compound was determined as: % Cell viability = [(average OD of test compound)/(average OD of control)] × 100. A general procedure is summarized in Fig. 1.1.

1.2 FLOW CYTOMETRIC ASSAY FOR CELL APOPTOSIS[11]

Trypsinized cancer cells (1×10^6 cells/mL) were successively washed twice in ice-cold Hanks solution, fixed in 75% ice-cold ethanol for at least 1 h, washed twice with phosphate-buffered saline (PBS), stained in 100 μg/mL propidium diiodide (PI, Sigma, St. Louis, MO, USA), 10 μg/mL Rnase in PBS, and incubated at 37°C for 30 min. After excitation of the fluorescent dye, fluorescence emitted from the propidium–DNA complex was quantitatively analyzed on the FACSCalibur system (BD Biosciences, San Jose, CA, USA).

1.3 DNA FRAGMENTATION ASSAY[11]

To identify DNA ladder using DNA fragmentation assay, the cancer cells (3×10^4 cells/mL) were plated into a 60-mm culture dish. The cells were treated with 50 μM and 100 μM quercetin, from which DNA samples can be extracted at 12, 24, and 48 h. Cells of different treatment groups were collected, washed once with ice-cold PBS, centrifuged, and the supernatants were removed carefully. The cells were dispersed in 50 μL lysis buffer [10 mM Tris pH 7.4, 100 mM NaCl, 25 mM ethylenediamine tetraacetic acid (EDTA), and 1% sarkosyl] and incubated at 50°C for 3 h. Mixing 4 μL DNA sample dye with bromophenol blue buffer, the solution was electrophoresed on 1.2% agarose gel at 50 V for 3 h until bromophenol blue dye reaching the foreland of the gel, and the DNA was visualized under UV light.

1.4 Bcl-XL/BH3 INTERACTION ASSAY[13]

In the presence of 50 μL test compound, His-tagged Basal cell lymphoma-extra large protein (His-Bcl-XL, 3 μM) was preincubated with 50 μL Ni^{2+}-sepharose magnetic beads. The total assay volume was 100 μL at 5% DMSO in buffer B (1 mM $MgCl_2$, 100 mM KCl, 10 mM Tris-HCl, pH 7.9) containing protease inhibitors and

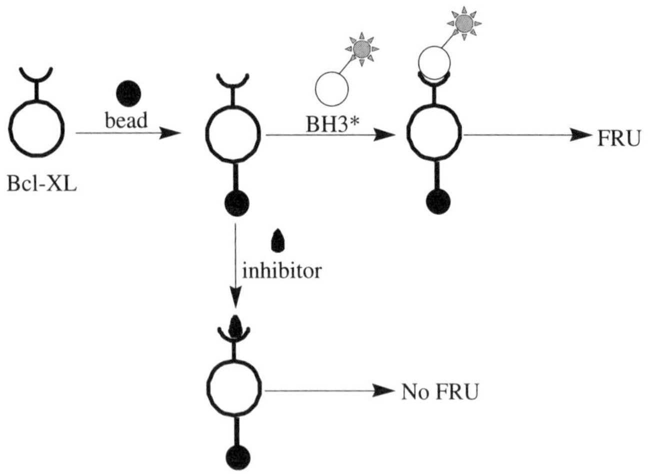

Figure 1.2 Principle of Bcl-XL/BH3 interaction assay.

was mixed for 1 h at room temperature. Subsequently, 50 μL fluoresceinated GQVGRQLAIIGADINR (BH3*; 6 μM) was added, followed by further incubation for 1 h in the dark. Reagents were removed, and the beads were washed with 150 μL 5 mM imidazole, 0.5 M NaCl, 20 mM Tris-HCl pH 7.9, 2 mM EDTA, and Proteases Inhibitors Mix (Sigma, St. Louis, MO, USA). Then the protein was eluted from beads with 0.5 M imidazole. Fluorescence was directly determined by measuring the relative fluorescence units (RFU) at 485-nm excitation and 530-nm emission (detector: Fusion; Packard Bio-Science, Perkin-Elmer, Inc., Boston, MA, USA). The assay was performed in 96 wells, blocked with bovine serum albumin (BSA) 3%. The protein:protein displacement was evaluated with the equation $\text{Displacement} = [1 - (\text{RFUs} - \text{RFUb})/(\text{RFUc} - \text{RFUb})] \times 100$, where RFUs, RFUb, and RFUc refer to the RFU in the presence of the tested samples, of the average of four wells without BH3*, and of the average of eight wells without added sample. Intra- and interexperiment variability was evaluated as standard deviation (SD) and variability coefficient (CV). The principle of this assay is represented in Fig. 1.2.

1.5 DISSOCIATION-ENHANCED LANTHANIDE FLUORO-IMMUNOASSAY (DELFIA)[14]

To each well of 96-well plates coated with streptavidin (Perkin-Elmer Inc., Boston, MA, USA), 100 μL of a variable concentration of biotin-labeled BH3 peptide, biotin-$(CH_2)_6$-GGGQVGRQLAIIGDDINR, was added (concentration, 2 to 40 nM). After 1 h (15 min to 1 h) incubation and three washing steps, the unbound biotin–BH3 peptide was removed. To each well, 89 μL anti-His Europium (Eu)-antibody conjugate (7.2 nM), 1 μL DMSO containing test compound, and 10 μL His6-Bcl-XL (or any other antiapoptotic His6-Bcl-2 family protein; concentration, 1–150 nM)

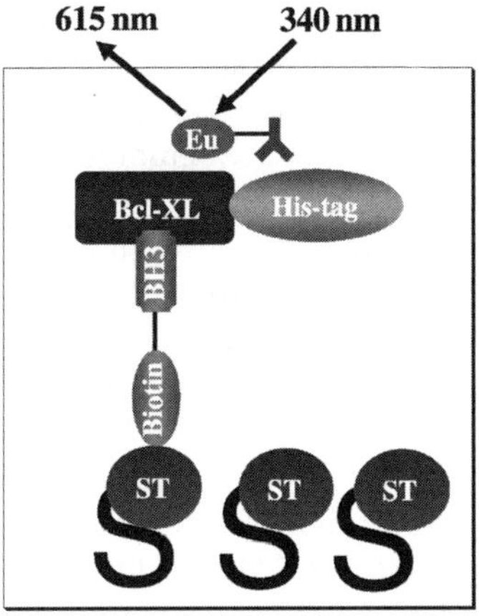

Figure 1.3 Representation for detecting BH3–Bcl-2 interactions with DELFIA.

were added. After 1 h incubation, the well was washed five times to remove the unbound protein (and so the Eu-antibody if displaced by test compound), then 200 μL enhancement solution was added. After 30 min incubation, the fluorescence was measured (excitation wavelength, 340 nm; emission wavelength, 615 nm). To give the relaxation properties of Eu, the measurements were made in time-resolved mode; controls included unlabeled peptide and blanks receiving no compounds; the DELFIA could also be obtained by using glutathione s-transferase (GST)-fusion Bcl-2 proteins and anti-GST Eu-antibody (Perkin-Elmer Inc., Boston, MA, USA); Z0 factor measurements were obtained by repeating the experiments (positive and negative controls) multiple times; and the assay buffer from Perkin-Elmer was used in each step. The representation of detecting BH3–Bcl-2 interactions with DELFIA is given in Fig. 1.3.

1.6 ISHIKAWA CELL AND RAT ASSAY FOR DETECTING ANTIESTROGENS[15–23]

Ishikawa cells (50,000 cells/well, in triplicate, Tsukuba University, Japan) were grown in 96-well plates in estrogen-free medium (phenol red free, with charcoal-stripped calf serum) in the presence of (stimulatory assay) test compounds as well as antiestrogens at concentrations that were varied over several log orders. For the antiestrogen assay, a range of concentrations of samples was added concurrently with 1 nM antiestrogens. After growth for 3 days, to determine alkaline phosphatase (AP) activity, the cells were frozen, defrosted, and incubated with the chromogenic substrate *p*-nitrophenylphosphate at room temperature. The hydrolysis product, *p*-nitrophenol, was measured kinetically at 405 nm.

The specificity of the antiestrogenic activity was determined using the estrogen receptor (ER) subtypes, ERα and ERβ, in ER element (ERE)-transfected human choriocarcinoma JAR cells transfected with plasmids containing a consensus ERE fused to a firefly luciferase reporter gene and separately with the expression vectors for either human ERα or human ERβ. JAR cells were routinely cultured in RPMI 1640 supplemented with 10% fetal bovine serum (FBS), 0.5% nonessential amino acids, and 1% PEST (100 U penicillin/mL and 100 μg streptomycin/mL). After seeding in 6-well plates for 24 h, cells were transfected using the Mirus Trans IT reagent in a serum- and antibiotic-free mixture of phenol red-free OptiMEM with 0.1–0.4 μg pC × N2 human ERα or pC × N2 h-ERβ and 0.75 μg 3× ERE-TATA-Luc reporter constructed by introducing an HpaI/BglII fragment containing 3× ERE-TATA into SmaI/BglII of the pGL3-Luc basic vector.[13–15] Medium was replaced with a phenol red-free RPMI containing 10% dextran-coated charcoal-treated calf serum, 0.5% nonessential amino acids, and no PEST. Antiestrogens were added 24 h later. Cells were incubated at 37°C in 5% CO_2 for 12 h and then harvested in 10 mM Tris-HCl/10 mM EDTA/150 mM NaCl and centrifuged at 4000 × *g* for 4 min. The supernatant was removed, and cell pellets were lysed in Lysis Buffer 2. Using the GenGlow system (Promega, Madison, WI), luciferase activity was measured.

To detect *in vivo* estrogenic/antiestrogenic activity, ERα was determined by Western blotting. After the AP assay, the Ishikawa cells were washed three times with PBS and lysed using 1% Nonidet P-40 and 0.1% sodium dodecyl sulfate in the presence of protease inhibitors. Proteins (25 μg/well) were separated by sodium dodecyl sulfate-polyacrylamide gel electrophoresis (SDS-PAGE) on ice using 10% polyacrylamide gel and transferred to nitrocellulose membranes stained with Ponceau red before the antibody incubation to ensure proper transfer. After blocking the membranes with 5% powdered milk in water, immunoblotting was performed. The blots were incubated overnight at 4°C with ERα monoclonal antibody clone 6F11. Using peroxidase-labeled horse anti-mouse secondary antibody and Chemiluminescence Reagent Plus (PerkinElmer, Wellesley, MA), ERα was detected. Using a digital imaging analysis system, the intensity of the signal was analyzed. β-Actin was used as an internal control to normalize the amount of protein loaded in the gels.

The uterotropic assay was performed with stimulation in immature rats. Female Sprague–Dawley rats (22 days old) were injected subcutaneously daily for 3 days with antiestrogens. Control animals received vehicle (0.1 mL sesame oil). Then, animals were killed, and uteri removed, dissected, blotted, and weighed. To determine whether antiestrogens had tissue-selective effects in cholesterol, uteri, and bone, ovariectomized female Sprague–Dawley rats (250 g) were injected with antiestrogens subcutaneously for 35 days. Then the animals were killed by exsanguination under ether anesthesia. The total cholesterol concentration of the serum was determined by a commercial chromogenic assay. The uteri were dissected, weighed, fixed in formalin, and embedded in paraffin to prepare for 5-μm sections. Using the Openlab image analysis system (Improvision, Lexington, MA), endometrial luminal epithelium and glandular cell height were measured. The tibia were dissected free of extraneous tissue and then histomorphometrically analyzed. The bones were fixed in 70% ethanol, dehydrated in graded ethanol, and cleared in toluene. The specimens were then infiltrated with increasing concentrations of methylmethacrylate (MMA) and embedded in MMA. After polymerization, MMA blocks were cut to size, sanded, and polished to the appropriate level to prepare for 4- to 5-μm sections. The sections were mounted on gelatin-coated slides and stained with toluidine blue and measured using the computerized Osteomeasure analysis system (Osteometrics, Atlanta, GA).

1.7 ATP ASSAY FOR EIGHT CELLS[24]

The primary cryopreserved human hepatocytes (lot HH227 in Li's Universal Medium), human renal proximal tubule cells (RPTCs in REGM), human small airway epithelial cells (SAECs in SAGM), human aortic endothelial cells (HAECs in EGM-2), normal human astrocytes (AGM), breast cancer cells (MCF-7 and MDA-MB-468 in MEME), and human colon tumor cell line (HCT-116 in MEME) in 25-cm^2 flasks were subcultured into 75-cm^2 flasks after approximately 4 days (75% confluent). Before trypsinizing and plating, the cells were cultured for another 7 days. The cells were plated into 96-well plates (1×10^4 to 1×10^5 cells per well) with 100 μL medium and incubated at 37°C and a highly humidified atmosphere of 95% air and 5% carbon dioxide for 24 h before the treatment with the tested compound. On the treatment day, the media were removed from all wells and replaced with 100 μL medium containing various concentrations of the tested compound. The treatment was performed in triplicate for an incubation period of 24–96 h. After the treatment period, the cells in individual wells were processed to determine viability. After removal of the treatment media, each of the individual wells was washed twice with isotonic PBS (pH 7.4), and then 50 μL mammalian cell lysis solution was added. The cell lysate was combined with the substrate solutions of the ATPlite system (Perkin-Elmer Inc., Boston, MA) and analyzed for chemiluminescence on a Wallac Victor 1420 Multilabel Counter (Wallac, Turku, Finland). Results were expressed as relative viability and were obtained according to the equation: Relative viability = $(ATP_{treatment}/ATP_{control}) \times 100\%$.

1.8 AP ACTIVITY ASSAY[25,26]

AP is a glycoprotein-structured metallophosphatase and presents in many tissues of all living beings from bacteria to mammals. AP may catalyze the hydrolysis of various monophosphate esters at alkaline pH. Based on the conversion of *p*-nitrophenylphosphate (*p*-NPP) to *p*-nitrophenol, the colorimetric determination of the resulting colored product was performed. In the assay, the cells (mouse myeloma cell lines P3, FO, SP2; erythroleukemia cell K562 and B cell hybridoma 1G2, 1.5×10^5 to 2.5×10^5 cells/well) were cultured in a humidified atmosphere of 5% CO_2 in air at 37°C in 96-well plates or 25-cm^2 Falcon tissue culture flasks containing DMEM supplemented with 10% fetal calf serum (FCS), 100 U/mL penicillin, 100 μg/mL streptomycin, and 2 mmol/mL glutamine for 4 days. Then cells were washed twice with PBS. Cell viability was assessed by exclusion of trypan blue dye (0.4%). The plates or tubes were centrifuged for 5 min at 3000 rpm. The supernatant was discarded, and the pellet was resuspended in 100 μL substrate buffer (10 mM diethanolamine, 0.5 mM $MgCl_2$ pH 10.5) containing 1 mg/mL *p*-NPP and incubated at room temperature for 20 min. The reaction was stopped with 50 mL 2 N NaOH. OD at 405 nm was determined using a Titertek Uniscan II ELISA Reader. According to the standard curve, AP activity was identified. For this standard curve, various concentrations of commercially provided AP from 0 to 400 ng/mL in a volume of 100 μL were added to flat-bottom 96-well plates, with 100 μL *p*-NPP (2 mg/mL) at room temperature incubated for 20 min. AP activity (OD) was determined spectrophotometrically at 405 nm. Each concentration of AP was measured in triplicate.

1.9 TUMOR ENDOTHELIAL CELL TUBE FORMATION ASSAY[27]

Murine tumor endothelial cells (3B11) were cultured in a humidified incubator with 5% CO_2, DMEM supplemented with 10% FBS, and 1× penicillin/streptomycin. Two days before the confluence of 3B11 cells reaching 80% to 95%, the cells (6×10^5) were cultured in a T75 flask. One day before performing this assay, a Matrigel matrix was incubated on ice overnight. On the day this assay was performed, 0.15 mL Matrigel matrix (basement membrane, BD Biosciences, Bedford, MA) was transferred onto a 48-well plate (or 0.25 mL to 24-well plate or 0.3 mL to 4-well chamber slides). The plates or slides were incubated at 37°C and in 5% CO_2 for 30 min, during which 3B11 cells in T75 flasks were washed with PBS, digested with 1X trypsin solution for about 3 min, and suspended in 8 mL DMEM complete medium. The cells were counted and diluted to 4×10^5/mL in DMEM complete medium. The diluted testing compound (120 μL) was transferred to a 1.5-mL tube and mixed with equal volume of endothelial cells. To each well containing the Matrigel matrix, this mixture (200 μL) was added. The plates or slides were incubated at 37°C and in 5% CO_2 for 4–16 h depending on the tube size. Under a phase-contrast inverted microscope at 5× objective magnification (or 2.5× and 10× objectives) using computer-controlled ProMax software (Bio-TEC Instruments Inc., Winooski, VT), endothelial cell tube formation could be continuously monitored and imaged without fixation and staining.

For quantitative analysis, after carefully removing the cell culture medium in the well, the tubes were fixed for 30 min in 1 mL 3.6% formaldehyde at room temperature, washed with PBS three times, permeated via incubation in 1 mL 0.1% Triton X-100–PBS for 5 min, washed with PBS another two times, and blocked in 1 mL 1% BSA–PBS for 30 min at room temperature, stained with rhodamine phalloidin (200 μL) in 1% BSA–PBS for 60 min, washed twice with PBS again, and imaged by 2D wide-field fluorescence or 3D confocal fluorescence microscopy.

Two-dimensional wide-field fluorescence image-based quantitative analysis was performed on a Leica DC350F CCD camera attached to an inverted Leica (Wetzlar, Germany) DMIL microscope and captured using ProMax software. Images were then converted to TIFF files of high resolution for analysis. Rhodamine phalloidin was used for both fluorescently labeling of the actin cytoskeleton, which is abundant in the endothelial cell cytoplasm, and monitoring endothelial cell cytoplasm tube formation under various conditions. The spatial location of the branched and clustered endothelial cells in the Matrigel were defined by use of the fluorescent rhodamine signal. The analysis was two-dimensional due to the single wide-field fluorescent images. To account for the 2D feature of the images, area calculations were important. The threshold tool Image Pro Plus software (Media Cybernetics Inc., Bethesda, MD) was used to trace the fluorescent signal in each 2D image. In these 8-bit images, a signal with fluorescent intensity between 50 and 256 was considered a real signal. This threshold was pseudo-colored in green by the software and displayed as a layer over the 8-bit gray-scale original image that showed the spatial area of interest. Once traced, the area colored in green was outlined with the software, and then these outlined regions were used to calculate 2D parameters including area (as a uniform height), perimeter, and average length of each connected vessel.

Three-dimensional images were performed on a Rainbow Radiance 2100 Laser Scanning Confocal system attached to a Nikon TE2000-U inverted microscope (BioRad-Carl Zeiss Inc., Thornwood, NY), optical image slices (8-bit) were obtained with a 10× objective to capture as much area as possible (3-μm-interval step slices). Using Laser Sharp 2000 software (BitPlane Inc., Saint Paul, MN), the images were imported into the Imaris software and the macro "filament tracker" used to generate a 3D solid mask of the original fluorescence signal. As mentioned above, the rhodamine phalloidin fluorescent signal of the 3D image was also used to define the spatial location of the clusters and branches of endothelial cells in the Matrigel. In these 8-bit images, a threshold between 70 and 256 was considered a true signal. The filament tracker tool in Imaris automatically traced the threshold signal and created a 3D cylindrical outline of the endothelial cell network. With the outline of the filament track, the software can extract parameters including average vessel length, area, volume, number of free unassociated termination points, and number of branch points. In addition, the "spot tracking" tool in Imaris automatically located the 4,6-diamidino-2-phenylindole dihydrochloride (DAPI)-stained nuclear fluorescence signal and marked the center of each nucleus with a solid sphere. The Imaris software calculated the parameters including the number of nuclei per branch, per tube, or per field and their fluorescent intensities and locations and displayed them in Excel (Microsoft, Redmond, WA) for statistics and graphing.

1.10 ANTIANGIOGENIC ASSAY[28]

Human RPE cells (ARPE-19) were cultured (37°C, 5% CO_2) to confluence in 56.7-cm^2 culture dishes with DMEM/F-12 growth medium (15 mL; ATCC) supplemented with 10% heat-inactivated FBS and penicillin (100 units/mL) and streptomycin (100 μg/mL). In hypoxia experiments, the medium was replaced with 15 mL DMEM/F-12 growth medium supplemented with penicillin (100 units/mL) and streptomycin (100 μg/mL) only, and the cells were incubated in a hypoxic incubator (37°C, 3% O_2) for 24 h. The culture medium was removed and centrifuged to remove any cellular debris, and the supernatant was frozen at −80°C until further use. Conditioned medium from two separate experiments was pooled together before freezing at −80°C. All experiments were carried out using cells within the first five passages. Observation of the cells by light microscopy after more than 36 h of hypoxia demonstrated that they were viable and healthy.

In short interfering RNA (siRNA) experiments, ONTARGETplus SMARTpool hypoxia-inducible factor (HIF)-1α siRNA or vascular endothelial growth factor (VEGF) siRNA was used with fluorescent siGLO RISC-free siRNA as a negative control. Cells with 90% confluence were washed with OptiMEM I reduced-serum medium and then incubated (37°C, overnight) in reduced-serum medium (15 mL) containing 600 pmol siRNA and 30 μL Lipofectamine 2000. The transfection solution was aspirated. The cells were washed with culture medium, given 36 h to recover from transfection, and then hypoxic experiments were performed. The amount of siRNA and transfection agent used was optimized; under the conditions used, 100% transfection was observed under fluorescent microscopy, and little to no cell toxicity was observed under light microscopy. With Western blot techniques, it was confirmed that HIF-1α and VEGF levels were barely detectable up to 72 h after siRNA transfection.

Conditioned medium was assayed for angiogenic factors [VEGF, angiogenin, placental growth factor (PlGF), erythropoietin (EPO), leptin, tumor growth factor (TGF)-β1, basic fibroblast growth factor (bFGF), interleukin (IL)-6, IL-8, and monocyte chemoattractant protein (MCP)-1] using Quantikine human immunoassay kits and for antiangiogenic factor pigment epithelium-derived factor (PEDF) using Chemikin pigment epithelium-derived factor sandwich ELISA kit according to the manufacturers' instructions and in triplicate.

1.11 *IN VIVO* HOLLOW FIBER ASSAY[29]

Cell confluent monolayers were collected by centrifugation and then resuspended in conditioned medium at a density of 10^6 or 5×10^5 cells/mL. Fibers were filled with the cells and incubated in 6-well plates at 37°C in a 5% CO_2 atmosphere for 24 h. Female athymic NCr *nu*/*nu* mouse (5–6 weeks old) hosted up to six fibers, which were implanted into two physiologic compartments. For intraperitoneal implants, the mouse was given a small incision through the skin and musculature of the dorsal abdominal wall. After inserting the fiber samples into the peritoneal cavity in a craniocaudal direction, the incision was closed with skin staples. For subcutaneous implants, the mouse was given a small skin incision at the nape of the neck for inserting an 11-gauge tumor implant trocar containing the hollow fiber samples, which was inserted caudally through the subcutaneous tissues. The fibers were deposited during withdrawal of the trocar. The incision was closed with skin staples.

Cell growth was assessed with fibers containing 5×10^5, 10^6, or 5×10^6 cells/mL. For treatment protocols, to increase solubility the test compounds were coprecipitated with polyvinylpyrrolidone (PVP, MW 360,000) and then dissolved in PBS. Mice were randomized into PBS vehicle control groups (6 mice/group) and test compound treatment groups (3 mice/group for each dose tested). From days 3 to 6 after implantation, test compounds were administered by intraperitoneal injection once daily, and body weights of mice were recorded daily. On day 7, mice were anesthetized by intraperitoneal injection of a mixture of ketamine (13.3 mg/100 g body weight) and xylazine (1.3 mg/kg body weight) and the fibers were retrieved. The fibers were placed into 6-well plates, each well containing 2 mL fresh and prewarmed culture medium (with 20% calf serum or FBS) and equilibrated for 30 min at 37°C. Using MTT assay, the survival of the cell mass in the intact hollow fibers was defined. Briefly, to each dish 1 mL of prewarmed culture medium (with 20% calf serum or FBS) containing 1 mg MTT was added, and the dishes were incubated at 37°C for 4 h. The culture media were removed by aspiration. To each well, 2 mL normal saline containing 2.5% protamine sulfate solution was added. The plates were stored at 4°C for 24 h, and protamine sulfate solution was removed by aspiration and replaced by 2 mL fresh protamine sulfate solution. The plates were stored at 4°C for at least 2–4 h. The fibers were transferred to 24-well plates, cut in half, and allowed to dry overnight. The residual formazan was extracted with DMSO (250 μL/well) for 4 h at room temperature on a rotation platform. Aliquots (150 μL) of DMSO containing formazan were transferred to individual wells of 96-well plates and assessed for optical density at a wavelength of 540 nm. The effect of the treatment regimen was determined by the net growth percentage of the cells relative to changes in body weight.

1.12 VX2 RABBIT LUNG ASSAY[30]

After the addition of VX2 cells isolated from freshly excised lung tumors of a donor rabbit (ca. 750,000 live VX2 cells/μL) to fresh rabbit plasma, thrombin was added. Before it clotted, using a syringe this plasma mixture was quickly drawn into 5-cm lengths of PE-90 tubing. A single 5-cm newly coagulated clot (volume ca. 30 μL, containing ca. 25,000 VX2 cells) was recovered in 0.5 mL saline and injected intravenously (ear) into a recipient rabbit. At 28 d after seeding, the rabbit was anesthetized by intravenously administered 35 mg/kg sodium pentobarbital and exsanguinated through a cannula (PE-190 tubing) placed in a carotid artery. From lung tumors, VX2 cells were harvested and used to seed the lungs of another rabbit. Twenty-four hours later, the rabbits were given test compounds. At 28 d after seeding, each rabbit with VX2 tumor was anesthetized, exsanguinated through a cannula (PE-190 tubing) placed in a carotid artery, and positioned on its back. At the sternum, the chest cavity was opened, and by syringe any obvious fluid within the interpleural space was drawn off with minimal contamination by blood. The fluid was weighed and centrifuged at $1200 \times g$ for 5 min. From a clear and pale yellow supernatant the pellet [equivalent to $(9.5 \pm 3.2) \times 10^6$ live VX2 cells/mL of effusion] was separated. The supernatant (pleural effusion) was snap-frozen (liquid N_2) and stored at -40°C. The lungs and only their tumors larger than 0.02 g were excised and individually weighed.

1.13 INSULIN-LIKE GROWTH FACTOR-I-INDUCED KINASE RECEPTOR ACTIVATION ASSAY[31]

Human breast adenocarcinoma (MCF-7) cells were cultured in 150-cm^2 tissue culture flasks, 1.5×10^6/flask for 4–7 days. For the assay, cells (2×10^5/well) were cultured in flat-bottom microtiter plates in 100 μL F-12/DMEM 50:50 at 37°C in 5% CO_2 for 24 h. The medium was supplemented with 10% FBS, 25 mM HEPES, and 2 mM L-Gln. After removal of the supernatants, the plates were lightly blotted on paper towels. To each well, medium containing either test compounds or the recombinant insulin-like growth factor I (IGF-I) standards were added, kept for 15 min, decanted and blotted. To each well, 100 μL lysis buffer containing 0.5% Triton X-100, 2 mM sodium orthovanadate (to prevent dephosphorylation of the receptors), and a cocktail of protease inhibitors was added to generate crude lysates that were transferred to an ELISA plate that had been coated overnight with an insulin-like growth factor-I receptor (IGF-IR)-specific antibody (3B7, 5.0 μg/mL) and "blocked" with 0.5% BSA. With incubation the IGF-IR in the crude lysates was effectively banded and thus directly purified in the ELISA well. The plate was extensively washed to remove unbound material, and with biotinylated anti-phosphotyrosine monoclonal antibody (4G10) and horse radish peroxidase (HRP)-conjugated dextran-streptavidin the degree of anti-phosphotyrosine binding was visualized with the development of a tetramethyl benzidine (TMB) substrate. The absorbance at 450 nm was read with a reference wavelength of 650 nm (*A*450).

1.14 INSULIN-LIKE gD.trkA-INDUCED KINASE RECEPTOR ACTIVATION ASSAY[32]

gD.trk-transfected CHO cells were seeded (5×10^4/well) in 96-well plates containing 100 μL medium and cultured at 37°C in 5% CO_2 for 16 h. The well supernatants were decanted, and the plates were blotted on a paper towel. To each well, 50 μL test compounds or the recombinant purified rhNGF, rhNT4/5, rhBDNF, or rhNT3 standards diluted in stimulation medium were added. The cells were stimulated at 37°C for 20 min, and the well supernatants were once again blotted on a paper towel. To each well, 100 μL lysis buffer was added to lyse the cells and solubilize the receptors. Lysis buffer consisted of 150 mM NaCl containing 50 mM HEPES, 0.5% Triton X-100, 0.01% thimerosal, 30 U/mL aprotinin, 1 mM 4-(2-aminoethyl)-benzenesulfonyl fluoride hydrochloride, and 2 mM sodium orthovanadate. The plate was then agitated gently on a plate shaker for 60 min at room temperature.

1.15 UV SPECTRA-BASED CALF THYMUS DNA INTERCALATION ASSAY[33,34]

Spectra were recorded on a UV-3100 spectrophotometer with a 1-cm quartz cuvette. In the absorbance titrations, the concentration of the test compound was kept constant while varying the concentration of calf thymus DNA. To take the influence from the absorbance of DNA into account, an equal amount of DNA was added into the sample cell and the reference cell. All absorbance measurements were made at 25°C and 286 nm. The absorption spectrum was recorded after each addition of DNA until no further decrease in absorbance was observed. Before measurements, all solutions were allowed to equilibrate for 5 min. By the equation $[DNA]/(\varepsilon_A - \varepsilon_F) = [DNA]/(\varepsilon_B - \varepsilon_F) + 1/K_b/(\varepsilon_A - \varepsilon_F)$, the binding constant ($K_b$) may be calculated, where ε_A, ε_B, and ε_F correspond with the apparent extinction coefficient of test compound, the extinction coefficient for free test compound, and the extinction coefficient for test compound in the fully bound form, respectively. Using the plot of $[DNA]/(\varepsilon_A - \varepsilon_F)$ versus [DNA], K_b was defined by the ratio of the slope to the y intercept, and its value reflected the intercalation strength.

1.16 FLUORESCENCE SPECTRA-BASED CALF THYMUS DNA INTERCALATION ASSAY[33,34]

Spectra were recorded on a Perkin-Elmer LS50B luminescence spectrometer with a 1-cm quartz cuvette and the temperature maintained with a CS501 Superthermostat. In the fluorescence titrations, the concentration of the test compound was kept constant while varying the concentration of calf thymus DNA. The experiments were carried out in buffer. The concentration of test compound typically was 13.8 mM, DNA concentration varying from 0 to about 0.5 mM until no further decrease in the fluorescence intensity for the test compound was found. Throughout the course of the titration, the solution system was continuously stirred. Before measurements, all solutions were equilibrated for 5 min. For fluorescence spectra, the samples were excited at 290 nm, and a slit width of 5 nm was used for the excitation and emission beams, and from 200 to 700 nm the emission spectra were recorded. According to the equation $I_0/I = 1 + K_{vs}[DNA]$, the fluorescence quenching data were plotted, where I_0 and I are the fluorescence intensities in the absence and presence of DNA, respectively, and K_{vs} is the Stern–Volmer quenching constant, which measures the efficiency of quenching by DNA. The data from fluorescence titrations were also useful for determining the binding constant of test compound with DNA. The concentration of the free test compound was calculated according to the equation $C_F = C_T(I/I_0 - P)/(1 - P)$, where C_T is the initial concentration of the test compound, C_F is the concentration of the free test compound, and P is the percentage of the observed quantum yield of the fluorescence of the totally bound test compound. P-value was obtained from the plot of I/I_0 versus 1/[DNA] as the limiting fluorescence yield given by y intercept. The amount of bound test compound can be calculated from $C_T - C_F$. According to the equation $r/C_F = K_i(1 - nr)\{(1 - nr)/[1 - (n - 1)r]\}^{n-1}$, the plot of r/C_F versus r was constructed, where r equals $C_T - C_F$ /[DNA], K_i is the binding constant, and n is the binding site size in base pairs. With Origin graph software (7.0, OriginLab.) on an Ascend computer, fitting the binding data into the equation $r/C_F = K_i(1 - nr)$ $\{(1 - nr)/[1 - (n - 1)r]\}^{n-1}$, the values of K_i and n can be extracted.

1.17 P-GLYCOPROTEIN PUMP IN MCF-7R CELLS ASSAY[35]

To assess the inhibitory effect of test compounds on the P-gp in MCF-7R cells (human breast adenocarcinoma cells, resistant to Adriamycin or doxorubicin) maintained at

37°C in humidified atmospheric air containing 5% CO_2 in nutrient mixture (F10/HAM) containing L-glutamine, with 10% fetal calf serum, 1% nonessential amino acids, 60 μg/mL tylosin, and 1% antibiotic/antimycotic solution, the cells were seeded in 96-well plates at 3×10^4 cells in 200 μL medium per well. The plates were incubated at 37°C for 24 h in humidified atmospheric air containing 5% CO_2. The medium was replaced with fresh medium containing 0.3 μM rhodamine 6G together with test compounds and thoroughly mixed. Each 50 μL of the mixture was diluted with 450 μL fresh medium containing 0.3 μM rhodamine 6G. After incubation (in the presence of rhodamine 6G) of test compound, reserpine (positive control), or rhodamine 6G alone (negative control), the plates were further incubated at 37°C for 3 h. The cells were washed twice with 200 μL ice-cold PBS and trypsinized with 100 μL phenol red-free trypsin solution for 15 min, which was transferred onto empty wells. To the cells was added 100 μL of 4% SDS in PBS, and the plates were shaken for 2 h to dissolve the cells and release rhodamine 6G, which was determined by measuring the dye fluorescence on a microplate fluorescence reader at excitation and emission wavelengths of 530 ± 25 nm and 590 ± 35 nm, respectively.

For fluorescence imaging, MCF-7R cells were seeded on Lab-Tek chamber slides at 5×10^4 cells per chamber and incubated at 37°C for 24 h in humidified atmospheric air containing 5% CO_2. The medium was replaced with fresh medium containing 0.3 μM rhodamine 6G with or without 50 μM reserpine, with 100 μg/mL test compounds in separate chambers of the slide. The cells were incubated again at 37°C for 3 h and imaged using fluorescence microscopy with a 510- to 560-nm band-pass excitation filter and a 590-nm long-pass emission filter set using a light-sensitive charge-coupled device digital camera.

1.18 P-GLYCOPROTEIN PUMP-RELATED EFFLUX CARRIERS ASSAY[36]

The inhibition by test compounds of P-gp-related efflux carriers was measured with the Caco-2 system via intestinal absorption of test compounds, in which P-gp was overexpressed. Caco-2 cells were harvested with trypsin–EDTA and seeded onto the Caco-2 assay system (MultiScreen) at a density of 1×10^4 cells/well. In the first 6 days, the culture medium was replaced at intervals of 48 h, thereafter the culture medium was replaced at intervals of 24 h. After 21 days in culture, the Caco-2 monolayer was used for transport assay. The transepithelial electrical resistance (TER) of the monolayers was measured daily before and after the assay on an epithelial volt-ohmmeter. After a 21-day culture, the TER values were generally greater than 1000 Ω. Apical to basolateral (P_{app} A-B) and basolateral to apical (P_{app} B-A) permeability of various concentrations of test compounds (1–100 μM) were measured at 120 min. For this, the test compounds were dissolved in Hanks balanced salt solution (HBSS; pH 7.4) and sterile filtered. After 21 days of Caco-2 cell growth, the medium was removed from the filter wells and receiver plate. The filter well and receiver plate were filled with 75 μL and 250 μL fresh HBSS buffer, respectively. This procedure was repeated twice. The plates were incubated at 37°C for 30 min, then the HBSS buffer was removed, and the solutions of test compounds were added to the filter well (75 μL). HBSS without test compounds was added to the receiver plate (250 μL). The plates were incubated at 37°C for 120 min. After incubation, samples were removed from the apical (filter well) and basolateral (receiver plate) side of the monolayer and stored in a freezer (−20°C) pending analysis. The concentrations of test compounds were analyzed using UV-vis spectroscopy. The apparent permeability (P_{app}, nm/s) was obtained by the equation $P_{app} = (V_A/\text{area} \times \text{time}) \times ([\text{Compd.}]_{acceptor}/[\text{Compd.}]_{initial})$, where V_A is the volume (μL) in the acceptor well, area is the surface area of the membrane (0.11 cm^2 of the well), time is the total transport time (7200 s), $[\text{Compd.}]_{acceptor}$ is the concentration of tested compounds measured by UV spectroscopy, and $[\text{Compd.}]_{initial}$ is the initial concentration of tested compounds (1×10^{-4} M) in the apical or basolateral wells.

1.19 [^{3}H]SUBSTRATE TRANSPORT INHIBITION ASSAY[36]

Caco-2 cells (1×10^4 cells/well) were seeded onto MultiScreen Plates for 21 days, and the cell monolayer exhibited a TER >800 $\Omega \times cm^2$. For assaying test compounds to basolateral compartment with and without P-gp inhibitors (from 200 nM to 400 μM), 20 nM [^{3}H]vinblastine or [^{3}H]mitoxantrone was added at 37°C for 120 min, and its appearance in the apical compartment was monitored. At 120 min (the end of the assay), 20 μL sample was taken from the basolateral compartment to determine the concentration of radioligand on an LS6500 Beckman Counter. For each compound, [^{3}H]vinblastine or [^{3}H]mitoxantrone transport inhibition was calculated as radioactivity difference between radioligand with and without test compound. These differences were expressed as inhibition percentage at single concentration of test compound.

1.20 LACTATE DEHYDROGENASE RELEASE ASSAY[36]

Cell release of lactate dehydrogenase (LDH) representing cell death was performed with the CytoTox-One kit (Promega Corp., Madison, WI, USA), which was calculated relative to the LDH release from total lysis of cells in the untreated control. It was assumed that test compound-treated wells and the control wells contained the same total number of cells (dead plus live cells) at the end of the assay, and the cytotoxic effect of test compound was unaffected by any underestimation of cytotoxicity that could occur due to decreased total number of cells in the treated samples compared with the untreated control. Cells were seeded into 96-well plates for optical performance in the fluorescent cell-based assay in 100 μL complete medium with or without various concentrations of test compound. The plate was incubated for 24 h in a humidified atmosphere containing 5% CO_2 at 37°C, and then 100 μL substrate mix in assay buffer was added. Ten microliters of lysis solution from the CytoTox-One kit was added to untreated wells in order to estimate total LDH. Plates were protected from light for 10 min at room temperature, and 50 μL stop solution from the CytoTox-One kit was added to all wells. The fluorescence was recorded on a LS55 Luminescence Spectrometer (PerkinElmer) at 560-nm excitation wavelength and 590-nm emission wavelength. The cytotoxicity percentage was estimated by the equation (LDH in medium of treated cells − culture medium background)/(total LDH in untreated cells − culture medium background) × 100.

1.21 FUNCTIONAL ASSAY OF MITOCHONDRIAL P-gp[37]

To evaluate mitochondrial autofluorescence as well as the uptake and efflux of doxorubicin into and out of organelles, whole isolated mitochondria (50 μg in hypertonic buffer) from doxorubicin-sensitive and resistant K562 cells were divided in test tubes either without any other drug or in the presence of specific monoclonal antibodies (4 μg UIC2 or 1.5 μg F4) or in the presence of inhibitor (5 μM cyclosporin A or 10 μM dexverapamil or 50 μM quinine) as positive control or test compounds (in various concentrations to measure IC_{50}). The tubes were preincubated at 20°C for 1 h (monoclonal antibodies) or 30 min (positive control and test compounds). To the tubes, doxorubicin (10 μM final) was added, and the tubes were incubated at 20°C avoiding light exposure for another 1 h and kept on ice for a few minutes until flow cytometric analysis. In order to estimate doxorubicin efflux from mitochondria, 2 mL hypertonic buffer was added to the tubes containing test compounds, the tubes were centrifuged for 5 min at 450 × *g* and 4°C, the residue was washed once more with 2 mL buffer and diluted in 500 μL buffer at 20°C. All tubes were incubated for 1 h at 20°C. After elimination of debris and aggregates, doxorubicin fluorescence (for uptake and retention evaluation) in each tube was measured on 10,000 isolated mitochondria with a flow rate of 500 events/s on a FACSVantage (Becton-Dickinson, Grenoble, France) flow cytometer.

1.22 RESISTANCE INDEX VALUE ASSAY[38]

(1) Determining doxorubicin sensitivity of MES-SA and MES-SA/Dx5 cells: Doxorubicin-sensitive MES-SA cells and doxorubicin-resistant MES-SA/Dx5 cells (during the exponential phase of growth with final concentration at 1×10^5/mL) in McCoy's 5 A medium containing 10% (v/v) fetal calf serum, penicillin (100 μg/mL), and streptomycin (100 μg/mL) were seeded in 96-well plates (100 μL/well) coated with poly-L-lysine. The cultures were propagated at 37°C in a humidified atmosphere containing 5% CO_2 for 4 h. To the plates, 25 μL solution (final concentrations, 10.00, 5.00, 2.50, 1.25, 0.63, 0.31, 0.16, 0.08, and 0.04 μM) of doxorubicin in the growth medium was added and the cells were propagated for 48 h. For the control well, 25 μL growth medium was added and the cells were also propagated for 48 h. To the blank well, which contains only the growth medium without cells, 25 μL growth medium was added and was also propagated for 48 h. The medium was removed, the residue was washed with PBS and replaced by fresh medium, to

each well 25 μL MTT (5 mg/mL) was added, and the plates were incubated for 4 h. After removing the growth medium, the residue was dried in the air, the residues were dissolved in 10 μL DMSO, and the absorption values of light of the formed purple solutions were recorded on a Bio-Rad 450 microplate reader (Bio-Rad, Ontario, Canada). Survival % = $(D_{Dx} - D_{blank})/(D_{control} - D_{blank})$, where D_{Dx} represented the light absorption values of medium of cells plus doxorubicin and growth medium, $D_{control}$ represented the light absorption values of medium of cells plus growth medium, and D_{blank} represented the light absorption values of growth medium alone. The survival and concentrations of doxorubicin were plotted to define the IC_{50} of doxorubicin against both MES-SA cells and MES-SA/Dx5 cells. Accordingly, the resistance index value was given by the equation (IC_{50} against MES-SA/Dx5)/(IC_{50} against MES-SA).

(2) Determining the effect of test compounds on the sensitivities of doxorubicin-resistant MES-SA/Dx5 cells: MES-SA/Dx5 cells (during the exponential phase of growth with final concentration of 1×10^5/mL in McCoy's 5 A medium containing 10% (v/v) fetal calf serum, penicillin (100 μg/mL), and streptomycin (μg/mL) were seeded in 96-well plates (100 μL/well) coated with poly-L-lysine. The cultures were propagated at 37°C in a humidified atmosphere containing 5% CO_2 for 4 h. To the plates, 25 μL solution (final concentrations, 10.00, 5.00, 2.50, 1.25, 0.63, 0.31, 0.16, 0.08, and 0.04 μM) of doxorubicin in the growth medium was added and the cells were propagated for 48 h. To the plates, the solution (final concentration, 1.00 μM) of test compounds in the growth medium was added and the cells were propagated for 48 h. For the control well, 25 μL growth medium was added, and the cells were also propagated for 48 h. To the blank well, which contained only the growth medium without cells, 25 μL growth medium was added and propagated also for 48 h. The medium was removed, the residue was washed with PBS and replaced by fresh medium, to each well 25 μL MTT (5 mg/mL) was added, and the plates were incubated for 4 h. After removing the growth medium, the residue was dried in the air, dissolved in 10 μL DMSO, and the absorption values of light of the formed purple solutions were recorded on a Bio-Rad 450 microplate reader. Survival % = $(D_{Dx} - D_{blank})/(D_{control} - D_{blank})$, where D_{Dx} represented the light absorption values of cells plus doxorubicin plus growth medium plus test compounds, $D_{control}$ represented the light absorption values of cells plus growth medium, and D_{blank} represented the light absorption values of growth medium alone. The survival and concentrations of doxorubicin were plotted to define the IC_{50} of doxorubicin against MES-SA/Dx5 cells. Accordingly, the resistance index value was given by the equation (IC_{50} of doxorubicin alone against MES-SA/Dx5)/(IC_{50} of doxorubicin with test compounds against MESSA/Dx5).

REFERENCES AND NOTES

1. A. Jang, C. Jo, K. Kang, M. Lee. Antimicrobial and human cancer cell cytotoxic effect of synthetic angiotensin-converting enzyme (ACE) inhibitory peptides. Food Chem 107 (**2008**) 327–336.

2. J. Li, Y. Yin, F. Yao, L. Zhang, K. Yao. Effect of nano- and micro-hydroxyapatite/chitosan-gelatin network film on human gastric cancer cells. Mater Lett 62 (**2008**) 3220–3223.
3. J. Chen, C. Wu, H. Kao, H. Chang, A.F.Y. Lic, T. Liu, C. Chi. Effects of COX-2 inhibitor on growth of human gastric cancer cells and its relation to hepatocyte growth factor. Cancer Lett 239 (**2006**) 263–270.
4. Y. Zhang, S.K. Vareed, M.G. Nair. Human tumor cell growth inhibition by nontoxic anthocyanidins, the pigments in fruits and vegetables. Life Sci 76 (**2005**) 1465–1472.
5. B. Jayaprakasam, Y. Zhang, N.P. Seeram, M.G. Nair. Growth inhibition of human tumor cell lines by withanolides from Withania somnifera leaves. Life Sci 74 (**2003**) 125–132.
6. F.S. Kimball, B.J. Turunen, K.C. Ellis, R.H. Himes, G.I. Georg. Enantiospecific synthesis and cytotoxicity of 7-(4-methoxyphenyl)-6-phenyl-2,3,8,8a-tetrahydro-indolizin-5(1H)-one enantiomers. Bioorg Med Chem 16 (**2008**) 4367–4377.
7. G.B. Jones, J.E. Mathews. Tricarbonyl η6 arene chromium (0) based antitumor agents. Bioorg Med Chem Lett 5 (**1995**) 93–96.
8. B. Zarranz, A. Jaso, I. Aldana, A. Monge. Synthesis and anticancer activity evaluation of new 2-alkylcarbonyl and 2-benzoyl-3-trifluoromethyl-quinoxaline 1,4-di-N-oxide derivatives. Bioorg Med Chem 12 (**2004**) 3711–3721.
9. J. Provencher-Mandeville, C. Descoteaux, S.K. Mandal, V. Leblanc, E. Asselin, G. Berube. Synthesis of 17β-estradiol-platinum(II) hybrid molecules showing cytotoxic activity on breast cancer cell lines. Bioorg Med Chem Lett 18 (**2008**) 2282–2287.
10. M. Pan, Q. Wei, F. Cao, Y. Lu, Y. Zhu, Y. Shu, X. Cao. Inhibition of cell proliferation by siRNA targeting hPRLR in breast cancer MCF-7 cell line. Journal of Nanjing Medical University 21 (**2007**) 372–376.
11. Z. Chen, Y. Liu, S. Yang, B. Song, G. Xu, Pi.S. Bhadury, L. Jin, D. Hu, F. Liu, W. Xue, X. Zhou. Studies on the chemical constituents and anticancer activity of Saxifraga stolonifera (L) Meeb. Bioorg Med Chem 16 (**2008**) 1337–1344.
12. X. Feng, Z. Gao, S. Li, S.H. Jones, S.M. Hecht. DNA polymerase beta lyase inhibitors from Maytenus putterlickoides. J Nat Prod 67 (**2004**) 1744–1747.
13. C. Pisoni, G. Cimoli, A. Resconi, D. Losi, R. Lorenzetti, S. Parodi, L. Carrano. A new assay for the discovery of Bcl-XL inhibitors. Il Farmaco 60 (**2005**) 938–943.
14. M.F. Rega, J.C. Reed, M. Pellecchia. Robust lanthanide-based assays for the detection of anti-apoptotic Bcl-2-family protein antagonists. Bioorg Chem 35 (**2007**) 113–120. *Note*: In time-resolved fluorescence resonance energy transfer (TR-FRET) assay, 29 μL detection buffer, 5 μL variable concentration of biotin-labeled BH3 peptide, biotin-$(CH_2)_6$-GGGQVGRQLAIIGDDINR (concentration, 2 to 5 nM), 5 μL antiapoptotic Bcl-2 protein (1 to 200 nM), and 1 μL DMSO containing test compound were mixed in each well of 96 wells (or 384) black plates (Perkin-Elmer). After 30 min (15 min to 1 h) incubation, to each well 5 μL Eu-antibody conjugate (3 to 25 nM) and 5 μL APC-streptavidin (1 to 4 nM) were added. The detection buffer was used. Fluorescence measurements were performed after 10 min or 2 h incubations at 340-nm excitation wavelength and 665-nm emission wavelength, controls included unlabeled peptide and blanks receiving no compounds, Z0 factor measurements were obtained by repeating the experiments (positive and negative controls) multiple times, and TR-FRET, as the DELFIA, could also be obtained using GST-fusion Bcl-2 proteins and anti-GST Eu-antibody.

15. M. Wormke, E. Castro-Rivera, I. Chen, S. Safe. Estrogen and aryl hydrocarbon receptor expression and crosstalk in human Ishikawa endometrial cancer cells. J Steroid Biochem Mol Biol 72 (**2000**) 197–207.

16. A.D. Darnela, T.K. Archerb, K. Yanga. Regulation of 11b-hydroxysteroid dehydrogenase type 2 by steroid hormones and epidermal growth factor in the Ishikawa human endometrial cell line. Steroid Biochem Mol Biol 70 (**1999**) 203–210.

17. J. Ashby, W. Owens, R. Deghenghi, J. Odum. Concept evaluation: An assay for receptor-mediated and biochemical antiestrogens using pubertal rats. Regul Toxicol Pharmacol 35 (**2002**) 393–397.

18. J.X. Zhang, D.C. Labaree, G. Mor, R.B. Hochberg. Estrogen to antiestrogen with a single methylene group resulting in an unusual steroidal selective estrogen receptor modulator. J Clin Endocrinol Metab 89 (**2004**) 3527–3535.

19. B.A. Littlefield, E. Gurpide, L. Markiewicz, B. McKinley, R.B. Hochberg. A simple and sensitive microtiter plate estrogen bioassay based on stimulation of alkaline phosphatase in Ishikawa cells: estrogenic action of Δ^5 adrenal steroids. Endocrinology 127 (**1990**) 2757–2762.

20. G.G. Kuiper, J.G. Lemmen, B. Carlsson, J.C. Corton, S.H. Safc, P.T. van der Saag, B. van der Burg, J.A. Gustafsson. Interaction of estrogenic chemicals and phytoestrogens with estrogen receptor. Endocrinology 139 (**1998**) 4252–4263.

21. G. Mor, E. Sapi, V.M. Abrahams, T. Rutherford, J. Song, X.Y. Hao, S. Muzaffar, F. Kohen. Interaction of the estrogen receptors with the Fas ligand promoter in human monocytes. J Immunol 170 (**2003**) 114–122.

22. H. Niwa, K. Yamamura, J. Miyazaki. Efficient selection for high-expression transfectants with a novel eukaryotic vector. Gene 108 (**1991**) 193–199.

23. A.M. Parfitt, M.K. Drezner, F.H. Glorieux, J.A. Kanis, H. Malluche, P.J. Meunier, S.M. Ott, R.R. Recker. Bone histomorphometry: Standardization of nomenclature, symbols, and units. Report of the ASBMR Histomorphometry Nomenclature Committee. J Bone Miner Res 2 (**1987**) 595–610.

24. A.P. Li, C. Bode, Y. Sakai. A novel in vitro system, the integrated discrete multiple organ cell culture (IdMOC) system, for the evaluation of human drug toxicity: Comparative cytotoxicity of tamoxifen towards normal human cells from five major organs and MCF-7 adenocarcinoma breast cancer cells. Chem-Biol Interact 150 (**2004**) 129–136.

25. H. Akcakaya, A. Aroymak, S. Gokce. A quantitative colorimetric method of measuring alkaline phosphatase activity in eukaryotic cell membranes. Cell Biol Int 31 (**2007**) 186–190.

26. P.J. Lamas-Ardisana, P, Queipo, P. Fanjul-Bolado, A. Costa-Garcia. Multiwalled carbon nanotube modified screen-printed electrodes for the detection of p-aminophenol: Optimisation and application in alkaline phosphatase-based assays. Anal Chim Acta 615 (**2008**) 30–38.

27. Q. Zhou, W.B. Kiosses, J. Liu, P. Schimmel. Tumor endothelial cell tube formation model for determining anti-angiogenic activity of a tRNA synthetase cytokine. Methods 44 (**2008**) 190–195.

28. F. Forooghian, B. Das. Anti-angiogenic effects of ribonucleic acid interference targeting vascular endothelial growth factor and hypoxia-inducible factor-1α. Am J Ophthalmol 144 (**2007**) 761–768.

29. Q. Mi, D. Lantvit, E. Reyes-Lim, H. Chai, W. Zhao, I. Lee, S. Peraza-Sanchez, O. Ngassapa, L.B.S. Kardono, S. Riswan, M.G. Hollingshead, J.G. Mayo, N.R. Farnsworth, G.A. Cordell, A.D. Kinghorn, J.M. Pezzuto. Evaluation of the potential cancer chemotherapeutic efficacy of natural product isolates employing in vivo hollow fiber tests. J Nat Prod 65 (**2002**) 842–850. *Note*: KB cell confluent monolayers were collected by centrifugation and suspended in Matrigel (5×10^5 cells/0.19 mL Matrigel, using a pre-cooled pipette or syringe or needle). Cell suspensions (0.19 mL) were administered subcutaneously into the right flank region of female athymic NCr *nu/nu* mice (the same characteristics as those used for the *in vivo* hollow fiber assays) on day 0. When the tumors reached a palpable mass at day 10, treatment was initiated for the mice of the experimental groups distributed randomly with either PBS (control) or test compounds on days 10, 13, 16, and 20. Test compound preparation was the same as for the *in vivo* hollow fiber assay. Using a digital caliper in two dimensions, the tumor size was measured twice weekly. With $V = [\text{length} + (\text{width})^2]/2$, individual tumor volumes (V) were calculated for evaluating the efficacy of test compounds. Body weights (as percent change after initiation of treatment) were determined twice weekly.

30. M.W.C. Hatton, S.M.R. Southward, B.L. Ross, B.J. Clarke, G. Singh, M. Richardson. Relationships among tumor burden, tumor size, and the changing concentrations of fibrin degradation products and fibrinolytic factors in the pleural effusions of rabbits with VX2 lung tumors. J Lab Clin Med 147 (**2006**) 27–35.

31. M.D. Sadick, A. Intintoli, V. Quarmby, A. McCoy, E. Canova-Davis, V. Ling. Kinase receptor activation (KIRA): A rapid and accurate alternative to end-point bioassays. J Pharm Biomed Anal 19 (**1999**) 883–891.

32. M.D. Sadick, A. Galloway, D. Shelton, V. Hale, S. Weck, V. Anicetti, W.L.T. Wong. Analysis of neurotrophin/receptor interactions with a gD-flag-modified quantitative kinase receptor activation (gD. KIRA) enzyme-linked immunosorbent assay exp. Cell Res 234 (**1997**) 354–361.

33. W. Zhong, J. Yu, W. Huang, K. Nl, Y. Liang. Spectroscopic studies of interaction of chlorobenzylidine with DNA. Biopolymers (Biospectroscopy) 62 (**2001**) 315–323.

34. J. Wu, G. Cui, M. Zhao, C. Cui, S. Peng. Novel N-(3-carboxyl-9-benzyl-carboline-1-yl)ethylamino acids: Synthesis, anti-proliferation activity and two-step-course of intercalation with calf thymus DNA. Mol BioSystems 3 (**2007**) 855–861.

35. M.K. Owusu, A.R. Kamuhabwa, C. Nshimo. Investigation of fractions present in the stem bark of Annickia kummeriae on their P-glycoprotein inhibitory pump activity. Phytother Res 18 (**2004**) 652–657.

36. N.A. Colabufo, F. Berardi, M. Cantore, M.G. Perrone, M. Contino, C. Inglese, M. Niso, R. Perrone, A. Azzariti, G.M. Simoneb, A. Paradiso. 4-Biphenyl and 2-naphthyl substituted 6,7-dimethoxytetrahydroisoquinoline derivatives as potent P-gp modulators. Bioorg Med Chem 16 (**2008**) 3732–3743. *Note*: (a) Based on the fact that P-gp belongs to the superfamily of adenosine triphosphate (ATP)-binding cassette (ABC) transporters, cell ATP availability assay was performed on Victor3 (from Perkin-Elmer Life Sciences) according to the instructions in the technical sheet of the ATPlite Kit for luminescence ATP detection. Caco-2 cells (2×10^4 cells/well) were seeded into a 96-well plate in 100 μL complete medium and incubated at 37°C in a humidified atmosphere containing 5% CO_2 overnight. After removal of the medium, 100 μL complete medium with or without different concentrations (from 1 to 100 μM) of test compound was added with incubation for another 2 h. To all wells, 50 μL mammalian cell lysis solution from the ATPlite kit was added and the plate

stirred for 5 min in an orbital shaker. To all wells, 50 μL substrate solution was added, and the plate was stirred for 5 min in an orbital shaker. The plate was dark adapted for 10 min, and the luminescence was measured on Victor3. (b) Determination of cell growth was performed using MTT assay at 24 and 48 h. On day 1, 10,000 cells/well were plated in 96-well plates in a volume of 200 μL, and on day 2, the various compounds alone or in combination were added. In all assays, the solvents (ethanol, DMSO) were added in each control to evaluate a possible solvent cytotoxicity. After the established incubation time with test compounds, 0.5 mg/mL MTT was added to each well and, after 1 h incubation at 37°C, the supernatant was removed. The formazan crystals were dissolved in 100 μL DMSO, and the absorbance values at 570 and 630 nm were determined on the SpectraCount microplate reader. (c) Effect of antiproliferative drug combination: In MCF7/Adr, test compounds and verapamil were used at 2 μM and 20 μM; doxorubicin at 5 μM (IC_{50} after 3 days test compounds exposure when the P-gp was not overexpressed); verapamil was used as reference compound. The schedule of test compound administration was the P-gp inhibitors plus doxorubicin for 2 days followed, after two wash steps with complete medium, by doxorubicin for 1 day. The analysis of cell growth inhibition was performed using the MTT assay.

37. E. Munteanu, M. Verdier, F. Grandjean-Forestier, C. Stenger, C. Jayat-Vignoles, S. Huet, J. Robert, M. Ratinaud. Mitochondrial localization and activity of P-glycoprotein in doxorubicin-resistant K562 cells. Biochem Pharmacol 71 (**2006**) 1162–1174.

38. J. Liu, G. Cui, M. Zhao, C. Cui, J. Ju, S. Peng. Dual-acting agents that possess reversing resistance and anticancer activities: Design, synthesis, MES-SA/Dx5 cell assay and SAR of 2,2-dimethyl-3-[1′-substituted acetyloxybenzyl]-4-oxoimidazo-[3′,5′:2,3]tetrahydro-β-carbolines. Bioorg Med Chem 15 (**2007**) 7773–7778.

2

METHODS AND APPLICATIONS OF ANTIVIRAL ASSAYS

Shiqi Peng, Ming Zhao, and Chunying Cui

Virus infection is an important public health concern. For instance, according to the World Health Organization, there are currently 33.2 million people living with human immunodeficiency virus (HIV) worldwide, and HIV/acquired immunodeficiency syndrome (AIDS) is considered to be uniformly fatal; hepatitis B virus (HBV) infection is a major cause of chronic liver disease worldwide, and in regions where hepatitis B is highly endemic such as Southeast Asia and the Western Pacific, cirrhosis and hepatocellular carcinoma are common causes of death; an estimated 3% of the world's population (170+ million persons) is infected with the hepatitis C virus (HCV), 55% to 80% with chronic infection, and HCV is a significant cause of global morbidity and mortality, responsible for approximately 25% of both chronic liver disease and hepatocellular carcinoma. Therefore, persons at high risk for HIV infection are also likely to be at risk for other infectious pathogens, including HBV or HCV. On the other hand, a wide variety of rheumatic diseases has been documented in the presence of HCV infection and in HIV infection, which includes inflammatory arthritis, reactive arthritis, myositis, vasculitis, psoriasis vulgaris, as well as pustular psoriasis. Many assays have been used to evaluate the antiviral activity of agents against HIV, HBV, and HCV infections, to explore the infection mechanisms, and to establish the corresponding therapies. In this chapter, 13 assays are described: nonradioactive HIV-1 reverse transcriptase (RT) activity assay,[1] respiratory syncytial virus assay,[2] influenza virus types A and B assay,[3,4] nasal exhaled NO concentration

Pharmaceutical Bioassays: Methods and Applications. By Shiqi Peng and Ming Zhao

assay,[5,6] nasal NOS2 mRNA quantity assay,[7–13] RT-polymerase chain reaction (PCR) and swine assay for anti-HEV antibody,[14–20] HIV-1 protease and RT kinetic assay,[21] anti-HIV assay,[22] robust antiviral assays,[23] HIV/SIV fusion assay,[24] rapid DNA hybridization assay,[25] antiviral screening assay for HepAD38 cell cultures,[26] and trak-C HCV core assay.[27]

2.1 NONRADIOACTIVE HIV-1 RT ACTIVITY ASSAY[1]

Nonradioactive HIV-1 RT activity assay was performed with HIV-1 RT nonradioactive assay kit. In the assay, 20 mL of the reaction mixture containing a homogenous template/primer hybrid [(rA)n(dT)15, final concentration 750 mA_{260nm}/mL] and a triphosphate substrate (dUTP/dTTP, final concentration 10 mM) was added to the wells of a streptavidin-coated microtiter plate containing 20 μL test compound solution and 4 ng HIV-1 RT in 20 μL lysis buffer. The reaction mixture was at 37°C for 1 h, to which 200 μL solution of anti-digoxigenin-peroxide and ABTS [2,29-azino bis(3-ethylbenzothiazoline-6-sulfonic acid) diammonium salt] substrate were added for coloring reaction. The absorbance of each well was measured at 405 nm with the reference wavelength at 490 nm. Nevirapine (Viramune, Boehringer-Ingelheim Pharma KG, Germany) was used as a reference compound. Enzyme kinetic assay was carried out as outlined above except for the concentration of enzyme solution (1 ng of the HIV-1 RT), incubation time (30, 52, 80, 105, 130 min), and various concentrations of either the template/primer (1500, 750, 187.5 mA_{260nm}/mL) or the triphosphate substrate (20, 15, 10, 5, 2.5 μM) in the presence of the inhibitor, 8,89-bieckol.

2.2 RESPIRATORY SYNCYTIAL VIRUS ASSAY[2]

In 6-well plates and at room temperature, the monolayer cultures of Vero cells were infected in DMEM (2% serum) with serially diluted respiratory syncytial virus (RSV) for 1.5 h. The culture medium was removed, replaced with DMEM containing methylcellulose, and incubated for 5 days. The cells were fixed with methanol/acetone (1/1), and viral plaques were visualized by immunoperoxidase staining. Primary antibody was mouse anti-RSV (1:1000 in 1X PBS), and the secondary antibody was peroxidase-conjugated anti-mouse IgG (1:1000 dilution in 1X PBS). Using 3,3′-diaminobenzidine/peroxide solution, plaques were developed, and plaque numbers were quantified by visual counting.

2.3 INFLUENZA VIRUS TYPES A AND B ASSAY[3,4]

The influenza virus types A and B rapid assay of all specimens was parallel with viral culture. In rapid assays, the manufacturer's instructions were followed, which consisted of two simple steps and took 30 min or less. The results are available to clinicians within 2 h of specimen receipt. Measures of quality controls including internal kit positive and negative controls, as well as external laboratory controls were performed for each test run. Specimens were simultaneously inoculated onto three cell culture monolayers [human foreskin fibroblasts, human lung carcinoma (A549), and rhesus monkey kidney cells] and examined daily for 21 days under light microscopy for evidence of viral cytopathic effect. On days 2, 5, and 14 of cell culture, incubation hemadsorption with guinea pig red blood cell suspension was tested. To identify all hemadsorption-positive cell cultures, virus-specific immunofluorescence assays were used. Using functional acid lability assays, picornaviruses were typed as enteroviruses or rhinoviruses.

2.4 NASAL EXHALED NO CONCENTRATION ASSAY[5,6]

Direct online measurement of exhaled NO (eNO) was made according to American Thoracic Society guidelines by using a chemiluminescence analyzer with a lower limit of detection of 1 ppb. For lower airway and nasal eNO measurement, to maintain constant flows of 9, 27, 41, and 78 mL/s, a mouthpiece or nasal mask through needle restrictors at pressures of 10 or 20 cm H_2O was used. Because of nasal eNO measurements following this method including lower airway NO, the wall concentration of NO from the nose should be extrapolated by performing a linear regression of nasal eNO versus flow by using three or four flows where reproducible measures are available. By extrapolating this regression to 0 flow, nasal NO wall concentration can be defined. Similarly, lower airway NO wall concentration could be determined by means of linear regression to 0 flow. Values for nasal and lower airway eNO extrapolated to 0 flow were used for data analysis.

2.5 NASAL NOS2 mRNA QUANTITY ASSAY[7–13]

With a Qiagen RNeasy Kit (Hilden, Germany), total RNA was extracted from nasal scrapings and nasal lavage fluid. On an Applied Biosystems Model 7700 Sequence Detector, RNA (100 ng) was transcribed to cDNA. In the presence of specific primers and a fluorescently labeled probe selected with Primer Express software, cDNA was given to PCR amplification. The forward primers of 59GGGTTTTCA TCTATGCTA TTGAG ATCA T39, reverse primers of 59AGT TGG CGT ATG TAT TGA GGG TAA G39, and probe primers of FAM-CTGAACAACAA-GGCCAAATTCAAATCTCTAAACATAMRA (Perkin Elmer Applied Biosystems Division) were used as the primers of nitric oxide synthase 2 (NOS2) and human rhinoviruses (HRV)-16. Absolute quantity of NOS2 mRNA was determined by using relative standard curves and was prepared by using a first-strand NOS2 cDNA (NOS2 amplicon, a 122-bp sequence in the NOS2 gene from 1457 to 1578 that included the binding sites for the forward and reverse NOS2 primers). The NOS2 amplicon was used in a range of 0.01 fg to 10 pg. Quantities of NOS2 mRNA in nasal scraping cells were expressed relative to glyceraldehyde-3-phosphate dehydrogenase

(GAPDH). A standard curve for HRV-16 was prepared by using RNA isolated from dilutions of a standard HRV-16 preparation (log $TCID_{50}$ ranged from 5 to 2).

2.6 RT-PCR AND SWINE ASSAY FOR ANTI-HEV ANTIBODY[14–20]

Swine hepatitis E virus (HEV) IgG antibodies were examined with enzyme-linked immunosorbent assay (ELISA) to find seronegative pigs from 75 specific-pathogen-free (SPF) pigs (2 weeks old) for inoculation. Prior to inoculation, the pigs were allowed to acclimate to the research facilities for 1 week. Tissues (liver, heart, pancreas, or skeletal muscle) and feces of HEV-infected SPF pigs were collected at 3–7, 14–20, and 27–55 days postinoculation (DPI), pooled, and stored at −70°C for inoculation. Each inoculum was prepared as a tissue homogenate or fecal suspension (10%, w/v) in PBS and tested by a semiquantitative RT-PCR for swine HEV RNA. The positive control inoculum was a standard infectious pool of swine feces with $10^{4.5}$ 50% pig infectious doses (PID_{50}) of swine HEV/mL.

Total RNA was extracted from 100 μL of each sample by Trizol reagent and tested by a nested PCR with primers located in the putative capsid gene (ORF2). The first-round PCR produced an expected fragment of 404 bp with the forward primer F1 (5′-AGCTCCTGTACCTGATGTTGACTC-3′) and the reverse primer R1 (5′-CTACAGAGCGCCAGCCTTGATTGC-3′) and for the second round an expected fragment of 266 bp with the forward primer F2 (5′-GCTCACGTCATCTGTCGCT-GCTGG-3′) and the reverse primer R2 (5′-GGGCTGAACCAAAATCCTGACA-TC-3′). Total RNA was reverse transcribed with R1 reverse primer and SuperScript II reverse transcriptase at 42°C for 1 h. The resulting cDNA was amplified by PCR with ampliTaq Gold DNA polymerase. The PCR reaction was carried out for 39 cycles of denaturation at 94°C for 1 min, annealing at 52°C for 1 min, extension at 72°C for 1.5 min, and incubation at 72°C for 7 min. Ten microliters each round of PCR were mixed as the template. The amplified PCR products were examined by gel electrophoresis. The virus titer of inocula was calculated and expressed as genome equivalent (GE)/mL. Serum samples of inoculated pigs collected from 7 to 21 days after inoculation were also tested by RT-PCR for swine HEV RNA.

Anti-HEV IgG antibodies in swine sera were detected using a standardized ELISA. A purified 55-kDa truncated recombinant putative capsid protein of human HEV strain Sar-55 was used as the antigen that cross-reacts well with the swine HEV. Peroxidase-labeled goat anti-swine IgG was used as the secondary antibody. Duplicates per serum sample were used.

2.7 HIV-1 PROTEASE AND REVERSE TRANSCRIPTASE KINETIC ASSAY[21]

Nonradioactive HIV-1 reverse transcriptase activities were evaluated with an HIV-1 RT nonradioactive assay kit (Roche Diagnostics GmbH, Germany). To the wells of a streptavidin-coated microtiter plate containing 20 μL test sample solution and 4 ng HIV-1 RT in 20 μL lysis buffer, 20 μL reaction mixture containing a homogenous template/primer hybrid, (rA)n(dT)15 (750 mA_{260nm}/mL final concentration), and a triphosphate substrate, dUTP/dTTP (10 μM final concentration), was added. The reaction was carried out at 37°C for 1 h and followed by the addition of 200 μL solution of anti-digoxigenin-peroxide and ABTS substrate for coloring reaction. The absorbance of each well was recorded at 405 nm with the reference wavelength at 490 nm. Nevirapine was used as a reference compound.

Enzyme kinetic assay was carried out as outlined above except for the concentration of enzyme solution (1 ng HIV-1 RT), incubation time (30, 52, 80, 105, 130 min), and various concentrations of either the template/primer (1500, 750, 187.5 mA 260 nm/mL) or the triphosphate substrate (20, 15, 10, 5, 2.5 mM) in the presence of the inhibitor.

HIV-1 protease assay was performed with slight modification. One microliter of a dimethyl sulfoxide solution of test compound was mixed with 10.5 μL substrate solution (His-Lys-Ala-Arg-Val-Leu-(*p*-NO_2)-Phe-Glu-Ala-Nle-Ser-NH_2, 0.1 mg/mL in HIV-1 protease assay buffer, Bachem AG, Switzerland) and then 0.5 μL recombinant protease solution (0.3 mg/mL) was added. After a 37°C and 15 min incubation, the reaction was stopped by the addition of 1.2 μL 10% trifluoroacetic acid and diluted with 20 μL water. The hydrolysate and the remaining substrate were quantitatively analyzed using HPLC with an Inertsil ODS-3 column (4.6 × 150 mm, GL Sciences Inc., Japan), a linear gradient of CH_3CN (15% to 40%) in 0.1% trifluoroacetic acid (TFA) elution, 20 μL injection volume, 1.0 mL/min flow rate, and 280 nm detection. The hydrolysate and substrate were eluted at 8.61 and 10.84 min, respectively.

2.8 ANTI-HIV ASSAY[22]

In the anti-HIV activity of the test compound at varying concentrations (from 100 nM by double dilution technique) against replication of HIV-1 (III B) assay, HIV-1-infected MT-4 cells were grown in RPMI-1640 DM medium supplemented with 10% (v/v) heat-inactivated calf serum and 20 μg/mL gentamicin. HIV-1 (III B) from the culture supernatant of normal MT-4 cells or HIV-1-infected MT-4 cells was stored at −70°C as the virus stocks until used. Anti-HIV assays were carried out in microtiter plates filled with 100 μL medium and 25 μL medium containing test compound in triplicate so as to allow simultaneous evaluation of their effects on HIV and mock infected cells. To either the HIV-infected or mock-infected part of the microtiter tray, 50 μL HIV in 100 μL CCID50 medium was added. At 37°C, the cell cultures were incubated in a humidified atmosphere of 5% CO_2 in air. After 5 days infection, the viability of mock- and HIV-infected cells were examined by the MTT method. The EC_{50} (effective concentration of compound achieving 50% protection in MT-4 cell lines against the cytopathic effect of HIV-1, nM) and CC_{50} (cytotoxic concentration of compound required to reduce the viability of mock-infected MT-4 cells by 50%, mM) values were calculated.

2.9 ROBUST ANTIVIRAL ASSAYS[23]

Vero cells were cultured in minimum essential medium (MEM; Invitrogen) supplemented with 10% FCS, 2 mM L-glutamine, and 0.04% gentamicin (50 mg/mL). The identical medium supplemented with 2% FCS was used for test compound treating culture medium. Vero cells (50 μL, 2.5 × 10^4 cells/well) were added to a 96-well Whiteview plate containing 25 μL of fourfold serially diluted testing compound in cell culture medium with 2% FCS. To each well, 25 μL of virus stock was added [0.01 multiplicity of infection for yellow fever virus (YFV), 0.001 multiplicity of infection for west nile virus (WNV), sindbis virus, and coxsackie B virus, respectively]. Cell controls received only cells and medium, and virus controls received virus without test compound. Plates were incubated at 37°C until the viral cytopathic effect (CPE) in the virus control wells reached about 100%. According to the supplier's instructions, ATPLite was added to all wells. Briefly, to each well 100 μL reconstituted lyophilized substrate solution was added, the plate was shaken at 700 rpm for 2 min, and the luminescence was measured on a Viewlux apparatus (Perkin Elmer Applied Biosystems Division) by taking a 0.1- to 0.5-s integrated reading of each test plate. The results were expressed as EC_{50}, the concentration of test compound having 50% inhibition of the virus-reduced luminescence signals compared with the uninfected cell control. The signal-to-noise ratio of an assay is the ratio between the mean luminescent signals of the cell controls and the virus controls. The dynamic range is defined as the ratio between the signals at the last (maximal signal) and first points in the linear range of the dose-response curve.

2.10 HIV/SIV FUSION ASSAY[24]

In the HIV-1 envelope-mediated fusion assay, the U87 cell lines were serum-starved for 36 h at 37°C and infected overnight with vCB21R at multiplicity of infection of 10, or in some cases with recombinant vaccinia virus encoding the constitutively active GTPase RacV12 mutant (vRacV12). Fusion partner BSC40 cells were infected overnight with vPT7-3 and vCB-39 encoding the adenosine deaminase activity (ADA) envelope. In some assays, 10^6 BSC40 cells were transfected with 5–10 μg of proviral plasmids for 1 h and infected with vPT7-3. Cells were then lightly trypsinized, washed, and at 37°C with or without the Rac GEF inhibitor incubated for 1 h. Cells (10^5) of each type were then mixed 1/1 in triplicate wells, incubated at 37°C for 3 h, and the fusion was stopped by adding Nalidet P-40 (NP-40) to 1% final concentration and freeze thawing at −20°C. β-Galactosidase activity was measured using chloro-phenol red-β-D-galactopyranoside and at 579 nm determining absorbance of each sample.

Free virus particles obtained 48 h after transfection of 293T cells were used for the virus-dependent fusion assay. In the presence of 20 μg/ml diethylaminoethyl (DEAE)-dextran virus was incubated with 10^5 U87.CD4.CCR5 cells, infected with vPT7-3 for 18 h, while 10^5 U87.CD4.CCR5 cells were infected with vCB21R for 18 h. Assays were performed in triplicate wells for fusion activity.

In Rac and Rho activation assays, 2×10^6 serum-starved U87.CD4.CCR5 cells were mixed with BSC40 cells infected with vaccinia virus expressing Env or vWT at 1/1 ratio. In some cases, one population of U87.CD4.CCR5 cells was infected with vRacV12. At the indicated concentrations, Rac GEF inhibitor was added to the target cells 1 h prior to mixing, incubated at 37°C for 30 min, washed two times with ice-cold PBS, the cells were lysed, the lysates were snap frozen, and, using a G-LISA Rac activation or Rho-A activation assay kit (Cytoskeleton, Denver, CO) according to the manufacturer's instructions, equal amounts of protein per well were analyzed.

2.11 RAPID DNA HYBRIDIZATION ASSAY[25]

To all wells except the cell control wells of a round-bottom 96-well plate, growth media containing 50 μg/mL goat anti-human IgG antibody was added. Serial dilutions of test compounds in duplicate with concentrations ranging from 100 to 0.032 μM were prepared. To each well, 100 μL of dilution of infected Akata cells was added at a final concentration of 4×10^5 cells/mL, the plates were incubated for 72 h, and using the blot hybridization assay, antiviral activity was assessed. After incubation, 100 μL and 50 μL denaturation buffer (1.2 M NaOH, 4.5 M NaCl) were added to each well and the wells of a Biodot (Bio-Rad, Hercules, CA) apparatus containing an Immobilon nylon membrane, respectively. From the cells that were induced to undergo a lytic infection, a 50-μL denatured DNA aliquot was aspirated through the membrane and then 50 μL of denaturation buffer was added. Before beginning the hybridization process, the membrane was allowed to dry, and incubated in DIG Easy Hyb (Roche Diagnostics, Indianapolis, IN) at 56°C for 30 min. According to the manufacturer's instructions, using PCR primers EBNA probe forward 5′-CCCAGGAGTCCCAGTAGTCA-3′ and EB virus nuclear antigen (EBNA) probe reverse 5′-CAGTTCCTCGCCTTAGGTTG-3′, a specific digoxigenin (DIG)-labeled probe was prepared. The resulting probe corresponding with coordinates 96802-97234 in Epstein-Barr virus (EBV) genome (AJ507799) was hybridized at 56°C to the blots overnight, blots were washed at 56°C with 0.2× SSC, with 0.1% SDS and 0.1× SSC, and with 0.1% SDS. With anti-DIG antibody, according to the manufacturer's instructions, specifically bound DIG probe was detected. With QuantityOne software (Bio-Rad Hercules, CA), an image of the film was captured and quantified. Using standard methods, test compound concentrations sufficient to reduce the accumulation of viral DNA by 50% (EC_{50}) were interpolated from the dose response.

2.12 ANTIVIRAL SCREENING ASSAY FOR HepAD38 CELL CULTURES[26]

In 60-mm dishes, HepAD38 cells were cultured with DMEM/F-12 medium supplemented with 10% FBS, 100 IU/mL penicillin, and 100 μg/mL streptomycin. In some cases, 1 μg/mL tetracycline was routinely added to suppress HBV pre genome RNA (pgRNA) transcription. At indicated time points, cells and culture media were harvested. By centrifugation at 1000 × *g* for 10 min, media were clarified, cells were washed with cold PBS then stored at −70°C until used.

Cells from a 60-mm dish were lysed with 1 mL lysis buffer containing 10 mM Tris-HCl (pH 8.0), 10 mM EDTA, 1% NP-40, and 8% sucrose at 37°C for 10 min; by centrifugation cell debris and nuclei were removed; the supernatant was mixed with 250 μL 35% polyethylene glycol (PEG)-8000 containing 1.5 M NaCl; incubated

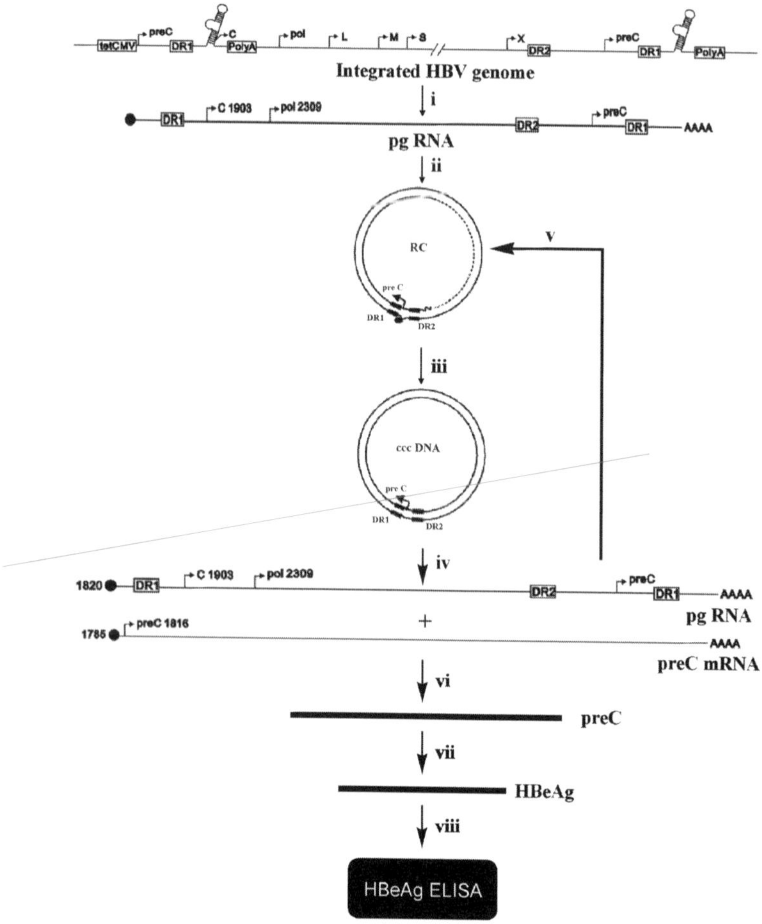

Figure 2.1 Schematic representation of procedure. (i) The integrated HBV DNA in HepAD38 cells consisted of a 1.1 overlength HBV genome and beginning from nt 1807 is placed downstream of a tetCMV promotor. (ii) Removing tetracycline from the cultures and pgRNA packaging results in viral pgRNA transcription and replication. (iii) Repair for cccDNA formation. (iv) Templates for new rounds of viral DNA synthesis. (v) Transcription from a cccDNA template to both pre-core mRNA and pgRNA. (vi) Translation from pre-core mRNA and pgRNA into preC both HBeAg and pgRNA. (vii) Posttranslation processing and secretion. (viii) HBeAg was measured by using the EKB-PLUS ELISA kit.

in ice for 1 h; viral nucleocapsids were pelleted at 4°C by centrifugation at 12,000 × *g* for 10 min; digested for 1 h at 37°C in 400 μL digestion buffer containing 0.5 mg/mL pronase, 0.5% SDS, 150 mM NaCl, 25 mM Tris-HCl (pH 8.0), and 10 mM EDTA; the digestion mixture was extracted twice with phenol; and DNA was precipitated with ethanol and dissolved in TE buffer (10 mM Tris-HCl, pH 8.0, 1 mM EDTA). From each plate, one sixth of the DNA sample was resolved by electrophoresis into a 1.5% agarose gel. The gel was denatured in a solution containing 0.5 M NaOH and 1.5 M NaCl, and the gel was neutralized in a buffer containing 1 M Tris-HCl (pH 7.4) and 1.5 M NaCl. In 20× SSC buffer, DNA was blotted onto Hybond-XL membrane (GE Health care).

Following the manufacturer's instructions, total cellular RNA was extracted with Trizol reagent, 10 μg total RNA was resolved in 1.2% agarose gel containing 2.2 M formaldehyde and then transferred onto Hybond-XL membrane in 20X SSC buffer.

Using a modified Hirt extraction procedure, covalently closed circular DNA (cccDNA) was extracted. Briefly, cells from a 60-mm dish were lysed for 30 min in 3 mL 10 mM Tris-HCl (pH 7.5), 10 mM EDTA, and 0.7% SDS at room temperature, the lysate was transferred into a 15-mL tube, 0.8 mL 5 M NaCl was added, then incubated overnight at 4°C. The lysate was clarified by centrifugation at 12,000 × *g* for 30 min at 4°C and extraction twice with phenol and once with phenol/chloroform. At room temperature, with two volumes of ethanol DNA was precipitated overnight and dissolved in 60 μL TE buffer (10 mM Tris-HCl, pH 8.0, 1 mM EDTA). At 85°C, one third of the cccDNA sample was denatured for 5 min, separated in a 1.2% agarose gel, and transferred onto Hybond-XL membrane.

For detecting HBV DNA and RNA, membranes were probed either with an α-^{32}P-UTP (800 Ci/mmol) labeled minus or plus strand specific full-length HBV riboprobe. In 5 mL EKONO hybridization buffer with 1 h prehybridization at 65°C and overnight hybridization at 65°C, 1 h wash with 0.1× SSC and 0.1% SDS at 65°C, hybridization was carried out. The membrane was exposed to a phosphoimager screen, and hybridization signals were quantified with QuantityOne software.

By adding polyethyleneglycol (PEG)-8000 to a final concentration of 10%, incubating on ice for 1 h and centrifugation at 4°C and 6000 × *g* for 10 min, viral particles (including virions, subviral particles, and nucleocapsids) in culture medium were precipitated, dissolved in TNE buffer containing 10 mM Tris-HCl (pH 8.0), 100 mM NaCl, and 1 mM EDTA, resolved in a 1% agarose gel, and transferred onto nitrocellulose membrane. By hybridization with a α-^{32}P-UTP (800 Ci/mmol) labeled minus strand specific full-length HBV riboprobe, viral DNA was detected (Fig. 2.1).

2.13 TRAK-C HCV CORE ASSAY[27]

Growth of Huh-7 clone 21-5, which harbored the bicistronic, full-length HCV I_{389}neo/Core-3′ replicon with the cell culture adapted HCV 1b sequence NK5.1, in the presence of G418 (at 250 μg/mL) in DMEM with Earle's salts, 10% FCS, 2 mM L-glutamine, 1X nonessential amino acids, 100 U/mL penicillin, and

100 μg/mL streptomycin, was strictly dependent on the replication of the HCV RNA (introduced into the cell by transfection), which harbored a neomycin resistance gene located 5′ to the HCV genome sequence. Non-transfected Huh-7 human hepatoma cells harboring no replicon RNA were used as control and maintained in the medium except the G418 was omitted.

Huh-7 clone 21-5 replicon cells or control Huh-7 cells were split from subconfluent monolayers by trypsin addition, counted with a hemocytometer by Trypan blue exclusion, seeded into microplates, grown at 37°C for 24 h, growth medium was collected from cell monolayer aliquots of the tissue culture, centrifuged at 1700 × *g* for 5 min to remove any cell debris, and the clarified supernatants were then stored at −80°C prior to the trak-C assay.

For transient replication assays, *in vitro* RNA transcripts of Con1/wt and Con1/5.1 (10 μg each) were generated from the unmodified full-length HCV 1a genomes harboring the wild-type (Con1/wt) or the cell culture-adapted Con1/5.1 genome containing three adaptive mutations (E1202G, T1280I, and S2197P) and electroporated into 8×10^6 Huh-7 cells essentially. Two independent electroporations of the same transcript were pooled and seeded into parallel tissue culture dishes, cells were lysed in 1 mL PBS supplemented with 1% Triton-X-100 at 4, 12, 24, and 48 h after electroporation, lysates were diluted 1/50 in PBS and probed with the trak-C assay.

Homozygous transgenic Alb-uPa/SCID mice harboring chimeric livers populated with primary human hepatocytes were generally generated and subsequently inoculated with patient sera harboring genotype 1a HCV. From HCV-infected mice on day 31 postinfection or a control noninfected mouse, serum samples were prepared, and aliquots (10 μL) of these serum samples and of the patient inoculum were tested in the trak-C assay. Only for core detection from the inoculum was the trak-C pretreatment step (to dissociate antibody–antigen complexes) necessary.

Using trak-C assay, core antigen was measured quantitatively. The 96-well microplates containing cell extracts (or cell culture supernatants) prepared as described above were thawed and at room temperature incubated for 10 min. In the trak-C assay for serum samples, 50 μL ortho-ELISA pretreatment buffer was added to 100 μL samples, and the contents were heated to 56°C for 30 min to dissociate any antigen–antibody immune complexes, which would otherwise mask the presence of HCV core in the sample, whereas in the trak-C assay for extracts from HCV replicon cells or cells transfected with HCV genomes, this step was unnecessary (as no antibody is present in these *in vitro* systems) and therefore omitted. The mixture of 10 μL cell extract (10% of the starting material) or cell culture supernatant and 75 μL trak-C standard diluent buffer was added to microwells coated with multiple HCV anti-core monoclonal antibodies containing 100 μL ortho-ELISA reaction buffer. When test serum samples were prepared from the murine model, 10 μL mouse serum was mixed with 80 μL trak-C standard diluent buffer without any pretreatment step. The only sample that did require the pretreatment step in the murine assay was the patient inoculum. The microwells were incubated at 25°C for 60 min; washed six times with 1× ortho-wash buffer with PBS and 0.05% Tween 20 [(polyoxyethylene sorbitan mono laurate)urea] to remove any unbound material; to each well 200 μL solution containing anti-core monoclonal antibody Fab fragments conjugated to horseradish peroxidase was added; the contents were incubated at 25°C for 30 min, washed, and aspirated using 1 × wash buffer with urea to remove unbound conjugate; to each well 200 μL substrate solution containing *o*-phenylenediamine (OPD) and hydrogen peroxide were then added; and at room temperature in the dark for 30 min, to each well 50 μL stop solution (4 N sulfuric acid) was added. The plate was read at 492 nm with a standard plate reader using 600- to 650-nm reference absorbance. To estimate the core antigen concentration, the assay included a series of recombinant HCV core protein dilutions containing 0, 1.5, 5, 10, 50, 100 pg of core and was run on each plate without the pretreatment step to generate a standard curve. Prior to evaluation of the standard curve and sample core determinations, the absorbance measured from blank wells (without cell extract) processed with all subsequent reagents was background subtracted from all other absorbance values. HCV core concentration in test samples was determined by interpolation to the standard curve of recombinant core protein obtained using linear regression analysis based on the equation $y = (OD_{sample} - y\text{-axis intercept})/\text{slope}$, where y was the sample concentration (in pg).

REFERENCES AND NOTES

1. M. Ahn, K. Yoon, S. Min, J.S. Lee, J.H. Kim, T.G. Kim, S.H. Kim, N.G. Kim, H. Huh, J. Kim. Inhibition of HIV-1 reverse transcriptase and protease by phlorotannins from the brown alga Ecklonia cava. Biol Pharm Bull 27 (**2004**) 544–547. *Note*: In HIV-1 protease assay, 1 μL of a dimethyl sulfoxide solution of test compound was mixed successively with 10.5 μL of the substrate solution (His-Lys-Ala-Arg-Val-Leu-(*p*-NO_2)-Phe-Glu-Ala-Nle-Ser-NH_2, 0.1 mg/mL in HIV-1 protease assay buffer; Bachem AG, Bubendorf, Switzerland) and 0.5 μL of the recombinant protease solution (0.3 mg/mL; Bachem AG). After incubation at 37°C for 15 min, the reaction was stopped by adding 1.2 μL 10% TFA and diluted with 20 μL water. The hydrolysate and the remaining substrate were quantitatively analyzed by HPLC (column, Inertsil ODS-3, 4.6 × 150 mm; GL Sciences Inc., Tokyo, Japan; elution, a linear gradient of CH_3CN (15–40%) in 0.1% TFA; injection volume, 20 μL; flow rate, 1.0 mL/min; and detection, 280 nm) and the hydrolysate and substrate were eluted at 8.61 and 10.84 min, respectively. % Inhibition $= 100 \times (A_{control} - A_{sample})/(A_{control})$, where A is a relative peak area of the hydrolysate. Acetyl pepstatin (Bachem AG) was used as a positive control in this assay.
2. V. Razinkov, C. Huntley, G. Ellestad, G. Krishnamurthy. RSV entry inhibitors block F-protein mediated fusion with model membranes. Antiviral Res 55 (**2002**) 189–200.
3. A.T. Cruz, A.C. Cazacu, J.M. Greer, G.J. Demmler. Rapid assays for the diagnosis of influenza A and B viruses in patients evaluated at a large tertiary care children's hospital during two consecutive winter seasons. J Clin Virol 41 (**2008**) 143–147. *Note*: The procedures were also used for identifying adenoviruses, cytomegalovirus (CMV), enteroviruses, herpes simplex virus (HSV), influenza A and B viruses, parainfluenza viruses types 1–4, RSV, and rhinoviruses. In these assays, standard laboratory techniques were used and viral culture was considered the reference standard for the diagnosis of influenza infection. Standard 2 × 2 tables were used for the calculation of the sensitivity, specificity, and likelihood ratios. Confidence intervals for proportions were calculated with a confidence likelihood of 95%. Performance of the tests during periods of peak influenza activity was analyzed *a priori* to investigate differences in test performance due to viral prevalence.

4. M. Hindiyeh, C. Goulding, H. Morgan, B. Kenyon, J. Langer, L. Fox, G. Dean, D. Woolstenhulme, A. Turnbow, E. Billetdeaux, S. Shakib, C. Gordon, A. Powers, G. Vardeny, M. Johnson, L. Skodack-Jones, K. Carroll. Evaluation of BioStar® FLU OIA® assay for rapid detection of influenza A and B viruses in respiratory specimens. J Clin Virol 17 (**2000**) 119–126. *Note*: Primary rhesus monkey kidney (PRMK) cells, Buffalo green monkey kidney (BGM) cells, rhabdomyosarcoma (RD) cells, human lung carcinoma (A549) cells, and human diploid fibroblasts (MRC-5) cells grown in shell vials were maintained in Eagle's minimal essential medium (EMEM) supplemented with 10% fetal bovine serum at 37°C in a CO_2 incubator. Once inoculated, the cells were maintained in EMEM supplemented with 2% fetal bovine serum, 50 U/mL penicillin, 50 μg/mL streptomycin, 20 μg/mL vancomycin, 20 μg/mL gentamicin, and 1 μg/mL fungizone in a 5% CO_2 incubator at 34°C. The specimens submitted for influenza virus culture were inoculated into three PRMK, while those submitted for respiratory virus cultures were inoculated into three PRMK, one BGM, one RD, one MRC-5 and one A549. Cell monolayers were inoculated, centrifuged at 2800 × *g* at 20°C for 15 min, and the vials were incubated in a CO_2 incubator at 34°C. On every other day, possible cytopathic effects (CPEs) in the cell monolayers were examined. In addition, the PRMK cells were hemadsorbed on days 3–4, 6–7, and 14. Using Bartels fluorescent antibody stain, all samples with positive hemadsorption were tested for influenza virus A and B, parainfluenza viruses 1–3, as well as RSV. Respiratory cultures were kept a total of 14 days. For Bartels fluorescent antibody stain, to 1–2 mL of the thoroughly vortexed respiratory sample aliquot in a holding tube, approximately 3 mL PBS was added and centrifuged at 500 × *g* for 10 min. The supernatant was poured off for a second wash. The cell pellet was resuspended in two to eight drops of PBS, spotted onto an acetone-cleaned 3-well slide, and allowed to air dry under the biohazard hood. Then the slides were fixed with acetone and allowed to dry completely. To the appropriate wells, adequate amounts of mouse anti-influenza A and B antibodies were added. The third well was incubated with polyclonal respiratory viral pool. In a humid chamber, the slides were incubated at 37°C for 30 min. The primary antibody in each well on the slide was rinsed with PBS, to the residue secondary antibody (anti-mouse IgGFITC) was added, and the culture was incubated for 30 min at 37°C. After the secondary antibody was washed off and a coverslip was placed over the sample wells, fluorescent staining was observed using a fluorescence microscope at 250× and 400× magnification.

5. S.P. Sanders, D. Proud, S. Permutt, E.S. Siekierski, R. Yachechko, M.C. Liu. Role of nasal nitric oxide in the resolution of experimental rhinovirus infection. J Allergy Clin Immunol 113 (**2004**) 697–702.

6. P.E. Silkoff, J.T. Sylvester, N. Zamel, S. Permutt Airway nitric oxide diffusion in asthma: Role in pulmonary function and bronchial responsiveness. Am J Respir Crit Care Med 161 (**2000**) 1218–1228.

7. P. McErlean, L.A. Shackelton, S.B. Lambert, M.D. Nissen, T.P. Sloots, I.M. Mackay. Characterization of a newly identified human rhinovirus, HRV-QPM, discovered in infants with bronchiolitis. J Clin Virol 39 (**2007**) 67–75.

8. K.E. Arden, P. McErlean, M.D. Nissen, T.P. Sloots, I.M. Mackay. Frequent detection of human rhinoviruses, paramyxoviruses, coronaviruses, and bocavirus during acute respiratory tract infections. J Med Virol 78 (**2006**) 1232–1240. *Note*: Clinical specimens (n = 1244) were predominately nasopharyngeal aspirates (NPAs; 92%). NPAs and purified nucleic acid extracts were stored at −70°C. Extracts were tested by RT-PCR (OneStep; Qiagen) for human metapneumovirus (HMPV), human rhinoviruses (HRVs),

and enteroviruses (HEVs), all four non-SARS viruses human coronaviruses (HCoVs), human bocavirus (HBoV), respiratory syncytial virus (HRSV), influenza A and B viruses, parainfluenza viruses, and human adenoviruses. Real-time RT-PCR was performed on the Rotor-Gene 3000 (Corbett Research) by use of 0.2–0.4 pmol oligonucleotide. Amplicons were purified when necessary and both strands sequenced (Big Dye v3.1; Applied Biosystems). First-strand cDNA synthesis (Transcriptor RT; Roche) using (T) 17ATA or gene-specific primers was followed by second-strand synthesis and amplification (Platinum *Pfx*; Invitrogen). HeLa-Ohio, A549, MRC-5, and WI-38 cells inoculated with patient respiratory secretions stored at −70°C were rolled at 33°C, watched daily for cytopathic effect, and HRV-QPM RNA levels monitored by real-time RT-PCR.

9. S. Moretti, F. Reinier, O. Poirot, F. Armougom, S. Audic, V. Keduas, C. Notredame. PROTOGENE: Turning amino acid alignments into bona fide CDS nucleotide alignments. Nucl Acids Res 34 (**2006**) W 600–603; S. Kumar, K. Tamura, M. Nei. MEGA3: Integrated software for molecular evolutionary genetic analysis and sequence alignment. Briefings Bioinform 5 (**2004**) 150–163. *Note*: With the program MUSCLE (Edgar, 2004) and Protogene, nucleotide or amino acid sequences were aligned and back translated, respectively. Maximum likelihood trees were constructed using a GTR + I + Γ_4 model of nucleotide substitution in PAUP and visualized with Tree Edit. The nucleotide distance matrix for neighbor-joining trees was generated using the Kimura two-parameter estimation in MEGA 3.1. Nodal confidence values indicate the results of bootstrap resampling ($n = 1000$).
10. K. Loens, H. Goossens, C. de Laat, H. Foolen, P. Oudshoorn, S. Pattyn, et al. Detection of rhinoviruses by tissue culture and two independent amplification techniques, nucleic acid sequence-based amplification and reverse transcription-PCR, in children with acute respiratory infections during a winter season. J Clin Microbiol 44 (**2006**) 166–171.
11. V. Caro, S. Guillot, F. Delpeyroux, R. Crainic. Molecular strategy for 'serotyping' of human enteroviruses. J Gen Virol 82 (**2001**) 79–91.
12. M.S. Oberste, K. Maher, D. Schnurr, M.R. Flemister, J.C. Lovchik, H. Peters, et al. Enterovirus 68 is associated with respiratory illness and shares biological features with both the enteroviruses and the rhinoviruses. J Gen Virol 85 (**2004**) 2577–2584.
13. P. Simmonds, J. Welch. Frequency and dynamics of recombination within different species of human enteroviruses. J Virol 80 (**2006**) 483–493. *Note*: To design new PCR primers permitting determination of the continuous coding sequence from overlapping fragments, the initial HRV-QPM sequences were used and bracketed by partial untranslated regions. Additional primers from HRV and HEV studies supplemented the process. To confirm the primary sequence was from a distinct virus and not the result of mixed template in the patient's specimen, a second round of RT-PCR and sequencing of overlapping fragments should be done using newly designed HRV-QPM-specific primers targeting RNA from the original specimen extract. A third round of RT-PCR and sequencing could be used to conform discrepant sequence results.
14. C. Kasorndorkbua, P.G. Halbur, P.J. Thomas, D.K. Guenette, T.E. Toth, X.J. Meng. Use of a swine bioassay and a RT-PCR assay to assess the risk of transmission of swine hepatitis E virus in pigs. J Virol Methods 101 (**2002**) 71–78.
15. P.G. Halbur, C. Kasorndorkbua, C. Gilbert, D. Guenette, M.B. Potters, R.H. Purcell, S.U. Emerson, T.E. Toth, X.J. Meng. Comparative pathogenesis of infection of pigs with hepatitis E virus recovered from a pig and human. J Clin Microbiol 39 (**2001**) 918–923.

16. X.J. Meng, P.G. Halbur, J.S. Haynes, T.S. Tsareva, J.D. Bruna, R.L. Royer, R.H. Purcell, S.U. Emerson. Experimental infection of pigs with the newly identified swine hepatitis E virus (swine HEV), but not with human strains of HEV. Arch Virol 143 (**1998**) 1405–1415.

17. X.J. Meng, P.G. Halbur, M.S. Shapiro, S. Govindarajan, J.D. Bruna, I.K. Mushahwar, R.H. Purcell, E.U. Emerson. Genetic and experimental evidence for cross species infection by swine hepatitis E virus. J Virol 72 (**1998**) 9714–9721.

18. N. Jothikumar, T.L. Cromeans, B.H. Robertson, X.J. Meng, V.R. Hill. A broadly reactive one-step real-time RT-PCR assay for rapid and sensitive detection of hepatitis E virus. J Virol Methods 131 (**2006**) 65–71. *Note*: Primers and probes for a rapid and sensitive real-time RT-PCR assay detecting HEV RNA were selected based on the multiple sequence alignments of 27 sequences of the ORF3 region. Thirteen HEV isolates corresponding with genotypes 1–4 were used to standardize the real-time RT-PCR assay. This assay specifically detected four genome equivalent (GE) copies of HEV plasmid DNA and sensitively detected 0.12 50% pig infectious dose (PID_{50}) of swine HEV. Different concentrations of swine HEV (120–1.2 PID_{50}) spiked into a surface water concentrate were detected in the real-time RT-PCR assay.

19. J. Inoue, M. Takahashi, Y. Yazaki, F. Tsuda, H. Okamoto. Development and validation of an improved RT-PCR assay with nested universal primers for detection of hepatitis E virus strains with significant sequence divergence. J Virol Methods 137 (**2006**) 325–333. *Note*: The nested universal primers were designed based on the analysis of 70 entire or nearly entire HEV sequences and within the ORF2/ORF3 overlapping region for PCR capable of amplifying HEV sequences of genotypes 1–4, and the amplified products could be used for HEV genotyping. The newly designed primers in this assay were sense primers HE361 (5′-GCR GTG GTT TCT GGG GTG AC-3′ with R = A or G) and HE366 (5′-GYT GAT TCT CAG CCCTTC GC-3′) antisense primers HE363 (5′-GMY TGG TCD CGCCAA GHG GA-3′ with H = A, T, or C) and HE364 (5′-CTGGGM YTG GTC DCG CCA AG-3′ with M = A or C; Y = T or C; D = G, A, or T).

20. Y. Zhou, R.H. Purcell, S.U. Emerson. An ELISA for putative neutralizing antibodies to hepatitis E virus detects antibodies to genotypes 1, 2, 3, and 4. Vaccine 22 (**2004**) 2578–2585. *Note*: At 4°C, the microwell plates were coated with open reading frame 2 (ORF2, 1 μg/mL) proteins in carbonate–bicarbonate buffer (pH 9.6) overnight, washed with 0.05% Tween 20 in PBS (PBS-T) three times, and then blocked with 5% skim milk in PBS-T at 37°C for 1 h. The plates were washed with PBS-T three times. After addition of the diluted solution of serum samples in skim milk (5%), the plates were incubated at 37°C for 1 h. The plates were washed with PBS-T five times and the bound antibodies detected by adding the solution of HRP-conjugated anti-human IgGF(ab′)2 (Sigma) in 2000-fold PBS-T. The plates were incubated at 37°C for 1 h, washed as above, and the substrate tetramethylbenzidine solution (Sigma) was added to the wells. The plates were incubated at room temperature for 15 min and then 2 M H_2SO_4 was added to stop the color development. Optical density (OD) at 450 nm was determined with an ELISA reader.

21. M.J. Ahn, K.D. Yoon, S.Y. Min, J.S. Lee, J.H. Kim, T.G. Kim, S.H. Kim, N.G. Kim, H. Huh, J. Kim. Inhibition of HIV-1 reverse transcriptase and protease by phlorotannins from the brown alga Ecklonia cava. Biol Pharm Bull 27 (**2004**) 544–547.

22. D. Sriram, P. Yogeeswari, M. Dinakaran, M. Sowmya. Synthesis, anti-HIV and antitubercular activities of nelfinavir diester derivatives. Biomed Pharmacother 62 (**2008**) 1–5.

23. E. Gong, T. Ivens, C. Van den Eynde, S. Hallenberger, K. Hertogs. Development of robust antiviral assays for profiling compounds against a panel of positive-strand RNA viruses using ATP/luminescence readout. J Virol Methods 151 (**2008**) 121–125.

24. S. Pontow, B. Harmon, N. Campbell, L. Ratner. Antiviral activity of a Rac GEF inhibitor characterized with a sensitive HIV/SIV fusion assay. Virology 368 (**2007**) 1–6.

25. M.N. Prichard, S.L. Daily, G.M. Jefferson, A.L. Perry, E.R. Kern. A rapid DNA hybridization assay for the evaluation of antiviral compounds against Epstein-Barr virus. J Virol Methods 144 (**2007**) 86–90.

26. T. Zhou, H. Guo, J.T. Guo, A. Cuconati, A. Mehta, T.M. Block. Hepatitis B virus e antigen production is dependent upon covalently closed circular (ccc) DNA in HepAD38 cell cultures and may serve as a cccDNA surrogate in antiviral screening assays. Antiviral Res 72 (**2006**) 116–124.

27. L. Cagnon, P. Wagaman, R. Bartenschlager, T. Pietschmann, T. Gao, N.M. Kneteman, D.L.J. Tyrrell, C. Bahl, P. Niven, S. Lee, K.A. Simmen. Application of the trak-C^{TM} HCV core assay for monitoring antiviral activity in HCV replication systems. J Virol Methods 118 (**2004**) 23–31.

3

METHODS AND APPLICATIONS OF ANTITUBERCULAR ASSAYS

Shiqi Peng, Ming Zhao, and Chunying Cui

In the past few decades, though significant progress has been made in treatment and control strategies for tubercular infections by introducing new diagnostic and monitoring tools and combination therapy, tuberculosis (TB) is still the leading infectious cause of death in the world today: Approximately 3 million patients die every year; nearly one third of the world's population is infected with *Mycobacterium tuberculosis*, accounting for more adult deaths worldwide than does any other single infectious agent; and the WHO estimates that about 30 million people will be infected within the next 20 years. In addition to these, the development of resistance to current antimycobacterial therapy also endorses the search for more effective agents and assays. In this chapter, 6 assays are introduced: *Mycobacterium tuberculosis* assay,[1–4] DNA polymerase β lyase assay,[5–7] agar dilution assay for *in vitro* antitubercular activity,[8,9] microplate Alamar blue assay for *in vitro* antitubercular activity,[10] radiometric respiratory assay for *in vitro* antitubercular activity,[11] and *Mycobacterium bovis* BCG inhibition assay.[12]

3.1 *Mycobacterium tuberculosis* ASSAY[1–4]

With the BACTEC system, the radiometric respiratory technique was used for testing susceptibility against *Mycobacterium tuberculosis*. Concentration of test compounds

Pharmaceutical Bioassays: Methods and Applications. By Shiqi Peng and Ming Zhao

in DMSO was 5×10^5 mg/L, stored at 4°C until use. In BACTEC 12B vials containing 4 mL 7H12 medium broth, the solution of test compound was diluted with DMSO to achieve the desired final concentrations of 5000.0, 2500.0, 1000.0, and 500.0 mg/L together with an antimicrobial supplement PANTA. In evaluation, solution of DMSO in the medium without test compound (1%) was used as black control and had no adverse effect on the growth of *Mycobacterium tuberculosis*; streptomycin, isoniazid, rifampicin, and ethambutanol were used as positive controls. In the vials containing test compounds as well as in the control vials, a homogenous culture (0.1 mL) of *M. tuberculosis* yielding 1×10^4 to 1×10^5 colony-forming units per milliliter (CFU/mL) was inoculated. Three test compound-free vials were used as controls (medium + 1% DMSO), two vials (V1) were inoculated the same as for the vials containing test compounds, and one vial (V2) was inoculated with a 1 : 100 dilution of the inoculum (1 : 100 control) to produce an initial concentration representing 1% of the bacterial population (1×10^2 to 1×10^3 CFU/mL). The minimum inhibitory concentration (MIC) represented the lowest concentration of the test compound that inhibited more than 99% of the bacterial population. In 7H12 medium containing ^{14}C-labeled substrate (palmitic acid), mycobacteria used the substrate and produced $^{14}CO_2$. The detected amount of $^{14}CO_2$ (reflecting the rate and amount of growth occurring in the sealed vial) was expressed in terms of the growth index (GI). At 37°C, the inoculated bottles were incubated, and each was assayed every day to measure GI at the same time point until cumulative results were interpretable. The difference in the GI values of the last 2 days was defined as ΔGI. GI values of the vials containing the test compounds were compared with the control vials (V2). Values were tested until the control vials, containing a 100-times lower dilution of the inoculum than that of the test vials, had a GI of 30 or more. If the ΔGI values of the vials containing test compounds were less than that of the control vials, the population was susceptible to test compound. All assays were repeated three times. Large and rapid increase in GI may suggest contamination; in this case bottles were inspected and the organisms were stained with Ziehl–Neelsen stain to determine whether the visible microbial growth was a mycobacterial organism, with which the bacilli appear as brilliantly stained red rods against a deep sky-blue background. Organisms often had a beaded appearance due to polyphosphate content and unstained vacuoles.

3.2 DNA POLYMERASE β LYASE ASSAY[5–7]

Using [α-^{32}P]ddATP and terminal deoxynucleotidyltransferase, the 3′-end of a 36-nucleotide oligodeoxyribonucleotide containing uridine at position 21 was labeled. The product was subjected to 20% denaturing polyacrylamide gel electrophoresis for purification. Using autoradiography, the interesting band was visualized and excised. After removal by the "crush and soak" method, the DNA substrate was annealed to its complementary strand by heating to 70°C for 3 min, followed by slow cooling to 25°C. To 200 μL 354 nM [α-^{32}P]-labeled double-stranded oligodeoxynucleotide with uridine at position 21, 10 mM HEPES–KOH (pH 7.4), 5 mM $MgCl_2$, 50 mM KCl, 10 mg/mL bovine serum albumin, 2.4 units of uracil-DNA glycosylase, and 3 units of apurinic (AP) endonuclease were added. After incubation at 37°C for 20 min, an AP site was created in the [α-^{32}P]-labeled double-stranded oligodeoxynucleotide for deoxyribose phosphate (dRP)-excision assay. To 5 μL reaction mixture, various concentrations of test compound and 0.172 unit of rat DNA polymerase β were added. After incubation at room temperature for 30 min, the reaction was terminated by adding 0.5 M $NaBH_4$ (50 mM final concentration) and incubated at room temperature for 10 min. The reaction products were further incubated at 70°C to 80°C for 20 min, separated on a 20% denaturing polyacrylamide gel, and visualized by autoradiography.

3.3 AGAR DILUTION ASSAY FOR *IN VITRO* ANTITUBERCULAR ACTIVITY[3,8,9]

(1) *In vitro* antibacterial activity assay: Minimum inhibitory concentration (MIC), the lowest concentration that yields no visible growth on the plate, was tested in side-by-side comparison with ciprofloxacin and norfloxacin against gram-positive bacteria (*Staphylococcus aureus*, *Streptococcus faecalis*, *Bacillus subtilis*) and gram-negative bacteria (*Klebsiella pneumoniae*, *Escherichia coli*, *Pseudomonas aeruginosa*) by broth microdilution method. Serial dilutions were prepared by progressively diluting the solution of 10 mg test compounds or ciprofloxacin or norfloxacin in 1 mL DMSO to 1, 2, 4, 8, 16, 31.25, 62.5, 125, 250, and 500 mg/mL with melted Müller–Hinton agar. The tubes were inoculated with 10^5 CFU/mL and were incubated at 37°C for 18 h. To ensure solvent had no effect on the bacterial growth, a control was performed with the test medium supplemented with DMSO alone.

(2) *In vitro* antitubercular H37Rv agar dilution assay: With broth dilution assay, MIC of each test compound was determined and defined as its lowest concentration that inhibited $\geq$99% bacterial population present at the beginning of the assay. A frozen culture in Middlebrook 7H9 broth supplemented with 10% albumin–dextrose–catalase and 0.2% glycerol was thawed, diluted to 10^5 CFU/mL drug-susceptible strain of *M. tuberculosis*, H37Rv, with broth and used for inoculation. To each U-tube, 1–500 μg/mL dilutions of compounds in DMSO was added and diluted with broth twice to the desired concentration with 1.3% final DMSO concentration. Each U-tube was inoculated with 0.05 mL standardized culture, incubated at 37°C for 21 days, and the growth was compared with visibility against positive control (without drug), negative control (without drug and inoculum), and standard isoniazid.

3.4 MICROPLATE ALAMAR BLUE ASSAY FOR *IN VITRO* ANTITUBERCULAR ACTIVITY[10]

Test compound was assayed at 6.25 mg/mL against *Mycobacterium tuberculosis* H37Rv (ATCC 27294) in BACTEC 12B medium using a broth microdilution method, the microplate Alamar blue assay (MABA). Test compounds exhibiting fluorescence were assayed in the BACTEC 460 radiometric system. In the primary evaluation, test compounds having more than 95% inhibition were considered as active and reassayed at lower concentrations against *M. tuberculosis* H37Rv to determine their actual MIC using MABA. MIC was defined as the lowest concentration effecting a reduction of 95% fluorescence with respect to the controls. Rifampin (MIC = 0.015 to 0.125 mg/mL) was used as the reference compound. With Vero-C-1008 cell line using neutral red uptake assay, the cytotoxicity of test compound at various concentrations (6.25–50 mg/mL) was measured to analyze the contribution of cytotoxicity to antitubercular activity.

3.5 RADIOMETRIC RESPIRATORY ASSAY FOR *IN VITRO* ANTITUBERCULAR ACTIVITY[11]

The BACTEC system was used for this radiometric respiratory assay. Test compounds were dissolved in DMSO at 5×10^5 μg/mL and stored at 4°C until use. Dilutions were made in BACTEC 12B vials with 7H12 medium, and the final concentration of DMSO was 1%. Into the vials, 0.1 mL of homogenous culture of *M. tuberculosis*, yielding 1×10^4 to 1×10^5 CFU/mL, was inoculated. Streptomycin, isoniazid, rifampin, and ethambutol were used as positive controls. Three compound-free vials were used as controls (medium + 1% DMSO); among them two vials (V1) were inoculated in the same way as the vials containing test compounds, and one (V2) was inoculated with a 1 : 100 dilution of the inoculum (1 : 100 control) to produce an initial concentration representing 1% of the bacterial population (1×10^2 to 1×10^3 CFU/mL). The MIC was defined as the lowest concentration of test compound that inhibited >99% of the bacterial population. Mycobacteria were grown in 7H12 medium containing ^{14}C-labeled substrate (palmitic acid), used the substrate, and produced $^{14}CO_2$. The amount of $^{14}CO_2$ was detected (reflecting the rate and amount of growth occurring in the sealed vial) and expressed in terms of the growth index (GI). Bottles were incubated at 37°C, each bottle was assayed at about the same hour(s) every day to measure GI until cumulative results were interpretable, and the difference in the GI values of the last 2 days was designated as ΔGI. The GI readings of the vials containing test compounds were compared with those of control vials (V2). Readings were taken until the control vials reached a GI of 30 or more, which contained a 100-times lower dilution of the inoculum than that of the test vials. If the ΔGI values of the vials containing test compounds were less than those of control vials, the population was considered to be susceptible to the compound. Each test was replicated three times.

3.6 *Mycobacterium bovis* BCG INHIBITION ASSAY[12]

Subculturing of *Mycobacterium bovis* BCG was routinely done in Dubos albumin agar slants or plates. *Mycobacterium bovis* BCG was incubated at 37°C and 150 revolutions per minute (rpm) in Dubos Tween albumin broth medium to prepare liquid inoculum. For all assays, 1% 1.0 A_{620} of the culture was used as standard inoculum size to yield a final inoculum of approximately 10^5 CFU/mL, and viable cell counts were measured.

In order to adapt the *in vitro* 0.5 HSR (head space ratio) model, which was based on gradual depletion of oxygen from mycobacterial cells to achieve the nonreplicating dormant stage, in microplate format, to each well of 375-μL-capacity 96-well plates, 250 μL culture containing approximately 10^5 cells/mL was added, and the head space was exactly maintained to 0.5 culture volume. By applying an adhesive type of microplate sealer to allow self-generation of hypoxia in the culture, the air supply of the culture in the microplate was blocked. Sealing of plates could be performed manually or using a sealer already attached through SAMI software placed on the Beckman Coulter HTS platform. Plates were incubated at 37°C in a CO_2 incubator, and fading/decolorization of methylene blue (1.5 μg/mL final concentration) was used to determine hypoxia/anoxia.

By reading A_{620} as well as by determining CFU/mL of the culture at different time intervals, the growth of *M. bovis* BCG in microplate format of dormancy model was measured. The lowest concentration of test compounds yielding a differential absorbance (A_{620}) of approximately zero or less was defined as the MIC. Nitrate reductase (NR) activity of *Mycobacterium bovis* BCG in the microplate dormancy model was measured using the Griess reaction method. Briefly,1% solution of sulfanilic acid (in 20% HCl) and 0.1% naphthyl ethylene diamine hydrochloride solution (in distilled water) was added to whole cell culture in 1/1/1 ratio, incubated for 15 min, and the absorbance was read at 540 nm.

REFERENCES AND NOTES

1. S.P.N. Mativandlela, N. Lall, J.J.M. Meyer. Pelargonium reniforme (CURT) and Pelargonium sidoides (DC) (Geraniaceae) root extracts. South African J Botany 72 (**2006**) 232–237.
2. D. Sriram, P. Yogeeswari, M. Dinakaran, M. Sowmya. Synthesis, anti-HIV and antitubercular activities of nelfinavir diester derivatives. Biomed Pharmacother 62 (**2008**) 1–5. *Note*: Using Middlebrook 7H10 medium supplemented with Middlebrook OADC medium, the agar dilution method for different concentrations of test compounds from 6.25 mg/mL was performed. After solidification of the agar, the plates were inoculated with 0.1 mL 10^{-2} and 10^{-4} dilutions of a McFarland 1.0 concentration of a suspension of organism. The plates in duplicate were incubated at 37°C for 4 weeks. MIC, the minimum inhibitory concentration, was considered as the lowest concentration that completely inhibited growth on agar plates, disregarding a single colony or a faint haze caused by the inoculums.
3. S.D. Joshi, H.M. Vagdevi, V.P. Vaidya, G.S. Gadaginamath. Synthesis of new 4-pyrrol-1-yl benzoic acid hydrazide analogs and some derived oxadiazole, triazole and pyrrole ring systems: A novel class of potential antibacterial and antitubercular agents. Eur J Med Chem 43 (**2007**) 1989–1996.
4. M. Shiradkar, G.V.S. Kumar, V. Dasari, S. Tatikonda, K.C. Akula, R. Shah. Clubbed triazoles: A novel approach to antitubercular drugs. Eur J Med Chem 42 (**2007**) 807–816. *Note*: The primary evaluation was performed using 6.25 mg/mL test compound against *Mycobacterium tuberculosis* H37Rv (ATCC 27294) in BACTEC 12B medium with the microplate Alamar blue assay (MABA). Test compounds exhibiting fluorescence were tested in the BACTEC 460 radiometric system. In the primary evaluation, test compounds

having more than 95% inhibition were considered as active and retested at lower concentrations against *M. tuberculosis* H37Rv for determining their actual MIC using MABA. MIC was defined as the lowest concentration effecting a reduction in fluorescence of 95% with respect to the controls. Rifampicin (RMP) was used as the reference compound (RMP, MIC = 0.015 to 0.125 mg/mL). The cytotoxicity analysis of test compounds was performed using neutral red uptake and Vero-C-1008 cell line at various concentrations (6.25–50 μg/mL).

5. X. Feng, Z. Gao, S. Li, S.H. Jones, S.M. Hecht. DNA polymerase beta lyase inhibitors from Maytenus putterlickoides. J Nat Prod 67 (**2004**) 1744–1747.
6. Z. Gao, D.J. Maloney, L.M. Dedkova, S.M. Hecht. Inhibitors of DNA polymerase β: Activity and mechanism. Bioorg Med Chem 16 (**2008**) 4331–4340.
7. V.S.P. Chaturvedula, Z. Gao, S.M. Hecht, S.H. Jones, D.G.I. Kingston. A new acylated oleanane triterpenoid from Couepia polyandra that inhibits the lyase activity of DNA polymerase β. J Nat Prod 66 (**2003**) 1463–1465.
8. D. Sriram, P. Yogeeswari, M. Dinakaran, M. Sowmya. Synthesis, anti-HIV and antitubercular activities of nelfinavir diester derivatives. Biomed Pharmacother 62 (**2008**) 1–5. *Note*: Using Middlebrook 7H10 medium supplemented with Middlebrook OADC medium, the agar dilution method for different concentrations of test compounds from 6.25 mg/mL was performed. After solidification of the agar, 0.1 mL 10^{-2} and 10^{-4} dilutions of a McFarland 1.0 concentration of a suspension of organism were inoculated onto the plates in duplicate and incubated at 37°C for 4 weeks. The MIC was defined as the lowest concentration that completely inhibited growth on agar plates, disregarding a single colony or a faint haze caused by the inoculums.
9. S. Talath, A.K. Gadad. Synthesis, antibacterial and antitubercular activities of some 7-[4-(5-amino[1,3,4]thiadiazole-2-sulfonyl)-piperazin-1-yl] fluoroquinolonic derivatives. Eur J Med Chem 41 (**2006**) 918–924.
10. M. Shiradkar, G.V.S. Kumar, V. Dasari, S. Tatikonda, K.C. Akula, R. Shah. Clubbed triazoles: A novel approach to antitubercular drugs. Eur J Med Chem 42 (**2007**) 807–816.
11. S.P.N. Mativandlela, N. Lall, J.J.M. Meyer. Antibacterial, antifungal and antitubercular activity of (the roots of) Pelargonium reniforme (CURT) and Pelargonium sidoides (DC) (Geraniaceae) root extracts. South African J Botany 72 (**2006**) 232–237.
12. A. Khan, D. Sarkar. A simple whole cell based high throughput screening protocol using *Mycobacterium bovis* BCG for inhibitors against dormant and active tubercle bacilli. J Microbiol Methods 73 (**2008**) 62–68.

4

METHODS AND APPLICATIONS OF THROMBUS-RELATED ASSAYS

Ming Zhao

Despite great advances being made in the prevention of thrombotic disorders by the development of new pharmacologic and surgical treatments and well-established application of antithrombotic drugs, such as anticoagulants, antiplatelet drugs, and thrombolytic drugs, thromboembolic diseases are still the leading cause of mortality and morbidity worldwide, and there is still an urgent need to gain insight into the pathologic mechanism, to evaluate the benefit of surgical treatments, and to find more potent and safer compounds in the prevention and treatment of ischemic symptoms. In recent years, a series of events such as that of platelet aggregation, fibrinogenesis, fibrin adhesion, fibrin aggregation, and vascular inner wall injury is known to be implicated in different stages corresponding with the categories of disease free, disease onset, and disease progression. This knowledge is widely used in constructing evaluation methods for understanding the pathologic mechanism, supporting surgical treatments, and screening new antithrombotic and/or thrombolytic drugs. Based on the facts that thrombosis, antithrombosis, and thrombolysis may relate to anti-platelet aggregation, chemical- and electrical-induced blood vessel injury, thread-induced fibrin or platelet adhesion, euglobulin clot lysis time, fibrinolytic area, and reduction of thrombus mass, a lot of *in vitro* and *in vivo* assays have been reported and have largely contributed to the discovery and to the validation of original treatments of

Pharmaceutical Bioassays: Methods and Applications. By Shiqi Peng and Ming Zhao

thrombotic disorders. The purpose of this chapter is to introduce 29 original thrombus-related models including *in vitro* anti-platelet aggregation assay,[1–6] fibrinolytic area assay,[1] thromboxane B_2 (TXB_2) and prostaglandin D_2 (PGD_2) TLC assay,[2] thromboxane A_2 (TXA_2) synthase activity assay,[2] $[Ca^{2+}]_i$ measuring assay,[2] arachidonic acid liberation assay,[2] serotonin secretion assay,[2] cyclic adenosine monophosphate (cAMP) release assay,[3] *ex vivo* anti-platelet aggregation assay for patients,[3] ATP release assay,[4] platelet-activating factor (PAF)-induced mice mortality assay,[4] prostaglandin E_2 (PGE_2) and TXB_2 ELISA,[4] thrombelastograph assay,[6,7] image monitored $FeCl_3$-induced thrombosis assay for rat,[8] weight monitored $FeCl_3$-induced thrombosis assay for rat,[9,10] occlusion time monitored $FeCl_3$-induced thrombosis assay for pig,[11] Doppler blood flow monitored $FeCl_3$-induced thrombosis assay for mouse,[12] rat groin flap assay,[13] ferret acute thrombosis assay,[14] rat acute thrombosis assay,[15] arteriovenous shunt assay,[16,17] plasma clotting time assay,[17] thromboembolic photochemical assay for repeated stroke in mice,[18] euglobulin clot lysis assay,[19–21] clot formation and lysis (CloFAL) assay,[22] fibrin microplate assay,[23] fibrinolytic activity assay,[24] fibrinolysis assay,[25] and thrombolytic assay.[26,27]

4.1 *IN VITRO* ANTI-PLATELET AGGREGATION ASSAY[1–6]

A general procedure is summarized in Fig. 4.1. Blood from the marginal ear vein of male New Zealand White rabbits was collected directly into 0.15 mL (v/v) anticoagulant citrate dextrose (ACD) solution that contained 0.8% citric acid, 2.2% trisodium citrate, and 2% dextrose (w/v). By centrifugation at room temperature and $200 \times g$ for 15 min, platelet-rich plasma was obtained. The platelet-rich plasma was mixed with 1/40 volume of EDTA (final concentration 5 mM) and recentrifuged at $1000 \times g$ for 12 min. The supernatant was discarded, and the platelet pellet was suspended in modified Ca^{2+}-free Tyrode's buffer (137 mM NaCl, 2.8 mM KCl, 2 mM $MgCl_2$, 0.33 mM NaH_2PO_4, 5 mM glucose, 10 mM HEPES) with 0.35% bovine serum albumin, heparin (50 unit/mL), and apyrase (1 unit/mL). The platelet counts were adjusted to $2 \times 10^5/\mu L$ by addition of autologous plasma. Platelet aggregation tests were conducted in an aggregometer using the standard turbidimetric technique. Platelet was pretreated with vehicle (negative control), aspirin (positive control), and test compounds (treating groups) at 37°C for 3 min in the aggregometer with stirring at 1000 rpm before aggregation was challenged by the addition of platelet-activating factor (PAF; final concentration 0.1–10 μM), adenosine diphosphate (ADP; final concentration 0.1–10 μM), thrombin (final concentration, 0.05 U/mL), collagen (final concentration, 10 μg/mL), arachidonic acid (AA; final concentration, 100 μM), U46619 (TXA_2 mimic; 1 μM), thapsigargin (Ca^{2+} ATPase inhibitor; final concentration, 0.5 μM), and A23187 (calcium ionophore; final concentration, 5 μM). All experiments were carried out within 4 h of isolation of platelets. Aggregation was expressed as percent change in transmission of light, the transmission of blank sample (buffer without platelets) was defined as 100%, and the transmission of washed platelet suspension was defined as 0%. The effects of tested compounds on agonists-induced platelet aggregation were observed. The maximal rate of platelet aggregation (Am%)

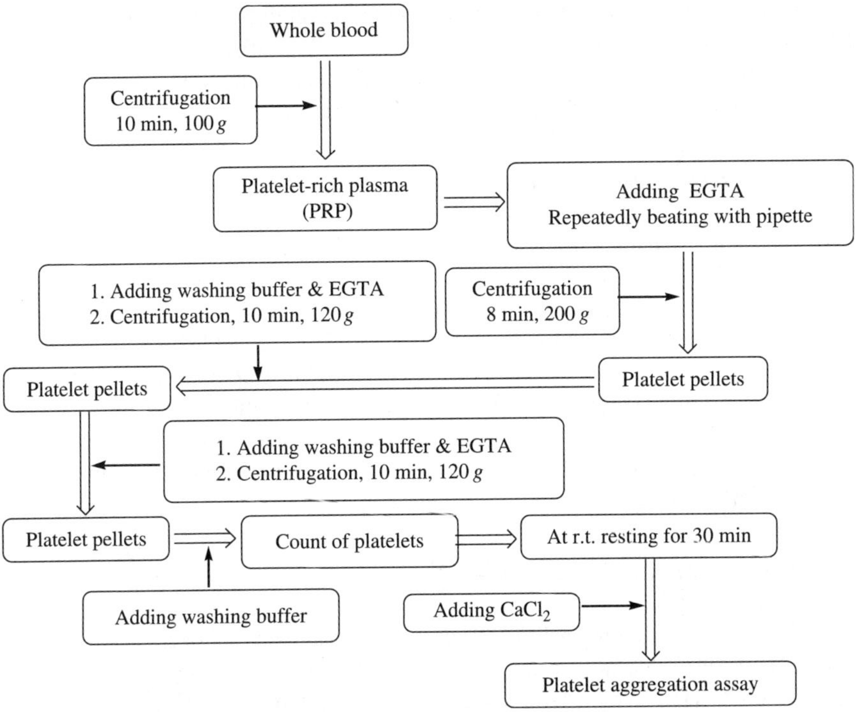

Figure 4.1 Flowchart of *in vitro* anti-platelet aggregation assay.

was represented by the peak height of the aggregation curve. The extent of inhibition of platelet aggregation was expressed as percentage of negative control.

4.2 FIBRINOLYTIC AREA ASSAY[1]

The fibrinogen–agarose mixture was prepared and coagulated with thrombin in plastic dishes accordingly. The fibrinogen–agarose mixture was prepared by mixing equal volumes of 0.3% rabbit fibrinogen and 0.95% agarose solutions, both dissolved in 50 mM sodium barbiturate buffer (pH 7.8). The fibrinogen–agarose mixture was coagulated with 100 μL thrombin (100 IU/mL) in plastic dishes (90 mm diameter × 1 mm depth). After 30 min at 4°C, an adequate number of wells, 5 mm in diameter, was perforated. To determine fibrinolytic activity, 30 μL normal saline (NS) (blank control), urokinase (UK) (positive control), and test compound were added to the corresponding well. The plate was incubated, and areas of lysis were quantified.

4.3 TXB_2 AND PGD_2 TLC ASSAY[2]

After preincubation with various concentrations of test compounds at 37°C for 3 min, washed platelets (4×10^8 cells/mL) were incubated with a mixture of [^{3}H]arachidonic acid ([^{3}H]AA) and unlabeled AA (2 μM, 1 μCi/mL) for 5 min. After extraction, the soluble lipids were separated by thin-layer chromatography (TLC) on silica gel G plates with ethyl acetate/iso-octane/acetic acid/H_2O (9/5/2/10, v/v/v/v) as development system. The area corresponding with each lipid was scraped off, and the radioactivity was determined by liquid scintillation counting.

4.4 TXA_2 SYNTHASE ACTIVITY ASSAY[2]

After pretreatment with indomethacin (50 μM), the platelets were incubated with vehicle (negative control), test compounds (final concentration, 10, 25, and 50 μM), or imidazole (typical TXA_2, synthase inhibitor, positive control, final concentration 5 mM) for 3 min, then 5 μM prostaglandin H_2 (PGH_2) was added, and the platelets were incubated for an additional 5 min. After addition of cooling ethylenebis(oxy-ethylenenitrilo)tetraacetic acid (EGTA) (2 mM), the incubations were centrifuged at 13,000 × *g* and 4°C for 4 min. With a commercial enzyme immunoassay kit and according to the manufacturer's instructions, the amount of TXB_2 in the supernatants was measured, which reflected TXA_2 synthase activity.

4.5 $[CA^{2+}]_i$ MEASURING ASSAY[2]

To 4 mL cuvettes containing a Teflon-coated stirrer bar, calcium green-1/AM-labeled platelets (2.5 mL) were added and treated with test compound (50 μM) or vehicle (negative control), to which EGTA was added to the buffer to a final concentration of 4 mM. $[Ca^{2+}]_i$ measurement was performed at room temperature on a MSIII fluorimeter (Photon Technology International, S. Brunswick, NJ, USA) using excitation wavelengths of 506 nm as well as an emission wavelength of 533 nm and calculated using the SPEX dM3000 software package (Photon Technology International, S. Brunswick, NJ, USA).

4.6 ARACHIDONIC ACID LIBERATION ASSAY[2]

After preincubation of $[^3H]$AA-prelabeled platelets (4×10^8 cells/mL) with 50 μM BW755C at 37°C for 5 min and in the presence of 1 mM $CaCl_2$, vehicle (negative control) or test compounds (10, 25, 50 μM) or arachidonyltrifluoromethyl ketone ($AACOCF_3$; phospholipase A_2 inhibitor, 50 μM) were added, incubated for 3 min, and stimulated with collagen (50 μg/mL). After terminating the reaction by adding chloroform/methanol/HCl (200/200/1, v/v/v), the soluble lipids were extracted and separated by TLC on silica gel G plates with petroleum ether/diethyl ether/acetic acid (40/40/1, v/v/v) as development system. The area corresponding with each lipid was scraped off, and the radioactivity was determined by liquid scintillation counting.

4.7 SEROTONIN SECRETION ASSAY[2]

In the pretreatment of washed platelet suspension with test compounds (10, 25, and 50 μM) or vehicle (negative control) at 37°C for 3 min, imipramine (5 μM) was used to prevent the reuptake of serotonin and then collagen (10 μg/mL) was added. Five minutes after incubation, the reaction was terminated by the addition of 5 mM EDTA on ice. The supernatant was centrifuged at 12,000 × *g* for 2 min. The supernatant was mixed with 6 M trichloroacetic acid and further centrifuged at 12,000 × *g* for 2 min. The mixture of 0.6 mL aliquot of trichloroacetic acid supernatant and 2.4 mL of the solution (0.5% *o*-phthalaldehyde in ethanol diluted 1:10 with 8 N HCl) was heated in boiling water bath for 10 min and cooled in ice. The excess trichloroacetic acid in this mixture was extracted with chloroform, and the fluorophore was measured at the wavelength of excitation (360 nm) and emission (475 nm). Using serotonin creatinine sulfate as standard solution, the extent of serotonin release was calculated. One hundred percent release of serotonin was determined by the treatment of platelets with Triton X-100.

4.8 cAMP RELEASE ASSAY[3]

In cAMP measurements, the washed rabbit platelets were adjusted to a density of 1×10^8 platelets/mL per reaction. The platelets were incubated with increasing concentrations of edema factor (EF) ranging from 2.2 to 11 nM, for which time-dependent incubations were also done (for 0.5, 1, 1.5, 2, 2.5, and 3 h), along with 11 nM protective antigen (PA). Adenylyl cyclase activator (forskolin, 12 μM) was incubated with platelets for 5 min. Calmodulin inhibitor (calmidazolium chloride, 10 μM) was incubated with platelets for 20 min prior to incubation with edema toxin (11 nM PA and 11 nM EF) for 2 h. EF mutant was incubated at a concentration of 50 nM along with wild-type PA. All incubations were carried out at 37°C and terminated by adding 5 mM EDTA. After freeze thawing platelets three times, the samples were heated at 80°C for 20 min and centrifuged at 14,000 × *g* for 10 min. The supernatant was assayed for cAMP (pM cAMP/10^8 platelets) using the cAMP assay kit (Amersham, Buckinghamshire, UK).

4.9 *EX VIVO* ANTI-PLATELET AGGREGATION ASSAY FOR PATIENTS[3]

On the next day of the last administration of the test compound before coronary intervention and at 3 h and 18 to 24 h after stenting, 20 mL of baseline blood sample was withdrawn from the right arm of a patient in a fasting state in the catheterization laboratory through the indwelling femoral vessel sheath and transferred to separate Vacutainer blood-collecting tubes containing 3.8% trisodium citrate or lithium heparin. Vacutainers were inverted (five to seven times) to ensure full mixing of the blood and the anticoagulant. The first 2 to 3 mL of free-flowing blood was discarded, and the Vacutainer tubes were filled to capacity and gently inverted three to five times to complete mix with trisodium citrate or lithium heparin. The blood-citrate or heparin tubes were centrifuged at 120 × *g* for 5 min and further centrifuged at 850 × *g* for 10 min to recover platelet-rich plasma (PRP) and platelet-poor plasma (PPP), respectively, and stored at room temperature for use within 2 h. Platelet were stimulated with 5 μM ADP and 1 mM AA. Aggregation was measured on a Chronolog Lumi-Aggregometer with the Aggrolink software package (Chronolog, Havertown, PA, USA) and expressed as the maximum percent change in light transmittance from baseline, with PPP as a reference.

4.10 ATP RELEASE ASSAY[4]

With a bioluminescence method, ATP released from platelets was detected. After centrifugation at 7000 × *g* for 1 min, an aliquot of the platelet suspension was resuspended in Tyrode's buffer to give 10^9 platelets/mL in a final volume of 4.5 mL, which was stirred in a plastic tube in a water bath (37°C) for 5 min to equilibrate. The mixture of 100 μL of samples were taken as indicated, and 900 μL of precooled (0°C) Tyrode's buffer was centrifuged at 7000 × *g* for 1 min. For determining ATP, 20 μL supernatant was mixed with 200 μL 1 M perchloric acid and 50 mM EDTA for 1 min to precipitate soluble proteins. By addition of 316 μL 1 M HEPES buffer (pH 7.75, 25°C), the solution was adjusted to pH 7.5–8.0, incubated at 0°C for 30 min, and centrifuged at 14,000 × *g* for 10 min to remove potassium perchlorate. To 500 μL HEPES buffer, 500 μL supernatant was added, and ATP content was detected by employing the kit HSII (Boehringer Mannheim GmbH, Mannheim, Germany) on a luminometer and calculated based on a calibration curve prepared using different ATP concentrations as standards.

4.11 PAF-INDUCED MICE MORTALITY ASSAY[4]

Male ICR mice were intravenously administered 500 mg/100 g or 1000 mg/100 g PAF solution. All mice were observed for at least 24 h after the PAF administration. Each test compound was intraperitoneally administered 15 min before the PAF administration. The results were given as 24-h survival rates.

4.12 PGE_2 AND TXB_2 ELISA[4]

In prostaglandin E_2 (PGE_2) and TXB_2 enzyme-linked immunosorbent assay (ELISA), a test kit from Promega (CA, USA) was employed. An aliquot of platelet suspension was centrifuged at 7000 × *g* for 1 min, resuspended in Tyrode's buffer aimed at 1.5×10^8 platelets/mL on a final volume of 4 mL, and transferred into a plastic tube located in a water bath (37°C). After 5 min equilibration, 200 μL stirred suspension were taken as indicated. For determination of PGE_2 alone, 1 μM thromboxane synthase (TXS)-inhibitor *E*/*Z*-7-phenyl-7-pyridyl-6-heptenoic acid was added to the suspension at the onset of equilibration. For determination of both PGE_2 and TXB_2, 200 μL suspension was mixed with 1800 μL ice-cooled stop solution (50 μM indomethacin, 25 mM HEPES, pH 7.4). By sonification at 0°C for 15 min platelets were disrupted, and the resulting homogenates were centrifuged at 5000 × *g* for 10 min to allow complete release of both eicosanoids. PGE_2 and TXB_2 contents in the supernatant were determined according to the ELISA kit supplier's instructions.

4.13 THROMBELASTOGRAPH ASSAY[6,7]

A blood sample was withdrawn from the right arm of a volunteer, transferred into Vacutainer tubes anticoagulated with 14.7 U/mL heparin and used for thrombelastograph (TEG) assay within 2 h to determine the maximum amplitude (MA) of clot shear elasticity, which was proportionate to the extent of platelet activation elicited by 1 mmol AA and represented the interaction of the activated platelet glycoprotein IIb/IIIa complex with a fibrin network formed by a reptilase-factor XIIIa mixture. Vacutainers were inverted (five to seven times) to ensure full mixing of the blood and the anticoagulant. This assay of platelet response to 1 mmol AA was analogous to optical platelet aggregation (OPA) measured aggregation used as an assay for nonsteroidal anti-inflammatory drug (NSAID) inhibition. In accordance with the manufacturer's instructions, responses were compared with the maximal MA of the same blood clotted by thrombin with the addition of heparinase and kaolin (MAKH). By subtracting the MA without platelet activator (MA_0) from the MA with AA (MAAA), dividing by MAKH minus MA_0, and then multiplying the result by 100%, a percent MA response (TEG %MA) was obtained, which correlated well with percent aggregation as determined by OPA.

4.14 IMAGE-MONITORED $FeCl_3$-INDUCED THROMBOSIS ASSAY FOR RAT[8]

Male Sprague-Dawley (SD) rats (weight, 245–331 g) were anesthetized with a mixture (1.5/1/1) of xylazine (50 mg/mL), acepromazine (50 mg/mL), and ketamine (50 mg/mL) at a dose of 0.1 mL/100 g, intramuscularly administered respectively, placed on a surgical board specifically designed for this procedure, and catheterized in the left femoral, and body temperature was maintained at 37°C by a silicone heating mat. After exposing the right testicle and removing the connective tissue from the cremaster muscle, the muscle was separated from the testicle by use of baroque-handled iris scissors and pinned flat over a transparent section of the surgical board. Using an industrial-scale scope with a Cool Snap ES camera and fluorescent (100 W mercury) light sources, images were made. Using MetaMorph software (Universal Imaging, West Chester, PA, USA), images were analyzed off-line. Rats were injected with 0.5 mL PBS or test compounds in PBS by tail veins, and the surface of their cremaster muscle vessels was proximally and distally covered by Whatman filter papers saturated with PBS or 30% $FeCl_3$ or 50% $FeCl_3$. Rats were given a solution of 0.5% rhodamine-6G-dextran (525-nm excitation wavelength and 555-nm emission wavelength). A video image before the tracer injection was recorded for 1 min. The blood flow was observed and, after the application of the chemicals, images were saved 5 min and every 15 min for 90 min or whenever the blood flow stopped.

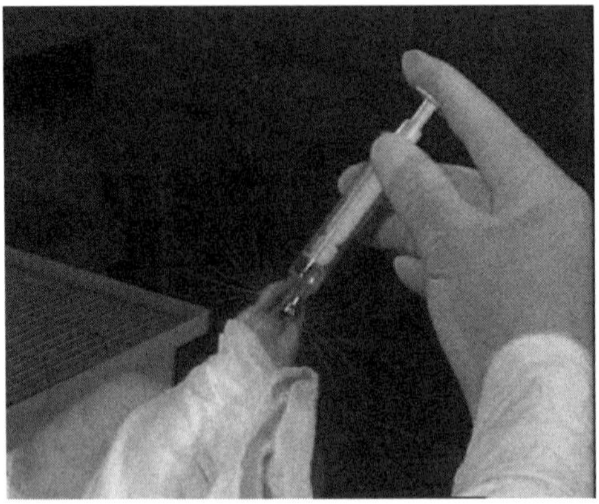

Figure 4.2 Oral administration for rat.

4.15 WEIGHT-MONITORED $FeCl_3$-INDUCED THROMBOSIS ASSAY FOR RATS[9,10]

After overnight fasting, male SD rats (weight, 320–380 g) were orally administered water (blank control), test compounds (15, 30, 60 mg/kg), and aspirin (positive control, 30 mg/kg), and anesthetized with dial urethane (1.5 g/kg, intraperitoneal) at 90 min after the oral administration (Fig. 4.2), or a lateral tail vein was cannulated for test compound or vehicle infusion. With a rectal thermometer, heat lamp, and temperature control unit, body temperatures of the mice were monitored and maintained at 37°C. The left common carotid artery was isolated carefully, and a plastic sheet was placed under the vessel to separate it from the surrounding tissue. The surface of the carotid artery was covered with a 4 × 0.5 cm cotton sheet saturated with $FeCl_3$ solution (25% w/v) for 15 min. The injured carotid artery segment (0.6 cm) was cut off, from which the formed thrombus was taken out. After drying for 24 h at room temperature in a dehumidifier, the dried weight of the thrombus was measured.

4.16 OCCLUSION TIME-MONITORED $FeCl_3$-INDUCED THROMBOSIS ASSAY FOR PIG[11]

Healthy, pure pietran pigs were premedicated with intramuscular administration of ketamine (20 mg/kg) and diazepam (1 mg/kg) to induce anesthesia, which was maintained by a continuous infusion of sufentanil (0.5 $\mu g\ kg^{-1}\ h^{-1}$) and pentobarbital (3 $mg\ kg^{-1}\ h^{-1}$). With pancuronium bromide (0.2 mg/kg), spontaneous movements were prevented. The pigs were given endotracheal intubation through a cervical tracheostomy and connected to a volume-cycled ventilator for delivering a tidal volume of 15 mL/kg with a fraction of inspired oxygen (FiO_2) of 0.4 and at a respiratory rate of 20 breaths/min. Ventilation adequacy was monitored by end-tidal CO_2 (ET CO_2) measurements, which was maintained between 30 to 35 mm Hg. Via closely monitoring and adjusting the FiO_2, arterial oxygen saturation was maintained above 95%. With a rectal probe and by means of a heating blanket, central temperature was measured and maintained at 37°C. Heart rate (HR) was monitored using a standard lead electrocardiogram. Through median sternotomy, the chest was entered, the incised pericardium was sutured to the chest wall to form a cradle for the heart, and the adherent fat and connective tissue were removed from the root of the aorta. Through the right carotid artery, a micromanometer catheter was inserted and advanced into the left ventricle. Through the right femoral artery, a micromanometer-tipped catheter was inserted and advanced into the ascending aorta. Around the aorta 2 cm downstream, a 14-mm-diameter perivascular flow-probe was closely adjusted to the aortic valve. By manipulating the micromanometer-tipped catheter, the pressure sensor was positioned just distal to the flow probe. With a micromanometer-tipped catheter inserted into the cavity through the superior vena cava, right atrial pressure was measured. Through a right femoral venotomy, a 6-F Fogarty balloon catheter was advanced into the inferior vena cava. Inflation of this balloon produced a titratable leftward shift in pressure. For eventual test compounds administration as well as blood samples analysis, the jugular vein and the left femoral artery were cannulated. For topical ferric chloride application, a segment of the left anterior descending (LAD) coronary artery was isolated distal to the first diagonal branch. Around the LAD coronary artery 3 cm distal to the site of $FeCl_3$ application, an electromagnetic flow probe was placed to measure coronary artery blood flow. After 30 min stabilization and drawing of a blood sample (time 0), a tissue strip (3 mm width) saturated with ferric chloride solution (20% or 50% w/v) was rolled around the surface of the LAD coronary artery for 45 min. On removal of the tissue strip, pigs were kept alive during 6 h. To evaluate the occlusion time and eventual reperfusion events, the blood flow in the LAD coronary artery was monitored throughout the assay.

4.17 DOPPLER BLOOD FLOW-MONITORED $FeCl_3$-INDUCED THROMBOSIS ASSAY FOR MOUSE[12]

C57BL/6 mice (weight, 18–25 g) were anesthetized by inhalation of a gas composed of 30% oxygen (300 mL/min) and 70% nitrous oxide (700 mL/min), which was passed through an isoflurane vaporizer set to deliver 3% to 4% isoflurane during initial induction and 1.5% to 2% during surgery. An incision was made in the skin of anesthetized mice directly on top of the right common carotid artery region. To expose a segment of the left common carotid artery, the fascia was bluntly dissected. With a miniature Doppler flow probe, carotid blood flow was measured. Thrombosis was induced by placing two pieces of filter paper (1 × 2 mm) saturated with 2.5% to 5% $FeCl_3$ on the opposite sides of the carotid artery (one beneath and one above) in contact with the adventitial surface of vessel for only 3 min. One minute prior to applying $FeCl_3$, mice were infused through the jugular vein with vehicle (negative control) or clopidogrel (10 mg/kg, positive control) or test compounds. The carotid blood flow was monitored at time 0 (prior to the ferric chloride paper application) and at 5, 10, 20, 30 min or sometimes up to 60 min (following the application of the filter paper). Thrombus formation was indicated by reductions in Doppler blood flow.

4.18 RAT GROIN FLAP ASSAY[13]

Male SD rats (weight, 300–350 g) were housed in cages with free access to food and water and maintained in room temperature of 24 ± 1°C with a 12-h light/dark cycle. Rats were anesthetized by intraperitoneal injection of sodium pentobarbital (50 mg/kg), which was maintained by intravenous injections of sodium pentobarbital (10 mg kg^{-1} h^{-1}) as needed. To obtain hairless skin, the groins were shaved with an electric razor and then treated with a depilatory cream. During surgery, to avoid temperature loss, the rats were kept on a heating pad, and the core temperature was thermostatically controlled. To directly observe the superficial epigastric vessels, the flap was resutured to its anatomic bed by two stitches. On a laser Doppler perfusion monitoring (LDPM) system (with an extended bandwidth up to 24 kHz for the artery measurements), flap microcirculation and blood flow in the feeding artery was monitored. Before incision, a laser Doppler probe holder with a standard probe PF408 for the skin measurements was attached to the flap with double adhesive tape. Using a needle probe with a custom-made vascular fixation device, artery blood perfusion was recorded, reading corpuscular movements in a volume of 1 mL, thus monitoring flow in the vessel lumen, with the vessel probe firmly kept in place by a micromanipulator. To stimulate the vessel wall, a bipolar stimulating electrode with two platinum contacts was used (electrode diameter 0.5 mm, electrode length 5 mm, electrode separation 1 mm). Via a constant-current unit (CCU 1A) and a stimulus isolation unit, the electrode was connected to a Grass stimulator. To systemically monitor the blood pressure, a pressure transducer was introduced into the left common carotid artery via a catheter. For data processing, LDPM and blood pressure data were stored in a Toshiba T3200SXC desktop computer with special software. On videotape with a video camera, microscopic images were recorded and monitored on a television screen. Thirty minutes after flap elevation, to avoid stimulation inducing artery contraction, Xylocaine was locally applied to the feeding artery, which was repeated every 30 min. Once the blood perfusion was stabilized and a 10-min baseline was recorded using monophasic pulses (frequency 4 Hz, pulse width 200 ms, voltage 100 mV), the superficial epigastric artery was stimulated proximal to the laser Doppler flowmeter (LDF) reading area, and a stimulus train of six impulses was used. In the case that no thrombi occurred 20 min after the first stimulation, another stimulation train was applied in the same way until thrombosis formation was observed. Five minutes prior to stimulation, rats were infused through the jugular artery with vehicle (negative control) or clopidogrel (10 mg/kg, positive control) or test compounds. Using the video camera system and LDPM measurements from the supplying vessel and its flap, the course of thrombus formation was continuously monitored and recorded. From stimulated sections, flap artery specimens were collected, perfused with 2.5% glutaraldehyde in a 0.1 M sodium phosphate buffer containing 2 mM $MgCl_2$, postfixed in glutaraldehyde, osmicated (2% osmium tetroxide in phosphate buffer), rinsed for 15 min in buffer, dehydrated in ethanol, and embedded in agar 100. On a LKB ultratome (Type 4802A; LKB-Produkter AB, Sweden), thin cross-sectional slices were cut, which were stained with Toluidine blue, and mounted for light microscopy examination. With baseline recordings set to 100%, laser Doppler values were presented as percentage changes from baseline and expressed as the mean values ± SE. In statistical analyses, the Kruskal–Wallis test with Dunn's multiple comparisons test was used to compare percentage changes of blood flow and skin microcirculation at various time points of stimulation. $P < 0.05$ was considered significant. Immediately after the experiments, the rats were given a lethal dose of the barbiturate.

4.19 FERRET ACUTE THROMBOSIS ASSAY[14]

Male neutered ferrets (weight, 1.1–1.3 kg) were anesthetized by intraperitoneal injection of sodium pentobarbital (45 mg/kg), which was maintained by intraperitoneal injections of sodium pentobarbital (20 mg/h, 60 mg total). Their body temperatures were monitored and maintained at 38°C with a heating pad. For test compounds administration and blood withdrawal as well as to monitor arterial blood pressure (ABP) with a P23Db transducer (Gould Inc., Oxnard, CA), polyethylene (PE) catheters were inserted into jugular veins (PE-90) and right femoral artery (PE-50), respectively. After cannulating the trachea, mechanical ventilation was initiated with a model 665 dual phase respirator (Harvard Apparatus, South Natick, MA), which was adjusted to maintain arterial $PO_2 > 80$ mm Hg and PCO_2 between 35 and 40 mm Hg as measured on an ABL 500 blood gas analyzer (Radiometer, Copenhagen, Denmark). As indicated in Fig. 4.3, after exploring the right carotid artery, a piece of Parafilm M (Sigma-Aldrich Co., St. Louis, MO, USA) was inserted under the vessel for electrical isolation. An L-shaped stainless steel wire, through which a 1-mA or 0.5-mA anodal current was delivered for 10 min using a constant DC current stimulator, and an electromagnetic flow probe were placed on the artery and attached to a model MDL 1401 flowmeter (Skalar, Delft, Netherlands) for measuring carotid artery blood flow (CBF). An alligator clip (the cathode) was attached to the hind limb. Fifteen minutes prior to stimulation, ferrets were intravenously injected vehicle (negative control) or aspirin (10 mg/kg, positive control) or test compounds. After termination of stimulation, the electrode was removed. On a R611 recorder (Sensor Medics, Anaheim, CA), CBF and ABP were continuously monitored, while to prevent overheating of the vessel the flow probe was switched off after about 20 min of zero flow and turned on intermittently thereafter. The observation usually lasted for 2 h, in the case where the vessel remained patent it was maintained for up

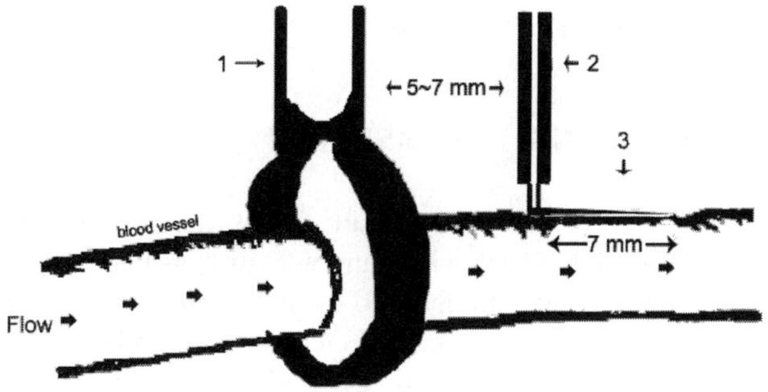

Figure 4.3 Measuring carotid artery blood flow.

to 3 h, at which the artery was isolated over its exposed length with two microaneurysm clips and dissected free. Under a stereomicroscope, the vessel segment was opened lengthwise, the thrombus was removed, and its wet weight was determined.

4.20 RAT ACUTE THROMBOSIS ASSAY[15]

After overnight fasting, male SD rats (weight, 250–350 g) were anesthetized with urethane (1.5 g/kg, intraperitoneal). The left carotid artery was isolated carefully. A plastic sheet was placed under the vessel to separate it from the surrounding tissue. Thrombus formation was induced generally. The holder contained two electrodes, and the temperature sensor was fixed under the exposed carotid artery. The rats were administered intravenously with NS (blank control), test compound (5, 10, 20 mg/kg), and aspirin (4 mg/kg), respectively. After 5 min, a current of 3 mA was delivered for 3 min. With thrombus formation, the temperature of the carotid blood was abruptly decreased. The occlusion time (OT) was measured through the temperature sensor and timer on the electric thrombosis stimulator. The rate of thrombosis inhibition was calculated according to the equation: Thrombosis inhibition $(\%) = (A1 - A)/A \times 100\%$, where A was the OT of the control group and A1 was the OT of agent groups.

4.21 ARTERIOVENOUS SHUNT ASSAY[16,17]

Male SD rats (weight, 320–380 g) were orally administered water (blank control), test compounds (15, 30, 60, 120 mg/kg), and aspirin (positive control, 30 mg/kg, twice daily for 2.5 days i.g.) and anesthetized by intraperitoneal injection of urethane (1.5 g/kg) 75 min after the last dose. The arteriovenous shunt operation was carried out; namely, the left jugular vein and the right carotid artery were cannulated with a 4-cm-long polyethylene tube (o.d. 1 mm) with heparin (50 U/kg) injection intravenously for anticoagulation. These catheters were connected to the ends of a 15-cm-long polyethylene tube (o.d. 2 mm) containing a 5-cm-long suture silk thread (no. 4). The blood was allowed to flow through the shunt for 15 min, and the silk thread with its associated thrombus was gently removed from the shunt, blotted twice on paper, and weighed immediately. The thrombus on the thread was calculated by subtracting the premeasured silk thread.

4.22 PLASMA CLOTTING TIME ASSAY[17]

Male SD rats (weight, 320–380 g) were orally administered water (blank control), test compounds (15, 30, 60, 120 mg/kg), and aspirin (positive control, 30 mg/kg, twice daily for 2.5 days i.g.) and anesthetized by intraperitoneal injection of urethane (1.5 g/kg) 75 min after the last dose. The collected arterial blood samples with sodium citrate as anticoagulant were centrifuged. On a fibrometer, by incubating 100 μL plasma with 100 μL activated partial thromboplastin time (aPTT) reagent for 3 rein and by adding 100 μL 25 mM $CaCl_2$, the plasma clotting time (aPTT) was measured, of which the data points were the mean of duplicate measurements and expressed as a ratio of treated versus baseline control.

4.23 THROMBOEMBOLIC PHOTOCHEMICAL ASSAY FOR REPEATED STROKE IN MICE[18]

Male C57BL/6J mice weighing 19–25 g and 12–16 weeks old were provided with a standard diet and tap water *ad libitum*, anesthetized with 3% isoflurane, and anesthesia was maintained with 1.5% isoflurane and a mixture of 30% oxygen and 70% nitrous oxide. The mice were placed in a supine position with a rectal probe connected to a heating pad, and their body temperatures were maintained at $37 \pm 0.5°C$. In surgical preparation, a commercial hair on the left inguino-femoral area and the midline of the neck from the lower jaw to the sternum was removed. Five minutes later, depilatory cream was removed and mice were administered (subcutaneous injection) 30 mL/kg saline NS to ensure their proper hydration before surgery. Mice were returned to their cages without food but with water *ad libitum* for recovery from the anesthesia, and they would further clean and groom themselves, leaving the planned site of surgery in pristine condition for surgery the following day. This maneuver may dramatically decrease postsurgical wound infection.

On the day of surgery, mice were anesthetized with 3% isoflurane, and anesthesia was maintained with 1.5% isoflurane and a mixture of oxygen 30% and nitrous oxide 70%. The mice were placed in a supine position and an 0.5-mm incision was made in the left inguinal area. The femoral vein was bluntly dissected, a 32-gauge catheter was inserted, advanced 4 mm, and secured using 6-0 nylon sutures. At room temperature, the surgical cavity was filled with NS. The mice received a rostro-caudal midline incision from the lower jaw to the inferior aspect of the sternal manubrium. By blunt dissection technique, the right common carotid artery (CCA), the proximal portion of the adjacent internal carotid artery (ICA), and the external carotid artery (ECA) were dissected.

In order to direct emboli to the cerebral vasculature a 6-0 nylon suture was used to permanently ligate the ECA. A Transonic Doppler flow probe (Model 0.5 VB, Transonic Systems, Ithaca, NY) was coupled with a temperature probe and placed on the right CCA approximately 3 mm proximal to its bifurcation. The surgical cavity was filled with a 1:1 mixture of Signa Gel (Parker Laboratories, Inc., Fairfield, NJ) and NS. The probe connected to a flowmeter was interpreted with a computerized data acquisition program. Isoflurane was decreased from 1.5% to 1%, and the CCA blood flow was monitored for 10–15 min.

Using erythrosin B (EB) as the photosensitizing dye, common carotid artery thrombosis (CCAT) was produced. Via the femoral venous catheter with an infusion pump (PHD2000; Harvard, Holliston, MA), a solution (2 mg/100 mL) of EB in saline was injected at a rate of 17.5 mg kg^{-1} min^{-1} for 2 min (total dose, 35 mg/kg). A tunable argon laser operating at 514.5 nm with a power of 165 mW (yielding an average intensity of 130 W/cm^2) was simultaneously focused through a 5× beam expander with focus range 1.2 M to ∞ placed 68.5 cm from the laser source. With a mirror, the laser beam was perpendicularly redirected onto the subject CCA (ca. 400 μm in diameter). A technique including gentle mechanical assistance for releasing the thrombus formed within the CCA was also employed for 1 min. Mice were returned to their cages. Twenty-four hours later, the same procedure was performed, while the laser was focused 2 mm distal to the initial site of the original irradiation on the CCA. Animals were again returned to their cages with food and water and sacrificed 72 h later.

4.24 EUGLOBULIN CLOT LYSIS ASSAY[19–21]

In a 13 × 100 mm glass test tube, the euglobulin fraction was prepared by adding 350 μL citrated plasma to 6.3 mL working acetic acid (1.5 mL 1% acetic acid in 90 mL deionized water), which was placed on ice for 10 min and centrifuged at 2000 × *g* at 25°C for 5 min. After discarding the supernatant, the tubes were drained via a cotton swab for 1 min to remove the residual liquid. The euglobulin fraction was resuspended in a solution consisting of 154 mM NaCl and 2.6 mM sodium borate (350 μL, pH 9.0), stirred with a glass rod, placed in a 37°C water bath for 90 s, restirred, and placed in a water bath for another 90 s. The sample was in duplicate transferred into a prewarmed 37°C 96-well Nunc-untreated polystyrene flat-bottom microtiter plate (Nalge Nunc International, Rochester, NY, USA, 150 μL/well). To one well of each sample, a 150-μL solution of test compounds was added, and the other well served as an individual blank control. Before first reading, the plate was shaken in the reader for 5 s. Temperature remained at 37°C, and a Power Wave HT microplate spectrophotometer (BIO-TEK, Winooski, VT, USA) was used, which was set up in kinetic mode to measure absorbance at 405 nm every 3 min for 720 min. Maximum absorbance was measured as the peak absorbance at 405 nm. The lysis time was defined as the time at which the curve, corrected for the individual blank, reaches an absorbance of 0.05 or less. A representative curve is shown in Fig. 4.4.

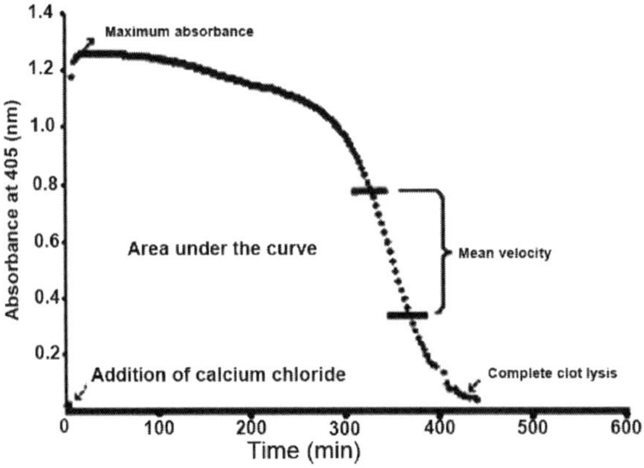

Figure 4.4 Formation and lysis curve of euglobulin clot.

4.25 CLOT FORMATION AND LYSIS (CloFAL) ASSAY[22]

Plasma samples were maintained in an ice-water bath until use, not to exceed 30 min. Recombinant lipidated human tissue factor (TF, 0.5 μg/mL stock solution prepared according to the manufacturer's instruction) or two-chain recombinant tPA or test compounds were added to a stock solution of Tris-buffered saline (TBS; 66 mM Tris, 130 mM NaCl, pH 7.0) containing 34 mM $CaCl_2$ to prepare reactant solution of 10 pM TF or and 900 ng/mL tPA or 1 M test compounds, and with TBS alone as the assay blank. TBS stock solutions were stored at 4°C for up to 1 month, and reconstituted stock solutions of tissue-plasminogen activator (tPA) or TF or test compounds were stored at −70°C for up to 1 month (and at least 24 h) for preparing fresh reactant solutions, which were maintained at room temperature not to exceed 30 min. Into each of four wells in a round-bottom 96-well plate, plasma was dispensed and prewarmed at 37°C for 3 min. Using a multitip automated pipette, to the first well of each quadruplicate 75 μL TBS blank was added, and to the remaining wells 75 μL reactant solution (prewarmed for 3 min) was simultaneously added. After an initial 5-s mixing step prior to the first reading, the plate was immediately placed in an eight-channel microplate spectrophotometer measuring 405 nm and 630 nm dual kinetic absorbance at 45-s intervals for 3 h. To eliminate artifact in baseline absorbance due to plate inconsistencies and lipemic or icteric plasma, absorbance data were blanked with wavelength (405 nm minus 630 nm) and reagent (reagent well minus TBS blank well). Absorbance data were averaged at each time point for the triplicate plasma wells containing reagent. From the absorbance data, maximum amplitude (MA), times to maximum absorbance (T_1), and completion of the first phase of decline in absorbance (T_2) were directly obtained. Using the area under the curve (AUC) over the course of the initial 30 min of the assay, a coagulation index $CI = [(AUC_{0\text{-}30\ mine})_{sample}/(AUC_{0\text{-}30\ mine})_{standard}] \times 100$ and fibrinolytic indices $FI = [(T_2/T_{1sample})/(T_2/T_{1standards})] \times (MA_{standard}/MA_{sample}) \times 100$ were obtained.

4.26 FIBRIN MICROPLATE ASSAY[23]

In the preparation of fibrinogen gels in 96-well microplates, fibrinogen 1 mg/mL was mixed with paranitroaniline (p-NA) and distilled water in a suitable concentration to give an optical density of approximately 0.800 at 405 nm. To each well, 190 μL mixture was added. On a microplate autoreader, the microplates were mixed for 1 s and measured at 405 nm. To the wells, 6 μL thrombin (20 NIH units/mL) was added, and the plates were stored at 37°C overnight.

In the assay, urokinase was used as reference (final concentrations 750, 375, 188, 94, 47, and 23 mIU/mL). With plate buffer I (0.05 M sodium diethylbarbiturate, 0.093 M NaCl, and 1 mM EDTA-Na), plasma samples (CPLa) were diluted 1:2. These dilutions were preincubated at 37°C with plate buffer I (3:1) or goat-anti-human urokinase IgG (3:1) for 1 h. Onto the fibrin gel, 20 μL test mixture was applied together with 5 μL plasminogen (0.375 mg/mL) in triplicate. Each microplate contained the standard references of urokinase. After 6 h incubation at 37°C, the converted substrate, the lysate, was removed with rinse buffer (20 mM NaH_2PO_4

and 150 mM NaC1, pH 7.4). To each of the remaining clots, 20 μL plasmin (1 U/mL) was added. Complete lysis was obtained overnight at 37°C. The wells were filled with rinse buffer, mixed for 10 s, and measured at 405 nm. Fibrinolytic activity was determined as absorption difference before and after lysis and dilution of a dye-containing fibrin clot and expressed as urokinase international μIU/mL by means of a urokinase dose-response curve based on a serial dilution of standard urokinase applied to each microplate. Assay values were the median value of three parallels applied.

4.27 FIBRINOLYTIC ACTIVITY ASSAY[24]

At room temperature, 1 NIH units/mL human thrombin and 2.5 mg/mL human fibrinogen in 500 μL HEPES buffered saline (HBS; 20 mM HEPES, 0.13 M NaCl, pH 7.4) with 20 mM $CaCl_2$ cross-linked fibrin clots were mixed in Eppendorf tubes (1 mL), and 100 μL polymerizing solution was immediately added. This mixture was stored at room temperature for 2 h to form clots. To each tube, 10 μL (50 ng) of the fibrinolytic enzyme in HBS was added. The tube was incubated at 37°C for 15 h, and then 50 μL denaturing solution (8 M urea, 4% β-mercaptoethanol, 4% SDS) was added for SDS-PAGE (12% gel) analysis. Fresh human plasma was used to form clots for visual analysis of clot lysis by test compounds. Plasma (500 μL) was clotted with thrombin (0.3 U), and the resultant clots were washed. Test compound was added to clots (25, 50, and 100 μg), incubated at 37°C for 2 h, and photographed with a Sony Cyber-shot DSC-W7 digital camera. Using the Chemidoc Imaging System with the QuantityOne Analysis software (Bio-Rad, Hercules, CA, USA), quantitative evaluation of clots was performed. The density of control was normalized to 100%, and the density (INT/mm^2) of clots treated with test compound given by the software was presented in the graph as "relative density (%)."

4.28 FIBRINOLYSIS ASSAY[25]

To 1.2 mL hypotonic phosphate buffer, platelets (165×10^6) were added. This solution was mixed with 150 μL whole blood supplemented with ca. 120 nCi ^{125}I-labeled fibrinogen, which was followed by adding test compound with suitable concentration. In a 17 × 100 mm polypropylene tube containing thrombin (0.80 IU/mL), these reaction components were gently mixed at room temperature for 5 min to form clots. To dislodge the clots from the sides of the tubes and hasten the reaction, the tubes were gently shaken. After 90 min, by diluting the reaction mixture with 10 mL phosphate buffer, the reaction was terminated. The diluted mixture was immediately filtered through a 25-mm Whatman GF/C filter, the filter residue was washed with another 5 mL phosphate buffer, dried at 55°C for 30 min, and the bound ^{125}I was measured by liquid scintillation counting on an LKB Wallac Rackbeta scintillation counter with 96% efficiency for ^{125}I. Fibrinolysis activity was represented by the difference between total counts in the reaction mixture and the filter residue.

4.29 THROMBOLYTIC ASSAY[26,27]

Male Wistar rats weighing 200–300 g were anesthetized with pentobarbital sodium (80.0 mg/kg, intraperitoneal). The right carotid artery and left jugular vein of the rats were separated. To the glass tube containing 1.0 mL blood collected from the right carotid artery of the rat, a stainless steel filament helix (12 circles; 0.8×1.0 mm) was added immediately (Fig. 4.5). Fifteen minutes later, the helix with thrombus was carefully taken out and weighted. It was then put into a polyethylene tube that was filled with heparin sodium (50 IU/mL NS), and one end was inserted into the left jugular vein. Heparin sodium was injected via the other end of the polyethylene tube for anticoagulation, following which the test compound was injected (Fig. 4.6). The blood was allowed to circulate through the polyethylene tube for 90 min, after which the helix was taken out and weighed (Fig. 4.7). The reduction of thrombus mass was recorded as thrombolytic activity.

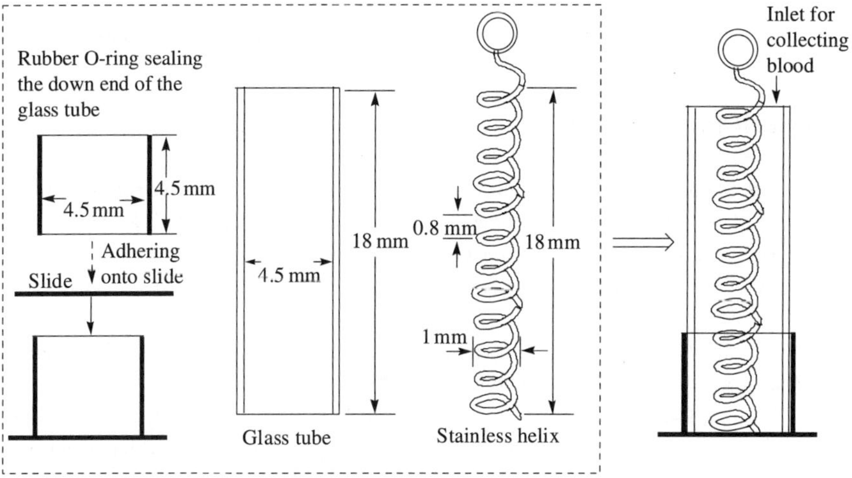

Figure 4.5 A device to prepare thrombus.

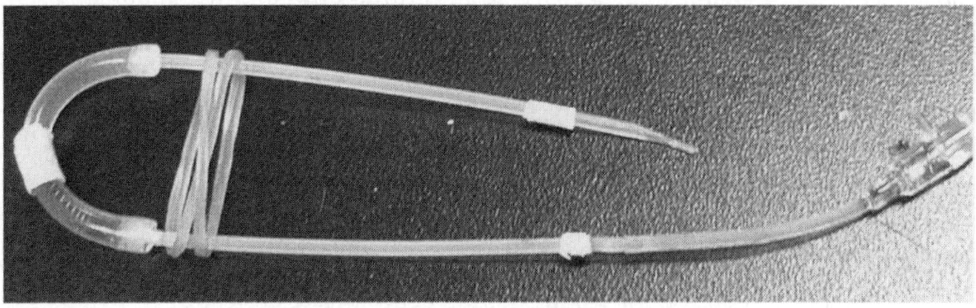

Figure 4.6 Polyethylene tube with thrombus.

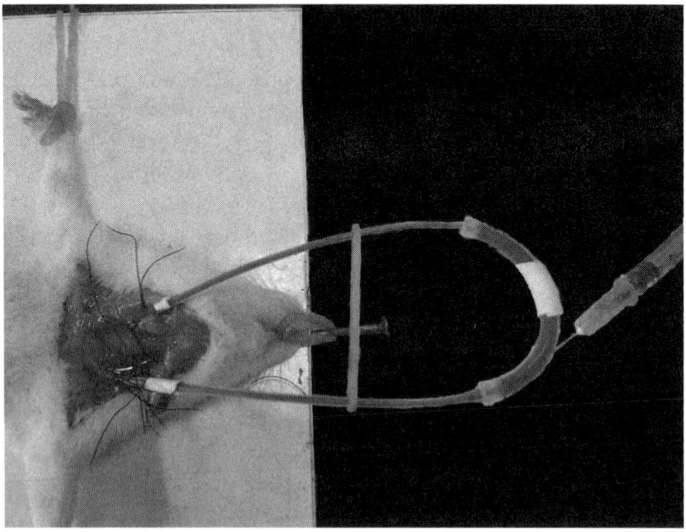

Figure 4.7 Blood circulating through the thrombus in the polyethylene tube.

REFERENCES AND NOTES

1. C. Kluft. Studies on the fibrinolytic system in human plasma: quantitative determination of plasminogen activators and proactivators. Thromb Haemost 41 (**1979**) 365–383.
2. K. Lee, Y. Jin, J. Lee, Y. Lim, D. Son, C. Lee, K. Yi, S. Yoo, H. Shin, Y. Yun. Anti-platelet activity of KR-32560, a novel sodium/hydrogen exchanger-1 inhibitor. Pharmacol Res 53 (**2006**) 265–270.
3. S. Alam, M. Gupta, R. Bhatnagar. Inhibition of platelet aggregation by anthrax edema toxin. Biochem Biophys Res Commun 339 (**2006**) 107–114.
4. G. Chang, S. Kang, J. Kim, K. Chung, Y. Chang, C. Kim. Inhibitory effect of the Korean herbal medicine, Dae-Jo-Whan, on platelet-activating factor-induced platelet aggregation. J Ethnopharmacol 102 (**2005**) 430–439. *Note*: PRP was prepared by centrifugation of fresh rabbit blood ($190 \times g$, for 10 min) with a 1/10 volume of 3.8% sodium citrate solution. Platelet numbers were adjusted to 6.0×10^5/mL. PRP was prepared by centrifugation of whole blood of apparently healthy volunteers ($100 \times g$, for 15 min at room temperature) with a 1/9 volume citrate–citric acid dextrose (100 mM trisodium citrate, 7 mM citric acid, 140 mM dextrose, pH 6.5). By centrifugation at $1000 \times g$ for 10 min, the platelets of the supernatant were sedimented. Platelet pellet was washed two times with citrate buffer (0.35% w/v BSA, 108 mM NaCl, 2.8 mM KCl, 1.0 mM $CaCl_2$, 1.6 mM $MgCl_2$, 0.3 mM NaH_2PO_4, 9.5 mM $NaHCO_3$, 10.1 mM trisodium citrate, 4.6 mM citric acid) by centrifugation at $555 \times g$ for 10 min, and resuspended in citrate buffer to obtain about 10^9 platelets/mL titer. The suspension was shaken in a Petri dish until the assays were performed. For human platelet aggregation assay, an aliquot of the platelet suspension was centrifuged at $7000 \times g$ for 1 min, and the pellet was resuspended in Tyrode's buffer (135 mM NaCl, 3.5 mM KCl, 1.2 mM $CaCl_2$, 2.0 mM $MgCl_2$, 0.3 mM NaH_2PO_4, 0.35% w/v BSA, 11.9 mM $NaHCO_3$, pH 7.4) to obtain about 10^9 platelets/mL titer.

5. C. Hung, W. Tsai, L.Y. Kuo, Y. Kuo. Evaluation of caffeic acid amide analogues as anti-platelet aggregation and anti-oxidative agents. Bioorg Med Chem 13 (**2005**) 1791–1797.

6. E. Nurtjahja-Tjendraputra, A.J. Ammit, B.D. Roufogalis, V.H. Tran, C.C. Duke. Effective anti-platelet and COX-1 enzyme inhibitors from pungent constituents of ginger. Thromb Res 111 (**2003**) 259–265.

7. R.C. Carroll, R.M. Craft, J.J. Chavez, C.C. Snider, S.J. Bresee, E. Cohen. A thrombelastograph whole blood assay for clinical monitoring of NSAID-insensitive transcellular platelet activation by arachidonic acid. J Lab Clin Med 146 (**2005**) 30–35. *Notes*: (1) By differential centrifugation, heparin anticoagulated blood was also used to isolate PRP and PPP for assay of OPA on a Chrono-Log Aggregometer with AggroLink software determining the maximum aggregation of the platelets for 10 min after adding 1 mmol AA and calculating percent aggregation. OPA was assayed without platelet count adjustment. (2) After removing the PRP layer, further cell fractionation was carried out. The leukocyte-rich buffy coat was also collected separately for some experiments. The blood cell fractions were diluted to 1/1 with a balanced salt solution containing 6 IU/mL heparin, centrifuged at 100 $\times g$ for 20 min at room temperature, and the top layer containing platelets was discarded. This procedure was repeated twice more until platelet counts were reduced to less than 10% of the original whole blood sample. To further reduce white blood cell count to less than 15% of the original whole blood counts, red blood cell fraction minus the buffy coat was also removed. Blood cell counts in various fractions were determined on an Ichor/Platele-tworks analyzer. (3) After MA values were obtained, TEG samples were collected into Microfuge tubes, centrifuged at 10,000 $\times g$ for 5 min at room temperature, and the sera was collected and frozen at $-70°C$. TXB_2 concentrations in serum samples were assayed with an enzyme immunoassay kit following the manufacturer's directions.

8. H. Du, J.A. Zawaski, M.W. Gaber, T.M. Chiang. A recombinant protein and a chemically synthesized peptide containing the active peptides of the platelet collagen receptors inhibit ferric chloride-induced thrombosis in a rat model. Thromb Res 121 (**2007**) 419–426.

9. K.D. Kurz, B.W. Main, G.E. Sandusky. Rat model of arterial thrombosis induced by ferric chloride. Thromb Res 60 (**1990**) 269–280.

10. M.A. Robinson, D.C. Welsh, D.J. Bickel, J.J. Lynch Jr., E.A. Lyle. Differential effects of sodium nitroprusside and hydralazine in a rat model of topical $FeCl_3$-induced carotid artery thrombosis. Thromb Res 111 (**2003**) 59–64.

11. J.M. Dogne, S. Rolin, V. Tchana-Sato, M. Petein, A. Ghuysen, B. Lambelmont, J. Hanson, P. Segers, V. Dorio, P.H. Kolh. Characterization of an original model of myocardial infarction provoked by coronary artery thrombosis induced by ferric chloride in pig. Animal models and experimentation. Thromb Res 116 (**2005**) 431–442.

12. X. Wang, L. Xu. An optimized murine model of ferric chloride-induced arterial thrombosis for thrombosis research. Thromb Res 115 (**2005**) 95–100.

13. Y. Qi, B. Gazelius, B. Linderoth, T. Lundeberg. Arterial blood flow and microcirculatory changes in a rat groin flap after thrombosis induced by electrical stimulation of the artery. Microvasc Res 63 (**2002**) 179–185.

14. W.A. Schumacher, T.E. Steinbacher, J.R. Megill, S.K. Durham. A ferret model of electrical-induction of arterial thrombosis that is sensitive to aspirin. J Pharmacol Toxicol Methods 35 (**1996**) 3–10.

15. J. Hladovec. Experimental arterial thrombosis in rats with continuous registration. Thromb Diath Haemorrh 26 (**1974**) 407–410.

16. T. Umetsu, K. Sanai. Effect of 1-methyl-2-mercapto-5-(3-pyridyl)imidazol (KC-6141), an antiaggregation compound, on experimental thrombosis in rats. Thromb Haemost 39 (**1978**) 74–83.
17. P.C. Wong, E.J. Crain Jr., O. Nguan, C.A. Watson, A. Racanelli. Antithrombotic actions of selective inhibitors of blood coagulation factor Xa in rat models of the thrombosis. Thromb Res 83 (**1996**) 117–126. *Note*: For venous thrombosis, model rats were anesthetized. A carotid artery, a jugular vein, and a femoral vein were cannulated for blood sampling, test compounds infusion, and hypotonic saline injection, respectively. The abdominal vena cava was isolated, and between the left renal and femoral veins, all side-branches of it were ligated. By a rapid injection of 1 mL hypotonic saline (0.225%) into the vena cava, thrombus formation was induced for 15 s, and a 15-min stasis in a segment (about 1 cm) of the vena cava was isolated. The formed thrombus in the vena cava was removed, immediately blotted twice on paper, and weighed.
18. J.D. Lozano, D.P. Abulafia, G.H. Danton, B.D. Watson, W.D. Dietrich. Characterization of a thromboembolic photochemical model of repeated stroke in mice. J Neurosci Methods 162 (**2007**) 244–254.
19. A.A. Smith, L.J. Jacobson, B.I. Miller, W.E. Hathaway, M.J. Manco-Johnson. A new euglobulin clot lysis assay for global fibrinolysis. Thromb Res 112 (**2003**) 329–337.
20. T. Urano, K. Sakakibara, A. Rydzewski, S. Urano, Y. Takada, A. Takada. Relationships between euglobulin clot lysis time and the plasma levels of tissue plasminogen activator and plasminogen activator inhibitor 1. Thromb Haemost 63 (**1990**) 82–86.
21. J.M. Wardlaw, C.P. Warlow, C. Counsell. Systematic review of evidence on thrombolytic therapy for acute ischaemic stroke. Lancet 350 (**1997**) 607–614.
22. N.A. Goldenberga, W.E. Hathaway, L. Jacobson, M.J. Manco-Johnson. A new global assay of coagulation and fibrinolysis. Thromb Res 116 (**2005**) 345–356.
23. S. Fossum, N.O. Hoem. Urokinase and non-urokinase fibrinolytic activity in protease-inhibitor-deprived plasma, assayed by a fibrin micro-plate method. Immunopharmacology 32 (**1996**) 119–121.
24. L.H. Gremski, O.M. Chaim, K.S. Paludo, Y.B. Sade, M.F. Otuki, M. Richardson, W. Gremski, E.F. Sanchez, S.S. Veiga. Cytotoxic, thrombolytic and edematogenic activities of leucurolysin-a, a metalloproteinase from Bothrops leucurus snake venom. Toxicon 50 (**2007**) 120–134.
25. A.P. Gadbut, J.R. Schullek, A.M. Hanel, D.R.E. MacAllan. A reconstituted dilute blood clot lysis assay for the medium throughput screening of thrombolytic compounds. Anal Biochem 270 (**1999**) 24–32.
26. M. Zhao, N. Lin, C. Wang, S. Peng. Synthesis and thrombolytic activity of fibrinogen fragment related cyclopeptides. Bioorg Med Chem Lett 9 (**2003**) 961–964.
27. M. Zhao, J. Liu, C. Wang, L. Wang, H. Liu, S. Peng. Synthesis and biological activity of nitronyl nitroxide containing peptides. J Med Chem 48 (**2005**) 4285–4292.

5

METHODS AND APPLICATIONS OF ANTICOAGULATION ASSAYS

Ming Zhao

The hypercoagulable state is associated with a number of risk factors leading to cardiovascular complication and is defined as an increasing tendency in developing thrombosis. Patients with either cardiovascular disorders or thrombus often need anticoagulation for surgical procedures. In the revised coagulation theory, the extrinsic pathway is considered to play an integral role. This pathway is initiated by a clotting signal tissue factor resulting from trauma (tissue injury), sepsis, inflammation, and many other conditions and is often upregulated. For instance, by damaging the endothelium and splitting the vessel wall, angioplasty allows the blood constituents to be exposed to the deeper tissues of the wall. In the wall, cells including macrophages express tissue factor, which activates the extrinsic coagulation pathway and generates thrombin and the activation and aggregation of platelets and leukocytes at the site of wall injury. Together with reduced fibrinolytic activity in the angioplastied area, these postthrombotic changes result in the development of intramural thrombosis. In the past decade, anticoagulation development has witnessed exponential progress, but the risks of anticoagulation-related hemorrhagic complications and development of chronic venous insufficiency are significant over time. To understand the coagulation mechanism in detail and develop new anticoagulation drugs, a series of procedures for anticoagulation assays have been established. In this chapter, 25 models are described: ecarin chromogenic assay,[1] anticoagulation activity assay in an *in vitro* system,[2] anticoagulation activity assays in an *in vivo* system,[2] rat thrombosis assay,[3,4]

Pharmaceutical Bioassays: Methods and Applications. By Shiqi Peng and Ming Zhao

in vivo microvascular inferior vena cava (IVC) blood flow assay,[3] Owren prothrombin time (PT) assay,[5] rabbit double-balloon injury assay,[6–8] rapid point-of-care assay for enoxaparin,[9] thromboplastin clotting assay,[10] rat assay for reproducible stasis-induced venous thrombosis,[11] *in vivo* activated clotting time (ACT) II/ecarin clotting time assay,[12] *ex vivo* and *in vivo* anticoagulation assay,[13] plasma-based ecarin clotting time assay for r-hirudin,[14] protamine titration for heparin in whole blood,[15] automated assay evaluating response of kaolin ACT to heparin,[16] platelet/monocyte interaction-based particulate matter (PM) exposure mouse assay,[17] mouse tail-bleeding time assay,[18,19] MRI assay for rabbit atherosclerotic lesions,[20] spiral computed tomography assay,[21] duplex ultrasound assay,[21] standard Hemochron assay,[22–24] platelet serotonin release assay,[24] activated partial thromboplastin time (aPTT) assay,[25–27] and rat stroke outcome assay.[28,29]

5.1 ECARIN CHROMOGENIC ASSAY[1]

By clean cubital vein puncture, whole blood from 10 healthy volunteers, 5 patients under oral anticoagulation therapy (OAT) and 5 patients with liver dysfunction (LD), was collected into plastic vials containing 3.2% sodium citrate (plasma/citrate, 9 : 1 v/v). All patients under OAT were anticoagulated for treatment of atrial fibrillation with international normalized ratio (INR) values ranging from 1.5 to 2.2 (1.73 ± 0.3) and matched to patients with LD regarding their INR. LD was classified according to Child–Pugh criteria: patients with Child B or C and increased INR ranging from 1.4 to 1.7 (1.52 ± 0.2) were involved. The means of baseline INRs between OAT and LD patients had no statistical significance. Among the five patients with LD, four had alcoholic liver cirrhosis Child B ($n = 2$) or Child C ($n = 2$); the cause of liver cirrhosis in one patient (Child C) remained unclear. To avoid inter-lot differences after centrifugation ($2500 \times g$, 10 min), plasma samples were immediately frozen, stored at −80°C, and analyzed within 6 weeks with only one reagent lot.

To achieve final argatroban plasma concentrations (125–2000 ng/mL) dilution series was prepared by a two-fold dilution series of an individual sample spiked to 2000 ng/mL argatroban (Argatra®; Mitsubishi Pharma, Düsseldorf, Germany) and then measured immediately.

On a BCS Coagulation Analyzer (Dade Behring, Marburg, Germany), all clotting time measurements were performed. For the activated partial thromboplastin time (aPTT) measurement, Pathromtin SL (Dade Behring Inc., USA), Actin FS (Dade Behring Inc., USA), aPTT-SP (Instrumentation Laboratory, Munich, Germany), and KCT KaoClot (Haemochrom Diagnostica GmbH, Essen, Germany) were used. Using Innovin reagent (Dade Behring Inc., USA, ISI value = 1.09), the PT (INR) was measured. By ecarin chromogenic assay (ECA), reaction time was tested. According to the manufacturer's original settings, all tests were applied. For ECA, into prewarmed cuvettes 100 μL ECA prothrombin buffer, 25 μL plasma sample, and 25 μL ECA substrate were added, gently mixed at 37°C for 1 min, 50 μL ECA ecarin reagent was added, and measurement was started. Normal values were

26–36 s for Pathromtin SL, 25.8–33.2 s for Actin FS, 29–41 s for aPTT-SP, and less than 65 s for KCT.

5.2 ANTICOAGULATION ACTIVITY ASSAY IN AN *IN VITRO* SYSTEM[2]

With citrated human PPP as test system, *in vitro* anticoagulation assay was performed. PPP was prepared according to the guidelines for preparing citrated plasma for hemologic analyses. From individual healthy donors, human blood was pooled, immediately mixed with 3.8% trisodium citrate in volume ratio 9 : 1, and centrifuged at 2500 × *g* for 15 min to provide PPP. By the classic coagulation assay of aPTT, PT, and thrombin time (TT), with heparin (150 IU/mg) as reference the *in vitro* anticoagulant activity of sodium microcrystalline cellulose (Na-MCS) was evaluated. For aPTT assay, 50 μL test sample was mixed with 500 μL PPP, incubated at 37°C for 1 min, 100 μL cephatin was added, incubated at 37°C for another 3 min, 100 μL 0.025 M aqueous $CaCl_2$ was added, and the clotting time (CT), that is aPTT, was with recorded with a CA-1500 automatic blood coagulation instrument (Diagnostica Stago, France). Using their relative assay reagents thrombin containing calcium and thrombin, respectively, instead of cephatin TT and PT were determined as for aPTT.

Activities of coagulation factor (F) IIa or FXa were analyzed based on the absorbance at 405 nm (*A*405) of released *p*-nitroaniline, their specific chromogenic substrates. *A*405 was positively related to FIIa and FXa activities, and in its analysis chromogenic FXa substrate CBS31.39 or chromogenic FIIa substrate chromozym P was added into 100 μL test system, after mixing and incubating at 37°C for 4 min, 50 μL interrupting agent was added to stop the reaction, and *A*405 was measured on an ELX800 microplate reader. The inhibition on FIIa or FXa by Na-MCS was studied by comparing the residual FIIa or FXa activities in PPP after adding Na-MCS, which were calculated according to the equation: Residual activity (%) = (A405/A′405) × 100%, wherein A′405 and A405 were the absorbance values at 405 nm before and after the addition of Na-MCS.

5.3 ANTICOAGULATION ACTIVITY ASSAYS IN AN *IN VIVO* SYSTEM[2]

Clotting time (CT) in mice was as the *in vivo* anticoagulation activity evaluated by Na-MCS in comparison with that of heparin. The test samples were dissolved in saline and given by subcutaneous administration of 5 mL/kg. The mice were divided randomly into 26 groups, and each group contained eight mice. In the assays (a) 11 groups of mice received Na-MCS or heparin at doses of 0.3, 0.6, 0.9, 1.2, and 1.5 mg/kg or saline, respectively, and 2 h later their blood was collected; (b) 15 groups of mice received Na-MCS or heparin at doses of 1.2 mg/kg, and 12, 13, 13.5, 14, 14.5, 15, and 18 h later their blood was collected from their hearts under intramusclar anesthesia of 10 mg/kg ketamine. To record the CT, each blood sample from the heart was directly drawn into a plastic test tube.

5.4 RAT THROMBOSIS ASSAY[3,4]

Thrombosis in male SD rats (weight, 350–450 g) was induced with proximal ligation of the inferior vena cava (IVC) below the renal veins and ligation of all visible tributaries. In more than 90% of rats, a clot developed. Rats were sacrificed on day 4 or day 8. At sacrifice, through the previous incision the thrombosed IVC was exposed and the block was removed. For histologic analysis, a small consistent portion was divided intact and preserved, while most of the specimen was divided into wall and thrombus portions with blunt or sharp dissection. The net thrombus was weighed and snap-frozen for later biochemical analysis.

Rats receiving NS were used as control, and rats at each time point receiving test agent were used as experimental groups. Six and 10 rats were assigned to each group at the 4-day time point and to each group at the 8-day time point, respectively. Rats received 1 μg study agent daily in 0.1 mL NS with 1:500 bovine serum albumin, with dosing extrapolated from other experimental studies. No specific pharmacokinetic studies were involved. The first dose was administered at the time of IVC ligation by directly injecting the agent into the ligated IVC using a 30-gauge needle and cotton-tipped applicator pressure for hemostasis. The subsequent doses were via tail vein injection given daily until sacrifice. An arterial blood sample was taken from the dorsal tail artery before both laparotomies. By means of automated peripheral blood differential, peripheral leukocyte quantification was determined.

5.5 *IN VIVO* MICROVASCULAR IVC BLOOD FLOW ASSAY[3]

To assess *in vivo* microvascular IVC blood flow of rats, a laser Doppler scanner was used. Laser Doppler scan (5-s scans with 30 s between over 4 cycles) was carried out at laparotomy, the pre-IVC ligation, immediate postligation, and harvest of the exposed IVC region of interest. To ensure the best estimation of the mid-coronal IVC, section depth was constant and adjusted for each rat. These scans were saved, and to estimate the mean color flow, accompanying image software was used. The color flow was summed and averaged for the 4 cycles. At the various time points for each rat, mean intensity was determined and represented as intra-IVC blood flow velocity, not volume flow. The intensities were reported as percent of baseline flows specific to each rat to ensure consistency (each served as its own control).

5.6 OWREN PT ASSAY[5]

The manufacturer calibrated the system, and the calibration information was stored in a lot-specific microchip code included in every pack of test strips. All patients were tested with the same batch of test strips. Two expert technicians performed all capillary blood sampling and the analysis. Capillary blood was obtained by a finger prick using an Accu-chek Softclix (Roche Diagnostics), and one drop of blood was applied to the CoaguChek S system.

The standard Owren prothrombin time (PT) test, the SPA Prothrombin complex assay, was calibrated using calibrant plasmas. The locally established international sensitivity index (ISI) was 0.99, and the mean normal PT was 21.8 s. The total imprecision was calculated as coefficient of variation to be 2.2% at international normalized ratio (INR) level of 1.0 and to be 4.2% at the INR level of 3.1. Prothrombin, factor V, factor VII, and factor X analyses were all performed with one-stage coagulation assays using Thromborel S as thromboplastin source. With a coagulation-based Clauss method, fibrinogen was determined, with the reagent Platelin LS, aPTT was determined, and with the thromboplastin reagent Neoplastine, traditional Quick PT determination was performed. The ISI value was 1.28 and the mean normal PT was 11.6 s. All coagulation assays were performed on the BCS instrument. The presence of antiphospholipid antibodies was detected with two different ELISAs, namely Asserachrom APA IgG with IgM and Anti-h2-glycoprotein IIgG with IgM. The reference interval recommended by the manufacturer was used for the antiphospholipid antibodies.

5.7 RABBIT DOUBLE-BALLOON INJURY ASSAY[6–8]

As controls, eight rabbits were injected with saline, as treatment group, eight rabbits received subcutaneous injections of tissue plasminogen activator (r-TPA) (1 mg/kg), and eight rabbits were injected with subcutaneous PEG-hirudin (0.7 mg/kg). In a consecutive manner, cohorts rabbits were operated on. Two hours prior to the second balloon injury, injections were given and repeated at 6-h intervals for 18 h. Previously, this dosing schedule had been shown to achieve adequate plasma concentrations over 6 h in hypercholesterolemic rabbits. Five minutes before the second injury and at 2 and 24 h, blood samples were taken. Polyethylene glycol-hirudin (PEG-H) blood levels were measured with ELISA. Using a lyophilized ecarin assay, with ecarin the specific activity of PEG-H was measured. Using a bio-immunoassay, blood levels of functionally active tissue plasminogen activator (tPA) were measured. Six weeks after the second injury, the rabbits were killed by an overdose of sodium pentobarbitone (100 mg/kg, intravenous) and through the right femoral artery via a 21-guage needle at a continuous pressure of 80 mm Hg immediately perfusion fixed with 10% buffered formalin. The contralateral femoral vein was exposed and opened to allow the effluent to escape until it was clear. The right and left common carotid arteries were harvested, oriented, and allowed to fix overnight in 10% buffed formalin. The arteries were paraffin embedded and cut to 5-μm-thick sections at 1-mm intervals. A minimum four sections per angioplastied segment was taken and stained with Millers elastic stain. To measure the total vessel area, the luminal area, to provide the cross-sectional area of the walls, computerized planimetry was used. To enable the software to translate bitmaps into units of square millimeters, the system was calibrated with a graticule. Two blinded observers performed these measurements and used the mean of two measurements.

5.8 A RAPID POINT-OF-CARE ASSAY FOR ENOXAPARIN[9]

A microprocessor-based analyzer and single-use assay-specific test cards were involved in the Rapidpoint Coag system (Pharmanetics Inc., Morrisville, NC) (Fig. 5.1). In a flat capillary chamber on the surface of a test card, the Rapidpoint phot-mechanically monitors fibrin clot formation. After the test card was inserted into the reader, the test card was warmed to 37°C and the operator added a single drop of citrated whole blood to the sample well. As blood was drawn into the reaction chamber, the reagents were rehydrated, and by an oscillating magnetic field, paramagnetic iron particles were stimulated to move in the test chamber. With the help of a specific factor X activator in blood sample, factor Xa was rapidly produced to initiate the clotting cascade. As a clot began to form, fibrin strands attached themselves to the iron particles and impeded their movement. Particle movement was monitored by an infrared optical system, while a preset reduction in particle movement signaled the test's end point. A complex of enoxaparin (ENOX) from the patient's blood and antithrombin formed, inhibited factor Xa, and proportionally lengthened the clotting time. The ENOX card was optimized for the ENOX monitor but was affected by other heparins. Thienopyridines or glycoprotein (GP) IIb/IIIa inhibitors did not affect the ENOX times. Among 120 normal volunteers, ENOX times ranged from 106 to 160 s and among 166 un-anticoagulated citrated blood

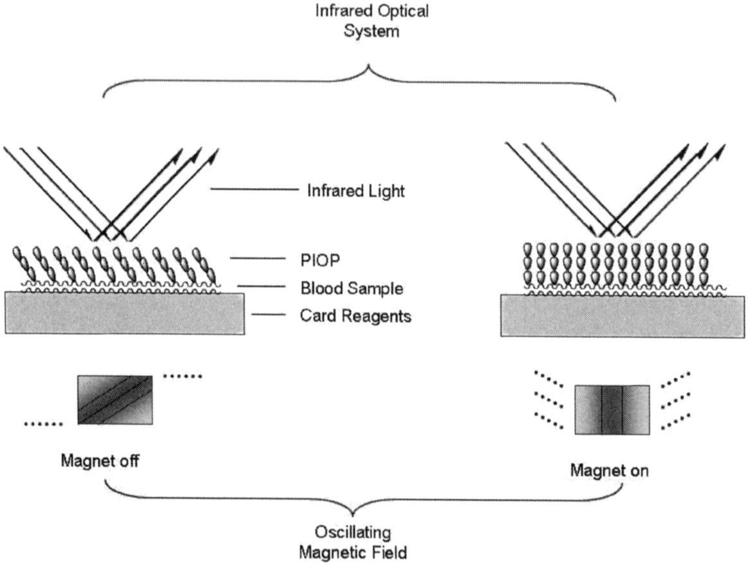

Figure 5.1 Diagrammatic sketch of Rapidpoint Coag system.

samples of patients, ENOX times ranged from 70 to 180 s. The correlation between ENOX times and plasma anti-Xa levels was high with a value of 260 s approximating a plasma anti-Xa level of 1.0 IU/mL.

5.9 THROMBOPLASTIN CLOTTING ASSAY[10]

Human plasma was prepared from blood donated by healthy volunteers. PPP was prepared from blood anticoagulated in 0.13 M trisodium citrate. After centrifugation of whole blood at 900 × *g*, 10,000 × *g*, and finally 23,000 × *g*, plasma from a minimum of three donors was pooled and stored at −80°C prior to use. Over the course of the described experiments, a number of different pools were used. By centrifugation (1100 × *g*) of anticoagulated blood collected from New Zealand White rabbits, citrated rabbit plasma was prepared (Fig. 5.2). Using actin FS aPTT reagent, and calcium on an automated coagulation timer, aPTT was determined.

5.10 RAT ASSAY FOR REPRODUCIBLE STASIS-INDUCED VENOUS THROMBOSIS[11]

Forty-two male SD rats (weight, 250–300 g) were anesthetized with an inhalation mixture of 1% to 2% isoflurane and 100% oxygen during the procedure. After induction on day 1, rats were given either 0.1 mL/day vehicle (0.1% serum albumin in normal saline) or IL-8 (1 mg in 0.1% serum albumin in normal saline) daily via tail vein injection for 4 or 8 days. This dosing was chosen empirically based on IL-8's extrapolated average rat serum volume and the known *in vitro* angiogenic/chemotactic activity (e.g., 10 nM). Aseptic right groin was exposed, and the right femoral vein was dissected and with 3-0 silk ties looped. Laparotomy was performed, the IVC

Figure 5.2 Collecting blood from the marginal ear vein of male New Zealand White rabbits.

was isolated, and major side branches were ligated. At this point with the abdomen open and the abdominal contents eviscerated, the right femoral vein was cannulated with a 21-gauge angiocatheter and the venous pressure was measured with a water manometer corrected to rat cardiac height. After the pressure column approximately equilibrated for 1 min, the reading was taken. For establishing thrombosis, the IVC was ligated below the renal veins. After the pressure column approximately equilibrated for 1 min, the postligation venous pressure was measured, the femoral vein was tied off, and the abdomen was closed. After either 4 or 8 days, the rat was again anesthetized in the same fashion, and the left femoral vein was isolated. A repeat laparotomy was performed, and the IVC was exposed. The left femoral vein was cannulated, and a venous pressure was taken with the water manometer. Through the left ventricle, the rat was administered 10 mL 50% India ink solution, which was allowed to circulate for several cardiac contractions. The rat was sacrificed, the thrombosed IVC segment was harvested below the renal veins and weighed, and the weight (mg) was corrected to vein length (cm). The vein was halved, the proximal portion was placed in 10% buffered formalin for 24 h, and then was placed in 70% EtOH for subsequent permanent section processing. The lower segment of the vein was snap-frozen in liquid N_2 and stored at $-70^{\circ}C$.

5.11 *IN VIVO* ACT II/ECARIN CLOTTING TIME ASSAY[12]

ACT II/ecarin clotting time (ECT) coagulation measurement was based on a mechanical detection of clot. Reagent was pipetted into the bottom of the cartridges containing kaolin for activated clotting time measurement. Plastic flags were inserted, and the blood sample was placed into the blood chamber. After an automatic incubation period, the flags were moved in the device. By inhibiting optical detection of the flag movement, clot formation was confirmed. According to a standard calibration curve, the measured clot formation time was related to a defined anticoagulant concentration. Assays were performed in duplicate in double cartridges.

The mixture of 50 mL 0.1 M $CaCl_2$, 40 mL 1 M HEPES buffer, 4 mL 5% BSA, and 106 mL H_2O was titrated to pH 7.0 to provide needed HEPES buffer. Then, 50 U lyophilized ecarin was dissolved in 2.5 mL HEPES buffer to result in an ecarin concentration of 20 U in the stock solution, and 10 U and 5 U solutions were obtained by further dilution of this solution. The reaction chamber of the blank cartridges was filled with 75 μL ecarin reagent, and the flags were inserted, while 75 μL standard human plasma and 75 μL citrated whole blood were placed into the blood chamber, and the test was started.

After informed consent, citrated whole blood samples were collected from 5 healthy volunteers (3 men and 2 women, aged 35–46 years), spiked with 0, 1, 2, 3, 4, and 5 μg/mL r-hirudin to establish a standard calibration curve. After determining ecarin's final concentration, in a second part of the *in vitro* setup, the within-assay variance of 5 IU/mL ecarin reagent final preparation was examined by 5 × duplicate measurements of the same sample for 1 μg /mL and 5 μg /mL r-hirudin. The stability of prepared and stored cuvettes was also investigated. Thus the ecarin was placed in the cartridge, and the flag was inserted and with the plasma overlaid. Afterward, the samples were stored at −80°C. After 1, 2, 4, and 12 weeks, the cartridges were defrosted. Measurements were performed in citrated whole blood samples spiked with 1 and 5 μg/mL r-hirudin and compared with the results of freshly prepared ecarin cartridges. Five duplicate measurements were performed for each measurement. In addition, the influence of variations in hematocrit value, platelet count, and plasmatic coagulation factors on the test results was clarified. Therefore by centrifugation and adjustment of the plasma fraction, 20%, 30%, and 60% hematocrit were obtained, by centrifugation of platelet-rich plasma and adjustment to the desired value with platelet-poor plasma 10×10^4/mL, 5×10^4/mL, and 3×10^4/mL platelets were obtained, and by substitution of platelet-poor plasma with adequate 5% albumin solution, 50%, 30%, and 10% of the initial procoagulants were obtained.

The *in vivo* investigation was performed with 5 and 10 IU/mL ecarin reagent, and five patients with heparin-induced thrombocytopenia type II (HIT II) who underwent cardiac surgery (coronary artery bypass grafting $n = 2$, aortic valve replacement $n = 2$, aortic valve replacement + bypass grafting $n = 1$) with r-hirudin as the anticoagulant were involved. For anticoagulation of cardiopulmonary bypass (CPB), 0.25 mg/kg bolus and 0.20 mg/kg priming solution of the r-hirudin regimen were used for the patients. When hirudin concentration achieved 3.5 to 4.0 μg/mL, CPB was initiated, which was maintained by an infusion of r-hirudin that was adjusted to the actual measured level. With the TAS analyzer (Cardiovascular Diagnostics Inc., Raleigh, NC), the r-hirudin level was monitored. A kallikrein inhibitor and antifibrinolytic agent, aprotinin, was used according to a high-dose regimen with 2×10^6 kIU bolus priming solution and 500,000 kIU/h constant infusion during CPB for all patients. Citrated whole blood samples were taken at 20-min intervals during CPB and for 2 h after CPB. To perform the chromogenic test in plasma, the plasma sample was incubated with tris-(hydroxymethyl)aminomethanehexadimethrine bromide buffer, S-2238 substrate, and thrombin reagent and measured with a COBAS MIRA analyzer (Behringwerke, Marburg, Germany) at 405 nm extinction.

5.12 *EX VIVO* AND *IN VIVO* ANTICOAGULATION ASSAY[13]

(1) *Ex vivo* assay: Ten patients with an elective coronary artery bypass procedure were involved. Cardiopulmonary bypass was performed with standard uncoated CPB circuits for perfusion. After starting CPB, mild hypothermia (32°C) was instituted. Heparin (pig mucosa, 5000 IU/mL) was used for anticoagulation, and 400 IU/kg heparin bolus as a standard dose was intravenously given. Activated clotting time (ACT) with kaolin as activating agent was at least 480 s before starting CPB. If ACT level was below this target level, additional heparin was given. Protamine sulfate was used for neutralizing 1.3 mg heparin per 100 IU. Before administering protamine, the extracorporeal bypass circuit was disconnected. If the postoperative ACT was more than 130 s, supplemental doses of protamine were considered. Blood samples were from the central venous cannula collected with a syringe containing no anticoagulant, discarding the first 10 mL (a) after inducing anesthesia before systemic heparinization, (b) 3 min after administering heparin, (c) 10 min after establishing CPB, (d) 30 min after establishing CPB, (e) 10 min after administering protamine, and (f) 2 h

postoperatively. Whole blood samples were (a) directly analyzed simultaneously (in dual-channel) for ACT determinations according to the manufacturer's instructions; (b) after addition of 1/10 volume of 0.129 M sodium citrate, sequentially analyzed using the thrombolytic assessment system heparin management test (TAS HMT) equipment according to the manufacturer's instructions; (c) at 4°C and 2700 $\times g$ centrifuged for 20 min to obtain plasma for subsequent heparin determination with chromogenic anti-FXa assay. The assay did not involve adding excess antithrombin to plasma samples. Using the same heparin (and lot number) for patients during cardiac operations, the standard curve was made. The plasma samples for heparin analysis were stored at −70°C until assayed in one batch.

(2) *In vitro* assay: A venous catheter placed in the cubital vein of one healthy donor was filled with saline and stoppered, blood samples were drawn through a syringe at intervals, after discarding the first 3 mL the blood was added to tubes containing either final concentration of 0, 0.15, 0.25, 0.5, 1.0, 1.5, 2.5, 3.5, 4.5, 5.5, 7.5, or 10.0 IU/mL unfractionated heparin (pig mucosa) for simultaneous ACT determinations (dual-channel) or the same amount of heparin plus sodium citrate (0.129 M) to 1/10 volume of total volume for sequential TAS HMT determinations. Plasma for heparin analysis was prepared from the remaining volume of the latter specimens by centrifugation at 4°C and 2700 $\times g$ for 20 min and stored at −70°C until assayed in one batch.

5.13 PLASMA-BASED ECARIN CLOTTING TIME ASSAY FOR r-HIRUDIN[14]

From 10 patients receiving heart surgery (6 with aorto-coronary bypass grafting, 4 with aortic and mitral valve replacement, 1 with heart transplantation) and r-hirudin anticoagulation during CPB and in the postsurgical period due to a present (2 patients) or past (8 patients) history of heparin-induced thrombocytopenia, the *ex vivo* blood samples containing different concentrations of r-hirudin were obtained. All patients included in a clinical study on the effectiveness of r-hirudin anticoagulant treatment suffered from heparin-induced thrombocytopenia. Before starting CPB, patients received intravenous 0.25 mg/kg bolus and 0.2 mg/kg via extracorporeal circuit prime. When r-hirudin plasma levels fell below 2.5 pg/mL, additional 5 mg r-hirudin boluses were administered.

Through a central venous line tier preflushing with 10 mL blood or during the CPB from the arterial line of the heart–lung machine, blood samples were collected into 0.1 M buffered trisodium citrate (9:1, blood:citrate), centrifuged at 3000 $\times g$ for 20 min, and prepared plasma samples were stored in 500-μL aliquots at −80°C until used.

For determining the aPTT, 0.1 mL plasma was mixed with 0.1 mL Pathromtin (Behringwerke), the mixture was incubated for 120 s, 0.1 mL 0.025 M calcium chloride was added, and the clotting time was recorded on a KC-1OA coagulometer (Amelung, Lemgo, Germany). The aPTT was tested in duplicate, and a normal control plasma was tested as every 25th sample.

For determining ECT, one vial containing 50 IU ecarin was reconstituted with 12.5 mL 0.2 M HEPES buffered saline containing 0.025 M calcium chloride, aliquoted at −80°C, stored and frozen. Alternatively, ecarin solution also could be stored at 4°C for up to 7 days without loss of activity. By adding 50 pL ecarin reagent to 100 pL plasma, the ECT was tested on a KC-1OA coagulometer, and the clotting time was recorded. By analyzing plasma samples of 50 healthy blood donors (24 males, 26 females, mean age 33 years, range 20–60 years), normal ECT values were defined.

To estimate the within-assay CV, plasma samples from 20 healthy blood donors were with r-hirudin spiked to achieve final 1, 2, 3, 4, 6, and 8 pg/mL r-hirudin. ECT values of each r-hirudin concentration were tested 10 times. The between-assay CV was calculated from the tested results on five different days using the same plasma samples. ECT values of each r-hirudin concentration were tested in triplicate.

With a thrombin-based chromogenic assay, the plasma concentration of free r-hirudin was determined on the coagulation analyzer Thrombolyzer (Organon, Heidelberg, Germany). Plasma samples to be analyzed were with TBS (100 mM NaCl, 50 mM Tris-HCl)/0.1% BSA diluted to 1:20. A 100-μL prediluted plasma sample was mixed with 0.05 mL solution containing 2 U/mL thrombin and 1 mg/mL Gly-Pro-Arg-Pro, incubated at 37°C for 1 min, 50 pL of 2 mM chromogenic thrombin substrate S2238 was added, and the change in absorbance was measured at 405 nm.

Using normal human plasma spiked with r-hirudin achieving final 0.5, 1.0, 1.5, 2.0, and 2.5 μg/mL r-hirudin, a calibration curve was constructed. The linear range of this assay was between 0.05 and 2.5 pg/mL r-hirudin. For plasma samples containing r-hirudin more than 2 μg/mL, the analysis was repeated using 1:40 and 1:80 plasma dilutions. Assay characteristics for within-assay CV and between-assay CV were <4.5% and <7.6%, respectively, indicating that the precision of this assay was comparable with that of the chromogenic assay.

5.14 PROTAMINE TITRATION FOR HEPARIN IN WHOLE BLOOD[15]

To achieve systemic anticoagulation during surgery, 24 cardiac surgery patients received 300 IU/kg bovine heparin. To maintain kaolin activated coagulation time (ACT) of greater than 480 s, additional bolus doses (10,000 IU) of heparin were administered as needed. All patients underwent moderate hypothermic CPB (lowest core temperature >28°C). To reverse the systemic heparinization, after successful separation from CPB all patients received protamine (1 mg protamine for each 100 IU heparin). Three patients received aprotinin (Trasylol, Bayer Pharmaceuticals, West Haven, CT), 15 patients received epsilon-aminocaproic acid, and 6 patients received neither.

The heparin sensor consisted of an Ag/AgCl internal reference electrode inside a polymer membrane tube impregnated with tridodecylmethylammonium chloride (TDMAC), and 0.12 M NaCl was used as the internal reference solution, while another Ag/AgCl electrode served as the external reference electrode. The sensor device's size was 2.5 mm $\times$ 60 mm. A potentiometric protamine titration was performed for each blood sample. To each of a series of 11 tubes, 250 μL whole blood was added, and each tube was coated with a known quantity of protamine sulfate. The electrochemical potential (EMF) determined with heparin sensor versus the external Ag/AgCl reference electrode was via a pH/mV meter for each tube sequentially recorded. To record the potentiometric response of each tube, 30 s were required. By plotting the 11 EMF readings against the amount of protamine in each tube, a protamine titration curve was constructed for each blood sample. Each titration curve consisted of two distinct regions: (a) A constant region where protamine was in excess and the potentiometric response was similar for each protamine amount. (b) Response of a variable region was related to the untitrated heparin. When the difference between one tube and the previous tube was less than 1 mV, the data point of that tube was assigned to the constant region. Any tube that varied from the previous tube by more than 1 mV was included in the variable region. To describe each region of the titration curve, linear regression lines were constructed. The intercept of the two extrapolated lines corresponded with the titration end point. Using the stoichiometry of heparin-protamine reaction as determined by potentiometric titration (100 U heparin/1 mg protamine), heparin levels were calculated. Heparin levels in whole blood were determined using the heparin-responsive sensor (HRS) and the Hepcon system (Medtronic Hemotec, Parker, CO) after heparin dosing, after CPB initiation at 30-min intervals during CPB, and after administering protamine. For HRS tests, 3 mL of whole blood was collected in citrated Vacutainer tubes and stored at 4°C until tested. Hepcon assays were performed at each time point in the operating room. A kaolin-activated ACT was at each time point performed. The samples where the Hepcon and HRS heparin levels differed by more than 1 U/mL had their anti-Xa activity determined. When comparisons were made between whole blood Hepcon and HRS and plasma anti-Xa measurements, results were corrected for the hematocrit of the whole blood samples. Bias and precision were calculated, and bias was defined as the mean difference of the two methods of measurement and represents the systematic error of one method versus another. Precision was the standard deviation of the differences and represents the random error inherent in the measurement methods. Correlation coefficients were determined by comparing the heparin levels measured by the Hepcon, the HSR, and the anti-Xa assay and by comparing each method with ACT values.

5.15 AUTOMATED ASSAY EVALUATING RESPONSE OF KAOLIN ACT TO HEPARIN[16]

Blood specimens were obtained from 41 adult patients undergoing cardiac operations requiring CPB. All patients were anesthetized with an opioid-based technique, which was supplemented with inhalational anesthetic agents, muscle relaxants, and benzodiazepines. With a Bio-Medicus centrifugal pump (Bio-Medicus, Eden Prairie, MN) and a Bentley membrane oxygenator (Baxter Healthcare, Bentley Division, Irvine, CA), extracorporeal circulation was accomplished, while during cardioplegia systemic hypothermia was maintained at 28°C. Systemic anticoagulation for CPB was accomplished with heparin, which was through use of the Hepcon instrument administered based on on-site measurements of heparin dose response, heparin concentration, and kaolin ACT. After the patient was rewarmed to 37°C, extracorporeal circulation was discontinued, and heparin was neutralized with protamine sulfate. Protamine dose was determined on the basis of the whole-blood heparin concentration measured prior to discontinuation of CPB (1.3 mg protamine per milligram of residual heparin).

Single blood specimens were collected from radial or femoral artery catheters or both after removal of 6 dead-space volumes and used for coagulation analysis by on-site whole-blood assays. Blood specimens were obtained prior to and 10 min after systemic administration of heparin, used to measure kaolin ACT values at various *in vitro* heparin concentrations with the heparin dose response (HDR) or manual titration, and the automated assay was used to evaluate the response of kaolin ACT to 0, 1.5, and 2.5 U/mL heparin using a Hepcon instrument. The manual titration technique consisted of adding incrementally greater doses of heparin to ACT cartridges and then measuring kaolin ACT using five separate automated clot timer instruments. Heparin concentrations required to prolong ACT values within the range of 0 to 500 s (low range) were estimated using the results derived from the heparin dose-response assay. Similarly, heparin was manually added to five ACT cartridges to prolong its values ranging from 500 to 1000 s (high range) using manually obtained low-range results. Kaolin ACT values were also measured after systemic administration of a heparin dose that was projected by the heparin dose-response assay to result in a greater than 480 s ACT value for kaolin. Obtained kaolin ACT values from both instruments at each heparin concentration were expressed as the mean of duplicate measurements. Activated clotting time that exceeded the detection limit of the instruments (999 s) was excluded from statistical analysis.

5.16 PLATELET/MONOCYTE INTERACTION-BASED PM EXPOSURE MOUSE ASSAY[17]

Mice were anesthetized with vapors from isoflurane/propanediol, and a suspension of 100 μg particulate matter (PM) in 100 μL sterile PBS (137 mM NaCl, 10 mM NaH_2PO_4, 1.47 mM KH_2PO_4, 2.7 mM KCl, pH 7.4) was delivered by intratracheal instillation. Blood was collected from a separate set of mice having 24-h PM exposure, placed in PGE_1-K_2EGTA anticoagulant tubes, and then divided for subsequent analyses. For flow cytometry analysis of CD41, red blood cells were lysed by adding Optilyse C in a 1:1 ratio, and 50 μL blood/Optilyse C (Beckman Coulter) solution was incubated with a fluorescent-conjugated antibody against CD41 or a matched isotype control antibody. By forward- and side-scatter parameters, the monocyte cell gate was determined, while fluorescence within this gate was analyzed by flow cytometry, and mean fluorescence intensity was calculated.

After complete blood count (CBC) analyses, whole blood remaining was centrifuged at 5000 × *g* for 10 min to remove plasma. Fibrinogen levels in fresh plasma were determined. At 37°C within the Coag-A-Mate XM coagulation timer

(Biomerieux, Durham, NC), 100 μL diluted plasma was incubated before adding 100 μL Fibriquik thrombin reagent (Biomerieux) to induce coagulation. Based on a human fibrinogen standard curve, clotting times were converted to fibrinogen level. Mean fibrinogen levels for each mouse group were calculated.

Twenty-four hours after exposure, with a human TAT complex-specific sandwich ELISA known to cross-react with mouse TAT antigen, according to kit instructions plasma thrombin/antithrombin (TAT) complex concentrations were determined. In 96-well plates coated with human thrombin-specific antibody, 50 μL citrated plasma was incubated, a peroxidase conjugated antibody to human antithrombin was added, color was developed, and OD value of the plate was read at 492 nm using a spectrophotometer. Negative control plate containing sample buffer was used to establish a background spectrophotometer reading at 492 nm, which was subtracted from sample readings, and based on a human TAT complex standard curve (2–60 ng/mL) supplied with the kit, TAT complex concentrations were determined. Mean plasma TAT concentrations were calculated.

5.17 MOUSE TAIL-BLEEDING TIME ASSAY[18,19]

NMRI male mice (approximately 30 g) were group-housed at 18°C to 22°C, 60% humidity, and 12-h light/dark cycle, fed rodent chow as well as fresh water *ad libitum*, and acclimatized for at least 7 days before assay. The mice were randomized to receive (intraperitoneal) either murine FVIIa blocked in the active site with phenylalanylphenylalanyl-arginyl-chloromethyl ketone (FFR-FVIIa, 0.1, 1.0, or 10 mg/kg), human FFR-FVIIa (10 or 25 mg/kg), or vehicle (10 mM glycyl-glycine, 150 mM NaCl, 10 mM $CaCl_2$, pH 7.5). The mice were anesthetized with sodium pentobarbital, placed on 38°C heating blankets, and the tail was placed in 14 mL 37°C saline 10 min before the cut. By cutting the tip of the tail (2 mm) with sharp scissors, the bleeding was initiated 1 h after FFR-FVIIa dosing. The tail was replaced in the saline and the bleeding time was monitored and defined as the sum of the duration of all bleeding episodes during a 45-min observation period. Blood samples from mouse heart were collected in 0.13 M trisodium citrate (1 : 10) at the termination of assay, centrifuged for 5 min at 4000 × *g*, and the prepared plasma samples were stored at −80°C until analyzed. After the blood sample was taken, the mice were euthanized with an overdose of sodium pentobarbital. GraphPad Prism statistical software (version 4.0, GraphPad Software, USA) was used for calculation and statistics and medians of bleeding time were calculated for each group of controls and treated mice, with use of nonparametric Kruskal–Wallis and Mann–Whitney tests for comparisons.

5.18 MRI ASSAY FOR RABBIT ATHEROSCLEROTIC LESIONS[20]

Aortic atherosclerosis was induced in 19 male New Zealand White rabbits (age, 3 months; weight, 3.4 ± 0.3 kg) by a combination of 9 months of high-cholesterol (HC) diet (0.2% cholesterol-enriched rabbit diet) and double aortic balloon denudation injury. One month after initiation of the atherogenic HC diet, using a 4-F Fogarty embolectomy catheter introduced through the iliac artery, the abdominal aorta's balloon denudation was performed from the top of the aortic arch to the iliac bifurcation. Three months after initiation of the HC diet, the same procedure was repeated again. All procedures were performed under general anesthesia by intramuscular injection of 20 mg/kg ketamine and 10 mg/kg xylazine. After inducing atherosclerosis, seven rabbits were sacrificed after magnetic resonance imaging (MRI) and served as histologic control. Complete plasma lipid profiles were analyzed by MRI.

Nine months after atherosclerosis induction, the rabbits underwent serial MRI of the aorta to assess the atherosclerotic burden degree, while at the end of month 15, the rabbits underwent serial MRI of the aorta to assess treating efficiency. In this design, each rabbit served as its own control, and true serial data on progression/regression of atherosclerosis in the rabbit aortas were obtained. A 1.5 T MRI system with enhanced gradients (40 mT/m), slow rates (150 mT/m/ms), and a custom-designed "birdcage" volume coil was used for assay. Sequential axial images (3-mm thickness) of the aorta from the celiac trunk to the iliac bifurcation were obtained using a fast spin-echo sequence with in-plane resolution of $352\ \mu \times 352\ \mu$ (PDW: TR/TE = 2300/17 ms; T2W: TR/TE = 2300/60 ms; field-of-view $9\ \text{cm} \times 9\ \text{cm}$, matrix 256×256, echo train length = 8, signal averages = 4). Fat suppression and flow saturation pulses were generally used.

For further analysis, the MR images were transferred to a Macintosh computer system, and using distances from the renal arteries and the iliac bifurcation, the MR images from the same rabbit at the two time points were matched for registration. With MRI allowing true serial follow-up of the course of atherosclerotic lesions over time, each segment of the abdominal aorta was compared across the two time points. For vessel wall measurements, 10 contiguous MR images immediately following the origin of the left renal artery were used. By an observer blinded to the treatments, cross-sectional areas of the lumen and outer boundary of each aortic section were determined by manual tracing with Image Pro-Plus (Media Cybernetics). As surrogates for atherosclerotic burden, the cross-sectional vessel wall area (VWA) was calculated by subtracting the area of the lumen from the total area of the vessel measured at the outer border of the vessel wall (VWA = total vessel area − lumen area). The outer wall was defined as the interface between vessel wall and adventitial fat. VWAs of 10 contiguous images were averaged, and their mean values were considered for statistical calculation.

5.19 SPIRAL COMPUTED TOMOGRAPHY ASSAY[21]

Consecutive patients with the first episode of symptomatic acute pulmonary embolism (APE) diagnosed, which was confirmed by contrast-enhanced spiral computed tomography, were included in the studied group. Patients suffering from chronic lung diseases or experiencing recurrence of venous thromboembolism (VTE) during the subsequent 6 months and received anticoagulation or did not complete this follow-up were not included into the study. At the diagnosis, all patients underwent detailed clinical assessment, such as transthoracic echocardiography and for right ventricular dysfunction (RVD), and APE diagnoses were accordingly given. On admission, plasma D-dimer concentration (normal value <500 ng/mL) was measured with VIDAS D-dimer ELISA method. Prior to introduction of oral anticoagulation, all patients were screened for malignancies. According to the European Society of Cardiology, patients after the first episode of APE with temporary or reversible risk factors were given at least 3 months of anticoagulation, and patients with idiopathic pulmonary embolism (PE) were given at least 6 months of anticoagulation. The included patients with unprovoked APE and with risk factors for VTE received 6 months anticoagulation and then were supervised in an outpatient clinic. During anticoagulant treatment, D-dimer was assessed.

To assess pulmonary parenchyma, initially high-resolution computed tomography (CT) was performed with 16-row scanner. Ten seconds after initiation of intravenous administration of 80 mL contrast medium (30 mL/5 mL/s and 50 mL/3 mL/s), angio-CT scans of pulmonary arteries were subsequently obtained. For the presence of filling defects, pulmonary arteries were assessed up to subsegmental pulmonary arteries.

5.20 DUPLEX ULTRASOUND ASSAY[22]

With an Ultramark 8 or 9 HDI color scanner (Advanced Technology Laboratory, Bothell, WA) using a 5- or 7.5-MHz linear transducer, duplex imaging were recorded, and using a 5-MHz probe, pulsed Doppler signals were obtained. Examined veins were the iliac, common femoral, superficial femoral, popliteal, and proximal tibial calf veins. All tests were blindly interpreted without the interpreter knowing the treatment protocol. Reflux assessment used a 5-MHz pulsed Doppler probe placed center stream at a 60-degree angle to the direction of the flow. In the proximal veins (common femoral and proximal superficial veins), reflux assessment was made by performing the Valsalva maneuver, while reflux assessment for mid and distal superficial femoral veins and the proximal and distal popliteal veins was made with compression over proximal muscle groups. For reflux diagnosis, reversed flow lasting longer than 2 s was required. Imaging of the anterior tibial veins and the peroneal veins was not routinely done particularly in the first several years of this study. Duplex interpretation criteria, including a return of the phasic Doppler signal with respiration and augmentation maneuvers, disappearance of the intraluminal echoes, and the ability to fully compress the vein in the transverse position by gentle pressure of the transducer, were used to consider the resolution of the deep venous thrombosis (DVT) process. To characterize chronic phlebitic changes or nonresolution of DVT during follow-up, criteria of abnormal thickening or increased echogenicity of the vein wall, abnormal valve motion with the presence of reflux, or an occluded vein were also used. When the previously mentioned criteria for DVT reappeared in a vein segment that had previously been considered normal or resolved, recurrence of the DVT process was diagnosed. According to duplex scanning, DVT outcome was divided into complete resolution (normal), chronic phlebitic changes, or recurrence of DVT.

5.21 STANDARD HEMOCHRON ASSAY[22–24]

From 30 patients undergoing cardiovascular surgical intervention, a total of 173 arterial samples before, during, and after reversal of anticoagulation with heparin were drawn, among which 19 patients had surgery using cardiopulmonary bypass (109 samples) and 11 were done without bypass (64 samples). The Hemochron system (HRFT CA510) used a celite-based ACT with 12 mg diatomaceous earth or a kaolin-based ACT with 12 mg kaolin for patients who received aprotinin. According to the manufacturer's instructions, 2 mL blood was drawn, mixed in a tube with the activator by rotating, the tube was tapped on a hard surface and heated by the machine to 37°C. When the fibrin monomers slowed the movements of a foreign body within the sample, the test was completed. Hemochron Jr. (HJr) used the ACT + kaolin and phospholipid mixture as not affected by aprotinin use. With the machine, the cartridges were prewarmed to 37°C, 15 μL blood sample was aspirated for evaluation, and no shaking or mixing was required. When the formation of clot slowed the movement of blood in a microcapillary tube below a preset rate, the test was completed using electronic optical detection. By means of quality control tests from the manufacturer to ensure reliability, all devices were calibrated daily. Using the standard Hemochron and simultaneously with HJr, ACT was determined in duplicate. A minimum of three samples was drawn per patient, and each sample was classified as a baseline or a heparinized value or a postprotamine value. Sample number varied from three to eight samples per patient depending on the length of the procedure and the number of additional heparin boluses required to achieve therapeutic anticoagulation. At the outset of the procedure before skin incision, baseline or pre-ACTs were measured. As a 300 U/kg intravenous bolus, heparin was administered with additional boluses given to achieve an ACT in excess of 480 s as measured by the Hemochron. During the cardiovascular surgical procedure, to monitor the anticoagulation level, the subsequent arterial samples were drawn. As necessary to maintain an ACT above 480 s for on-pump cases and above 400 s for off-pump cases, additional heparin was given. After either discontinuation of cardiopulmonary bypass or completion of the proximal anastomoses in off-pump procedures, to achieve an ACT less than 200 s protamine was administered as a 1 mg/100 U heparin infusion. All "post" values were accordingly recorded.

5.22 PLATELET SEROTONIN RELEASE ASSAY[24]

By performance of a serotonin release assay, heparin-induced platelet activation was measured *in vitro*. In the assay, the reacted platelets were centrifuged over a solution of 0.135 M formalin and 5 mM EDTA. Donor platelets used in this assay were chosen from a pool of persons, and their relative Fc receptor expressions equaled or was greater than that of a known donor that reacted to heparin–antibody complexes and were found to be responsive to heparin–antibody complexes with known plasmas containing antibodies reactive to adding heparin. By means of flow cytometry with a fluorescein isothiocyanate–labeled monoclonal antibody that reacts with FcγRII, expression of Fc receptors was determined.

5.23 ACTIVATED PARTIAL THROMBOPLASTIN TIME ASSAY[24–26]

For bedside examination and in the operating room, with kaolin or celite clotting tubes in a Hemochron ACT reader, automated ACTs were obtained by means of drawing blood directly into the clotting tube anticoagulated with 0.625 to 3.75 IU/mL danaparoid sodium. Clotting tubes containing either kaolin or celite were used in these examinations. With ACT, additional examinations were performed on samples anticoagulated with danaparoid sodium and aprotinin 10 to 50 U/mL. With Actin aPTT reagent (Baxter-Dade, Miami, FL) on an MLA 1600 instrument (Baxter-Dade, Miami, FL), aPTT was obtained. According to the manufacturer's instructions, with Coatest heparin (Chromogenix AB, Molndal, Sweden) anti-factor Xa, activity was measured. To determine the anti-factor Xa activity of unfractionated heparin or danaparoid sodium, the standard curve for the assay was prepared with each form of the anticoagulant in dilutions of 0.2 to 1.0 U/mL.

5.24 RAT STROKE OUTCOME ASSAY[27,28]

After induction, anesthesia of male SD rats (weight, 250–325 g) was maintained with 2% halothane and nitrous oxide:oxygen (60:40). A midline neck incision was performed to expose the left carotid artery, and the external carotid and pterygopalatine arteries were ligated with 5-0 silk. In the arterial wall, an incision was made, and from the bifurcation of external and internal carotid arteries, a 4-0 heat blunted nylon suture was advanced 18 mm; this distance reliably produced blockage of the origin of the middle cerebral artery (MCA). To generate a range of ischemia durations for each experimental group, the duration of occlusion for an individual rat was varied. Via a thermister probe linked to a controller, right temporalis muscle temperature was monitored and via a heat lamp brain temperature was maintained to be normothermic.

To determine if the filament was positioned correctly, each rat was injected via tail vein with 0.2 mL 2% Evan's blue in saline, which allowed to circulate for 30 to 60 min, the rate was perfused transcardially with 100 mL NS, the brain was removed and immediately examined under the operating microscope. For correctly placed filament, the damaged endothelium in the epicerebral internal carotid and anterior cerebral artery appeared blue. The MCA was never stained because the filament did not enter this vessel. Anterior cerebral artery not staining blue implied that the rat was not ischemic and was thus excluded from further study. If subarachnoid blood was seen, the filament was assumed to be puncturing an artery, and the rat was also excluded as a subject. After 4% paraformaldehyde immersion fixation for at least 48 h, each brain was sectioned in a coronal plane just anterior to the temporal lobe tip and under a stereomicroscope examined for intracerebral hemorrhage. Test compounds were intraperitoneally or intravenously administered. Controls received a vehicle. Hypothermia should occur spontaneously in all rats unless placed on a heating pad after surgery. To study the effect of hypothermia on outcome, a group of rats were kept off the heating pad. To measure stroke outcome, a quantal dose-response curve was fit as a function of occlusion duration versus the 48-h behavior rating, for which rat behavior was rated normal/abnormal at 48 h and any one (or more) of obtundation/reduced exploration, forepaw retraction on tail lifting, asymmetric forepaw grasp, forced circling, or death was used as abnormal criteria. The fraction of abnormal rats was calculated for each occlusion duration, including after short durations none or few rats were abnormal, after long durations all the rats were abnormal, and after intermediate durations rats gave intermediate results. By iteration the curve was fit, the best fit was given, and for each treatment group the ED_{50} and its standard deviation was calculated. Neuroprotection was demonstrated if the compound prolonged the ED_{50} compared with that seen in controls. The potency ratio for a treatment was the ED_{50} for that compound divided by the ED_{50} for the appropriate control group.

REFERENCES AND NOTES

1. R. Siegmund, K. Boer, K. Poeschel, G. Wolf, T. Deufel, M. Kiehntopf. Comparison of the ecarin chromogenic assay and different aPTT assays for the measurement of argatroban concentrations in plasma from healthy individuals and from coagulation factor deficient patients. Thromb Res 10.1016/j.thromres.2008.02.013.
2. Z.M. Wang, L. Li, B.S. Zheng, N. Normakhamatov, S.Y. Guo. Preparation and anticoagulation activity of sodium cellulose sulfate. Int J Biol Macromol 41 (**2007**) 376–382.
3. M.R. Varma, D.M. Moaveni, N.A. Dewyer, A.J. Varga, K.B. Deatrick, S.L. Kunkel, G.R. Upchurch Jr, T.W. Wakefield, P.K. Henke. Deep vein thrombosis resolution is not accelerated with increased neovascularization. J Vasc Surg 40 (**2004**) 536–542. *Notes*: (1) The specimen harvested for histologic analysis was embedded in paraffin, sliced into 10-μm sections, and stained. The tissue was deparaffinized, dehydrated through graded alcohols, and nonspecific binding sites were with species specific serum blocked; the sections were incubated with either 1/1000 anti-neutrophil antibody, 1/100 ED-1 antibody for monocytes, or 1/500 anti-laminin for neovascular channels. According to the manufacturer's instructions, species-specific secondary antibodies and ABC kits were used to develop slides, the slides were counterstained with hematoxylin and coverslipped. Positive staining cells or channels were in a blinded fashion counted for five randomly selected high-power fields around the vein wall thrombus interface. (2) By homogenization, sonication (30 s), and centrifugation (15 min, 10,000 × *g*) in buffer solution, thrombi were prepared. ELISAs were carried out as for keratinocyte cytokine (KC; a CXC rodent IL-8 analogue), monocyte chemotactic protein-1 (MCP-1), and basic fibroblast growth factor (bFGF), with species-specific primary antibodies quantified by using a double ligand technique. Plates were read at 450 nm. To account for differing protein composition of samples, total protein was quantified. Using the homogenate from ELISAs and with serial dilutions

of bovine serum albumin as standards, a modified Bradford assay was performed. Results are expressed as ng or pg per milligram of protein. (3) According to the manufacturer's instructions, thrombus collagen content was estimated with a commercially available kit. The Sircol collagen assay is a colorimetric assay, in which Sirius red dye binds to the side chains of amino acids of collagen. Plates were read at 590 nm. Specimen preparation and standardization of results were identical to those for ELISAs; units were micrograms per milligram of clot.

4. P. Thanaporn, D.D. Myers, S.K. Wrobleski, A.E. Hawley, D.M. Farris, T.W. Wakefield, P.K. Henke. P-Selectin inhibition decreases postthrombotic vein wall fibrosis in a rat model. Surgery 134 (**2003**) 365–371.

5. M. Jonsson, A. Hillarp, P. Svensson. Comparison between CoaguChek S- and Owren-type prothrombin time assay for monitoring anticoagulant therapy. Thromb Res 114 (**2004**) 83–89.

6. B. Alexander, K.G. Burnand, C.L. Lattimer, J. Humphries, P.J. Gaffney, D. Eastham, A. Smith. The effect of anticoagulation with subcutaneously delivered polyethylene glycol conjugated hirudin and recombinant tissue plasminogen activator on recurrent stenosis in the rabbit double-balloon injury model. Thromb Res 113 (**2004**) 155–161. *Notes*: (1) New Zealand White rabbits (weight, 3.0–3.5 kg) were given food and water *ad libitum* and before carotid dissection anesthetized with fentanyl/fluanisone (5 mg/kg, intramuscular) supplemented with a bolus of intravenous midazolam (2 mg/kg, intravenous). Prior to recovery, analgesia (0.02 mg/kg, subcutaneous) was given. With a balloon de-endothelialization, injury plaque formation was induced within the right common carotid artery. To expose a branch of the right external carotid artery, a midline neck incision was used, which was cannulated with a size 2CH Fogarty embolectomy catheter. Before being withdrawn along the common carotid artery for a length of approximately 6 cm, this catheter was passed retrograde to the aortic arch and the balloon was fully inflated to 0.2 mL. To ensure complete de-endothelialization, this process was repeated twice. The external carotid artery was then ligated with 5/0 black silk as a "marker." After the neck wound was closed, the rabbit was allowed to recover, fed a diet supplemented with 2% cholesterol for 2 weeks in order to produce hypercholesterolemia, the silk ligature was located through the same incision, and the origin of the external carotid artery was freed from surrounding scar tissue and cannulated as described above. The balloon was again inflated to 0.2 mL, however, before being deflated and withdrawn, it was only passed down the artery for 1 cm below the carotid bifurcation and was retained in position for only 20 s. This "angioplasty" was performed under direct vision, obviating the need for imaging to confirm the position of the catheter. The external carotid was religated, the neck wound was closed, and the rabbit was allowed to recover. The left common carotid artery was left uninjured and used as an internal control vessel. The cholesterol diet was continued for the remainder of the assay. (2) After 8 weeks, carotid arteries were harvested, all specimens were washed in PBS, divided into two segments, one segment was snap-frozen in liquid nitrogen for cryostat sectioning, the remaining segment was fixed in formalin overnight, embedded in paraffin wax, and cut to serial tissue sections. To locate the presence of 1-actin, macrophages, proliferating cells, and endothelial cells, immunohistochemistry was used. Using a three-stage avidin-biotin immunoperoxidase technique, sections were stained. As a negative control, nonimmune mouse serum was used in place of primary antibodies. For identifying laminated thrombus and to aid morphologic and cell characterization, sections were also stained with hematoxylin and eosin, Millers elastic stain, and Martius Scarlet blue.

7. H. Eto, M. Miyata, N. Kume, M. Minami, H. Itabe, K. Orihara, S. Hamasaki, S. Biro, Y. Otsuji, T. Kita, C. Tei. Expression of lectin-like oxidized LDL receptor-1 in smooth muscle cells after vascular injury. Biochem Biophys Res Commun 341 (**2006**) 591–598.

8. M. Sakurai, T. Nagata, K. Abe, T. Horinouchi, Y. Itoyama, K. Tabayashi. Oxidative damage and reduction of redox factor-1 expression after transient spinal cord ischemia in rabbits. J Vasc Surg 37 (**2003**) 446–452.

9. D.J. Moliterno, J.B. Hermiller, D.J. Kereiakes, E. Yow, R.J. Applegate, G.A. Braden, E.J. Dippel, M.I. Furman, C.L. Grines, N.S. Kleiman, G.N. Levine, T. Mann, R.N. Nair, R.A. Stine, S.J. Yacubov, J.E. Tcheng. A novel point-of-care enoxaparin monitor for use during percutaneous coronary intervention results of the evaluating enoxaparin clotting times (ELECT) study. J Am Coll Cardiol 42 (**2003**) 1132–1139.

10. S.T. Edwards, A. Betz, H.L. James, E. Thompson, S.J. Yonkovich, U. Sinha. Differences between human and rabbit coagulation factor X-implications for in vivo models of thrombosis. Thromb Res 106 (**2002**) 71–79.

11. P.K. Henke, T.W. Wakefield, A.M. Kadell, M.J. Linn, M.R. Varma, M. Sarkar, A. Hawley, J.B. Fowlkes, R.M. Strieter. Interleukin-8 administration enhances venous thrombosis resolution in a rat model. J Surg Res 99 (**2001**) 84–91.

12. A. Koster, M. Loebe, R. Hansen, M. Bauer, F. Mertzlufft, H. Kuppe, R. Hetzer. A quick assay for monitoring recombinant hirudin cardiopulmonary bypass in patients with heparin-induced thrombocytopenia type II: Adaptation of the ecarin clotting time to the act II device. J Thorac Cardiovasc Surg 119 (**2000**) 1278–1283.

13. H.I. Flom-Halvorsen, E. Øvrum, M. Abdelnoor, S. Bjørnsen, F. Brosstad. Assessment of heparin anticoagulation: Comparison of two commercially available methods. Ann Thorac Surg 67 (**1999**) 1012–1017.

14. B. Potzsch, S. Hund, K. Madlener, C. Unkrig, G. Muller-Berghaus. Monitoring of recombinant hirudin: Assessment of a plasma-based ecarin clotting time assay. Thromb Res 86 (**1997**) 373–383.

15. J.A. Wahr, J.-H. Yun, C. Yang, L.M. Lee, B. Fu, M.E. Meyerhoff. A new method of measuring heparin levels in whole blood by protamine titration using a heparin-responsive electrochemical sensor. J Cardiothoracic Vasc Anesth 10 (**1996**) 447–450.

16. G.J. Despotis, A.L. Alsoufiev, E. Spitznagel, L.T. Goodnough, D.G. Lappas. Response of kaolin ACT to heparin: Evaluation with an automated assay and higher heparin doses. Ann Thorac Surg 61 (**1996**) 795–799.

17. E. Cozzi, C.J. Wingard, W.E. Cascio, R.B. Devlin, J.J. Miles, A.R. Bofferding, R.M. Lust, M.R. Vanscott, R.A. Henriksen. Effect of ambient particulate matter exposure on hemostasis. Translational Res 149 (**2007**) 324–332.

18. L.C. Petersen, P.L. Nørby, S. Branner, B.B. Sørensen, T. Elm, H.R. Stennicke, E. Persson, S.E. Bjørn. Characterization of recombinant murine factor VIIa and recombinant murine tissue factor: a human-murine species compatibility study. Thromb Res 116 (**2005**) 75–85.

19. M. Zhao, Z. Li, L. Peng, Y. Tang, C. Wang, Z. Zhang, S. Peng. A new class of analgesic agents toward prostacyclin receptor inhibition: Synthesis, biological studies and QSAR analysis of 1-hydroxyl-2-substituted phenyl-4,4,5,5-tetra-methylimidazolines. Eur J Med Chem 43 (**2008**) 1048–1058.

20. R. Corti, J. Osende, R. Hutter, J.F. Viles-Gonzalez, U. Zafar, C. Valdivieso, G. Mizsei, J.T. Fallon, V. Fuster, J.J. Badimon. Fenofibrate induces plaque regression in

hypercholesterolemic atherosclerotic rabbits: In vivo demonstration by high-resolution MRI. Atherosclerosis 190 (**2007**) 106–113.

21. A. Kaczyńska, M. Kostrubiec, R. Pacho, J. Kunikowska, P. Pruszczyk. Elevated D-dimer concentration identifies patients with incomplete recanalization of pulmonary artery thromboemboli despite 6 months anticoagulation after the first episode of acute pulmonary embolism. Thromb Res 122 (**2008**) 21–25.
22. A.F. AbuRahma, D.L. Stickler, P.A. Robinson. A prospective controlled study of the efficacy of short-term anticoagulation therapy in patients with deep vein thrombosis of the lower extremity. J Vasc Surg 28 (**1998**) 630–637.
23. C.L. Aylsworth, F. Stefan, K. Woitas, R.H. Rieger, M. LeBoutillier III, V.J. DiSesa. New technology, old standards: Disparate activated clotting time measurements by the hemochron Jr compared with the standard hemochron. Ann Thorac Surg 77 (**2004**) 973–976.
24. G.J. Despotis, A.L. Alsoufiev, E. Spitznagel, L.T. Goodnough, D.G. Lappas. Response of kaolin ACT to heparin: Evaluation with an automated assay and higher heparin doses. Ann Thorac Surg 61 (**1996**) 795–799.
25. S.D. Gitlin, G.M. Deeb, C. Yann, A.H. Schmaier. Intraoperative monitoring of danaparoid sodium anticoagulation during cardiovascular operations. J Vasc Surg 27 (**1998**) 568–575.
26. M.S. Hulin, T.W. Wakefield, P.C. Andrews, S.K. Wrobleski, M.D. Stoneham, A.L. Doyle, G.B. Zelenock, L.A. Jacobs, C.J. Shanley, V.M. TenCate, J.C. Stanley. A novel protamine variant reversal of heparin anticoagulation in human blood in vitro. J Vasc Surg 26 (**1997**) 1043–1048.
27. M. Patteril, M. Stafford-Smith, J.G. Toffaletti, B.P. Bute, C.A. Milano, I.J. Welsby. Changing systems for measuring activated clotting times: Impact on the clinical practice of heparin anticoagulation during cardiac surgery. Clin Chim Acta 356 (**2005**) 218–224.
28. C. Jackson-Friedman, P.D. Lyden, S. Nunez, A. Jin, R. Zweifler. High dose baclofen is neuroprotective but also causes intracerebral hemorrhage: A quantal bioassay study using the intraluminal suture occlusion method. Exp Neurol 147 (**1997**) 346–352.
29. P.D. Lyden, C. Jackson-Friedman, C. Shin, S. Hassid. Synergistic combinatorial stroke therapy: A quantal bioassay of a GABA agonist and a glutamate antagonist. Exp Neurol 163 (**2000**) 477–489.

6

METHODS AND APPLICATIONS OF BLOOD PRESSURE-RELATED ASSAYS

Ming Zhao

In the past decade, hypertension onset and progress have been widely correlated with endogenous substances and the changes of their levels. For instance, new pressor protein (NPP) derived from human plasma increases blood pressure, heart rate, and plasma adrenal medullary catecholamines, and high plasma NPP activity has been found in hemodialysis patients with hypertension. As a potent pulmonary artery (PA) vasoconstrictor, 5-hydroxytryptamine (5-HT; also known as serotonin) binds to G protein-coupled receptors (primarily 5-$HT_{1B/1D}$ receptors) to induce vasoconstriction. Change in 5-HT receptor activity plays an important role in regulating PA vascular responsiveness, particularly in the setting of pulmonary hypertension. Superoxide mediates hypoxia/reoxygenation (H/R)-induced constriction of isolated mouse coronary arteries (CA). Under normoxia and H/R conditions, CAs from transgenic (Tg) mice overexpress human CuZn–superoxide dismutase (SOD). Cerebral microvessel extracellular matrix (ECM) contributes to both the permeability barrier and cerebral microvessel integrity and consists of type IV collagen, laminin, fibronectin, heparan sulfate proteoglycans (HSPGs), and other glycoproteins. During middle cerebral artery occlusion (MCAO) in the non-human primate, the gradual loss of microvessel type IV collagen, laminin, and fibronectin correlates with increased permeability and hemorrhagic transformation. Based on this knowledge, a series of

Pharmaceutical Bioassays: Methods and Applications. By Shiqi Peng and Ming Zhao

assays is reported and widely used for understanding the hypertension mechanism and for finding new antihypertension drugs. In this chapter, 14 assays are introduced: human plasma new pressor protein (NPP) assay,[1–6] pulmonary hypertension assay,[7–11] coronary arteries constriction assay,[12] MRI and brain natriuretic peptide assays,[13] competition enzyme-linked immunosorbent assay,[14] right ventricular pressure assay,[14] plasma nitrite/nitrate concentration assay,[15] adeno-associated virus vector-caused rat pulmonary artery pressure assay,[16] adeno-associated virus vector-caused rat protein and mRNA assay,[16] N-terminal pro-brain natriuretic peptide (N-T proBNP) assay,[17] middle cerebral artery occlusion (MCAO) assay,[18–23] vascular endothelial growth factor (VEGF) level assay,[24] antispasmodic agent *in vivo* action assay,[25] and temperature assay in awake subjects.[26–31]

6.1 HUMAN PLASMA NEW PRESSOR PROTEIN ASSAY[1–6]

The solution of test peptide was injected intravenously into adult male Wistar rats (housed in pairs, under a 12 : 12 h light/dark cycle, 0.4% NaCl rat chow and water available *ad libitum*, after 3–5 days of adaptation to the facility, about 300 g) at a proper dose. Inactin (100 mg/kg intraperitoneal, sodium ethyl-1-methylpropylmalonylthio-urea) anesthetized rats were treated with atropine (2.4 mg/kg, subcutaneous) and their ganglions were blocked with pentolinium (19.2 mg/kg, subcutaneous). To the rats in captopril group, 10 mg/kg captopril was administered intravenously acutely and inhibited fully the pressor effect of 60 ng/kg angiotensin I.

To assist ventilation, the trachea was incubated, and both vagi were severed to block reflex bradycardia. Into the right femoral veins and carotid artery, catheters were inserted for injecting test compound, other agonists, and antagonists for monitoring arterial blood pressure, heart rate, and blood sampling.

Responses, in terms of the increases in systolic (SBP) and diastolic (DBP) blood pressures, to test peptide and bradykinin were determined. After ganglion blockade with pentolinium (group GB) without or with adding captopril (group GB + Cap), heart rate, and plasma adrenaline and noradrenaline concentrations were determined. In some cases, the rats were not subjected to ganglion blockade and served as controls for rats in the GB group (control group).

To evaluate the contribution of bradykinin B2-receptor-mediated mechanisms, the selective bradykinin B2-receptor antagonist HOE-140 was given for ganglion blockade, the rats were captopril treated, and the increases in SBP, DBP, heart rate, and plasma adrenaline/noradrenaline induced by test peptide and bradykinin (group GB + Cap + HOE) were observed. To evaluate the contribution of AT1-receptor-mediated mechanisms, the selective angiotensin II type 1 (AT1) receptor antagonist, losartan, was given to group GB rats before and after treatment with captopril (1 mg/kg, intravenous), and the increases in SBP, DBP, heart rate, and plasma adrenaline/noradrenaline induced by test peptide were observed.

For determining plasma adrenaline and noradrenaline, approximately 1 mL of arterial blood was withdrawn via the carotid cannula at baseline (before injecting test peptide and bradykinin), at the peak of the SBP response, and after recovery from the response. Before each withdrawal, the rat was given a transfusion of approximately 1 mL blood from a similarly prepared donor rat. The SBP peak was observed at 2–3 min after injecting test peptide and at 1–2 min after injecting bradykinin; blood sampling coincided with these intervals. Using HPLC and fluorimetric detection, plasma catecholamines were determined.

6.2 PULMONARY HYPERTENSION ASSAY[7–11]

The chest cavity of mouse (10–20 weeks old, 22–30 g) anesthetized with pentobarbital sodium (80 mg/kg, intraperitoneal) was opened, the lungs were removed rapidly and placed in Krebs–Ringer bicarbonate solution (KRBS) containing 118.3 mM NaCl, 4.7 mM KCl, 1.2 mM $MgSO_4$, 1.2 mM KH_2PO_4, 2.5 mM $CaCl_2$, 25.0 mM Na_2CO_3, and 10.0 mM glucose, and bubbled with 21% O_2. From the intrapulmonary artery (3rd to 4th generation, PA) rings (50–100 μm internal diameter, 2–3 mm long) were isolated using a dissection microscope. PA rings were mounted as ring preparations by threading two steel wires into the lumen and securing the wires to two supports and placed in a small vessel wire myograph chamber. The support was attached to a micrometer for the control of ring circumference and to a force transducer for measuring isometric tension, respectively. After removal of endothelial cells by gently rubbing the intraluminal surface with a steel wire, some vessels were successively perfused with 2 mL air bubbles and 2 mL KRBS (perfusion pressure <5 mm Hg), before being mounted in the chamber. In the chamber filled with KRBS (pH 7.35–7.45), bubbled with 21% O_2–5% CO_2–balance N_2, the whole preparation was kept at 37°C. To control oxygen tension over the superfusate, Plexiglas was used as a cover over the chamber. The temperature and PA tension were recorded. At initial tension of 0 mN [1 g = 4.905 milliNewton (mN)], isolated murine PA rings in the chamber were allowed to equilibrate for 10–15 min, which was increased to 5 mN in 2.5-mN steps at 4–5 min intervals and held constant thereafter. Through preliminary assays, the resting tension for maximal constrictor response was optimized, in which the resting tension levels of 2.5, 5.0, and 7.5 mN were compared. In the evaluation of vascular viability, after the treatment of PA with 60 mM KCl, the tension was determined. The PA was washed extensively with KRBS, successively exposed to U-46619 (0.01 μM, thromboxane A_2 agonist) and 1 μM Ach, the resulting tension was recorded, stabilized for 5–10 min, and the agonists were washed out of the myograph chamber with KRBS.

Murine PA was isolated, placed in a confocal pressure myograph chamber, cannulated at both ends with glass micropipettes, and secured with 12-0 nylon monofilament suture. To control transmural pressure, both cannulas were connected to a reservoir that can be raised or lowered. In the chamber, PA was superfused at 37°C constantly with KRBS and gassed with 21% O_2–5% CO_2–balance N_2. To control oxygen tension over the superfusate, Plexiglas was used as a cover over the chamber. To the chamber, 0.1 mM dihydroethidium (DHE) was added, in which murine PA was incubated at 37°C for 45 min and washed for 30 min. The reaction of intracellular DHE and reactive oxygen species (ROS) produced a fluorescent oxidized product, and PA images were scanned (480-nm-line argon laser, fluorescence 620 nm). Superoxide anion levels in isolated murine PA was measured by lucigenin-enhanced chemiluminescence technique and scintillation counter, using solution of 5 μM lucigenin in Krebs–HEPES buffer (10.0 mM HEPES acid, 135.3 mM NaCl, 4.7 mM KCl, 1.2 mM $MgSO_4$, 1.2 mM KH_2PO_4, 1.8 mM $CaCl_2$, 0.026 mM Na-EDTA, and 11.1 mM glucose) at 1 mL total volume, after 5 min background stabilizing for chemiluminescence activity.

6.3 CORONARY ARTERIES CONSTRICTION ASSAY[12]

Mouse coronary artery rings were isolated, mounted, and their isotonic shortening contraction was measured. Mouse left main CA (70–90 μm in diameter, 1-mm length) was placed into a microvascular chamber, cannulated at both ends with glass micropipettes, and pressurized. The endothelial cells of some vessels were removed by gently rubbing the intraluminal surface with a steel wire. With a pressure transducer positioned at the level of the vessel lumen, the vascular intraluminal pressure (Ptm) was measured. In the chamber, vessels were superfused constantly with recirculating Krebs–Ringer bicarbonate solution containing 118.3 mM NaCl, 4.7 mM KCl, 1.2 mM $MgSO_4$, 1.2 mM KH_2PO_4, 2.5 mM $CaCl_2$, 25.0 mM $NaHCO_3$, and 11.1 mM glucose, which was gassed with 16% O_2, 5% CO_2, balance N_2 (pH 7.35–7.45) and maintained at 37°C. To control oxygen tension over the superfusate, Plexiglas was used as a cover over the chamber, through which port an oxygen electrode was passed into the superfusate and positioned near the vessel to provide continuous measurement of oxygen tension. The vascular intraluminal diameter (ID) was measured continuously. The oxygen tension, vascular ID, and Ptm were recorded.

In the chamber, the isolated CA was allowed to equilibrate for 30 min at 10 mm Hg of Ptm, which was increased to 60 mm Hg in 10 mm Hg steps at 5- to 7-min intervals and held constant thereafter. Five minutes after Ptm was increased to 60 mm Hg (time 0), the measurement of ID at 60 mm Hg of Ptm (ID60) was started and continued throughout the assay.

Using a lucigenin (bis-*N*-methyacidinium nitrate) enhanced chemiluminescence technique, the generation of superoxide anions in isolated murine CA was measured in 5 μM lucigenin buffer at 1 mL total volume; after background stabilizing chemiluminescence activity for 5 min, the CA was placed in the chemiluminometer, and photon emission was recorded continuously, and the chemiluminescence signal was recorded as relative light units per second (RLU/s).

6.4 MRI AND BRAIN NATRIURETIC PEPTIDE ASSAYS[13]

Before pulmonary endarterectomy (PEA), MRI was performed with a four-element body phased-array coil and a 1.5 T whole body system. MRI breath-hold cine imaging was triggered electrocardiographically and performed in the cardiac short-axis view within a stack of parallel imaging planes, which covered the left and right ventricles from base to apex. A spoiled gradient echo sequence was used and specified. From this stack of short-axis cine images, using the MR Analytical Software System (Medis, Leiden, The Netherlands), right ventricular (RV) and left ventricular (LV) volumes were calculated for each temporal frame in the cardiac cycle. From the stack of parallel short-axis images, the end-diastolic volume (EDV) and end-systolic volume (ESV) were assessed, and ejection fraction (EF) and stroke volumes (SV) were calculated. From the stack of parallel short-axis images, by manual detection of endocardial and epicardial borders on each slice and using the MR Analytical Software System, RV and LV myocardial masses were assessed. For body surface area, cardiac volume and mass were corrected. By the curvature, interventricular septal bowing was quantified. This curvature positive ratio denoted rightward septal bowing, and negative values denoted leftward ventricular septal bowing.

From the brachiocephalic vein, blood was tested at rest in a horizontal position for plasma, at 4°C and 3000 rpm centrifuged for 10 min, and stored at −80°C until analyzed. With an immunoradiometric assay, the brain natriuretic peptide (BNP) was determined.

6.5 COMPETITION ENZYME-LINKED IMMUNOSORBENT ASSAY[14]

Adult rabbits were immunized with 5×10^2 mg cis-4-hydroxy-L-proline-PEG (CHOP-PEG) in complete Freund's adjuvant and boosted at weeks 3 and 6 with 2.5×10^2 μg antigen in incomplete Freund's adjuvant. Microtiter plates were coated overnight with 100 μL CHOP-PEG (2 μg/mL in 0.2 M $NaHCO_3$, 4°C, pH 9.5); washed in PBS containing 1% BSA (pH 7.4); to the wells serial dilutions of sera were added for 9 additions for 10 minutes each; relative antibody concentrations were determined by adding horseradish peroxidase-conjugated goat-anti-rabbit IgG for 9 additions for 10 minutes each and 3,3′,5,5′-tetramethylbenzidine (TMB) substrate for 3 additions for 10 minutes each. Using a plate reader, the product was detected. By fluid-phase competition, pooled sera from several immunized rabbits (end titers of ca. $1:3 \times 10^4$) were assayed for specificity. To test competition of CHOP-PEG for the antibody, after mixing pretitered sera ($1:2 \times 10^4$ dilutions) with serial dilutions of CHOP-PEG in BSA-blocked plates for 2 h, the mixture was added to plates with immobilized CHOP-PEG. By fluid phase competition, specificity of the antiserum was determined for binding the rabbit IgG to immobilized CHOP-PEG by several chemical intermediates in the synthesis of CHOP-PEG. Competitor compounds were added in twofold serial dilutions with a 50 μg/mL top concentration. Uninhibited antiserum binding to immobilized CHOP-PEG produced 1.5–2.0 net OD. At about 1.0 μg/mL and 20 μg/mL, fluid phase CHOP-PEG inhibited antiserum binding by 20% and 80%, respectively. The antiserum preferentially recognized the unit consisting of CHOP-Lys with the α and ε amines of Lys linked via amide bonds to either PEG or Boc group.

6.6 RIGHT VENTRICULAR PRESSURE ASSAY[14]

At the end of the exposure periods, rats were anesthetized with an intraperitoneal injection of 0.5 mg/kg sodium pentobarbital, through the jugular vein a polyethylene catheter (PE-90) was inserted, and mean right ventricular pressure (RVP) was recorded. Prior to RVP measurements, rats breathed room air for at least 20 min, which was a sufficient time to eliminate hypoxic vasoconstriction. Rats were sacrificed by cutting the hepatic vein. Typically, experimental groups included controls, the untreated hypoxic animals group, and the treated hypoxic group. In some cases, a blood pressure-lowering effect of the test compound may be estimated as a reduction in RVP, or RedRVP, and calculated as RedRVP $= 100 \times [1 -$ (treated hypoxic $-$ control)/(untreated hypoxic $-$ control)]. Identical exposure conditions were used for all comparisons; the variables that differed were the dose of the test compound, the type of test compound, or the route of administration.

6.7 PLASMA NITRITE/NITRATE CONCENTRATION ASSAY[15]

Arterial blood samples of mongrel dogs were collected in tubes containing EDTA, centrifuged at $800 \times g$ for 5 min, plasma aliquots were removed and stored at -70°C, and their nitrite and nitrate (NOx) content was analyzed in duplicate using an ozone-based chemiluminescence assay. In the assay, the plasma samples were treated with a twofold volume of cold ethanol, centrifuged at $14{,}000 \times g$ for 5 min, 25 μL supernatant was treated with vanadium (III) in 1 N hydrochloric acid at 90°C in a glass purge vessel to reduce NO gas for quantitation of NOx. A nitrogen stream was bubbled successively through the purge vessel containing vanadium (III), 1 N NaOH, and NO analyzer, which detected NO released from NOx for chemiluminescent detection.

6.8 ADENO-ASSOCIATED VIRUS VECTOR-CAUSED RAT PULMONARY ARTERY PRESSURE ASSAY[16]

Fischer rats (12 weeks old, pathogen-free) housed in micro-isolator boxes (2–3 rats per box) were anesthetized with intraperitoneal injection of ketamine (50 mg/kg) and xylazine (10 mg/kg), intubated and ventilated using a model 683 Harvard rodent ventilator (Harvard Apparatus, Holliston, MA), the heart was exposed via a left anterolateral thoracotomy at the fourth intercostal space, 2×10^{10} genomic particles of adeno-associated virus-angiopoietin-1 (AAV-Ang-1) in 200 μL PBS/1 mM $MgCl_2$ was directly injected into the right ventricular outflow tract (just beneath the pulmonic valve) of the hearts of 30 rats, whereas the same amount of adeno-associated virus-lacZ (AAV-lacZ) was injected into 30 rats of the first control group; 30 sham rats of the second control group were injected with 200 μL PBS/1 mM $MgCl_2$ alone. At 1-month and 2-months post-gene delivery time points, rats from each group were sacrificed, and organs/blood was collected for tissue and molecular analysis. To observe the natural history of their pulmonary disease, four rats in each group were followed for 1 year or until death. All rats received care in accordance with the Guide for the Care and Use of Laboratory Animals.

Pulmonary artery systolic and diastolic pressures, heart rate, and systemic blood pressures were measured at two serial time points after the gene delivery operation just before the sacrifice of the rats, which was accomplished with a sternotomy performed after anesthetic induction and intubation in order to expose the heart. By inserting directly into the right ventricular outflow tract, a 22-gauge angiocatheter was advanced into the pulmonary artery, of which the pressures were measured using a standardized pressure transducer, whereas by direct aortic transduction systemic blood pressures were measured. Afterwards, the rats were sacrificed by exsanguination through the abdominal aorta. The pulmonary vasculature of five rats in each group at each time point was flushed with saline and then perfused using a syringe infusion pump with liquid silicone polymer Microfil (Flow Tech., Inc., Carver, MA) at 0.25 mL per min for 4 min. After 4°C and 12 h storage, the lungs were sequentially bathed in increasing concentrations of ethanol, placed in methyl salicylate, and photographed with a digital camera through a dissecting microscope.

6.9 ADENO-ASSOCIATED VIRUS VECTOR-CAUSED RAT PROTEIN AND mRNA ASSAY[16]

According to standard procedures, the protein extracts and Western blot analyses of the rats that had adeno-associated virus vector-caused pulmonary hypertension were performed using 100 mg of lung extracts. Using RNAzol B reagent, total RNA was isolated. According to the manufacturer's instructions, 2 μg total RNA was digested with DNase I and reverse-transcribed for all PCR analyses. In standard conditions, PCR amplifications were performed, and each PCR product was electrophoresed on 1% SeaKem LE (BioWhittaker Molecular Applications, Walkersville, MD) agarose gels stained with ethidium bromide. Using viral-specific Ang-1 5′-GGTACCCGAATGACAGTTTTCC-3′ and 5′-CTCCATTTCTAAGATTTGT-GCTC-3′, rodent Ang-1 5′-GAGAAGCAACTTCTCCAACAG-3′ and 5′-CTCGTT-CCCGAGCCAATATTC-3′, glyceraldehyde-3-phosphate dehydrogenase 5′-CATC-ATCTCTGCCCCCTCTG-3′ and 5′-CCTGCTTCACCACCTTCTTG-3′, each gene was analyzed. The viral-specific Ang-1 primer sequence overlapped with the multiple cloning site sequence in pAAV-shuttle vector and therefore only recognized the Ang-1 gene sequence delivered by AAV.

6.10 N-TERMINAL pro-BRAIN NATRIURETIC PEPTIDE (N-T proBNP) ASSAY[17]

Patients with pulmonary arterial hypertension (PAH) had a mean age of 57 years (range, 34–80 years), the control subjects had a mean age of 56 years (range, 27–78 years). Within 1 month of the N-T proBNP estimation, 6-min walk tests

(SMWTs) were performed on all patients with PAH. N-T proBNP values were compared with a patient's functional state as measured by the SMWT.

From each patient, a 10-mL whole blood sample was drawn into a heparinized (EDTA) plastic tube and assayed within 4 h. With the Elecsys Systems E170 proBNP assay kit (Sigma-Aldrich Co, St. Louis, MO, USA), the plasma N-T proBNP level was measured and the N-T proBNP concentration of each sample was calculated in values of pg/mL. Expected value for women and men at age <50 years and 50–65 years were 153 pg/mL, 334 pg/mL, 88 pg/mL, and 227 pg/mL, respectively.

6.11 MIDDLE CEREBRAL ARTERY OCCLUSION ASSAY[18–23]

Male baboons were randomly divided into treatment groups (a group underwent MCAO for 1 h or 2 h, a group underwent 3 h MCAO with subsequent reperfusion for 1 h or 4 h, and a group underwent 1.5 h MCAO with 24 h reperfusion), controls of group that did not undergo any preparation procedure, and a group that suffered from MCAO at surgical implantation and sustained hemiparesis for 7 days. After transcardiac perfusion with isosmotic heparinized perfusate, brain tissues were removed under thiopental Na and prepared for frozen and paraffin sections. To recipient tissues consisting of unfixed 10-μm-thick normal basal ganglia sections mounted on microscope slides, the samples of normal and 50 μL ischemic brain tissue or 100 μL purified reagents were applied.

For application of proteases to normal recipient tissues, PBS (pH 7.0) was used as the incubation buffer for type-7 bacterial collagenase. In a buffer containing 90 mM NaCl, 5 mM KCl, 1.5 mM $MgCl_2$, 23 mM Na gluconate, 27 mM Na acetate, 10 mM $CaCl_2$, and 40 mM $ZnCl_2$ (pH 7.4), the solution of matrix metalloproteinases (MMPs) and urokinase/plasminogen were prepared. In 200 mM sodium acetate containing 8 mM dithiothreitol, 1 mM EDTA, and 2.7 mM L-cysteine (pH 6.0), the solution of cathepsin B and L was prepared. Before application to recipient sections, the mixture of 5 IU/mL plasminogen and 5 IU/mL uPA was incubated at 37°C for 1 h, to which 1 μg/mL pro-MMP-2 or 1 μg/mL pro-MMP-9 was added and incubated for 1 h. On the recipient sections, 100 μL of each mixture or buffer alone was incubated at 37°C for 5 h (collagenase) or 18 h (MMPs and cathepsins), and the sections were washed with PBS and fixed. Donor samples were derived from approximately 1.0 cm × 1.0 cm × 10 μm frozen sections. Ten consecutive 10-μm sections of ischemic or normal basal ganglia were centrifuged (300 × *g*, 30 s) and mixed by repeated gentle pipetting. To the recipient sections, 50 μL of sample or PBS was applied, the mixture was incubated at 37°C for 18 h, the recipient sections were washed with PBS and fixed in acetone or paraformaldehyde (PFA).

Specific antigens were developed by immunoperoxidase methods. Acetone-fixed frozen sections were incubated overnight with the primary antibody at 4°C and developed for immunoperoxidase with 3,3-diaminobenzidine tetrahydrochloride. After deparaffinization, PFA-fixed paraffin-embedded sections were subject to the same procedures. dUTP incorporating into nuclear DNA was considered as evidence of nuclear DNA scission/repair, and at 2 h MCAO defined the ischemic core and peripheral regions of cellular neuronal injury. PFA-fixed cryosections were subject to the DNA polymerase I method. All ischemic samples contained ischemic core and ischemic peripheral regions. To detect MMP-related activities, gelatin zymography was performed. Protease activities were identified via incubating the gels in buffer containing GM6001 or 4-amidinophenyl methanesulfonyl fluoride (APMSF).

6.12 VASCULAR ENDOTHELIAL GROWTH FACTOR LEVEL ASSAY[24]

Consecutive acute myocardial infarction (AMI) patients received emergent coronary angiography and successful reperfusion therapy by primary percutaneous coronary intervention using sirolimus-eluting stents (SES) or bare-metal stent (BMS) after ballooning coronary angioplasty. To match the age, gender, and other traditional risk factors of the AMI, 12 control subjects were selected from 57 consecutive subjects with normal coronary angiograms and left ventriculograms and served as a reference group for the coronary vasomotor response and the plasma VEGF concentrations.

For all AMI patients, coronary angiography was performed twice immediately after admission during the acute phase of AMI and at week 2 after AMI. At the second coronary angiography after an overnight fast in all of the AMI patients, the measurements of coronary vasomotor response and left ventriculography were performed. Before systemic heparinization on the third day after AMI in 8 patients with SES implantation and at the time of the second coronary angiography in all of the AMI patients, blood sampling from the anterior interventricular vein (AIV) and the aortic root (AO) was performed. In all of the control subjects, the coronary vasomotor response measurement and blood sampling were performed only once. With a sandwich enzyme-linked immunosorbent assay, plasma levels of free VEGF were measured. With HPLC-MS, sirolimus levels in whole blood were determined at 0.1 ng/mL minimum detection limit. More than 3 days before the coronary angiographic assay in patients with AMI and in control subjects, all lipid-lowering drugs and vasodilators were withdrawn.

6.13 ANTISPASMODIC AGENT *IN VIVO* ACTION ASSAY[25]

Male C57/BL6 mice (9–18 weeks old) were anesthetized with a subcuticular injection of 0.3 mL Hypnoval® (500 μmg/mL, midazolam, Antigen Pharmaceuticals) and 0.3 mL Hypnorm® (10 μmg/mL fentanyl and 330 μg/mL fluanisone, Janssen Animal Health), the abdominal and thoracic cavities of the donor mouse were opened widely, systemic heparinization (100 U via inferior vena cava) was given, and exsanguinated and descending thoracic aorta was immediately harvested. For the anesthetized recipient mouse, a midline abdominal incision was made, and the abdominal aorta was exposed and mobilized by dissection of the periaortic fat. Between two vascular clips the aorta was transected, and from the cut ends blood was rinsed with heparinized saline (500 U/mL). By inserting 11-0 monofilament nylon proximal and distal interrupted stay sutures, a 2-mm donor aortic segment was positioned. With 8–10 additional interrupted sutures, the distal and proximal anastomoses were completed. After removal of the vascular clips, hemostasis was achieved. The viscera were replaced, and with a continuous 4-0 absorbable suture the abdomen was closed in two layers. In a warming cabinet, mice were recovered until fully mobile and given diet and water *ad libitum*. To evaluate the *in vivo* duration action of phenoxybenzamine and Verapamil and nitroglycerin (VG) solution after topical application to vascular grafts, donor aortic segments were incubated with topical antispasmodic agents (phenoxybenzamine 10 mM, or verapamil/nitroglycerin 30 mM each) or buffer alone (controls) for 15 min immediately prior to grafting. At approximately 2, 6, 13, and 23 h after surgery, the mice were sacrificed and the duration action was measured from removal of the vascular clamps and resumption of circulation.

By reopening the abdomen, exposing the graft, placing vascular clamps proximally and distally to the graft, and cutting through the suture lines, leaving the majority of the graft undamaged and suitable for analysis, the grafts were immediately harvested and placed in ice-cold buffer for vasomotor assays to measure contractile responses to K^+, phenylephrine, and prostaglandin F2α. For mice that died during the procedure or in the postoperative period, their grafts did not undergo vasomotor assays and were excluded from the assays.

6.14 TEMPERATURE ASSAY IN AWAKE SUBJECTS[26–31]

With an intraluminal filament occlusion of the middle cerebral artery (MCA), male SD rats (weight, 250–325 g) were caused reversible ischemia, after which anesthesia was maintained with 2% halothane and nitrous oxide : oxygen 60 : 40 by face mask. To expose the left carotid artery, a midline neck incision was made, and with 5-0 silk the external carotid and pterygopalatine arteries were ligated. In the arterial wall, an incision was made, and from the bifurcation of external and internal carotid arteries a 4-0 heat blunted nylon suture was advanced 18 mm. This distance reliably produced blockage of the origin of the MCA. The duration of occlusion for individual rats was varied to generate a range of ischemia durations for each experimental group. During surgery, the temporalis and brain temperature were monitored and maintained.

During the postanesthetic recovery period (4–5 h), the awaking rats were maintained at a temperature of 37.5°C. To verify the effect of this procedure on brain temperature in a separate set of nonischemic rats, an indwelling, radio telemetry thermister was implanted. After induction and using the identical anesthesia, the subject was placed in a stereotaxic head frame. The thermister probe was placed in the left parietal cortex (3 mm deep to dura and 1 mm anterior, 3 mm lateral to bregma). To secure the probe to the skull, a plastic cap was fitted using methyl methacrylate. After full recovery from the anesthesia, for 48 h telemetered brain temperature measurements were recorded at 15-min intervals. After a drug was administered during a baseline period of 24 h, temperatures were measured at 5-min intervals for 6 h. After drug treatment, temperature of some rats was measured at 15-min intervals for 2 h. Normal/abnormal behavior ratio was measured at 48 and 72 h. All obtundation/reduced exploration, forepaw retraction on tail lifting, asymmetric forepaw grasp, axial twist, forced circling, or death were defined to abnormal.

To determine that the filament was positioned correctly, each rat was injected with 0.2 mL 2% Evan's blue in saline via tail vein; 30 to 60 min after Evan's blue circulation, the rat was perfused transcardially with 100 mL NS, and the brain was removed and immediately examined. To assess therapeutic efficacy, one of the test therapies was injected 30, 60, 75, 120, 240, or 360 min after initiating occlusion. Allocation to treatment groups was random: each day two subjects were allotted to control and four or five to drug treatment. The test compounds were dissolved in saline and given via tail vein.

Stroke outcome was measured with the quantal assay. A quantal dose-response was time dependent. In separate groups, physiologic assays were performed to determine the effects of the test compound on blood gases, heart rate, and arterial blood pressure. In order to determine the interaction between test compound and anesthesia, the assays were performed in awaking or in anesthetized subjects. The ventral tail artery was cannulated with PE-50 tubing for recording the arterial pressures and sampling arterial blood. Once the rat was stable for 30 min without further adjustments needed to maintain homeostasis, steady-state mean arterial pressure (MAP) was 75–90 mm Hg, and halothane at a minimum, test compound was administered. To obtain similar data from awaking subjects, tail artery catheters were implanted under halothane inhalational anesthesia. After implantation, the rat was allowed to recover and lightly restrained. One hour after stopping the anesthesia, the cannula was attached to the transducer and vital signs recorded for a 30-min baseline period. After the subject was shown to be stable for 30 min, the test compound was given. Blood pressures and blood gases were recorded at 10-min intervals for 1 h and then hourly.

REFERENCES AND NOTES

1. A.A. Amfilochiadis, P.C. Papageorgiou, N. Kogan, F. Boomsma, D.H. Osmond. Role of bradykinin B2-receptor in the sympathoadrenal effects of 'new pressor protein' related to human blood coagulation factor XII fragment. J Hypertens 22 (**2004**) 1173–1181.
2. L. Mavrogiannis, D.M. Trambakoulos, F. Boomsma, D.H. Osmond. The sympathoadrenal system mediates the blood pressure and cardiac effects of human coagulation factor XII-related 'new pressor protein'. Can J Cardiol 18 (**2002**) 1077–1086.
3. L. Mavrogiannis, K.P. Kariyaisam, D.H. Osmond. Potent blood pressure raising effects of activated coagulation factor XII. Can J Physiol Pharmacol 75 (**1997**) 1398–1403.
4. P.C. Papageorgiou, A. Pourdjabbar, A.A. Amfilochiadis, E.P. Diamandi, F. Boomsma, D.H. Osmond. Are the cardiovascular and sympathoadrenal effects of human 'new pressor protein' preparations attributable to human coagulation {beta}-FXIIa? Am J Physiol Heart Circ Physiol 286 (**2004**) H837–H846.

5. D. Simos, F. Boomsma, D.H Osmond. Human coagulation factor XII-related 'new pressor protein': role of PACAP in its cardiovascular and sympathoadrenal effects. Can J Cardiol 18 (**2002**) 1093–1103.
6. F.A. Van der Hoorn, F. Boomsma, M.A.J. Veld, M.A. Schalekamp. Determination of catecholamines in human plasma by high-performance liquid chromatography: Comparison between a new method with fluorescence detection and an established method with electrochemical detection. J Chromatogr 487 (**1989**) 17–28.
7. J.Q. Liu, R.J. Folz. Extracellular superoxide enhances 5-HT-induced murine pulmonary artery vasoconstriction. Am J Physiol Lung Cell Mol Physiol 287 (**2004**) L111–L118.
8. M. Athar, C.A. Elmets, D.R. Bickers, H. Mukhtar. A novel mechanism for the generation of superoxide anions in hematoporphyrin derivative mediated cutaneous photosensitization. Activation of the xanthine oxidase pathway. J Clin Invest 83 (**1989**) 1137–1143.
9. L.B. Becker, T.L. vanden Hoek, Z.H. Shao, C.Q. Li, P.T. Schumacker. Generation of superoxide in cardiomyocytes during ischemia before reperfusion. Am J Physiol Heart Circ Physiol 277 (**1999**) H2240–H2246.
10. R.P. Brandes, M. Barton, K.M. Philippens, G. Schweitzer, A. Mugge. Endothelial-derived superoxide anions in pig coronary arteries: Evidence from lucigenin chemiluminescence and histochemical techniques. J Physiol 500 (**1997**) 331–342.
11. T.J. Guzik, S. Mussa, D. Gastaldi, J. Sadowski, C. Ratnatunga, R. Pillai, K.M. Channon. Mechanisms of increased vascular superoxide production in human diabetes mellitus: Role of NAD(P)H oxidase and endothelial nitric oxide synthase. Circulation 105 (**2002**) 1656–1662.
12. J.Q. Liu, I.N. Zelko, R.J. Folz. Reoxygenation-induced constriction in murine coronary arteries: The role of endothelial NADPH oxidase (gp91phox) and intracellular superoxide. J Biol Chem 279 (**2004**) 24493–24497.
13. H.J. Reesink, I.I. Tulevski, J.T. Marcus, F. Boomsma, J.J. Kloek, A.V. Noordegraaf, P. Bresser. Brain natriuretic peptide as noninvasive marker of the severity of right ventricular dysfunction in chronic thromboembolic pulmonary hypertension. Ann Thorac Surg 84 (**2007**) 537–543. *Note*: Consecutive male patients (18, age range 25–78 years) who were diagnosed with chronic thromboembolic pulmonary hypertension (CTEPH) and cardiopulmonary hemodynamics by pulmonary angiography and right heart catheterization received no specific medical treatment for pulmonary hypertension and received oral anticoagulants for at least 3 months. Mean pulmonary artery pressure (mPAP) greater than 25 mm Hg at rest or greater than 30 mm Hg during exercise was defined as pulmonary hypertension. Patients diagnosed with renal insufficiency (creatinine >115 μM), concomitant left-sided heart disease, uncontrolled systemic hypertension, or uncontrolled diabetes mellitus (DM) were excluded from the assay. As a part of the preoperative workup, coronary angiography was involved and routinely performed in all patients older than 50 years of age and in patients older than 40 years of age if they had a history of smoking. Ten nonsmoking healthy male volunteers (age range, 27–60 years) were age-matched and served as controls. For the included 32 patients a pulmonary endarterectomy (PEA) was performed; according to a protocol, five patients were considered to have distal, inoperable CTEPH. PEA was postponed for one additional patient with exercise-induced pulmonary hypertension. On the first or second day after PEA before removal of the Swan-Ganz catheter, postoperative hemodynamic characteristics were determined.

14. P.M. Simon, J. Pachence, B. Belinka, G.J. Poiani, S.-E. Lu, C.A. Tozzi, D.J. Riley. Prodrug of proline analogue reduces hypoxic pulmonary hypertension in rats. Pulm Pharmacol Ther 19 (**2006**) 242–250.

15. A.R. Souza-Silva, C.A. Dias-Junior, J.A. Uzuelli, H. Moreno Jr., P.R. Evora, J.E. Tanus-Santos. Hemodynamic effects of combined sildenafil and L-arginine during acute pulmonary embolism-induced pulmonary hypertension. Eur J Pharmacol 524 (**2005**) 126–131.

16. D. Chu, C.C. Sullivan, L. Du, A.J. Cho, M. Kido, P.L. Wolf, M.D. Weitzman, S.W. Jamieson, P.A. Thistlethwaite. A new animal model for pulmonary hypertension based on the overexpression of a single gene, angiopoietin-1. Ann Thorac Surg 77 (**2004**) 449–457.

17. D. Mukerjeea, L.B. Yap, A.M. Holmes, D. Nair, P. Ayrton, C.M. Black, J.G. Coghlan. Significance of plasma N-terminal pro-brain natriuretic peptide in patients with systemic sclerosis-related pulmonary arterial hypertension. Respir Med 97 (**2003**) 1230–1236.

18. S. Fukuda, C.A. Fini, T. Mabuchi, J.A. Koziol, L.L. Eggleston Jr., G.J. del-Zoppo. Focal cerebral ischemia induces active proteases that degrade microvascular matrix. Stroke 35 (**2004**) 998–1004.

19. G.F. Hamann, Y. Okada, R. Fitridge, G.J. del Zoppo. Microvascular basal lamina antigens disappear during cerebral ischemia and reperfusion. Stroke 26 (**1995**) 2120–2126.

20. G.J. del Zoppo, B.R. Copeland, L.A. Harker, T.A. Waltz, J. Zyroff, S.R. Hanson, E. Battenberg. Experimental acute thrombotic stroke in baboons. Stroke 17 (**1986**) 1254–1265.

21. M. Tagaya, H.P. Haring, I. Stuiver, S. Wagner, T. Abumiya, J. Lucero, P. Lee, B.R. Copeland, D. Seiffert, G.J. del Zoppo. Rapid loss of microvascular integrin expression during focal brain ischemia reflects neuron injury. J Cereb Blood Flow Metab 21 (**2001**) 835–846.

22. M. Tagaya, K.F. Liu, B. Copeland, D. Seiffert, R. Engler, J.H. Garcia, G.J. del Zoppo. DNA scission after focal brain ischemia: Temporal differences in two species. Stroke 28 (**1997**) 1245–1254.

23. J.H. Heo, J. Lucero, T. Abumiya, J.A. Koziol, B.R. Copeland, G.J. del Zoppo. Matrix metalloproteinases increase very early during experimental focal cerebral ischemia. J Cereb Blood Flow Metab 19 (**1999**) 624–633.

24. J. Obata, Y. Kitta, H. Takano, Y. Kodama, T. Nakamura, A. Mende, K. Kawabata, Y. Saitoh, D. Fujioka, T. Kobayashi, T. Yano, K. Kugiyama. Sirolimus-eluting stent implantation aggravates endothelial vasomotor dysfunction in the infarct-related coronary artery in patients with acute myocardial infarction. J Am Coll Cardiol 50 (**2007**) 1305–1309.

25. S. Mussa, T. Prior, N. Alp, K. Wood, K.M. Channon, D.P. Taggart. Duration of action of antispasmodic agents: Novel use of a mouse model as an in vivo pharmacological assay. Eur J Cardiothoracic Surg 26 (**2004**) 988–994.

26. P.D. Lyden, C. Jackson-Friedman, C. Shin, S. Hassid. Synergistic combinatorial stroke therapy: A quantal bioassay of a GABA agonist and a glutamate antagonist. Exp Neurol 163 (**2000**) 477–489.

27. E.Z. Longa, P.R. Weinstein, S. Carlson, R. Cummins. Reversible middle cerebral artery occlusion without craniectomy in rats. Stroke 20 (**1989**) 84–91.

28. F. Colbourne, G.R. Sutherland, R.N. Auer. An automated system for regulating brain temperature in awake and freely moving rodents. J Neurosci Methods 67 (**1996**) 185–190.

29. J.B. Bederson, L.H. Pitts, M. Tsuji, M.C. Nishimura, R.L. Davis, H. Bartkowski. Rat middle cerebral artery occlusion: Evaluation of the model and development of a neurologic examination. Stroke 17 (**1986**) 472–476.

30. J.A. Zivin, U. DeGirolami, A. Kochhar, P.D. Lyden, V. Mazzarella, C.C. Hemenway, M.E. Henry. A model for quantitative evaluation of embolic stroke therapy. Brain Res 435 (**1987**) 305–309.

31. J.A. Zivin, D.R. Waud. Quantal bioassay and stroke. Stroke 23 (**1992**) 767–777.

7

METHODS AND APPLICATIONS OF ASSAYS RELATED TO PARKINSON'S DISEASE AND GRAVES' DISEASE

Shiqi Peng

As a representative disease of the central nervous system, Parkinson's disease (PD) is characterized by bradykinesia, resting tremor, and gait disturbances. Pathologic characteristics of PD include degeneration of dopaminergic neurons in the substantia nigra compacta, locus caeruleus, hypothalamus, cerebral cortex, nucleus basalis, central and peripheral components of the autonomic nervous system, and the presence of Lewy bodies. In Graves' disease (GD), thyroid-stimulating antibody (TSAb), which binds to the thyrotropin receptor (TSHR) and stimulates thyroid hormone production of thyrocytes, causes hyperthyroidism. In primary atrophic hypothyroidism, thyroid stimulation blocking antibody (TSBAb), which binds to TSHR and inhibits the action of thyrotropin (TSH), induces thyroid atrophy and hypothyroidism. In the past decade, the onset and progress of both PD and GD are widely correlated with cytokines, enzymes, and receptors, based on which a whole set of assays is established and used to explain their biological mechanism and screen new drugs for their therapy. In this chapter, 24 assays are introduced: flow cytometric assay for cellular DNA content and caspase-3,[1] swine resuscitation assay,[2] human T-cell leukemia virus

Pharmaceutical Bioassays: Methods and Applications. By Shiqi Peng and Ming Zhao

type 1 (HTLV)-tax1 or pMuLV-SV-nlslacZ vectors transfected cell assay,[3] parkinsonian rat assay,[4] ELISA for nerve growth factor (NGF) antigen,[5] assay for β-nerve growth factor levels in cerebrospinal fluid,[6] Madin-Darby canine kidney (MDCK) scatter assay,[7] facial nerve and spinal root avulsions,[7] TSH receptor autoantibodies (TRAb) assay,[8–11] human thyrotropin receptor (hTSHR) assay,[12–15] soluble intercellular adhesion molecule-1 (ICAM-1), thyroid-stimulating antibodies (TSAb), and thyrotropin-receptor antibodies (TBIAb) activity assays,[16] microarray immunoassay for human thyrotropin receptor (hTSHr) production,[17] affinity assay for [^{35}S]GTPγS binding to gas/olf,[18] tissue segment binding assay for α_{1B}-adrenoceptor,[19] membrane binding assay with rat cerebral cortex for α_{1B}-adrenoceptor,[19] whole cell binding assay for α_{1B}-adrenoceptor,[19] functional assay for $\alpha_{1A,B}$-adrenoceptor,[19] rat neuroprotective assay,[20] γ-amino butyric acid (GABA)-benzodiazepine receptor assay,[21,22] forced swimming and tail suspension assays,[23] IGF-I kinase receptor activation (KIRA) assay,[24] gD.trkA kinase receptor activation (KIRA) assay,[24] gD-flag-modified quantitative kinase receptor activation (gD.trk KIRA)-ELISA,[25,26] and assays for cannabinoid receptors in rat cerebella or mouse brains.[27]

7.1 FLOW CYTOMETRIC ASSAY FOR CELLULAR DNA CONTENT AND CASPASE-3[1]

To assess cellular DNA content, the cells were cultured in 6-well plates, attached to the plate, collected with trypsin, mixed with detached cells present in the supernatant, centrifuged, resuspended in a solution containing 50 μg/mL propidium iodide, 1 μg/mL RNAse A, 0.01% NP-40 in PBS, incubated in the dark at 4°C for 30 min, and analyzed by flow cytometry on FACScan (Becton and Dickinson, Mountain View, CA, USA) using a 560-nm dichromatic mirror and 600-nm bandpass filter. The decreased percentage of DNA staining in apoptotic cells resulting from either fragmentation or decreased chromatin in a minimum of 20,000 cells per experimental condition was determined and expressed as the percentage of apoptotic (hypodiploid) nuclei. Cells with very low DNA content and uncertain death type were excluded from the analysis. As assayed by flow cytometry, the solvents used for preparing solutions could not induce any significant degree of apoptosis at any dose.

Caspase-3 was assayed using caspase-3 inhibitor (Z-DEVD-FMK) and caspase-8 inhibitor (Z-IETD-FMK) conjugated to fluorescein isothiocyanate (FITC) as the fluorescent *in situ* marker. Cell-permeable FITC-Z-DEVD-FMK and FITC-Z-IETD-FMK had no toxicity and irreversibly bound to activated caspase-8 and caspase-3 in apoptotic cells. FITC label allowed the activated caspases in apoptotic cells to be directly detected with flow cytometry. After stimulation, the cells were washed in saline solution and incubated for 1 h at 37°C in 5% CO_2 with 1 mL FITC-Z-DEVD-FMK or FITC-Z-IETD-FMK. In the analysis, a FACScan with a fluorescein channel detector was used.

7.2 SWINE RESUSCITATION ASSAY[2]

Domestic swine (15 males, mean weight 41 $\pm$ 4 kg, and 15 females, mean weight 39 $\pm$ 5 kg) were premedicated with ketamine (20 mg/kg) and xylazine (2 mg/kg), intubated endotracheally, received general anesthesia with isoflurane via nose cone, and maintained with inhaled isoflurane (MAC 1.0% to 2.5%) and nitrous oxide in a 1 : 1 mixture with oxygen. Using a side-stream capnometer, end-tidal carbon dioxide was continuously monitored, and minute ventilation was adjusted to maintain 35–45 mm Hg. The surface electrocardiogram's standard lead II was continuously monitored.

High fidelity and micromanometer-tipped catheters were positioned under fluoroscopic guidance in the ascending aorta via a femoral artery and in the right atrium via a jugular vein. A catheter with standard 7-F bipolar was placed in the jugular vein and advanced to the apex of the right ventricle. Using PowerLab Chart v. 5.2 (AD Instruments, Castle Hill, Australia), hemodynamic data were recorded and stored in a laptop computer. To the left and right lateral aspects of the thorax, standard adhesive defibrillation electrode patches were applied. Transthoracic impedance was measured using a tetrapolar constant current impedance measuring system. A 30-Ω noninductive resistor was placed in series with the impedance-compensating, truncated exponential biphasic defibrillation waveform defibrillator. After instrumentation, by passing 60 Hz ac current for approximately 0.5 s through the electrodes of a bipolar catheter positioned in the right ventricular apex, ventricular failure (VF) was electrically induced. After 7 min of untreated VF, swine in the supine position began to receive manual closed-chest compressions at a rate of approximately 100/min with force sufficient to depress the sternum 3.5–5.0 cm for 1 min and were given transthoracic countershock at 200 J. If VF persisted, additional shocks were administered at an escalating energy sequence (300 J and 360 J). VF termination was defined as successful defibrillation, regardless of the postshock cardiac rhythm or hemodynamic outcome, that is, spontaneous QRS complexes with or without associated arterial pressure pulses, determined 5 s after a defibrillation shock. Between shocks, chest compressions were performed while positive pressure ventilations ($FiO_2 = 1.00$) were performed at 8–10 ventilations/min. If VF persisted, 0.5 mg and 150 mg epinephrine was administered, cardiopulmonary resuscitation (CPR) continued, and shocks were repeated for 15 min or until VF was terminated. If shocks were followed by asystole or pulseless electrical activity (PEA), CPR and additional epinephrine were administered for 15 min or until spontaneous arterial pressures of 60 mm Hg appeared. At the end of 15 min of CPR, the swine remaining in VF, PEA, or asystole were considered failures in resuscitation, and resuscitative efforts were terminated. In those swine achieving return of spontaneous circulation (ROSC), defined as an arterial systolic pressure >60 mm Hg, hemodynamic and blood gas were measured at intervals for 3 h. Before electrical VF induction and at 15, 30, 60, 90, 120, and 180 min after ROSC, arterial blood was sampled, placed in 0°C sterile heparinized tubes, and centrifuged at 5000 rpm for 10 min. Plasma was immediately separated and stored at −80°C until analyzed. By quantitative sandwich ELISA, tumor necrosis factor (TNF)-α concentrations were determined using a commercially available kit specific for porcine TNF-α. With specific radioimmunoassay kit-based direct radioimmunoassays, plasma 17-estradiol and testosterone levels were measured in eight male and eight female swine. The lower limit of quantitation for 17β-estradiol (E2) and testosterone were 20 pg/mL and 0.025 ng/mL, respectively.

7.3 HTLV-tax1 OR pMuLV-SV-nlslacZ VECTORS TRANSFECTED CELL ASSAY[3]

Between the EcoRI and Hind III sites of the pBR327 plasmid by inserting a 0.8 kbp Sma I fragment containing LTR sequences of the MT2 provirus upstream of the 1acZ gene and the SV40 polyadenylation signal both derived from the pCHll0 vector the plasmid pA18G was constructed, while the plasmid pA20 was derived from pAl8G by inserting, at the pBR327 Sal I site, the neomycin gene under the control of the thymidine kinase promoter derived from the pMuLV-SV-tkneo plasmid. These constructs were introduced into the HTLV-I permissive human osteosarcoma cell line HOS and the Syrian Hamster kidney cell line (BHK-21) known to express a HTLV-I receptor. By double pulse (600 V–40 μF–74 Ω and 200 V–900 μF–74 W) electroporation in a Celliect apparatus (Eurogentec, Belgium), 30 μg of pA20 DNA was introduced into HOS cells (3×10^6 in 300 μL of DMEM medium). Using the calcium phosphate procedure with pAl8G plasmid and the neomycin expression vector PVV 1, BHK-21 cells were co-transfected. After transfection, by end point dilution geneticin G418-resistant cells were cloned and named pA20-HOS and pA18G-BHK-21 cells. With the tax expression vector HTLV-tax1 or the pBR327 plasmid as a negative control, pA20-HOS and pA18G-BHK-21 cells were assayed for the presence of a functional construct by transfection. In the same conditions on each cell line, by transfecting the pMuLV-SV-nlslacZ that constitutively expresses β-galactosidase, transfection efficiency was estimated. Two days after transfection, the cells were fixed and with 4-chloro-5-bromo-3-indolyl-β-D-galactoside (X-gal), lacZ expression was assayed. In pA20-HOS and pA18G-BHK-21 cells transfected with pBR327, lacZ gene expression was not observed. In contrast, in both cell lines after transfection with the HTLV-tax1 or the pMuLV-SV-nlslacZ vectors, β-galactosidase–positive cells were observed and in an almost equal amount, suggesting all cloned pA20-HOS and pA18G-BHK-21 cells had retained an LTR-1acZ construct inducible by tax.

7.4 PARKINSONIAN RAT ASSAY[4]

Hemiparkinsonian rats (adult male Wistar rats, weighing 200–250 g at the time of lesion) were used. By stereo taxic injection of 6-OHDA (8 μg/3 μL saline solution) into the right medial forebrain bundle (AP 4.4 mm, L 11.2 mm, and V 7.8 mm from dura, incisor bar 2.4 mm below the interaural line), dopaminergic lesions was performed. One month after lesion, all rats administered D-amphetamine (5 mg/kg in saline) intraperitoneally and apomorphine hydrochloride (50 μg/kg) subcutaneously were assayed for rotational behavior. In an automated rotomer, the circling behavioral was assessed and the number of full 360-degree turns/min were recorded automatically. Only rats that exhibited at least seven full ipsilateral turns/min with the D-amphetamine injection in 90 min and 3.3 full contralateral turns/min over the 45-min period with apomorphine were included in the assay.

7.5 ELISA FOR NERVE GROWTH FACTOR ANTIGEN[5]

Adult female SD rats (200–225 g) were perfused transcardially with ice-cold 10 mM PBS (pH 7.4). Their hippocampal formations (HF) were dissected out, stored at −70°C, weighed, supplied with 8 volumes of PBS, placed within Eppendorf tubes in an ethanol dry ice bath sonicated for 30 s, pooled, aliquots were supplied with 1/9 of their volume of freshly prepared PBS containing 30% BSA, filtered, adjusted to pH 7.4, incubated for 30 min at room temperature and pH 7.4 or pH 10.5, centrifuged for 1 h at 4°C and 100,000 × *g*, supernatant fraction was removed, pellet-containing tubes were rinsed once with 500 μL ice-cold PBS, pellets were resuspended via a 5-s sonication in a volume of PBS equal to 8/10 its precentrifugation volume, supplied with 1 volume of PBS containing 30% BSA, adjusted to a proper pH by adding very small amounts of 5 N NaOH or HCl (ca. 2.3 μL per 300 μL), and immediately vortexed. In Eppendorf tubes, final pH was measured with a micro pH-electrode. Samples before and after centrifugation were analyzed for NGF antigen content.

EIA/RIA high binding microtiter plates were coated overnight at 4°C with 50 μL carbonate buffer (pH 9.6) containing either goat antiserum against mouse β-NGF (1:500 dilution) or commercial goat serum (1:500 dilution), washed twice with about 100 μL washing buffer (PBS, 0.4 M NaCl, and 0.05% Triton X-100), blocked for 1 h at room temperature with 90 mL/well PBS containing 3% BSA, and washed once with washing buffer and once with PBS. Into the wells, samples (50 μL/well) containing sample buffer (3% BSA-PBS) alone, known quantities of purified mouse β-NGF (range 1–100 pg/well) or unknown material to be assayed (1:2 diluted sonicate fractions) were pipetted, at 37°C incubated for 3.5 h, at room temperature washed 3 × 20 min with washing buffer and 2 × 20 min with PBS, and incubated overnight at 4°C with 50 μL purified rabbit polyclonal antibodies against mouse β-NGF [diluted 1:1000 in PBS containing 0.4 M NaCl, 0.5% bovine serum albumin (BSA), 0.1 mM PMSF, 20 kI/mL aprotinin, 5 mM EDTA, and 0.1% Triton X-100]. Before adding 50 μL of peroxidase-conjugated affinity-isolated swine anti-rabbit antibody (diluted 1:1000 in PBS containing 0.1% BSA and 0.25% Triton X-100), unbound antibodies were removed, wells were washed twice with washing buffer and twice with PBS, incubated overnight at 4°C, and washed twice with washing buffer and twice with PBS alone. By adding 150 mL solution containing 0.033 M sodium phosphate, 0.016 M citric acid, 0.05% *o*-phenylenediamine, and 0.02% H_2O_2, a soluble colorimetric product was obtained, the plates were shaken and in the dark incubated for 15 min, the reaction was stopped with 50 mL 0.2 N H_2SO_4

and the OD was read using a Bio-Tek Instruments microplate reader at 490 nm. With the corresponding values from goat serum-coated wells, β-NGF containing samples in GAM-4A-coated wells were routinely corrected. With a purified mouse β-NGF stock containing 200 ng/mL β-NGF standard curves were run in each plate in every assay. β-NGF was serially diluted in sample buffer (3% BSA-PBS).

7.6 ASSAY FOR β-NERVE GROWTH FACTOR LEVELS IN CEREBROSPINAL FLUID[6]

NGF levels of cerebrospinal fluid (CSF) of 38 children with infantile spasms (IS) and 22 control patients were examined. All patients were carefully examined for blood/serum cell counts, vacuolated lymphocytes, glucose, TORCH antibodies, C-reactive protein, Na^+, K^+, Ca^{2+}, phosphate, alkaline phosphatase, urinary amino acids, glycosaminoglycans, oligosaccharides, and, when clinically indicated, organic acids and chromosome analysis, brain CT or nuclear magnetic resonance, and an ophthalmologic examination.

The mean age of onset of the spasms in the cryptogenic group, symptomatic noninfectious group, and the postinfectious group ranged from 5.5 to 6 months, from 2 to 10 months, and from 2.5 to 9 months, respectively. The duration of the spasms before admission to the hospital were 0 to 21 (mean 11) days, 2 to 132 (mean 40) days, and 5 to 49 (mean 26) days in the first group, the second group, and the third group, respectively. The patient where the delay before adrenocorticotropic hormone (ACTH) treatment had been exceptionally long (14 months) was excluded. Before ACTH therapy, the children with cryptogenic etiology did not receive any anticonvulsant medications. One of this group responded to pyridoxine and received no ACTH therapy. Fifteen of the 32 patients with symptomatic etiology had been treated early with nitrazepam (seven patients), nitrazepam plus valproate (one patient), nitrazepam plus phenobarbital (one patient), valproate (three patients), valproate plus fenemal (one patient), valproate plus vigabatrin (one patient), and phenobarbital (one patient) earlier. Due to response to anticonvulsants (one patient), infection (one patient), delayed diagnosis of WS and mental retardation (one patient), and severe mental retardation (two patients), five children from the whole symptomatic group did not receive any ACTH. The age at hospital admission for spasm treatment were 6 to 7 (mean 6.4) and 2 to 20 (mean 7.4) months in the cryptogenic and symptomatic groups. Treatment was started in all children (150 mg of pyridoxine daily for 3 to 4 days). ACTH treatment was based on interim evaluations of the response to treatment at weeks 2 and 4. A good treatment response meant there was total cessation of the spasms and disappearance of the hypsarrhythmia. On the basis of etiology (symptomatic or cryptogenic) and response to therapy, carboxymethylcellulose-prolonged corticotrophin (Acton prolongatum R, Ferring, Denmark) was started and the dose was adjusted for individuals in a stepwise manner. Initially, in every morning Acton prolongatum R 3 IU/kg was intramuscularly given and any other anticonvulsant was gradually withdrawn. Cessation of the spasms and hypsarrhythmia were evaluated at 2-week intervals. At week 2, in patients with response and cryptogenic etiology, Acton prolongatum R was gradually withdrawn, in patients with response and symptomatic etiology 3 IU/kg Acton prolongatum R was continued for another 2 weeks and then gradually withdrawn and in patients with no response the dose was doubled for 2 weeks. Similarly, at week 4 the dose was either tapered off (response) or doubled for 2 weeks (no response).

The maximal dose (12 IU kg^{-1} day^{-1}) was always gradually withdrawn after 2 weeks at l-week intervals halving each dose each time. If there was no response, other anticonvulsant drugs were introduced. The dose was withdrawn stepwise at 1-week intervals halving each dose. Thus, the therapy duration varied from 4 to 10 weeks. If there was a relapse, the highest preceding dose was resumed. In 32 cases, CSF samples were drawn before starting ACTH therapy, and in 11 cases CSF samples were drawn again during the maximal therapy. In 6 further cases, samples were drawn only during ACTH therapy. CSF samples were immediately frozen and stored at −70°C. With a sensitive two-site enzyme-linked immunosorbent assay, NGF content of the CSF was determined. Human NGF was used as standard. Twenty-two patients aged 1 month to 2 years with some other neurologic disease served as controls; for them the CSF examination was made to exclude a CNS infection or to reveal other etiology of the disorder.

7.7 MDCK SCATTER ASSAY[7]

From pBS-hHGF with deletion of 15 base pairs human hepatocyte growth factor (HGF) cDNA was excised and then cloned into SwaI cloning site of a cassette cosmid pAxCAwt (TaKaRa, Osaka, Japan), which carried an adenovirus type-5 genome lacking the E3, E1A, and E1B regions, to prevent the virus replication. The 5′-end and 3′-end of the cosmid pAxCAwt contained the CAG (cytomegalovirus-enhancer-chicken β-actin hybrid) promoter and a rabbit globin poly (A) sequence, respectively. With the adenovirus genome lacking the E3 region, the cosmid was then co-transfected to 293 cells. From 293 cells and by two rounds of CsCl centrifugation, a recombinant adenoviral vector encoding HGF (AxCAhHGF) was propagated, isolated, and purified.

COS1 cells were infected for 1 h with AxCAhHGF at a multiplicity of infection (moi) of 100 in serum-free DMEM and incubated at 37°C with serum-free DMEM in 5% CO_2. At day 3 postinfection, the conditioned media (CMs) were harvested for ELISA and Western blot analysis. The ELISA was generally performed. For Western blot analysis, concentrate HGF CM was treated with heparin beads, and CM or rhHGF was in reduced condition on 4% to 20% gradient SDS/polyacrylamide gels electrophoresed and transferred to PVDF membrane. The blotted membrane was blocked with 3% skim milk, incubated overnight with rabbit anti-HGF (1:500) and incubated with goat anti-rabbit IgG-HRP conjugate (1:1000, DAKO, Glostrup, Denmark). Reactions were visualized by enhanced chemiluminescence detection using an ECL Western blotting detection kit (Amersham, Piscataway, NJ).

MDCK cells were cultured in DMEM with 10% FBS, trypsinized, seeded on 24-well plate (5000 cells/well) in the presence or absence of AxCAhHGF-infected COS1

CMs or rhHGF in DMEM with 5% FBS, and incubated for 24 h at 37°C. Under a phase-contrast microscope, cell scattering was viewed.

7.8 FACIAL NERVE AND SPINAL ROOT AVULSIONS[7]

Fischer 344 male rats (12–14 weeks old, 200–250 g) were anesthetized with intraperitoneal injection of sodium pentobarbital (40 mg/kg). Under a dissecting microscope, right facial nerve of the rats was exposed at exit from the stylomastoid foramen, using microhemostat forceps by gentle traction the proximal facial nerve was avulsed and removed from the distal facial nerve, microsyringe was immediately inserted into the stylomastoid foramen, and through the facial canal 20 μL AxCAhHGF (1×10^8 pfu), AxCALacZ (1×10^8 pfu), or PBS was injected. Rats' wounds were covered with a small piece of gelatin sponge, sutured closed, and the rats were sacrificed at weeks 1 and 4 postoperation.

Under a dissecting microscope, the right seventh cervical segment (C7) nerve was exposed by separating the surrounding cervical muscles and connective tissue until the vertebral foramen was seen. Using microhemostat forceps, C7 ventral, dorsal roots, and dorsal root ganglia (DRG) were avulsed and removed from the peripheral nerve. Immediately after avulsion, a small piece of Gelfoam was presoaked with 10 μL AxCAhHGF (1×10^8 pfu), AxCALacZ (1×10^8 pfu), or PBS was placed in contact with the lesioned vertebral foramen. The wounds were sutured closed, and rats were sacrificed at week 4 postoperation.

7.9 TRAb ASSAYS[8–11]

Through 0.2-μm membrane, serum pools were filtered, and total IgG was extracted on solid phase–bound streptococcal protein G. Briefly, a 5 mL protein G agarose column was equilibrated with phosphate buffer (200 mM NaH_2PO_4, 50 mM NaCl, pH 7.2), on which 20 mL of 1 : 1 serum/buffer was applied twice, washed with PBS, and with freshly prepared 20 mM citrate buffer containing 50 mM NaCl (pH 2.1) to perform acidic elution.

The total IgG preparation was incubated with wild-type K562 cells to remove IgG by binding to K562 cells. Cells were produced in a fermenter of 30,000 mL, harvested by centrifugation at 4°C and 500 × *g* for 10 min, washed with PBS, resuspended with the IgG preparation at a concentration of 10% (v/v) in buffer (20 mM HEPES, 50 mM NaCl, 10% glycerin, 1% BSA, pH 7.5) with 5×10^8 cells per 1 IU TRAb, incubated overnight at 4°C and under slow agitation, centrifuged for 10 min at 4°C and 2000 × *g*, and the IgG containing supernatant was removed by aspiration to reduce contamination by cell debris. In identical conditions, the same procedure was used for K562 cells expressing the human TSH receptor. With 5×10^8 cells per 1 IU Graves' IgG an optimal purification can be obtained. Cells were incubated overnight at 4°C under slow agitation, washed twice with PBS, three times with HEPES buffer (20 mM HEPES, 500 mM NaCl, pH 7.5) and once with water, and the supernatant was removed by aspiration. By centrifugation with elution buffer (25 mM citric acid, 100 mM NaCl, pH 2.1), TRAb was eluted three times, fractions were combined, neutralized with 200 mM Tris-HCl (pH 7.5), centrifuged at 100,000 × *g* to remove any cellular debris, 1% (w/v) BSA was added to prevent adsorption of TRAb to surfaces, and the preparation was stored at −80°C.

After overnight coupling, 10 mg of monoclonal antibody BA8 specific for the TSH receptor was coupled to 2.5 mL Carbolink Gel (Pierce Biotechnology, USA) in 70% to 90% yield, and the gel was equilibrated with PBS. TSH-R extract was obtained from K562 cells, and 200 mL membrane extract, which represented the detergent preparation of around 5×10^9 cells, was at 4°C under mild agitation with 2.5 mL gel in batch incubated for 5 h. After 10 min careful centrifugation at 500 × *g*, the supernatant was decanted, gel material was resuspended and centrifuged with washing buffer (50 mM HEPES, 0.5% BSA, 0.5% Triton, 50 mM NaCl, pH 7.4) three times. After the last washing step, the gel material was resuspended in TRAb enriched buffer obtained in the first purification step at a ratio of 1 mL gel to 15 IU TRAb, at 4°C under mild agitation incubated over night to allow TRAb specifically binding to the solid phase bound hTSHR, at 500 × *g* centrifuged for 10 min, the supernatant was decanted, the gel material was washed three times with 50 mL washing buffer, eluted with 15 mL freshly prepared elution buffer (50 mM citric acid, 1% BSA, 150 mM NaCl, pH 2.1) to perform acidic elution, the fraction was immediately neutralized with 15 mL PBS containing 1% BSA and additional 200 mM phosphate buffer (pH 10) to a final pH of 6.8. Prior to freezing, the buffer was centrifuged at 100,000 × *g* to remove precipitated TSHR and other cellular debris.

According to the manufacturer's instructions. TBII activity was measured with the DYNOtest TRAK human (BRAHMS AG, Hennigsdorf, Germany). This assay was calibrated 1 : 1 to the WHO standard 90/672 and data was expressed as international units (IU/L), with 1 IU/L, 8 IU/L, and 40 IU/L (the highest standard) representing about 10%, 50%, and >85% inhibition of TSH binding, respectively, while the samples with TRAb > 40 IU/L were diluted to obtain quantitative data. TSAb detection was performed in low-salt conditions using human TSHR transfected Chinese hamster ovary (CHO) cells, and cAMP was measured in a commercial radioimmunoassay (RIA).

Thyroid stimulation index (SI) was based on calculated SI (%) = 100 × (cAMP patient/cAMP euthyroid control). Bovine TSH (1 mU/mL) was added either with euthyroid control or with test serum for TBAb detection. Inhibition index (InI) was based on calculated InI (%) = 100 × [1 − (cpm patient/cpm euthyroid control)] .

7.10 HUMAN THYROTROPIN RECEPTOR (hTSHR) ASSAY[12–15]

From the plates with PBS containing EDTA and EGTA (5 mM each), CHO cells expressing hTSHR or control CHO cells were detached, transferred into Falcon 2052 tubes (200,000 cells per tube), centrifuged for 3 min at 4°C and 500 × *g*, the supernatant was removed by inversion, at room temperature with 100 μL PBS-BSA (0.1%) containing 5 μL serum incubated for 30 min, washed with 4 mL PBS-BSA

(0.1%), centrifuged for 3 min at 4°C and 500 × *g*, incubated for 30 min on ice in the dark with fluorescein-conjugated chain-specific goat anti-mouse IgG in the same buffer; propidium iodide (10 μg/mL) was used for detecting damaged cells, which were excluded from the assays, cells were washed once again and resuspended in 250 μL PBS-BSA (0.1%). Using a FACScan flow cytofluorometer (Becton-Dickinson, Erembodegem, Belgium), the fluorescence of 5000 cells per tube was assayed.

On a CHO cell line expressing high levels of the human thyrotropin receptor JP19, thyrotropin binding inhibition activity (TBII) was measured. At room temperature in 0.1 mL modified Hanks' buffer without NaCl (isotonicity maintained with 280 mM sucrose) supplemented with 2.5% low fat milk, ^{125}I-labeled TSH (30000 cpm) and 10 μL serum 5×10^4 cells/well in 96-well plates were incubated for 4 h, rapidly rinsed with the same ice-cold buffer, solubilized with 0.2 mL 1 N NaOH, and the radioactivity was measured in a gamma counter. All experiments were performed in duplicate, and results were expressed as cpm bound.

Using a CHO cell line expressing moderate levels of the human thyrotropin receptor JP26, thyroid stimulating (TSAb) activity was measured. In 5 mM KCl, 0.25 mM KH_2PO_4, 0.5 mM $MgSO_4$, 0.4 mM Na_2HPO_4, 1 mM $CaCl_2$, 0.1% glucose, 2 mM IBMX, 20 mM HEPES, and 0.3% BSA containing 20 μL serum (200 μL total volume per well), 5×10^4 cells/well in 96-well plates were at 37°C incubated for 4 h, and cAMP released into the medium was measured by RIA. All assays were performed in duplicate, and results were expressed as pmol cAMP/mL.

Hybridoma was generated with spleen cells of mouse 154D, 15 days later the supernatants from 1000 clones were collected, 10 μL supernatant was tested for antibodies against hTSHR by use of flow immunocytometry (FACS) on JP19 cells, 50 μL supernatant was tested for binding competition of ^{125}I-labeled TSH by use of DYNOtest TRAK human coated tubes, while 10 μL supernatant was stimulated cAMP production by use of JP26 CHO cells. By limiting dilution, clones scoring positive in one of above tests were cloned and expanded for further characterization. With a mouse monoclonal antibody isotyping kit, the Ig class of mAbs was determined. By Sepharose-protein A affinity chromatography, mAb IRI-SAb1 scoring positive in above tests was purified. Using a mouse IgG capture ELISA, the concentration of IgG was determined. Various concentrations' mAb IRI-SAb1 was tested for its ability to compete with ^{125}I-labeled TSH and using DYNOtest TRAK human coated tubes for binding to immobilized receptor. In normal isotonic medium or low-salt sucrose medium, its ability to stimulate cAMP production using JP26 CHO cells was also evaluated. In 24 wells, 3×10^5 cells were seeded 24 h before the assay, culture medium was discarded and replaced by Krebs-Ringer-HEPES buffer (KRH) or NaCl-free KRH supplemented with 220 mM sucrose for 30 min, cells were incubated in 200 μL fresh medium supplemented with 25 μM phosphodiesterase inhibitor Rolipram and various concentrations of purified IRI-SAb1 for 1 h, the medium was discarded and replaced with 0.1 M HCl, cell extracts were dried in a vacuum concentrator, resuspended in water and appropriately diluted for cyclic AMP measurements by RIA. All samples were assayed in duplicate and results were expressed as pmol cAMP/mL.

In Western blotting, purified ectodomain obtained from CHO cells expressing the hTSH receptor with a GPI anchor (ECD-10His-GPI) was used. To 5 ng of receptor protein, 10% laemmli sample buffer containing SDS and 1 M β-mercaptoethanol as a reducing agent was added and at 40°C denatured for 1 h. On 10% acrylamide gel, the samples were run and with monoclonal antibodies BA8, 28, or IRI-SAb1 (5 μg/mL) probed. With an anti-mouse IgG horseradish peroxidase conjugate and the ECL Plus Western blotting detection system, the proteins were visualized.

7.11 SOLUBLE ICAM-1, TSAb, AND TBIAb ACTIVITY ASSAYS[16]

The study involved 45 Graves' disease (GD) patients diagnosed from typical clinical signs, such as goiter, occasional exophthalmos (eight patients), and increased free thyroid hormone concentrations (free triiodothyronine >8.9 pM and free thyroxin >23.4 pM), while patients showing toxic nodules on computed tomography scan were excluded. All patients were treated for 36 months with 60 mg/day of 1-methyl-2-thio-3-carbethoxyimidazole. Thirty sex- and age-matched healthy blood donors served as controls. Before drug therapy, during the treatment course (at months 3, 6, and 18), and at the time of relapse or 3 years after the end of treatment for the patients in remission, thyroid antibodies and sICAM were assayed. Blood samples were collected into anticoagulant-free tubes, at 4°C and 1000 × *g* centrifuged for 10 min, and sera were decanted for storage at −20°C until assayed.

With a commercially available slCAM-1 ELISA kit, serum concentrations for slCAM-1 were determined, for which serum samples were diluted 1:100, applied to prepared polystyrene microwells precoated with murine monoclonal antibody against human slCAM-1, a horseradish peroxidase–conjugated anti-mouse monoclonal antibody that binds to the slCAM-1 captured by the primary antibody was added, at room temperature on a rotator incubated for 2 h, extensively washed, the reaction product was developed in tetramethylbenzidine substrate solution for 10 min, with 3.6 M sulfuric acid the enzyme reaction was terminated, and samples were determined by comparing the mean absorbance of duplicate samples with the standard curve for each assay concentrations for slCAM-1 in serum.

TSAb levels were assayed in human thyrocyte cultures according to a general method. Normal thyroid cells (1×10^6 cells) from patients free of thyroid disease (e.g., during excision of a cold nodule) were obtained after enzyme dispersion with Dispase II (5 mg/mL, Boehringer-Mannheim, Germany), and cultured in 24-multiwell plates with Ham's solution F12 containing 10% fetal calf serum, 100 IU/mL penicillin, 100 μg/mL streptomycin, and 5 mg/mL 3-isobutyl-l-methylxanthine. This incubation took 48 h in 95% air-5% CO_2 watersaturated atmosphere. After deproteinization by perchloric acid and supernatant neutralization with K_2CO_3, total cAMP both in cells and in incubation medium was measured by a radio-competition method with a kit. All TSAb tests were performed in triplicate, and its activity was expressed as a percentage of thyrocyte stimulation during 48 h culture compared with the stimulation observed for pooled equal volumes of the 30 control sera.

With a Henning radioreceptor kit (Rueil-Malmaison, France), TBIAb activity was assayed according to the manufacturer's instructions, for which 50 μL of standards or patient's serum was mixed with 50 μL of solution of porcine thyrotropin receptors in buffer (pH 7.4), at room temperature incubated for 15 min, 100 μL of ^{125}I-labeled bovine thyrotropin was added, at room temperature incubated for 2 h, at 4°C 2000 μL of polyethylene glycol was added, at 2000 × *g* centrifuged for 20 min, the supernatant was discarded, and the radioactivity of the sediment was counted.

7.12 MICROARRAY IMMUNOASSAY FOR hTSHr PRODUCTION[17]

This assay is summarized in Fig. 7.1. P-SAHz and biotin saturated SAHz were used as a randomly distributed specificity control comprising 10% of all spots and diluted to 0.15 mg/mL in 0.1 M MES buffer (pH 4.7) with adding glycerol (20%, v/v). Protein A was used as incubation control (IC) and streptavidin with a Dy-633 fluorescence label (Dyomics GmbH, Jena, Germany) was used as position control (PC), both at 0.15 mg/mL and as solution in 50 mM $NaHCO_3$ (pH 9.6) with adding glycerol (20%, v/v). At dew point temperature, solutions were with the sciFLEXARRAYER

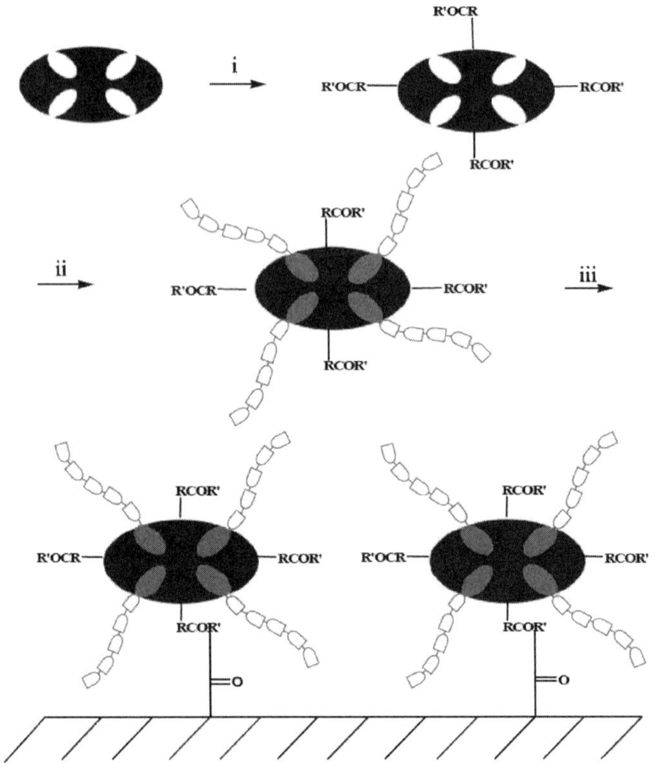

Figure 7.1 Schematic outline of the method.

piezoelectric dispenser printed onto glass slides with aldehyde surface coating. Printed slides were at 4°C in moist chambers incubated for 12 h. At room temperature, by washing the slides in 1 M TBS with adding Nonidet P40 (0.5%, v/v, Erie Scientific Company, Portsmouth, NH, USA) and nonfat dry milk (10%, w/v) for 2 h, surface passivation was achieved. Slides were rinsed in dH_2O, dried with nitrogen, and immediately used for antibody analysis.

Monoclonal anti-hTSHR-antibodies were used at 1–5 μg/mL in 12 mM PBS (pH 7.4) with 150 mM NaCl, Nonidet P40 (0.5%, v/v), and nonfat dry milk (3%, w/v). Arrays were overlaid with LifterSlips (Erie Scientific Company, Portsmouth, NH, USA), samples were injected into the reaction chambers, at room temperature or 4°C incubated for 2 h, the slides were thoroughly washed in 12 mM PBS (pH 7.4) with 150 mM NaCl, Nonidet P40 (0.5%, v/v) on a horizontal shaker at 250/min for 3 × 5 min, at room temperature the arrays were incubated for 1 h with a Cy5-labeled goat antimouse IgG secondary antibody (Jackson Immuno Research, Cambridgeshire, UK) at 10 μg/mL as solution in 12 mM PBS (pH 7.4) with 150 mM NaCl, Nonidet P40 (0.5%, v/v) and nonfat dry milk (3% w/v), and slides were washed, rinsed with dH_2O, and dried in nitrogen. The GenePix Professional 4200A microarray scanner (Molecular Devices Corporation, Sunnyvale, CA, USA) was used for scanned peptide microarrays and GenePix Pro 6.0 software was used for fluorescence image data acquisition.

7.13 AFFINITY ASSAY FOR [^{35}S]GTPγS BINDING TO GAS/OLF[18]

Binding affinity was determined by competition binding experiments from striatal at 1/300e (w/v) and cortical rat D1 receptor membranes at 1/150e (w/v) in presence of [^{3}H]-SCH23390 (0.5 nM) in buffer A containing 50 mM HEPES (pH 7.5), 100 mM NaCl, 5 mM KCl, 4 mM $CaCl_2$, 1 mM $MgCl_2$, and mesulergine (10 mM) to block 5-HT_2 receptors, for which using a Polytron, rat striata and cortices were homogenized in the assay buffer, at 37°C in the presence of 1 mM GTP incubated for 45 min to favor the dissociation of endogenous dopamine from receptors, washed three times by successive centrifugations at 4°C and 40000 × g for 20 min, in the assay buffer homogenized, the resulting pellets were resuspended in the same buffer at 1/300e (w/v), at 37°C in 96-well plates incubated for another 45 min, and nonspecific binding was defined by butaclamol (10 mM). By rapid filtration through Unifilter-96 GF/B filters pretreated with polyethyleneimine (0.3%) using a Filtermate harvester (PerkinElmer Life Science, Boston, MA), experiments were terminated. Using a Top-Count microplate scintillation counter (PerkinElmer Life Science, Boston, MA), by liquid scintillation counting radioactivity retained on the filters was determined. With nonlinear regression, isotherms were analyzed and yielded IC_{50} values, from which according to the Chenge-Prusoff equation inhibition constants (Ki values) were derived. In presence of 10 mM mesulergine, saturation binding experiments at rat D1 receptors were also performed to block the binding of [^{3}H]-SCH23390 to 5-HT_2 receptors, for which before rapid filtration through Unifilter-96 GF/B filters pretreated with polyethyleneimine (0.3%), striatal and cortical membranes were at 37°C in buffer A containing 12 concentrations of [^{3}H]-SCH23390 (0.1–12 nM, 85 Ci/mmol) incubated for 45 min, and radioactivity was counted.

Purified 25 ng of recombinant rat Gao, Gai1, Gai2, Gai3, Gaq, and Gas (short splice variant) or 10 μg of striatal or cortical membranes were loaded onto 10% polyacrylamide gel and transferred to nitrocellulose. Using polyclonal anti-Gas/olf (C18) and anti-Gaq/11 (C19) or anti-Gao (SA-280) and anti-Gai1/3 (SA-281) at 0.4 μg/mL (1/500), immunoblotting of Ga subunits was performed, and with horseradish peroxidase as secondary antibody (1/5000) chemiluminescence detection was enhanced.

By measuring the [^{35}S]GTPγS binding (1332 Ci/mmol) coupled to an antibody capture, the activation of native rat D1 receptor-linked specific G-proteins by agonists was assessed on rat striatal and cortical membranes. Rat striata and cortices were at 22°C in a buffer A′ containing 50 mM HEPES (pH 7.5), 100 mM NaCl, 0.2 mM DTT, and 0.2 mM EGTA homogenized, and pellets were resuspended in buffer B′ : A′ supplemented with 5 mM $MgCl_2$ and 0.1 mM GDP for Gas/olf, Gaq/11, Gai1/3, or GDP 50 mM for Gao. With agonists and antagonists in the buffer B′ rat D1 receptor membranes (ca. 10 μg per well) from striatum (1/300e w/v) or cortex (1/150e w/v) were preincubated for 30 min, which was started in 96-well optiplates with addition of ca. 0.4 nM [^{35}S]GTPγS in a final volume of 200 μL, at 35°C incubated for 60 min, to each well 20 μL of Nonidet P-40 was added and with gentle agitation incubated for 30 min. Before additional 30 min incubation, to each well 10 μL of rabbit polyclonal antibodies against Gas/olf and Gaq/11-proteins subtypes (ca. 1.74 μg/mL final dilution) or mouse monoclonal antibodies against Gao and Gai1/3 subunits (ca. 0.36 μg/mL final dilution) were added. At a dilution indicated by the manufacturer, 50 μL scintillation proximity assay beads coated with secondary anti-rabbit or anti-mouse antibody were added and the plates incubated for 3 h.

7.14 TISSUE SEGMENT BINDING ASSAY FOR α_{1B}-ADRENOCEPTOR[19]

After pentobarbital anesthesia, male Wistar rats (7–12 weeks old, about 300 g) were exsanguinated, their tail artery, thoracic aorta, and cerebral cortex were carefully isolated, and at 4°C immersed in a modified Krebs-Henseleit solution (120.7 mM NaCl, 5.9 mM KCl, 1.2 mM $MgCl_2$, 2.0 mM $CaCl_2$, 1.2 mM NaH_2PO_4, 25.5 mM $NaHCO_3$, and 11.5 mM D-glucose, pH 7.4) gassed with 95% O_2 and 5% CO_2, under a stereoscopic light microscope cut to approximately 3.5 mm in length for tail artery and thoracic aorta and 1- to 2-mm square and 1-mm thick for the cortex, at 4°C in 1 mL Kreb's incubation buffer (of which the composition was essentially the same as the modified Krebs-Henseleit solution except that the $NaHCO_3$ concentration was reduced to 10.5 mM to adjust the pH to 7.4 in air, with [^{3}H]-prazosin) incubated for 16 h. In saturation binding experiments, the concentrations of [^{3}H]-prazosin were 50–1000 pM and 50–300 pM for the tail artery and cerebral cortex and for the thoracic aorta, respectively; in competition binding experiments, the concentrations of [^{3}H]-prazosin were 200 pM and 100 pM for the tail artery and cerebral cortex and for the thoracic aorta, respectively. After incubation, the pieces were at 4°C in plastic tubes containing 1.5 mL incubation buffer blotted and gently rinsed by vortexing for 1 min, in 0.3 M NaOH solution again blotted, and solubilized to estimate the radioactivity and protein content. By subtracting the radioactivity bound in the presence of 30 μM phentolamine from the total binding radioactivity, the specific binding of [^{3}H]-prazosin was obtained. With a liquid scintillation counter (ULTIMA GOLD; Perkin Elmer Life and Analytical Sciences, Boston, MA, USA) using water-miscible scintillation fluid, radioactivity was measured. Using a Bio-Rad Protein Assay kit and bovine serum albumin, protein concentration in each tissue segment was measured.

7.15 MEMBRANE BINDING ASSAY WITH RAT CEREBRAL CORTEX FOR α_{1B}-ADRENOCEPTOR[19]

The isolated rat cerebral cortices stored at −80°C were with scissors minced and using a Polytron homogenizer (specify setting 8, 5 × 20 s at 4°C) homogenized in 50 volumes (v/w) of Krebs incubation buffer containing a proteinase inhibitor, at 4°C and 1000 × *g* centrifuged for 10 min, the supernatant was through four layers of gauge filtered, at 4°C and 80,000 × *g* centrifuged for 30 min, the pellets were resuspended in incubation buffer without proteinase inhibitors and as a crude membrane fraction directly used for binding assays. Binding was at 4°C in 1 mL of incubation buffer carried out for 4 h. In binding saturation and competition experiments, 50–1000 pM and 200 pM [^{3}H]-prazosin were used, respectively. To terminate the reactions, the mixtures were rapidly filtered using a Brandel cell harvester (Brandel Inc., Gaithersburg, MD, USA) and Whatman GF/C filters (Rickly Hydrological, Columbus, OH, USA) presoaked in 0.3% polyethyleneimine for 15 min, the filters were with 5 mL ice-cold incubation buffer washed three times, dried, and trapped radioactivity was quantified by liquid scintillation counter. Specific [^{3}H]-prazosin binding was defined by the difference between counts without and with 30 μM phentolamine. Protein contents of the total homogenates obtained before centrifugation and of the crude membrane fractions were tested in the same way as in the tissue segment binding experiments.

Human embryonic kidney 293 T (HEK 293T) cells were maintained as monolayer cultures at 37°C in DMEM supplemented with 10% FBS, 2 mM L-Gln, minimal essential medium with nonessential amino acids, and 100 IU/mL penicillin/streptomycin in a humidified atmosphere of 5% CO_2, 95% O_2, seeded at 2×10^6 cells/100-mm dish, and after 48 h transfected using FuGENE 6 transfection reagent. The pIRES-puro2-α_{1a}, pCR3-α_{1b} and pCR3-α_{1d} plasmids were used for transient expression of human α_{1A}-, α_{1B}-, and α_{1D}-adrenoceptors, respectively.

7.16 WHOLE CELL BINDING ASSAY FOR α_{1B}-ADRENOCEPTOR[19]

After 48 h transfection, the cells were washed twice and harvested with gentle pipetting without using trypsin. With [^{3}H]-prazosin in 1 mL Krebs-HEPES buffer (140 mM NaCl, 5.4 mM KCl, 1.2 mM $MgCl_2$, 2.0 mM $CaCl_2$, 0.3 mM NaH_2 PO_4,

11.1 mM glucose, 5.4 mM HEPES, and 4.6 mM Na-HEPES, pH 7.4) in a final volume of 1 mL at 4°C for 4 h, whole cell binding assays were performed in duplicate. Nonspecific binding was defined as the amount of radioligand bound in the presence of 30 μM phentolamine. In saturation binding experiments, 30–1500 pM [^{3}H]-prazosin was used. In the presence of 300 pM [^{3}H]-prazosin with adding and increasing concentrations of the unlabeled drugs, binding competition experiments were done. By rapid filtration with a cell harvester onto Whatman GF/C glass filters pre-soaked in 0.3% polyethyleneimine for 30 min, reactions were terminated, the filters were washed three times, dried, and the filter-bound radioactivity was measured by liquid scintillation counting.

7.17 FUNCTIONAL ASSAY FOR $\alpha_{1A,B}$-ADRENOCEPTOR[19]

Rat tail arteries and thoracic aortas were cut into rings of approximately 2 mm in length, at 37°C suspended in an organ bath containing 5 mL of a modified Krebs-Henseleit solution gassed continuously with 95% O_2 and 5% CO_2. To eliminate the possible involvement of endothelium-derived nitric oxide, 100 μM N^G-nitro-L-arginine methylester hydrochloride was present throughout the experiments. A resting tension of 0.5–1.0 g was applied, and changes in isometric force contractile responses were recorded on a transducer. By adding agonists phenylephrine or methoxamine in a cumulative fashion, concentration-response curves were drawn, and agonist trials were repeated three times for each ring preparation at 90-min intervals with the second concentration-response curve as a control. In tail arteries, in the presence of silodosin (3 nM), α_{1A}- and α_{1B}-adrenoceptor mediated response were produced by methoxamine and phenylephrine, respectively. To rule out the possible involvement of α_{1D}-adrenoceptor, BMY 7378 (30 nM) was also present throughout the experiments. In thoracic aortas, the contractile response was produced with phenylephrine. Antagonists were used 45 min before and during the third agonist trial.

7.18 RAT NEUROPROTECTIVE ASSAY[20]

After induction of anesthesia, male SD rats (250–325 g) were maintained with 2% halothane and nitrous oxide:oxygen (60:40). A midline neck incision operation was performed to expose the left carotid artery, and the external carotid and pterygopalatine arteries were ligated with 5-0 silk. In the arterial wall an incision was made, and from the bifurcation of external and internal carotid arteries a 4-0 heat blunted nylon suture was advanced 18 mm; this distance reliably produced blockage of the origin of the middle cerebral artery (MCA). To generate a range of ischemia durations for each experimental group, the duration of occlusion for each individual rat was varied. Via a thermister probe linked to a controller, right temporalis muscle temperature was monitored, and via a heat lamp brain temperature was maintained to be normothermic.

To assay neuroprotective efficacy, 5 min after initiating occlusion the test compounds were administered. Using the rodent examination scale, behavior was rated normal/abnormal at 48 and 72 h. Any one (or more) of obtundation/reduced exploration, forepaw retraction on tail lifting, asymmetric forepaw grasp, forced circling, or death was used as abnormal criteria. The fraction of abnormal rats was calculated for each occlusion duration, including after short durations none or few rats were abnormal, after long durations all the rats were abnormal, and after intermediate durations rats gave intermediate results. By iteration the curve was fit, the best fit was given, and for each treatment group the ED_{50} and its standard deviation was calculated. Neuroprotection was demonstrated if the compound prolonged the ED_{50} compared with that seen in controls. The potency ratio for a treatment was the ED_{50} for that compound divided by the ED_{50} for the appropriate control group.

7.19 GABA-BENZODIAZEPINE RECEPTOR ASSAY[21,22]

After decapitation, cerebral cortices of rats were at 0°C to 4°C immediately removed, homogenized in 20 volumes of 50 mM Tris-citrate buffer (pH 7.1) by ultra-turraxing, at 27,000 × *g* centrifuged for 15 min, pellets were washed three times with 20 volumes of Tris-citrate buffer by centrifugation at 27,000 × *g* for 15 min, and resuspended in 20 volumes of Tris-citrate buffer, at 37°C incubated for 30 min to remove endogenous GABA, centrifuged at 27,000 × *g* for 10 min, final pellets were resuspended to yield 2 mg of fresh brain tissue/mL buffer, and as membrane preparation frozen in aliquots at −20°C until used. The membrane preparation was thawed, resuspended in Tris-citrate buffer, at 0°C to 4°C and 27,000 × *g* centrifuged for 10 min, the pellet was resuspended in Tris-citrate buffer at 500 mL of buffer per gram of original tissue and used for binding assays. At 0°C to 4°C, 500 μL of membrane suspension was mixed with 25 μL of test solution and 25 μL of ^{3}H-Ro 15-1788 (flumazenil, 0.5 nM in assay), and the mixture was incubated for 40 min. Using 1 μM diazepam, nonspecific binding was determined. After incubation, to each sample 5 mL ice-cold Tris-citrate buffer was added, under suction filtered through Advantec GC-50 glass fiber filters, and immediately washed with another 5 mL buffer. Radioactivity on the filters was determined by scintillation counting. Specific binding was defined as total binding minus nonspecific binding.

7.20 FORCED SWIMMING AND TAIL SUSPENSION ASSAYS[23]

In a glass cylinder (20 × 14 cm) containing fresh water up to a height of 10 cm at 25 ± 1°C, 30 min after orally administering vehicle or test compounds mice were forced to swim individually for 6 min. During the final 4 min of the test, the duration of immobility was recorded.

In a box (25 × 25 × 30 cm) with the head 5 cm from the bottom, 30 min after orally administering vehicle or test compounds, a mouse was individually suspended by its tail using a clamp (2 cm from the end) for 6 min. Testing was performed in a darkened room with minimal background noise. During the 6-min test period, the duration of immobility was recorded.

7.21 IGF-I KINASE RECEPTOR ACTIVATION (KIRA) ASSAY[24]

Adherent MCF-7 cells deriving from a human breast adenocarcinoma and expressing endogenous IGF-I receptor were seeded in each well of a flat-bottomed 96-well culture plate, cultured overnight, supernatants were decanted, the plates were lightly blotted on paper towels, and medium containing either experimental samples or the recombinant IGF-I standards were added, the cells were exposed to ligand for 15 min, and the plates were decanted and blotted. To each well by adding lysis buffer containing Triton X-100, sodium orthovanadate, and a cocktail of protease inhibitors, crude lysates were generated, transferred to an ELISA plate that had been coated overnight with 5.0 μg/mL 3B7 antibody (IGF-I receptor-specific antibody), and blocked with 0.5% BSA. Incubation of the crude lysates in the ELISA plate inducing effective affinity directly in the ELISA well purified the IGF-IR. The plate was extensively washed to remove unbound material, and with biotinylated anti-phosphotyrosine monoclonal antibody the degree of receptor tyrosine phosphorylation was quantified by HRP-conjugated dextran-streptavidin. With the development of a tetramethyl benzidine (TMB) substrate, the degree of anti-phosphotyrosine binding was visualized. With a reference wavelength of 650 nm, the absorbance was read at 450 nm.

7.22 gD.trkA KINASE RECEPTOR ACTIVATION (KIRA) ASSAY[24]

Full-length recombinant human receptor was stably transfected into CHO cells to allow cloning a polypeptide flag onto the N-terminus or C-terminus of the receptor. A 26-amino-acid polypeptide derived from HSV glycoprotein D (gD) and, as a capture reagent in the ELISA phase of the kinase receptor activation (KIRA), a monoclonal antibody specific for the gD flag were used. Based on this format, as the capture antibody, 3C8 mAb was used for all KIRA assays. The receptor for the neurotrophin, NGF, was trkA. The gD.trkA KIRA assay was essentially identical to the IGF-I KIRA. The only differences were (a) CHO cells, (b) recombinant human trkA receptor, (c) an N-terminus gD-flag, and (d) mAb 3C8.

7.23 gD.trk KIRA-ELISA[25,26]

The procedure is represented as flow diagram of Fig. 7.2. In a flat-bottom 96-well culture plate, gD.trk-transfected CHO cells were seeded (5×10^4/well) in 100 μL medium, at 37°C in 5% CO_2 cultured for 16 h, the well supernatants were decanted, the plates were blotted on a paper towel, to each well 50 μL experimental samples or the recombinant purified rhNGF, rhNT4/5, rhBDNF, or rhNT3 standards diluted in stimulation medium were added, the cells were stimulated at 37°C for 20 min,

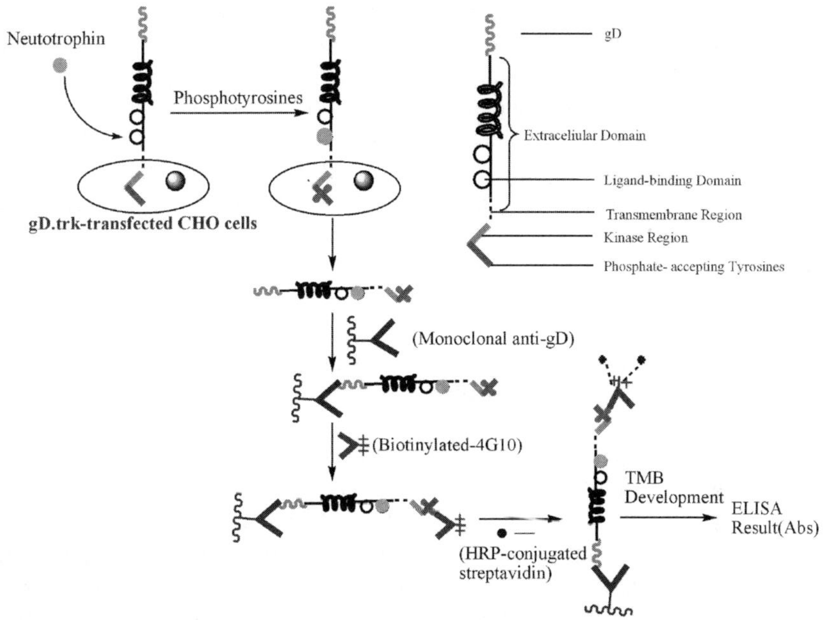

Figure 7.2 Flow diagram of gD.trk KIRA-ELISA.

the well supernatants were decanted, the plates were once again blotted on a paper towel, to each well 100 μL lysis buffer consisting of 150 mM NaCl containing 50 mM HEPES, 0.5% Triton X-100, 0.01% thimerosal, 30 U/mL aprotinin, 1 mM 4-(2-amithenoethyl)benzenesulfonyl fluoride hydrochloride, 50 nM leupeptin, and 2 mM sodium orthovanadate was added to lyse the cells and solubilize the receptors, and at room temperature on a plate shaker the plate was gently agitated for 60 min.

While the cells were being solubilized, an ELISA microtiter plate was at 4°C with 3C8 monoclonal anti-gD (0.5 mg/mL in 50 mM carbonate buffer, pH 9.6, 100 mL/well) coated overnight, decanted, blotted on a paper towel, at room temperature with gentle agitation using 150 μL/well block buffer (PBS containing 0.5% BSA and 0.01% thimerosal) blocked for 60 min, and with wash buffer (PBS containing 0.05% Tween-20 and 0.01% thimerosal) using an automated plate washer washed six times.

From the cell culture microtiter well, the lysates (85 μL/well) containing solubilized gD.trk were transferred to an anti-gD-coated and blocked ELISA well, at room temperature with gentle agitation incubated for 2 h, washed with wash buffer to remove the unbound receptor, to each well 100 μL biotinylated 4G10 (anti-phosphotyrosine) diluted to 0.1 μg/mL in dilution buffer (PBS containing 0.5% BSA, 0.05% Tween 20, 5 mM EDTA, and 0.01% thimerosal) was added, at room temperature incubated for 2 h, washed, to each well 100 μL of HRP-conjugated streptavidin diluted 1:40,000 in dilution buffer was added, at room temperature with gentle agitation incubated for 30 min, to each well the free avidin conjugate (tetramethyl benzidine, 2-component substrate kit) was added, the reaction was allowed to proceed for 10 min, and the color development was stopped by adding 100 μL/well of 1.0 M H_3PO_4. On a Spectromax plate reader (Molecular Devices, Palo Alto, CA) controlled with a Macintosh Centris 650 (Apple Computers, Cupertino, CA) and SoftMax Pro software (Molecular Devices), the absorbance at 450 nm was read with a reference wavelength of 650 nm. By stimulating gD.trkA-, B-, or C-transfected CHO dp 12 cells with 100, 33, 11, 3.7, 1.2, 0.4, and 0 ng/mL rhNGF, rhBDNF, rhNT4/5, or rhNT3, the standard curve was generated. By interpolation of their absorbance on the standard curve using the SoftMax Pro program, sample concentrations were obtained as ng/mL.

7.24 ASSAYS FOR CANNABINOID RECEPTORS IN RAT CEREBELLA OR MOUSE BRAINS[27]

At 0°C to 4°C, cerebella of male Wistar rats (250–300 g) or brains of mice (18–27 g, OF1-IOPS) were carefully dissected on ice, homogenized in 50 mM Tris-HCl (pH 7.4) with a potter and a Dounce, at 400 × *g* the homogenates were centrifuged for 10 min, the supernatants were collected and at 39,000 × *g* centrifuged for 10 min. The pellets were resuspended in 50 mM Tris-HCl (pH 7.4), homogenized, at 39,000 × *g* centrifuged for additional 10 min, and the pellets were washed twice more in the same conditions. Using Bradford assay and Coomassie blue with bovine serum albumin as a standard, the protein concentration was measured.

At 0°C to 4°C, after peritoneal incision, spleens from rats and mice were carefully dissected on ice, cut in several pieces, placed in a 50 mM Tris-HCl (pH 7.4) containing 3 mM $MgCl_2$, 1 mM EDTA, and 0.5% bovine serum albumin, and the membranes were prepared.

With the cDNA sequences encoding either the human CB1 or CB2 cannabinoid receptors, CHO cells were stably transfected, grown in Nutrient mixture Ham's F12 supplemented with 10% FCS, 2.5 μg/mL amphotericin B (Fungizone), 100 IU/mL penicillin, 100 μg/mL streptomycin, and 400 μg/mL G418, and confluent cells were trypsinized and collected by centrifugation at 100 × *g* for 10 min. At 0°C to 4°C, pelleted cells were lysed in ice-cold 50 mM Tris-HCl (pH 7.4), homogenized, centrifuged at 400 × *g* for 10 min, resuspended in the same buffer, homogenized, centrifuged at 15,000 × *g* for 10 min, and washed twice more in the same conditions. The protein concentration was measured as indicated above.

At 30°C in siliconized plastic tubes, the competitive binding assays were performed in presence of 1 nM appropriate radioligands at ([^{3}H]-SR141716A, [^{3}H]-CP 55940, [^{3}H]-WIN 55212-2) on membranes from rat cerebellum (20 μg protein/tube), rat spleen (80 μg/tube), mouse brain (100 μg protein/tube), mouse spleen (80 μg/tube), or transfected CHO cells (40 μg protein/tube), resuspended in 1 mL of (final volume) binding buffer (50 mM Tris-HCl, 5 mM $MgCl_2 \cdot 6H_2O$, 1 mM disodium EDTA, BSA [0.5%, w/v], pH 7.4).

Competitors were added at varying concentrations, and in the presence of 10 μM HU 210, the nonspecific binding of the radioligands was determined. After 1-h incubation, through the 0.5% PEI pretreated GF/B glass fiber filters on a 48-well cell harvester, the suspension was rapidly filtered, washed twice with 3 mL ice-cold binding buffer without bovine serum albumin, and with a Pharmacia Wallac 1410 β-counter (Lumac, Schaesberg, The Netherlands) by liquid scintillation in 10 mL Aqualuma, radioactivity on filters was measured. In saturation binding assays with 0.5–50 nM [^{3}H]-SR 141716A, 0.05–10 nM [^{3}H]-CP 55940, or 0.9–30 nM [^{3}H]-WIN 55212-2, similar binding conditions were used. All assays were performed in triplicate.

At 30°C, to the plastic tubes containing 20 μg protein resuspended in 1 mL of (final volume) binding buffer (50 mM Tris-HCl, pH 7.4; 5 mM $MgCl_2 \cdot 6H_2O$; 1 mM disodium EDTA; 100 mM NaCl, 0.1%, w/v, BSA) supplemented with 20 μM GDP and 0.01 nM to 100 μM agonists or antagonists, [^{35}S]-GTPγS (0.05 nM final concentration) was added, incubated for 1 h, the reactions were terminated by adding 3 mL ice-cold washing buffer (50 mM Tris-HCl, pH 7.4; 5 mM $MgCl_2 \cdot 6H_2O$; 1 mM disodium EDTA; 100 mM NaCl), and using a 48-well cell harvester the suspension was immediately filtered through GF/B filters and washed twice with ice-cold binding buffer. Radioactivity trapped on the filters was counted. In the presence of 100 μM Gpp(NH)p, the nonspecific binding was measured. All assays were made in triplicate.

REFERENCES AND NOTES

1. B. Caramés, M.J. López-Armada, B. Cillero-Pastor, M. Lires-Dean, C. Vaamonde, F. Galdo, F.J. Blanco. Differential effects of tumor necrosis factor-α and interleukin-1β on cell death in human articular chondrocytes. Osteoarthritis Cartilage 16 (**2008**) 715–722.
2. J.T. Niemanna, J.P. Rosborough, S. Youngquist. Is the tumour necrosis factor-alpha response following resuscitation gender dependent in the swine model? Resuscitation 77 (**2008**) 258–263.
3. T. Astier-Gin, J.P. Portail, F. Lafond, B. Guillemain. Identification of HTLV-I- or HTLV-II-producing cells by cocultivation with BHK-21 cells stably transfected with a LTR-1acZ gene construct. J Virol Methods 51 (**1995**) 19–30.
4. L.L. Pedre, N.P. Fuentes, L.A. González, A. McRae, T.S. Sánchez, L.B. Lescano, R.M. González. Nerve growth factor levels in parkinson disease and experimental parkinsonian rats. Brain Research 952 (**2002**) 122–127.
5. M.C. Hoener, S. Varon. Reversible sedimentation and masking of nerve growth factor NGF antigen by high molecular weight fractions from rat brain. Brain Res 772 (**1997**) 1–8.
6. R.S. Riikonen, S. Söderström, R. Vanhala, T. Ebendal, D.B. Lindholm. West syndrome: Cerebrospinal fluid nerve growth factor and effect of ACTH. Pediatr Neurol 17 (**1997**) 224–229.
7. Y. Hayashi, Y. Kawazoe, T. Sakamoto, M. Ojima, W. Wang, T. Takazawa, D. Miyazawa, W. Ohya, H. Funakoshi, T. Nakamura, K. Watabe. Adenoviral gene transfer of hepatocyte growth factor prevents death of injured adult motoneurons after peripheral nerve avulsion. Brain Res 1111 (**2006**) 187–195.
8. N.G. Morgenthaler, W.B. Minich, M. Willnich, T. Bogusch, J.M. Hollidt, W. Weglöhner, C. Lenzner, A. Bergmann. Affinity purification and diagnostic use of TSH receptor autoantibodies from human serum. Mol Cell Endocrinol 212 (**2003**) 73–79.

9. C. Massart, J. Orgiazzi, D. Maugendre. Clinical validity of a new commercial method for detection of TSH receptor binding antibodies in sera from patients with Graves' disease treated with antithyroid drugs. Clin Chim Acta 304 (**2001**) 39–47.

10. Y. Watanabe, H. Tada, Y. Hidaka, T. Takano, N. Amino. Effect of solubilization of porcine thyrotropin (TSH) receptor on TSH binding and on radio-receptor assay for anti-TSH receptor antibodies. Biochem Biophys Res Commun 248 (**1998**) 110–114.

11. D. Botero, R.S. Brown. Bioassay of thyrotropin receptor antibodies with Chinese hamster ovary cells transfected with recombinant human thyrotropin receptor: Clinical utility in children and adolescents with Graves' disease. J Pediatr 132 (**1998**) 612–618.

12. S. Costagliola, J.D.F. Franssen, M. Bonomi, E. Urizar, M. Willnich, A. Bergmann, G. Vassart. Generation of a mouse monoclonal TSH receptor antibody with stimulating activity. Biochem Biophys Res Commun 299 (**2002**) 891–896.

13. T. Ando, R. Latif, T.F. Davies. Thyrotropin receptor antibodies: New insights into their actions and clinical relevance. Best Practice Res Clin Endocrinol Metab 19 (**2005**) 33–52.

14. C. Massart, J. Gibassier, F. Verite, P. Fergelot, D. Maugendre. Use of Chinese hamster ovary cell lines transfected with cloned human thyrotropin receptor for the measurement of thyroid-stimulating antibodies: Advantages and difficulties. Clin Chim Acta 291 (**2000**) 67–81.

15. M. Ueda, H. Sugawa, S. Ichiyama, T. Mori. Expansion of helper T-cell recognition in mice immunized with a synthetic peptide corresponding to the C-terminal thyrotropin receptor-specific insert. Peptides 20 (**1999**) 1085–1090.

16. C. Massart, E. Sonnet, J. Glbassler, D. Maugendre, B. Genetet. Clinical validity of intercellular adhesion molecule-1 (ICAM-1) and TSH receptor antibodies in sera from patients with Graves' disease. Clin Chim Acta 265 (**1997**) 157–168.

17. H. Andresen, K. Zarse, C. Grötzinger, J.M. Hollidt, E. Ehrentreich-Förster, F.F. Bier, O.J. Kreuzer. Development of peptide microarrays for epitope mapping of antibodies against the human TSH receptor. J Immunol Methods 315 (**2006**) 11–18.

18. C.M. la Cour, S. Vidal, V. Pasteau, D. Cussac, M.J. Millan. Dopamine D1 receptor coupling to Gs/olf and Gqin rat striatum and cortex: A scintillation proximity assay (SPA)/antibody-capture characterization of benzazepine agonists. Neuropharmacology 52 (**2007**) 1003–1014.

19. Z.S. Sathi, A.S.M. Anisuzzaman, S. Morishima, F. Suzuki, T. Tanaka, H. Yoshiki, I. Muramatsu. Different affinities of native α_{1B}-adrenoceptors for ketanserin between intact tissue segments and membrane preparations. Eur J Pharmacol 584 (**2008**) 222–228. *Note*: For membrane binding assays, the cells harvested 48 h after transfection and stored at −80°C were resuspended in ice-cold Krebs-HEPES buffer containing proteinase inhibitor, at 4°C and 1000 × *g* centrifuged for 5 min, resuspended in Krebs-HEPES buffer, sonicated, at 4°C and 1000 × *g* centrifuged for another 5 min, the supernatant was taken and used as a crude membrane preparation. Binding experiments were carried out following the same procedure described for whole cell binding.

20. C. Jackson-Friedman, P.D. Lyden, S. Nunez, A. Jin, R. Zweifler. High dose baclofen is neuroprotective but also causes intracerebral hemorrhage: A quantal bioassay study using the intraluminal suture occlusion method. Exp Neurol 147 (**1997**) 346–352.

21. G.I. Stafford, A.K. Jäger, J. van Staden. Activity of traditional South African sedative and potentially CNS-acting plants in the GABA-benzodiazepine receptor assay. J Ethnopharmacol 100 (**2005**) 210–215.

22. P.D. Lyden, C. Jackson-Friedman, C. Shin, S. Hassid. Synergistic combinatorial stroke therapy: A quantal bioassay of a GABA agonist and a glutamate antagonist. Experimental Neurology 163 (**2000**) 477–489.

23. W. Wang, X. Hu, Z. Zhao, P. Liu, Y. Hu, J. Zhou, D. Zhou, Z. Wang, D. Guo, H. Guo. Antidepressant-like effects of liquiritin and isoliquiritin from glycyrrhiza uralensis in the forced swimming test and tail suspension test in mice. Progr Neuro-Psychopharmacol Biol Psychiatry 32 (**2008**) 1179–1184.

24. M.D. Sadick, A. Intintoli, V. Quarmby, A. McCoy, E. Canova-Davis, V. Ling. Kinase receptor activation (KIRA): A rapid and accurate alternative to end-point bioassays. J Pharm Biomed Anal 19 (**1999**) 883–891.

25. M.D. Sadick, A. Galloway, D. Shelton, V. Hale, S. Weck, V. Anicetti, W. Lee, T. Wong. Analysis of neurotrophin/receptor interactions with a gD-Flag-modified quantitative kinase receptor activation (gD.KIRA) enzyme-linked immunosorbent assay. Exp Cell Res 234 (**1997**) 354–361.

26. M.D. Sadick, M.X. Sliwkowski A. Nuijens, L. Bald, N. Chiang, J.A. Lofgren, W. Lee, T. Wong. Analysis of heregulin-induced ErbB2 phosphorylation with a high-throughput kinase receptor activation enzyme-linked immunosorbant assay. Anal Biochem 235 (**1996**) 207–214.

27. S.J. Govaerts, E. Hermans, D.M. Lambert. Comparison of cannabinoid ligands affinities and efficacies in murine tissues and in transfected cells expressing human recombinant cannabinoid receptors. Eur J Pharm Sci 23 (**2004**) 233–243.

8

METHODS AND APPLICATIONS OF ALZHEIMER'S DISEASE ASSAYS

Shiqi Peng

As a progressive neurodegenerative disorder, Alzheimer's disease (AD) is a cause of dementia in the aged population. The decline in cognition and memory is accompanied by different pathologic signals with progressive deposition of senile plaques made of β-amyloid (Aβ) and other proteins such as glutamine synthetase, α-chain of hemoglobin, macrophage migration inhibitory factor, albumin, β-tubulin, as well as GAPDH and neurofibrillary tangles, the latter characterized by accumulation of an abnormally hyperphosphorylated form of microtubule binding protein tau, leading to neuronal degeneration. In the past decade, all of these important features have been considered as indicators of the onset and progress of AD and have been used to establish assay methods for improving therapy for AD. In this chapter, 20 assays are described: assay for oxidative stress in cerebral cortex of AD mice,[1] reporter assay for primary neuronal cultures,[2] electrophoretic mobility shift assay (EMSA),[3] binding assays using aggregated Aβ peptide in solution,[4] assay for muscarinic receptor 1 in Alzheimer's dementia model,[5] β-secretase activity assay,[6] Aβ fibril binding assay,[7] thin layer chromatography (TLC) and microplate assays for acetylcholinesterase inhibitors,[8] immunocapture assay measuring specific enzyme activity of neprilysin,[9] *in vivo* acetylcholinesterase (AChE) inhibition assay,[10,11] single particle assay for Aβ aggregates,[12] indirect immunofluorescence assay,[13]

Pharmaceutical Bioassays: Methods and Applications. By Shiqi Peng and Ming Zhao

mouse behavioral assays,[14,15] center of pressure (CoP) assay in mice,[16] assay for plasma levels of DJ-1,[17] membrane filter assay for tau aggregation,[18] assays for motor neuron degeneration,[19] HPLC assay for neuroprotective agent in mouse plasma,[20] tissue culture assays for SOD1 mutations,[21] and luciferase-based reporter assay.[22]

8.1 ASSAY FOR OXIDATIVE STRESS IN CEREBRAL CORTEX OF AD MICE[1]

Conditional double knock-out of presenilin-1 (PS1) and presenilin-2 (PS2) (PS cDKO) and age- and gender-matched mice with B6CBA background served as treated and control animals. The cerebral cortex of 2-, 4-, and 7-month-old mice were dissected, homogenized in lysis buffer (6 M urea; 2 M thiourea; 10% glycerol; 50 mM Tris-HCl, pH 7.8–8.2; 2% *n*-octyl glucoside (*n*-OG); 1 mM protease inhibitor) on ice, the cytosolic fractions were centrifuged for 15 min at 15°C and 20,000 × g, the protein concentration was determined using Bradford method with BSA as standard, using the MDA Assay Kit (Nanjing Jiancheng Corp., China), the SOD Assay Kit-WST (Dojindo Inc., Japan), and CAT and GSHpx assay kits (Nanjing Jiancheng Corp., China) according to the manufacturers' instructions, the lipid peroxidation product MDA, SOD activity, and CAT and GSH-px activities were determined, respectively. In the determination of total protein carbonyls, 80 μL of cortex samples were mixed with 80 μL of 2,4-dinitrophenylhydrazine (DNPH, 10 mM) in 2 M HCl, at room temperature incubated for 1 h, proteins were precipitated with an equal volume of 30% trichloroacetic acid (TCA), the pellets were with 1 mL of ethanol-ethyl acetate (1 : 1, v : v) washed three times, and the pellets were at 37°C with 80 μL of 6 M guanidine HCl treated for 30 min. Using a molar absorption coefficient of 22,000/M · cm, the carbonyl content was calculated from the absorbance at 366 nm.

Following the manufacturer's instructions using Oxyblot Kit (Millipore-Chemicon), on 20 μg of protein DNPH was derivatized, resolved in 12% SDS-PAGE, transferred to Hybond-P PVDF membrane (Amersham), at room temperature in TBST containing 5% BSA blocked for 60 min, the membranes were washed with tris buffered saline with tween (TBST), incubated overnight at 4°C with anti-dinitrophenyl (anti-DNP) (1 : 150), the membranes were washed with TBST, incubated at room temperature with horseradish peroxidase (HRP)-conjugated anti-rabbit immune globulin G (IgG) for 45 min, washed again with TBST and according to the manufacturer's instructions with the ECL chemiluminescence kit (Amersham) the proteins were visualized and normalized to GAPDH. For Western blot, each cortical protein sample was mixed with loading buffer, heated at 100°C for 4 min, resolved by SDS-PAGE, transferred to Hybond-P PVDF membrane, and incubated with antibodies against glial fibrillary acid protein (GFAP) (1 : 2000) and GAPDH (1 : 20,000).

To address the time course of oxidative stress with the pathologic progress of AD, oxidative status at both lipid and protein levels in the PS cDKO mice were assessed. By DNPH assay, protein oxidative modification levels were measured as protein carbonyl content. To examine oxidative stress status during AD development, endogenous antioxidant defense including CAT, SOD, and GSH-px in the cerebral cortices of PS cDKO mice was assessed. To find the possible source of the enhanced oxidative stress occurring in the brain in response to the loss-of-function of presenilin (PS), as a marker of inflammation of cortex, the level of cortical GFAP was measured for 2-, 4-, and 7-month-old transgenic and WT mice.

8.2 REPORTER ASSAY FOR PRIMARY NEURONAL CULTURES[2]

From E17 embryos, SD rat neocortex were prepared by mechanical dissociation of pure primary cortical cultures, seeded at a density of 6×10^4 cells/well in poly-lysine-coated 96-well plates, cultured for 5 days, and according to the manufacturer's instructions transfected with the different constructs utilizing a lipofectamine 2000-based protocol and milder conditions (i.e., lower ratio lipofectamine/DNA and short time of transfection). In RNAi knock-down experiments, cells were transfected with Firefly Luciferase (FL), Renilla Luciferase (RL) constructs, and siRNAs targeting luciferase (as controls) or endogenous rat neuronal genes. The functional effects modulating the expression of control genes on neuronal viability in the presence/absence were tested and compared with those obtained from standard morphometric/imaging readouts. Using the reporter or morphometric/imaging readouts, the effects of Aβ25-35 on the viability of primary neuronal cultures from rat embryonic cortex were analyzed, for which neuronal cultures were transfected with the luciferase reporter and an expression construct encoding a fluorescent protein and treated with 25 μM Aβ25-35 for 24 or 48 h.

8.3 ELECTROPHORETIC MOBILITY SHIFT ASSAY (EMSA)[3]

Wild-type CCAAT/enhancer binding protein (C/EBP) sense oligonucleotide 5′-TGCAGATTGCGCAATCTGCA-3′, mutated C/EBP sense oligonucleotide 5′-TGCAGAgaGtagtcTCTGCA-3′ (mutated nucleotides in lowercase letters), and wild-type NF-jB sense oligonucleotide 5′-AGTTGAGGGGACTTTCCCAGGC-3′ (MWG) were used for EMSA. To identify C/EBPβ (sc-150X) and C/EBPδ (sc-151X), for supershift analysis subunit-specific antibodies were incubated with the samples for 30 min prior to adding the radiolabeled probe. To quantify the supershifted samples, the remaining band's intensity was determined and subtracted from the intensity of the band derived from the corresponding prepared nuclear extract not exposed to the antibody.

8.4 BINDING ASSAYS USING AGGREGATED Aβ PEPTIDE IN SOLUTION[4]

A solid Aβ(1-42) was gently dissolved (0.25 mg/mL) in a buffer (pH 7.4) containing 10 mM sodium phosphate and 1 mM EDTA, and at 37°C with gentle and constant shaking incubated for 42 h. Using the standard iododestannylation reaction, 6-^{125}iodo-2-(4′-dimethylamino)phenyl-imidazo[1,2]pyridine (^{125}I-IMPY) with 2200 Ci/mmol specific activity and greater than 95% radiochemical purity was prepared. In 12 × 75 mm borosilicate glass tubes, binding assays were performed. Briefly, the reaction mixture contained 50 μL Aβ(1-42) aggregates (29 nM in the final assay mixture), 50 μL ^{125}I-6-iodo-2-(4′-dimethylamino)phenylimidazo[1,2]pyridine (IMPY, 0.02 nM diluted in 10% EtOH), 50 μL of inhibitor (8 pM-12.5 μM diluted in 10% EtOH), and 850 μL of 10% EtOH, at 37°C incubated for 3 h, and the bound and free radioactivities were separated by vacuum filtration through Whatman GF/B filters using a Brandel M-24R cell harvester and washed 2 × 3 mL with 10% EtOH at room temperature. Filters containing bound I-125 ligand were placed in a gamma counter (Aloka, ARC-380) to measure radioactivity. Nonspecific binding was defined when 400 nM IMPY presented in the same assay tubes, while in assay conditions the specific bonding fraction accounted for less than 15% of the total radioactivity. From displacement curves using GraphPad Prism 4.0 (GraphPad Software, San Diego, CA), IC_{50} was determined, and using the Cheng-Prusoff equation inhibition constant (Ki) was calculated.

8.5 ASSAY FOR MUSCARINIC RECEPTOR 1 IN ALZHEIMER'S DEMENTIA MODEL[5]

Crude membrane pellet obtained from male Wistar rat brain cortex was homogenized in 20 volumes of 50 mM Tris-HCl buffer (pH 7.4) containing 0.32 M sucrose, at 4°C and 1000 × *g* centrifuged for 10 min to remove cellular debris, supernatant was at 4°C and 15,000 × *g* centrifuged for 20 min, pellet was resuspended in 50 mM phosphate assay buffer (pH 7.4) containing 1 mM $MgCl_2$, and the protein concentration was estimated. Using [^{3}H] qunclidinyl benzillate (QNB) (0.2 nM, specific activity 48 Ci/mM, Amersham, Little Chalfont, Buckinghamshire, UK), the affinity of test compounds toward muscarinic 1 receptor was estimated. In 200 μL aliquot of synaptosomal membrane proteins (50 μg), different concentrations of test compounds (0.1–200 μM), [^{3}H]QNB (0.2 nM) and assay buffer was at 37°C incubated for 2 h, by adding ice-cold assay buffer the reaction for displacement assay was stopped, the reaction mixture was rapidly filtered through GF/B filters under vacuum, the filter was transferred to a vial containing 5 mL of scintillation fluid, allowed to equilibrate overnight, radioactivity was measured in a liquid scintillation counter (Tris-Carb 2100TR, Packard, USA) at 65% efficiency, the data were analyzed, and from the Ligand-Drug program IC_{50} and Ki values were obtained.

8.6 β-SECRETASE ACTIVITY ASSAY[6]

Commercially available fluorogenic β-secretase substrate I, MCA-EVKMDAEFK-(DNP)-NH_2, corresponding with positions 668–676 of the wild-type amyloid precursor protein (APP) sequence having its N-terminus modified with a fluorescent

7-methoxycoumarin-4-yl acetyl group (MCA) and C-terminal lysine residue modified with a 2,4-dinitrophenyl group (DNP), was used. By internal quenching from DNP group, the fluorescence of MCA group in the intact peptide was abolished. β-Secretase activity leading to cleavage and the MCA fluorescence could be detected. Using a SpectraMax Gemini XS Spectrofluorometer (Molecular Devices), 96-well black plates were assayed. At the concentrations indicated, platelet membrane protein and MCA-EVKMDAEFK-(DNP)-NH_2 in a final volume of 200 μL in assay buffer (50 mM sodium actetate, pH 4.5) were at 37°C incubated for 30 min allowing MCA-EVKMDAEFK-(DNP)-NH_2 to hydrolyze, and the accompanying fluorescence increase was measured (excitation 325 nm, emission 393 nm) at 60-s intervals. Over the 8- to 25-min linear phase, initial rates were calculated and were either expressed as relative fluorescence units (RFU) per min or converted to [MCA] liberated using a calibration curve on the plate, which was constructed from a range of dilutions of 7-methoxycoumarin-4-acetic acid. The initial rate were expressed as pM MCA detected per min per μg membrane protein. All platelet samples were assayed in triplicate.

8.7 Aβ FIBRIL BINDING ASSAY[7]

Specific binding affinity of test compounds to Aβ fibrils were evaluated with an *in vitro* Aβ fibril competitive binding assay, in which test compounds competed with preformed Aβ42 aggregates for radioligand [^{125}I]TZDM. Kd value of 0.13 nM [^{125}I]TZDM for Aβ42 aggregates was estimated. In inhibition assays, the reaction mixture of 50 μL of Aβ42 aggregates (11.5 nM final concentration), 50 μL of test compounds (10^{-6} to 10^{-12} M in DMSO), 50 μL of [^{125}I]TZDM (in 40% EtOH, 0.05 nM final concentration), and 10% EtOH in a final volume of 1 mL was incubated at room temperature for 3 h, the bound and free radioactivities were separated by a vacuum filtration through Whatman GF/B filters using a Brandel M-24R cell harvester, at room temperature washed with 2 × 3 mL of 10% EtOH, the bound radioligand on filters were counted in a gamma-counter and subjected to non-linear regression analysis using GraphPad Prism software, by which Ki values were calculated. Nonspecific binding was defined by adding 2 μM Th-T for [^{125}I]TZDM binding.

8.8 TLC AND MICROPLATE ASSAYS FOR ACETYLCHOLINESTERASE INHIBITORS[8]

Using *O*-dianisidinebis(diazotized)hydrochloride zinc double salt (Fast Blue B salt) as reagent, the inhibitory activity of test compounds was determined with a TLC assay. In TLC assay, according to their solubility the test compounds were dissolved either in ethyl acetate or in methanol, spotted first at 1 μg and then at decreasing quantities onto the TLC plate. As positive control (±)-huperzine-A (Sigma-Aldrich Chemical, St. Louis, MA, USA) (0.001 μg) was added to each plate. Migration was carried out with hexane/ethyl acetate (1/1, v/v). The plate was sprayed with an enzyme stock solution (6.67 IU/mL), at 37°C in a humid atmosphere preincubated for 20 min, onto the plate the mixture of 10 μL of α-naphthyl acetate solution and 40 mL of Fast Blue B salt solution was sprayed to give a purple coloration after 1–2 min. The minimum quantity of test compound was noted for which an inhibition spot was still visible.

Using Fast Blue B salt as reagent, the inhibitory activity of test compounds was determined with a microplate assay. In the microplate assay, the mixture of 235.7 μL or 188.4 μL of phosphate buffer (0.1 M, pH 7.4), 20 μL of solution of acetylcholinesterase inhibitors (AChE) (0.1 U/mL final concentration in 0.1 M phosphate buffer, pH 7.4), 2.7 μL of test compound in DMSO or 50 μL of test compound in buffer was treated with 1.6 μL of α-naphthyl acetate (final concentration equal to the Km value) in DMSO for 90 s, the reaction mixture was at 25°C incubated for 90 s, the reaction was stopped with 20 μL of sodium dodecyl sulfate (SDS, 5%) in water, and the color was developed with 20 μL of Fast Blue B salt solution (0.17 mM final concentration in water). After formation of the purple-colored diazonium dye, which turned to stable blue in presence of SDS, by determining the absorbance in end point lecture at 600 nm, enzyme activity and inhibition were quantified.

The microplate assay may also follow another method. Briefly, in plate wells the mixture of 227.3 μL or 210 μL of Ellman's reagent (5,5′-dithiobis-2-nitrobenzoic acid) in 0.1 M phosphate buffer (pH 7.4) at 0.15 mM final concentration, 20 μL of acetylcholinesterase in 0.1 M phosphate buffer (pH 7.4) at 0.037 U/mL final concentration and 2.7 μL of test compound in DMSO or 20 μL of test compound in buffer was treated with 20 μL of acetylthiocholine iodide (ATCI) in demineralized water (final concentration equal to the Km value) to initiate the enzymatic reaction at 270 μL of final assay volume; the plate was shaken for 2 s and using a microplate spectrophotometer the absorbance increase was at 25°C and 412 nm monitored for 180 s. In controls, test compound solutions were instead by the corresponding volume of DMSO or buffer. The kinetic parameters Km and Vmax and inhibitory potencies were measured in presence of 0 (control), 0.6%, 1.6%, and 2% DMSO.

8.9 IMMUNOCAPTURE ASSAY MEASURING SPECIFIC ENZYME ACTIVITY OF NEPRILYSIN[9]

Using a Precellys 24 automated tissue homogenizer with 2.3-mm silica beads, unfixed frozen cortex (200 mg) from the left mid-frontal region (Brodmann area 6) in either 1% SDS buffer containing 5 M NaCl, 1 M Tris (pH 7.6), protease inhibitors aprotinin (1 μg/mL), and PMSF (10 μM, Sigma-Aldrich, Dorset, UK) or 0.5% Triton X-100 made up in 20 mM Tris (pH 7.4), 10% sucrose (w/v), protease inhibitors aprotinin (1 μg/mL, Sigma-Aldrich, Dorset, UK), and PMSF (10 μM) was homogenized for 30 s, at 4°C and 13,000 rpm centrifuged for 15 min, and the supernatants were aliquoted and stored at −80°C until used. Cerebrospinal fluid (CSF) was collected in polypropylene tubes, centrifuged, and aliquoted and immediately frozen at −80°C until used.

1. *Fluorogenic Enzyme Assay*: To serial dilutions of recombinant human neutral endopeptidase (NEP), insulin degrading enzyme (IDE), ECE-1, and ACE in 100 mM Tris-HCl (pH 7.5), 50 mM NaCl and 10 μM $ZnCl_2$ fluorogenic peptide substrate (10 μM) was added, the reaction mixture was at 37°C in the dark incubated for 1 h to produce standard curves of enzyme activity, and with excitation at 320 nm and emission at 405 nm in a fluorescent plate reader (FLUOstar, BMG Labtech, UK) the fluorescence was measured. Prior to adding fluorogenic peptide in 100 mM Tris-HCl (pH 7.5), 50 mM NaCl, and 10 μM $ZnCl_2$, thiorphan was with either recombinant human standards or crude tissue homogenates at room temperature incubated for 10 min.

2. *Standard ELISA Protocol*: From serial dilutions of recombinant human NEP and with crude tissue homogenates prepared in 1% SDS lysis buffer using the Human Neprilysin Duoset ELISA kit (R&D Systems, Abingdon, UK), which had previously been calibrated against a highly purified Sf21-expressed recombinant human NEP and showed no cross-reactivity with ECE-1, ECE-2, or neprilysin-2, standard curves were produced according to the manufacturer's instructions.

3. *Immunocapture-Based NEP-Specific Enzyme Activity Assay*: At room temperature anti-human NEP (1.6 μg/mL) diluted in PBS (pH 7.4) was coated on Nunc MaxiSorp 96-well plates for 18 h, the plates were washed six times in PBS containing 0.5% Tween-20, at room temperature by adding 1% PBS-bovine serum albumin blocked for 3 h, washed six times, in wells recombinant human NEP and crude brain tissue homogenates in either 1% SDS lysis buffer or 0.5% Triton X-100 lysis buffer were diluted in PBS (pH 7.4), at room temperature with continuous shaking incubated for 2 h, washed for another six times, to each well 10 μM fluorogenic peptide diluted in 100 mM Tris-HCl (pH 7.0), 50 mM NaCl, and 10 μM $ZnCl_2$ was added, incubated at 37°C in the dark, and fluorescent readings were taken at 1, 2.5, and 18 h. Control wells contained PBS alone were run in parallel. Thiorphan inhibition of NEP-specific enzyme activity on serial dilutions of recombinant human NEP and crude tissue homogenates was tested. As a control prior to adding fluorogenic peptide substrate, thiorphan in 100 mM Tris-HCl (pH 7.0), 50 mM NaCl, 10 μM $ZnCl_2$, or buffer alone was added to the wells at room temperature for 10 min.

8.10 *IN VIVO* AChE INHIBITION ASSAY[10,11]

ICR mice (25–35 g) were randomly assigned into control and treating groups, their whole cortex was dissected on ice at 5, 15, 30, 60, 120, and 240 min after test compound administration, homogenized in ice-cold PBS (0.1 mM, pH 7.4), at 4°C and 3000 rpm centrifuged for 15 min, and at 37°C with 0.1 mM ethopropazine (selective inhibitor of BchE, Sigma Chemicals Ltd., St. Louis, MO, USA) preincubated for 5 min. A 2 mL mixture of 0.1 mL of acetylcholine iodide (12 mM, Sigma Chemicals Ltd., St. Louis, MO, USA), 1.8 mL of sodium phosphate buffer (0.1 mM, pH 7.4), and 0.1 mL of homogenate was at 37°C incubated for 8 min, the reaction was terminated by adding 1 mL of 3% (w/v) SDS and 1 mL of 0.2% (w/v) 5,5′-dithiobis(2-nitrobenzoic) acid (DTNB) to produce a yellow complex, the color production was measured spectrophotometrically at 420 nm (model D7401; Barnstead Thermolyne Co., Dubuque, IA, USA). At the same time, with the Coomassie blue protein-binding method and bovine serum albumin as standard, the protein content was measured. Based on AChE inhibition $= -(1 - OD_{test}/OD_{control}) \times 100\%$, the AChE inhibition was calculated, in which OD_{test} and $OD_{control}$ were the absorbance values of the samples from treating and control groups, respectively.

8.11 SINGLE PARTICLE ASSAY FOR Aβ AGGREGATES[12]

1. *Preparing Aβ-Aggregates*: Aβ(1-42) (Sigma-Aldrich, Hamburg, Germany) was dissolved in DMSO to prepare 400 μM solution, diluted in PBS (140 mM NaCl, 2.7 mM KCl, 10 mM Na_2HPO_4, pH 7.4) to 33 μM final concentration, at 37°C incubated for at least 2 days, thioflavin T (ThT) (10 mM, Sigma, Hamburg, Germany) was added to monitor aggregate formation, and fluorescence was monitored with a microplate reader at excitation wavelengths of 440 nm and emission wavelengths of 490 nm (Polarstar Optima, BMG, Offenburg, Germany). A dilution of 1:10 of the aggregates in pooled cerebrospinal fluids (CSF) of healthy people was used for surface-fluorescence intensity distribution analysis (FIDA).

2. *Collecting CSF*: By lumbar puncture samples were obtained, within 2 h centrifuged at 3000 rpm for 5 min, the supernatant was stored at −70°C before being analyzed, and repeated freeze/thaw cycles were avoided. Using FCS Olympus IX 50 equipped with a beam scanner unit (Evotec OAI, Hamburg, Germany) working in dual-color mode with argon ion laser (excitation wavelengths 488 and 514 nm) and helium–neon laser (excitation wavelength 633 nm), FCS measurements were performed. In 24-well assay (Evotec OAI, Hamburg, Germany), chips samples were scanned in horizontal and vertical dimensions using the beam scanner, for which the focus was moved 1 mm in one direction with 100-μm rectangular deviation, 50 Hz frequency, and 50 μs integration time. For each sample, these settings were applied seven times site by site. To allow a precise z-positioning of the focus in the 100-nm range in the optic of the FCS Olympus IX 50, a piezo element was integrated.

3. *Surface-FIDA*: Briefly, 0.25 μg of capture antibody (mAb 4G8, recognizing amino acids 17–24 of Aβ) was adhesively bound to activated glass surfaces. After blocking unspecific binding sites with 10% fetal calf serum, 20 μL of sample was added to activated glass surfaces adhesively binding 0.25 μg of capture antibody (mAb 4G8, recognizing amino acids 17–24 of Aβ), at 4°C incubated for at least 2 h allowing the potentially present Aβ-aggregates to bind to the capture, washed twice with PBS buffer (140 mM NaCl; 2.7 mM KCl; 10 mM Na_2HPO_4, pH 7.4), fluorescence labeled detection antibodies (0.1 μg/μL) were added, at 20°C incubated for 1 h, washed with PBS containing 0.1% (w/w) Tween-20 five times and with PBS two times, and the measurements were carried out.

8.12 INDIRECT IMMUNOFLUORESCENCE ASSAY[13]

The immortalized hybridoma cell line EAhy926 or the primary cultures HMVEC-L (Provitro GmbH, Berlin, Germany) or human neuroblastoma cell line SKNBE2 were grown to 60% to 70% confluence, seminated at 5×10^6 per well on glass coverslips, either untreated or treated with 20 ng/mL TNF-α and 10 mg/mL cycloheximide for 8 h, at 4°C permeabilized with acetone/methanol (1/1, v/v) for 10 min, at 25°C soaked in balanced salt solution for 30 min, at 25°C in the blocking buffer (2% bovine serum albumin in PBS containing 5% glycerol and 0.2% Tween-20) incubated for 30 min, washed with PBS, at 4°C with the human anti-RABPT5 antibody and with control human IgG (0.1 mg/mL, gchain specific, Sigma) in PBS containing 1% BSA incubated for 1 h, at room temperature with 50 μg/mL of recombinant antigens incubated overnight to block the reactivity of anti-RABPT5 antibody, fluorescein isothiocyanate–conjugated anti-human IgG were added and at 4°C incubated for 30 min. Using the Apoptest binding kit containing annexin V-FITC and binding buffer by propidium iodide staining and the binding fluorescein isothiocyanate (FITC)-conjugated annexin V, apoptosis was evaluated. After washing with PBS, Olympus U RFL microscope (Olympus, Hamburg, Germany) was used to analyze the fluorescence.

8.13 MOUSE BEHAVIORAL ASSAYS[14,15]

Mice from a second-generation cross between heterozygous APPK670 N, M671L and heterozygous PS1 transgenic line 6.2 were genotyped, singly housed after weaning, confirmed for genotyping 1 month prior to behavioral assay, maintained on a 14-h light and 10-h dark cycle for the duration of the assays, and behavioral assays were carried out during the light cycle.

1. *Open Field Assay*: To measure exploratory behavior and general activity, mice were individually placed into a 81 × 81 cm open black box with 28.5-cm-high walls. With white lines this area was divided into 16 squares measuring 20 × 20 cm. Over a 5-min period, lines crossed were counted for each mouse.

2. *Balance Beam Assay*: To measure balance and general motor function, the mice were placed on a 1.1-cm-wide, 50.8-cm-long beam, which was suspended 46 cm above a padded surface by two identical columns and attached to a 14 × 10.2 cm escape platform at each end in a perpendicular orientation and monitored for 60 s as a trial. The time spent by each mouse before falling or reaching one of the platforms was recorded for each of three successive trials. A time of 60 s was assigned for the trial that a mouse reached one of the escape platforms. All three trials were averaged.

3. *String Agility Assay*: To examine forepaw grip capacity and agility, mice were placed in the center of a taut cotton string suspended 33 cm above a padded surface between the same two columns as in the balance beam task, allowed to grip the string with only their forepaws and released for maximum 60 s. A rating system ranging from 0 to 5 was employed to assess string agility for a single trial of 60 s.

4. *Y-maze*: In the measurements of general activity and basic memory function, mice were placed in the center of a black Y-maze with three arms, of which each measured 21 × 4 cm with 40-cm-high walls, faced the center area, allowed to explore for 5 min, the number and sequence of arm choices were recorded. The total number of arm entries was defined as general activity, while the percentage of spontaneous alternation (the ratio of arm choices differed from the previous two choices divided by the total number of entries) was measured as basic mnemonic function.

5. *Elevated Plus Maze Assay*: To measure anxiety/emotionality, mice were placed in the 5 × 5 cm maze center of an elevated plus maze 82 cm above the floor, which consisted of two opposite "open" 30 × 5 cm arms and two opposite "closed" 30 × 5 cm arms surrounded by 15-cm-high black aluminum walls, faced a closed arm and allowed to explore for 5 min. The total number of open-arm entries, closed-arm entries, and total time (seconds) spent in the open arms were recorded.

6. *Morris Water Maze Assay*: This task was used to measure reference learning (acquisition) and memory retention, for which mice were placed in a 100-cm pool that was divided into four equal quadrants by black lines drawn on the floor of the pool. In the center of quadrant 2, a transparent 9-cm platform was placed 1.5 cm below the surface of the water. The pool was surrounded with an assortment of visual cues. Mice were started from a different quadrant for each of four successive trials of 1 min per day, and the same quadrant start pattern was used across 10 days of acquisition. For each trial, the latency to find the platform (60 s maximum) was recorded, and for statistical analysis the four daily trials were averaged. The mouse finding the platform was allowed to stay on it for 30 s, while the mouse failing to find the platform was gently guided to the platform and given the 30-s stay as well as a latency of 60 s. On day 11, a trial of 60-s memory retention (probe) was performed, for which the platform was removed and the trial was started from the quadrant opposite the platform-containing quadrant. From videotape recordings of this probe trial, the percentage of time spent in each quadrant, annulus crossings, and average swim speed were determined.

7. *Circular Platform Assay*: For testing spatial (reference) learning and memory and for the single trial administered on each of 8 test days, mice were placed in a 69-cm circular platform with 16 holes equally spaced on the periphery, of which only 1 was a box filled with bedding to allow the mouse to escape from aversive stimuli (e.g., two 150-W flood lamps hung 76 cm above the platform and one high-speed fan 15 cm above the platform), faced away from the escape hole, and each was given one 5-min trial per day to explore the area. Escape latency was defined as the total number of errors (e.g., the number of head pokes into nonescape holes) and was measured (maximum of 300 s).

8. *Platform Recognition Assay*: To measure the ability to search for and identify or recognize a variably placed visible platform, mice were placed in a 100-cm pool that was divided into four equal quadrants by black lines drawn on the floor of the pool. A 9-cm circular platform raised 0.8 cm above the surface of the water attaching a prominent 10 × 40 cm black ensign was employed in this task. Mice were given four 60-s maximum trials starting from the same location in the pool per day for

4 days, for each trial the platform location was changed to a different one of the four quadrants, allowed 60 s maximum to search/identify and ascend the platform and have a 30-s stay on the located platform. The mice that did not find the platform within 60 s were gently guided to the platform manually and allowed to stay for 30 s. Escape latencies for all four daily trials were averaged for statistical analysis.

9. *Radial Arm Water Maze (RAWM) Assay*: RAWM used the same clear platform and visual cues as Morris maze testing and evaluated working (short-term) memory, however, an aluminum insert was placed in the pool to create six radially distributed swim arms emanating from a central circular swim area, and the last of four consecutive acquisition trials (T4) as well as a 30-min delayed retention trial (T5) were used as the indices of working memory. On any testing day, the submerged clear platform was placed at the end of one of the six swim arms, its location was changed daily to a different arm in a semirandom pattern, and mice were tested for 9 consecutive days. On each day, from the remaining five swim arms different start arms for each of the five daily trials were selected in a semirandom sequence that involved all five arms. For any given trial, the mouse was placed in the start arm, allowed to face the center swim area, and given 60 s to find the platform with a 30-s stay. The mouse entering a non-platform-containing arm was gently pulled back into the start arm and given an error record, or the mouse failing to enter any arm within 20 s was returned to the start arm and given an error record, or the mouse entering the platform-containing arm but not finding the platform was given an error record. Within a 60-s trial, the mouse failing to find the platform was guided to the platform manually, allowed to stay for 30 s and assigned a latency of 60 s. For each daily trial, both the number of errors (incorrect arm choices) and the escape latency were recorded.

8.14 CENTER OF PRESSURE (CoP) ASSAY IN MICE[16]

Transgenic mice (APPSwDI) containing the Swedish, Dutch (E22Q), and Iowa (D23 N) amyloid precursor protein (APP) mutations, T-279 mice expressing the N279K human FTDP-17 tau mutation under the control of the human TAU promoter, as well as P301L mice expressing the P301L human tau mutation were generated, while bigenic mice were produced by crossing P301L mice with $NOS2^{-/-}$ mice, and control mice for the T-279 strain were generated from a back-crossed strain and aged in the same conditions and for the same time period as the genetic strain. C57/Bl6 littermate mice served as control mice for all other strains. All mice were genotyped using standard procedures. Male and female mice were bred and housed under standard temperature and light conditions in an AAALAC/NIH approved facility.

The CoP analysis system (Fig. 8.1) consisted of an AMTI Biomechanics Force Platform (Model HE6 × 6; Advanced Mechanical Technology Inc., Watertown, MA, USA), which was sensitive enough to detect changes in force and moment initiated by tremor in 10- to 50-g mice, automated data acquisition software, analysis software, and a digital video camera. Each mouse was placed onto the platform within

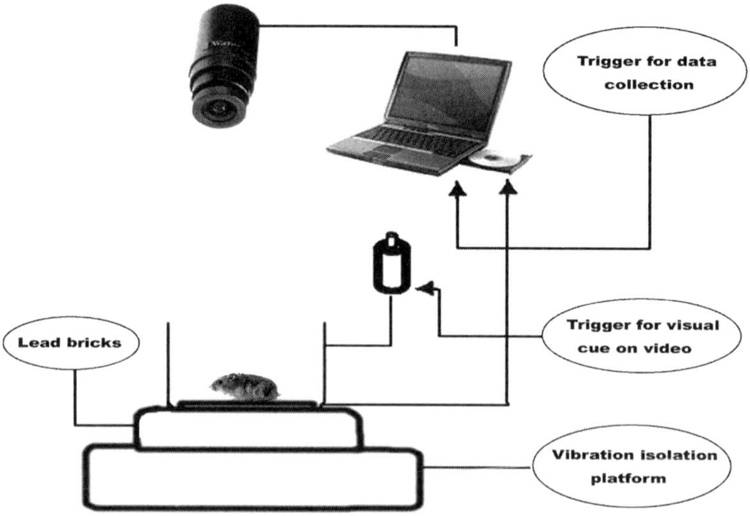

Figure 8.1 Typical apparatus for the CoP.

the box, allowed to explore for approximately 1–2 min, and watched via the operator and a video image recorded for each force measurement in order to avoid measuring voluntary movements such as walking or grooming, while the video camera and acquisition program were activated. When the mouse remained in a prolonged stationary position on all four legs without overt motion, a signal acquisition was initiated. The software was set to record the forces experienced by the force platform for the 1 s prior to signal initiation, which corresponded with the time that the animal was standing still. The video was activated with a remote trigger, and verbal descriptions of the mouse's activity were recorded to further confirm that mouse did not exhibit active movements of the limbs or head. In this manner, the video image of each force measurement could be used for confirming the "resting status" of each mouse. For each test (approximately 30 min), the measurements were repeated 8 to 10 times per mouse.

8.15 ASSAY FOR PLASMA LEVELS OF DJ-1[17]

The assay included 104 patients with sporadic PD (54 to 78 years old), 30 patients with sporadic dementia with Lewy body (68 to 79 years old), and 80 age-matched healthy controls (62 to 73 years old). To determine plasma DJ-1, immunoblotting was performed for plasma samples obtained from patients and normal controls. From all subjects, 5 mL of blood was collected in plastic tubes containing either sodium citrate or potassium EDTA, and plasma was collected in 0.5 mL plastic tubes and stored frozen at −80°C until used. Using a BioRad Protein Assay reagent (BioRad, Hercules, CA), plasma protein concentrations were determined. Immunoblot analysis was generally performed. Using SDS-PAGE (15%), 20 μg of plasma samples were separated and transferred to a nitrocellulose membrane. The membrane was blocked with 3% bovine serum albumin, incubated overnight at 4°C with mouse anti-human DJ-1 monoclonal antibody (1 : 1000) and detected with Immobilon Western Chemiluminescent HRP substrate (Millipore, Billerica, MA, USA). Recombinant GST-tagged DJ-1 protein (1.5 ng) was used as a positive control for DJ-1. Using anti-human fibronectin monoclonal antibody (1:500) as loading control, each plasma sample was also immunoblotted.

To quantitatively evaluate the plasma DJ-1 level, by use of CircuLex TM Human DJ-1/PARK7 ELISA Kit (Cat. No. CY-9050, CyLex Co., Ltd., Nagano, Japan), ELISA was performed at 0.92 ng/mL sensitivity with 4.5% intra-assay variation. Using recombinant GST-human DJ-1 provided by the ELISA kit solution at 100, 50, 25, 12.5, 6.25, 3.13, and 1.56 ng/mL, the standard curve for the ELISA was obtained. According to each standard curve and the dilution factor, the concentrations of the samples in each plate were calculated. To further determine the altered plasma DJ-1 levels during the course of disease progression, the plasma DJ-1 was also analyzed according to the disease stage.

8.16 MEMBRANE FILTER ASSAY FOR TAU AGGREGATION[18]

Tau protein (50 μL final volume) was incubated at 37°C in assembly buffer of 10 mM HEPES (pH 7.4), 100 mM NaCl, and 5 mM dithiothreitol with or without fibrilization inducer thiazine red (100 μM) or octadecyl sulfate (50 μM) for up to 24 h and immediately subjected to assay. For filter assay, tau fibrilization reaction products were with 2% SDS diluted up to 10-fold to prepare a series of descending tau filament concentrations. Before vacuum filtration through a 96-well dot blot apparatus containing nitrocellulose, polyvinylidene fluoride (PVDF, Billerica, MA, USA), or cellulose acetate membranes, all samples underwent a further 1 : 3 dilution in 2% SDS, the resultant membranes were washed twice with 2% SDS, blocked in 4% nonfat dry milk dissolved in 100 mM Tris-HCl (pH 7.4) and 150 mM NaCl for 2 h, incubated with primary antibody (1:1000) for 1.5 h, washed twice in blocking buffer, incubated with HRP-linked secondary antibody (Gaithersburg, MD, USA) for 1.5 h, washed twice in blocking buffer, and then developed with the ECL (enhanced chemiluminescence) Western Blotting Analysis System (GE Healthcare, Buckinghamshire, UK). On an Omega 12iC Molecular Imaging System, chemiluminescence was recorded and quantified using UltraQuant software (UltraLum, Claremont, CA, USA).

With adding 2% glutaraldehyde, tau fibrilization reactions were terminated and adsorbed onto 300-mesh Formvar/carbon-coated copper grids for 1 min. The grids were rinsed with water, with 2% uranyl acetate negatively stained for 1 min, washed again with water, images were captured on a Tecnai G2 Spirit Bio-TWIN transmission electron microscope (FEI, Hillsboro, OR, USA) operated at 80 kV and ×23,000 magnification, and analyzed with Image J software (National Institutes of Health). Average total filament length was generally determined.

Regression analysis was performed with SigmaPlot software (Systat Software, San Jose, CA, USA). Analyte concentration dependence of the filter assay was fit to the power function $y - y_0 = ax^b$, where y was the signal intensity produced in the presence of aggregation inducer at tau concentration x, y_0 was the background signal produced in the absence of tau aggregation inducer at tau concentration x, and a as well as b were constants. From the abscissa intercept after least squares linear regression, the critical concentration for fibrilization was estimated and reported ± standard error of the estimate.

8.17 ASSAYS FOR MOTOR NEURON DEGENERATION[19]

1. *Ethacrynic Acid (EA)-Mediated Glutathione (GSH) Depletion or L-Buthionine Sulfoximine (BSO)-Mediated GSH Synthesis Assay*: NSC34 cells were incubated with ethacrynic acid (EA) ranging from 20 μM to 100 μM for up to 12 h, or NSC34 cells were incubated with L-buthionine sulfoximine (BSO, Sigma Chemical Company, St. Louis, MO, USA) ranging from 20 μM to 100 μM for up to 48 h; cells were harvested and analyzed. For colorimetric analysis control, EA- or BSO-treated cells were incubated with 40 μM monobromobimane (MBM) for 30 min, the fluorescent intensity was directly measured with a fluorometer, and the relative levels of cellular GSH after EA or BSO treatments were expressed with an arbitrary unit of fluorescent intensity. For biochemical analysis, 1×10^7 EA- or BSO-treated cells were homogenized completely in MES buffer (200 mM 2-*N*-morpholinoethane-sulfonic acid, 50 mM phosphate and 1 mM EDTA, pH 6.0), at 4°C and 12,000 × *g* centrifuged for 10 min, and with Bioxytech GSH/GSSG-412 kit (Oxis International, Inc., Foster City, CA, USA) according to the manufacturer's instructions the supernatant was deproteinated and processed for GSH (or GSSG) analysis. The levels of GSH (or GSSG) in the spinal cord lumbar region of G93A-SOD1 transgenic mice and age-matched normal control mice at different stages corresponding with disease free, disease onset, and disease progression were also determined with this assay.

2. *Assay of Cellular Production of ROS*: In a 96-well plate (Corning Life Sciences, Corning, NY, USA), 2×10^5 control EA- or BSO-treated NSC34 cells/well were in incubated the dark with 100 μM 6-carboxy-2′,7′-dicholorfluorescin diacetate (DCFHDA) for 1 h and in a fluorometer (Molecular Device Inc., Sunnyvale, CA, USA) at the excitation wavelengths of 485 nm and emission wavelengths of 530 nm measured for the oxidation of DCFH-DA to represent the relative steady state of ROS generation in cells.

3. *Transient Transfection and Luciferase Activity Assay*: For transient transfection and luciferase activity analysis, 1×10^7 EA- or BSO-treated NSC34 cells were incubated at room temperature with 50 μg of vector control and vector containing 2 × AP-luciferase plasmid DNA in 0.4 mL of Opti-MEM (Invitrogen) for 10 min, electroporated using a low-voltage mode at 240 V, one pulse, and 25 ms/V of pulse length, and at room temperature kept for 30 min and cultured in complete medium.

4. *Adhesion Assay*: Twenty-four-well plates were in 1 × phosphate buffer solution (PBS) coated overnight with fibronectin, collagen, or laminin, washed, to each well 2×10^6 control or EA-treated NSC34 cells were added, incubated for 12 h, cell culture medium was removed, and the wells were briefly washed with 1 × PBS twice.

Trypsin cells adhering to the substrate were dissociated and counted. The number of cells adhered to each coated plate divided by the total number of cell seeded was defined as the percentage of adhesion.

5. *Quantitative Real-Time PCR Analysis of Gene Expression*: Vehicle control or EA-treated NSC34 cells were harvested, RNA was purified, total RNA purified was treated with RNase-free DNase (Ambion, Austin, TX, USA) to perform quantitative real-time PCR analysis, the concentration was measured, and using superscript first-strand synthesis kit 1 μg of total RNA from each sample was used to synthesize first strand cDNA. With MX4000 system real time PCR was performed. The forward real-time PCR primers for amplification of heme oxygenase-1 (HO-1) transcripts was 5′-CTCACTGGCAGGAAATCATCCC-3′ and the reverse real-time PCR primers for amplification of heme oxygenase-1 (HO-1) transcripts was 5′-GAGAGGTCAC-CCAGGTAGCG-3′, and the probe was 5′-6-FAM-(CACGCCAGCCACACAGC-ACTATGTAAAGC)BHQ-1-3′. Actin was used as endogenous control, the forward real-time PCR primers for amplification of its transcripts was 5′-TACAAT-GAGCTGCGTGTGGC-3′, the reverse real-time PCR primers for amplification of its transcripts was 5′-ATGGCTGGGGTGTTGAAGGT-3′ and the probe was 5′-6-FAM (CACCCTGTGCTGCTCACCGAGGC)BHQ-1 3′.

6. *Immunohistochemical Assay*: For immunohistochemical analysis, vehicle control and EA-treated NSC34 cells were harvested, washed twice in ice cold 1 × PBS, on ice in 4% paraformaldehyde (PFA) fixated for 1 h, permeabilized with 0.2% Triton X-100 in PBS, blocked with 10% goat serum, at 4°C with specific antibodies (cytochrome c and active caspase-3 antibodies were all at 1 : 300 dilution) incubated overnight, sections were washed five times with 0.2% Triton X-100 in PBS for 5 min each, at room temperature in the dark with specific secondary antibody conjugated with fluorescein or rhodamine incubated for 2 h, extensively washed, incubated with 4′-6-diamidino-2-phenylindole (DAPI) for nuclei staining, and mounted with anti-fade medium. Collected images were analyzed with a Nikon optical microscope equipped with Spot digital camera and Photoshop software (Adobe Systems, San Jose, CA, USA).

7. *Western Blotting Assay*: For Western blotting analysis, vehicle control or EA-treated NSC34 cells were scraped off the wells, harvested in ice-cold 1 × PBS, at 4°C and 2500 × *g* centrifuged for 5 min, the supernatant was discarded, the cell pellet was suspended in 400 μL of ice-cold lysis buffer (10 mM K_2HPO_4, pH 7.2/ 1 mM EDTA/5 mM EGTA/10 mM $MgCl_2$/50 mM glycerophosphate/1 mM Na_3VO_4/2 mM DTT/1% Triton X-100) with complete protease inhibitor, the cell lysate was at 4°C and 12,000 × *g* centrifuged for 10 min, the protein concentration in the supernatant was measured using BioRad protein assay kit (BioRad, Hercules, CA, USA), 20 μg of protein sample was separated in a 12.5% SDS-PAGE gel, transferred onto nitrocellulose membrane, at room temperature on a shaker plate the membrane was blocked with 5% dry milk dissolved in TBST (10 mM Tris-HCl, pH 7.5, 150 mM NaCl, 0.1% Tween-20) for 1 h, at 4°C with the specific antibody overnight incubated, at room temperature with TBST washed three times for 5 min each time, at room temperature with HRP-conjugated secondary antibody (KPL, Gaithersburg,

MD, USA) incubated for 1 h, washed three times with TBST for 5 min each, and visualized in ECL reagents (Amersham Bioscience, Piscataway, NJ, USA).

8. *Cell Viability Assay*: In Trypan blue exclusion assay of cell viability, treated NSC34 cells were at room temperature with 0.1% Trypan blue dye incubated for 10 min and counted on a hemocytometer with a microscope. The number of viable cells (dye-excluding) divided by total number of cells was defined as cell viability. To confirm the Trypan blue exclusion assay, cell viability was also measured by the LDH release assay based on the manufacturer's instructions.

9. *Cell Apoptosis Assay*: According to the manufacturer's instructions, apoptosis assays in control or EA-treated NSC34 cells were performed using an ELISA-formatted assay that detects histone-associated DNA fragments (Roche Biochemicals). To analyze apoptotic cell death, a terminal deoxynucleotidyl transferase–mediated nick-end labeling assay was also performed. Two independent apoptotic assays generated very similar results.

8.18 HPLC ASSAY FOR NEUROPROTECTIVE AGENT IN MICE PLASMA[20]

1. *Chromatographic Apparatus and Conditions*: A Waters system equipped with a Wisp-717 sample processor, a model 510 solvent delivery system, and a model 2487UV detector set at 263 nm was coupled with a model C-R6A Chromatopac Shimadzu integrator. At room temperature, on a μBondpack C18 column (3.9 mm × 300 mm; Waters) and with $CH_3CO_2H/CH_3CN/CH_3OH$/0.01 M KH_2PO_4 (1/5/43/51, v/v/v/v) as the mobile phase, which was degassed through a 0.45-μm filter before use and delivered isocratically at a flow rate of 1 mL/min, the separation was performed.

2. *Biological Samples*: Female SOD1-G93A mice and their nontransgenic littermates were used in a study to evaluate the neuroprotective activity of test compounds. The pharmacologic treatment started at the age of 7 weeks. Mice in the terminal stage of disease and their C57 littermates were killed by decapitation under deep anesthesia, 1 and 2 h after the last dose, respectively. Blood samples were collected in heparinized tubes, centrifuged, and the plasma was stored at −20°C. Brains and spinal cords were rapidly removed, blotted with paper to remove excess surface blood, and stored at −20°C.

3. *Extraction*: With test compound free plasma the plasma samples (0.1 mL) were adjusted to 0.5 mL, mixed with 25 μL of methanolic internal standard (IS, 10 μg/mL), with 0.01 M phosphate buffer (pH 7.4) diluted to1.0 mL, the final solution was vortex-mixed, at 4°C and 2000 rpm in a Sorval centrifuge centrifuged for 10 min, and the supernatant was applied on 3 mL Bakerbond spe Octyl (C8) disposable extraction columns activated previously with 2 mL of CH_3OH and 2 mL of H_2O.

In deionized water, brain tissue was homogenized (10 mL/g). To 1 mL aliquots of this homogenate, 25 μL of IS was added, vortex-mixed, at 4°C and 2000 rpm centrifuged for 10 min, the precipitates were redissolved in 0.5 mL of distilled water, centrifuged, and the combined supernatants were processed as plasma samples. After loading the brain tissue sample and equilibration with 0.01 M K_2HPO_4, the extraction cartridges were washed with 0.5 mL of H_2O and 0.5 mL of 10% CH_3OH in H_2O. Test compounds and IS were eluted with 2 mL of 1 M NH_4OH in CH_3OH, the eluent was dried under a stream of nitrogen, the residue was dissolved in 150 μL of mobile phase, and 100 μL aliquot was injected in to the HPLC system.

4. *Data Analysis*: By replicate analyses of quality control (QC) samples containing small, medium, and large known amounts of riluzole stored at −20°C, the precision and accuracy of the method were determined. On different days, these QC samples were assayed with standard samples and the calculated concentrations were compared (interassay variance), while by replicate analysis of QC samples on the same day intra-assay variance was checked.

With six concentrations over the working range, daily standard curves were plotted in duplicate with QC samples injected between the two sets of standards. As the analyte to the IS peak height ratio, the relative response factor was computed. Using least squares linear regression on the relative response factors against the nominal concentration of test compound, calibration lines were constructed. By back-solving $y = a + bx$ of the calibration line, QC concentrations and unknown samples were obtained.

8.19 TISSUE CULTURE ASSAYS FOR SOD1 MUTATIONS[21]

Rat C6 glioma cells maintained in DMEM (Gibco, UK) supplemented with 10% heat-inactivated FBS and 1% penicillin/streptomycin (Gibco, UK) were transfected with cDNA constructs containing normal human SOD1 or one of three mutant forms (G37R, G93A, and I113T) associated with familial motor neuron disease (MND), and cloned into the mammalian expression plasmid pCEP4 (Invitrogen). By selection with hygromycin, stable clones of cells were isolated. After transfection, SOD1 was uniformly expressed in the cells over 18 passages. Wild type, untransfected cells and cells transfected with the pCEP4 vector alone served as controls. For toxicity assays, cells were trypsinized, 8×10^4 cells were plated out per well into a 24-well tissue culture plate, incubated for 48 h, and toxicity was assayed. From transgenic mice overexpressing either wild type human SOD1 or G93A-mutant SOD1, highly enriched astrocyte cultures were prepared, and cultures from non-transgenic litter-mates were used as controls. Using human SOD1 primers, PCR of genomic DNA was performed to identify transgenic cultures. Astrocyte cultures from the cerebral cortices of mice on postnatal day 1 or 2 were prepared and in supplemented DMEM containing gentamicin (50 μg/mL) grown to confluence. Human SOD1 was expressed in the primary cultures, and the astrocytes were subcultured to maintain expression. Using an antibody that recognized both human and mouse SOD1, Western blotting was performed to confirm SOD1 expression in the cortical astrocytes cultures as well as the C6 cells. For toxicity assays, the primary astrocytes were trypsinized, 3×10^4 cells were plated out per well into a poly-L-lysine–coated

96-well microtiter plate, and after reaching confluence toxicity was assayed. Four test compounds were used as donors of different free radical species. After exposure to test compounds at 37°C for 24 h, cell viability was measured by reduction of the vital dye MTT. The concentrations of test compounds were titrated in initial experiments to produce similar amounts of cell death for each cell type (C6 cells versus primary astrocytes).

8.20 LUCIFERASE-BASED REPORTER ASSAY[22]

For transient transfection, 24 h prior to Lipofectamine-based transfection with pSVK3-APP695 and/or pCMV-Fe65 (2 μg/well) according to the manufacturer's instructions (Invitrogen, Carlsbad, CA), 3×10^5 CHO cells/cm^2 were seeded, lysed in 150 μL of sample buffer (deoxycholate 1%, pH 11.3, 5 mM PNT), and sonicated for 5 s. Thirty micrograms of the sample was loaded onto NuPage 10% to 20% Tricine gels, transferred to nitrocellulose membranes, saturated (5% skimmed milk in 0.05% Tween-20/PBS) for 30 min, washed, at 4°C with the primary antibody diluted in 0.05% Tween-20/PBS (0.5 μg/mL for the WO-2 antibody, 1:2000 for the anti-Fe65 antibody, and 1:5000 for the anti-APP C-terminal antibody) incubated overnight, washed, at room temperature with 1:10,000 of the appropriate HRP-conjugated secondary antibody incubated for 60 min, washed, and the immunoreactive bands were visualized by chemiluminescence. Using the Quantity One software coupled to the Gel Doc 2000 device (Bio-Rad, Hercules, CA), signal ratios were quantified. APPGal4 transactivation assays were performed in CHO cells. Using a co-transfected phRL-TK Renilla luciferase vector, luciferase activity was corrected for transfection efficiency. With 0.4 μg APPGal4 or APP695 vectors, 0.2 μg Gal4RE reporter gene, 0.2 μg Fe65 or empty pSV2 vector and 0.001 μg phRL-TK, cells (2×10^5) were transfected for 48 h. With the dual-luciferase assay, system luciferase activity was measured.

REFERENCES AND NOTES

1. F. Gu, M. Zhu, J. Shi, Y. Hu, Z. Zhao. Enhanced oxidative stress is an early event during development of Alzheimer-like pathologies in presenilin conditional knock-out mice. Neurosci Lett 440 (**2008**) 44–48.
2. G. Pollio, R. Roncarati, T. Seredenina, G.C. Terstappen, A. Caricasole. A reporter assay for target validation in primary neuronal cultures. J Neurosci Methods 172 (**2008**) 34–37.
3. M. Samuelsson, V. Ramberg, K. Iverfeldt. Alzheimer amyloid-β peptides block the activation of C/EBPb and C/EBPd in glial cells. Biochem Biophys Res Commun 370 (**2008**) 619–622.
4. M. Ono, M. Haratake, H. Saji, M. Nakayama. Development of novel β-amyloid probes based on 3,5-diphenyl-1,2,4-oxadiazole. Bioorg Med Chem 16 (**2008**) 6867–6872.
5. Y.C.S. Kumar, M. Malviya, J.N.N.S. Chandra, C.T. Sadashiva, C.S.A. Kumar, S.B.B. Prasad, D.S. Prasanna, M.N. Subhash, K.S. Rangappa. Effect of novel N-aryl

sulfonamide substituted 3-morpholino arecoline derivatives as muscarinic receptor 1 agonists in Alzheimer's dementia models. Bioorg Med Chem 16 (**2008**) 5157–5163.

6. J.A. Johnston, W.W. Liu, D.T.R. Coulson, S. Todd, S. Murphy, S. Brennan, C.J. Foy, D. Craig, G.B. Irvine, A.P. Passmore. Platelet β-secretase activity is increased in Alzheimer's disease. Neurobiol Aging 29 (**2008**) 661–668.
7. J.H. Lee, S.R. Byeon, S.J. Lim, S.J. Oh, D.H. Moon, K.H. Yoo, B.Y. Chungb, D.J. Kim. Synthesis and evaluation of stilbenylbenzoxazole and stilbenylbenzothiazole derivatives for detecting b-amyloid fibrils. Bioorg Med Chem Lett 18 (**2008**) 1534–1537.
8. S.D. Giovanni, A. Borloz, A. Urbain, A. Marston, K. Hostettmann, P.-A. Carrupt, M. Reist. In vitro screening assays to identify natural or synthetic acetylcholinesterase inhibitors: Thin layer chromatography versus microplate methods. Eur J Pharm Sci 33 (**2008**) 109–119.
9. J.S. Miners, M.M. Verbeek, M.O. Rikkert, P.G. Kehoe, S. Love. Immunocapture-based fluorometric assay for the measurement of neprilysin-specific enzyme activity in brain tissue homogenates and cerebrospinal fluid. J Neurosci Methods 167 (**2008**) 229–236.
10. H. Yu, W.M. Li, K.K.W. Kan, J.M.K. Ho, P.R. Carlier, Y.P. Pang, Z.M. Gu, Z. Zhong, K. Chan, Y.T. Wang, Y.F. Han. The physicochemical properties and the in vivo AChE inhibition of two potential anti-Alzheimer agents, bis(12)-hupyridone and bis(7)-tacrine. J Pharm Biomed Anal 46 (**2008**) 75–81.
11. Y.E. Kwon, J.Y. Park, K.T. No, J.H. Shin, S.K. Lee, J.S. Eun, J.H. Yang, T.Y. Shin, D.K. Kim, B.S. Chae, J.Y. Leem, K.H. Kim. Synthesis, in vitro assay, and molecular modeling of new piperidine derivatives having dual inhibitory potency against acetylcholinesterase and Aβ1-42 aggregation for Alzheimer's disease therapeutics. Bioorg Med Chem 15 (**2007**) 6596–6607.
12. S.A. Funke, E. Birkmann, F. Henke, P. Görtz, C. Lange-Asschenfeldt, D. Riesner, D. Willbold. Single particle detection of Aβ aggregates associated with Alzheimer's disease. Biochem Biophys Res Commun 364 (**2007**) 902–907.
13. F. Delunardo, P. Margutti, S. Pontecorvo, T. Colasanti, F. Conti, R. Riganò, E. Profumo, A. Siracusano, A. Capozzi, M. Prencipe, M. Sorice, A. Francia, E. Ortona. Screening of a microvascular endothelial cDNA library identifies rabaptin 5 as a novel autoantigen in Alzheimer's disease. J Neuroimmunol 192 (**2007**) 105–112.
14. G.W. Arendash, M.T. Jensen, N. Salem Jr., N. Hussein, J. Cracchiolo, A. Dickson, R. Leighty, H. Potter. A diet high in omega-fatty acids does not improve or protect cognitive performance Alzeheimer's transgenic mice. Neuroscience 149 (**2007**) 286–302.
15. J. Zou, Z. Yao, G. Zhang, H. Wang, J. Xu, D.T. Yew, E.L. Forster. Vaccination of Alzheimer's model mice with adenovirus vector containing quadrivalent foldable Aβ1-15 reduces Aβ burden and behavioral impairment without Aβ-specific T cell response. J Neurol Sci 272 (**2008**) 87–98. *Note*: The Morris water maze (MWM) assay was used at the end of the treatment to examine the differences of spatial and temporal navigation abilities between the treated Tg2576 mice and the controls. In MWM trial with a circular pool (1 m diameter and 0.47 m high) filled with water of 23 $\pm$ 2°C and virtually divided into four quadrants, which was made turbid with powdered milk, and using visual cues, mouse swam and found a hidden platform. A white plastic platform of 7 cm in diameter was submerged 0.5 cm below the water surface at the center of one of the four quadrants of the maze as an escape platform. Before the first training, all mice underwent nonspatial pretraining (NSP) to evaluate mouse swimming ability and allow mice to become accustomed the test. In NSP, each mouse swam freely for 60 s in the maze without the escape

platform. Two days after NSP, all mice underwent two tests per day as spatial learning test in the maze with escape platform for 4 days. In tests, each mouse was gently lowered into the water, faced the maze wall, and allowed to reach escape platform in 60 s for each trial. The trial was started randomly from one of the four quadrants. In a 60-s trial, the mouse failing to find the escape platform was manually guided to the platform and allowed to rest on the platform for 10 s. One hour after the 8th trial, the escape platform was removed from the pool and each mouse was given a 60-s swim probe trial. With a video camera suspended 2.5 m above the center of the maze and connected to a video tracking system, the movements of the mouse in the maze were recorded and analyzed with a computer.

16. D. Hutchinson, V. Ho, M. Dodd, H.N. Dawson, A.C. Zumwalt, D. Schmitt, C.A. Colton. Quantitative measurement of postural sway in mouse models of human neurodegenerative disease. Neuroscience 148 (**2007**) 825–832.
17. M. Waragai, M. Nakai, J. Wei, M. Fujita, H. Mizuno, G. Ho, E. Masliah, H. Akatsu, F. Yokochi, M. Hashimoto. Plasma levels of DJ-1 as a possible marker for progression of sporadic Parkinson's disease. Neurosci Lett 425 (**2007**) 18–22.
18. E. Chang, J. Kuret. Detection and quantification of tau aggregation using a membrane filter assay. Anal Biochem 373 (**2008**) 330–336.
19. L. Chi, Y. Ke, C. Luo, D. Gozal, R. Liu. Depletion of reduced glutathione enhances motor neuron degeneration in vitro and in vivo. Neuroscience 144 (**2007**) 991–1003.
20. M. Colovic, E. Zennaro, S. Caccia. Liquid chromatographic assay for riluzole in mouse plasma and central nervous system tissues. J Chromatogr B 803 (**2004**) 305–309.
21. R.E. Williams, M.R. Cookson, A.E. Fray, P.M. Manning, F.M. Menzies, D.A. Figlewicz, P.J. Shaw. Cultured glial cells are resistant to the effects of motor neurone disease-associated SOD1 mutations. Neurosci Lett 302 (**2001**) 146–150.
22. S. Huysseune, P. Kienlen-Campard, J.-N. Octave. Fe65 does not stabilize AICD during activation of transcription in a luciferase assay. Biochem Biophys Res Commun 361 (**2007**) 317–322.

For ALP activity assay, the primary osteoblasts were plated at 1×10^5 cells/well in 24-well dishes. The ALP activity was measured according to general procedure and expressed as nanomoles of *p*-nitrophenol liberated per minute per milligram protein, while total protein was generally assayed.

For osteoclastic TRAP activity assay, cells were washed twice with PBS and 10 μL 0.1% Triton X-100 was added for 10 min to lyse the cells. Then 100 μL substrate solution was added and incubated for 30 min at 37°C, for which 200 mL reactive solution was prepared by successively dissolving 0.4 g of disodium *p*-nitrophenyl-phosphate and 2.0 g of potassium tartrate in 150 mL of deionized water, and the pH was adjusted to 3.5 with 1M HCl, to each well 100 μL of NaOH (1 M) was added to stop the reaction, the samples and standards were diluted in 20 mM NaOH, and the absorbance was measured at 405 nm, the nanomolar number of *p*-nitrophenol liberated per minute per 100 osteoclasts in each well was calculated and was defined as TRAP activity. At the same time, the positive cells for TRAP were counted.

9.3 MURINE OSTEOBLASTIC MC3T3-E1 CELL CALCIFICATION AND VAN KOSSA ASSAYS[2]

Using Alizarin red S staining solution, the calcium deposition of MC3T3-E1 cell culture was assayed. By mixing 0.1 mL of 28% ammonia aqueous solution with aqueous Alizarin red S (10 mg/mL) and adjusting pH to approximately 6.4, fresh Alizarin red S solution was prepared. In 48-well plates containing DMEM medium and 10% FBS, MC3T3-E1 cells (10^4 cells/well) were seeded. After attaching onto the well bottom, the medium was changed to DMEM + 10% FBS medium containing 10 mM disodium β-glycerophosphate, 0.15 mM ascorbic acid, and 100 nM dexamethasone; to each well different concentrations of DL-3-hydroxybutyrate (3HB) sodium salt (Sigma Chemical Co., St. Louis, MO) were added and incubated for 21 days.

1. For calcification assay, the cells were washed three times with Dulbecco's PBS without calcium and magnesium salts, fixed by adding 10% formalin in Dulbecco's PBS without calcium and magnesium salts for 1 h, washed three times with distilled water, in Alizarin red S solution stained for 15 min, and washed twice with distilled water to remove the redundant stains. From Alizarin red S stained cultures, the digital images (DC 300F, Leica, Germany) were obtained and under the microscopic counting (DM IRB, Leica, Germany) by averaging six values of different sights, the number of the calcification nodules was calculated.
2. For Van Kossa assay, the cells were washed twice with 150 mM NaCl, fixed by adding 10% formalin in Dulbecco's PBS without calcium and magnesium salts for 1 h, under ultraviolet radiation with 100 μL of 5% $AgNO_3$ treated for 30 min, $AgNO_3$ solution was removed, washed twice with Dulbecco's PBS without calcium and magnesium salts, into each well 5% $Na_2S_2O_3$ was added and sustained for 10 min, washed twice with distilled water, and in Neutral red stained for 10 min, washed twice with distilled water to remove the redundant stains. Digital images of the stained cultures were obtained.

9.4 OSTEOCLAST GENERATION ASSAY FOR MALE SENILE RAT[2,3]

Male SD rats were subjected to a 12-h light and dark cycle with a temperature of $22 \pm 1°C$ and a humidity of 40% to 50%, housed individually in Macrolon cages (inside length × width × height, 25 × 19.5 × 14 cm). At the age of 23 months, blood was collected by heart puncture from overnight fasted and anesthetized rats. After sacrifice by isoflurane anesthesia, the femurs were aseptically removed, prepared free of adherent muscle, the bone ends were cut off, bone marrow cells were flushed out, seeded into wells of 24-well tissue culture plates (2×10^6 cells/well), at 37°C in a humidified incubator contained α-MEM (Gibco, Paisley, UK), which was supplemented with 10^{-8} M 1.25(OH)$_2$ vitamin D$_3$, 15% FCS, 100 U/L penicillin as well as 100 μg/mL streptomycin, and 5% CO_2 were cultivated for 7 days. Culture medium was changed at 2–3 day intervals. At culture end, adherent cells were fixed with 10% neutral buffered formalin and with acetone for 10 min and 30 s, respectively, washed twice with distilled water, and stained for tartrate-resistant acid phosphatase with a commercially available kit (Sigma-Aldrich, Steinheim, Germany).

9.5 BONE RESORPTION AND RECOVERY RELATED ASSAYS[4]

1. *Bone Resorption Assay*: To evaluate bone resorbing activity, 2-day-old mice were injected subcutaneously with $^{45}CaCl_2$ (2 μCi); 2 days later the parietal bones were taken out, in Ham's F-12 medium (Nissui Pharmaceutical Co., Ltd., Tokyo, Japan) containing 5% (v/v) heat-inactivated horse serum precultured for 1 day, and in fresh medium containing 0.1 nM VD$_3$ (Sigma-Aldrich, USA) and test compounds cultured for 6 days. The medium was changed at 3-day intervals. After finishing the culture, bones were extracted with 0.01 M EDTA-acetate buffer (pH 5.5) for ^{45}Ca contained in bones, while ^{45}Ca in the medium and in the EDTA solution were countered separately. The percentage of ^{45}Ca released into the medium to the total ^{45}Ca was defined as bone resorption.

2. *Bone Recovery Assays*: To evaluate bone recovery activity, 2-day-old mice were injected subcutaneously with $^{45}CaCl_2$ (2 μCi), 2 days later the parietal bones were taken out, in Ham's F-12 medium containing 5% (v/v) heat-inactivated horse serum precultured for 1 day, in fresh medium containing 0.1 nM VD$_3$ and test compounds cultured for 3 days, and in fresh medium containing 0.1 nM VD$_3$ alone cultured for 3 days. After finishing the culture, bones were extracted with 0.01 M EDTA-acetate buffer (pH 5.5) for ^{45}Ca contained in bones, while ^{45}Ca in the medium and in the EDTA solution were countered separately.

3. *Osteoclast-Like Cell Formation Assay*: Osteoblast-like cells (OCLs) and bone marrow cells were prepared from calvarias of 1-day-old mouse and tibias of 6-week-old male mice, respectively. OCLs (1×10^4/well) and bone marrow cells (1×10^5/well) were co-cultured for 6 days with 10^{-8} M VD$_3$ and test compounds; medium was replaced at 2-day intervals. After finishing the culture, the adherent

cells on the well surface were fixed with 10% formalin-PBS (pH 7.2), treated with ethanol/acetone (1/1, v/v), at room temperature with tartrate-resistant acid phosphatase (TRAP) stained for 12 min, and TRAP-positive cells containing three or more nuclei were counted as OCLs.

4. *OCLs Recovery Assays*: To evaluate OCLs-recovery activity, OLCs and bone marrow cells were prepared from calvarias of 1-day-old mouse and tibias of 6-week-old male mice, respectively. OCLs (1×10^4/well) and bone marrow cells (1×10^5/well) were co-cultured for 4 days with 10^{-8} M VD_3 and test compounds, medium was replaced at 2-day intervals, and in fresh medium containing 0.1 nM VD_3 alone cultured for 2 days. After finishing the culture, the adherent cells on the well surface were fixed with 10% formalin-PBS (pH 7.2), treated with ethanol/acetone (1/1, v/v), at room temperature with TRAP stained for 12 min, and TRAP-positive cells containing three or more nuclei were counted as OCLs.

5. *Pit Formation Assay*: In culture dishes precoated with collagen gel matrix, primary OCLs (2×10^6 cells/dish) and bone marrow cells (2×10^7 cells/dish) were with α-MEM containing 10% FBS and 10^{-8} M VD_3 co-cultured for 8 days, dishes were with 4 mL of collagenase (0.2%) treated for 20 min, and the OCLs were collected. Dentin's transverse slices (diameter 4 mm, ca. 200 μm thick) were placed in 96-well plates containing α-MEM (0.1 mL/well) with 10% FBS medium, 0.1 mL of OLCs preparations were transferred onto the slices to interact at 37°C for 90 min, the slices were transferred to 24-well plates containing α-MEM with 10% FBS (0.5 mL) and incubated for 48 h. At the beginning of the pit formation assay the test compounds were added. At the culture end, slices were placed in 1 M NH_4OH for 30 min, cleaned by ultrasonication to remove adherent cells, stained with Mayer's hematoxylin solution (1.0 g of hematoxylin, 0.2 g of $NaIO_3$, 50 g of $AlNH_4(SO_4)_2 \cdot 12H_2O$, 7.5 mL/L of CH_3CO_2H, pH 2.8) for 1 min, and washed with distilled water. With light microscopy and photography, the areas of resorption pits stained with the hematoxylin solution were identified.

9.6 MOUSE BONE MINERAL DENSITY ASSAY[4]

Female ddY mice (8 weeks old) were randomly divided into ovariectomized (OVX), OVX + test compound and sham-operated groups. On the OVX and OVX + test compound groups under pentobarbital anesthesia, bilateral ovariectomies were conducted. Sham operations were performed on the sham group with forceps touching their ovaries only. The mice in OVX + test compounds groups received proper doses of test compounds in solvent, while the mice in OVX and sham groups received the same solvent alone. The assay period was 5 weeks. On the 36th day, the mice were measured under pentobarbital anesthesia for bone mineral density (BMD) of the lumber spine using dual energy X-ray absorptiometry (DXA), killed by CO_2 asphyxiation, uteri were collected, trimmed of fat and connective tissue, and weighted immediately. During the assay, the mice were anesthetized with pentobarbital and lay in a fixed prone position with tape. Using the small animal program, the lumber spine was scanned and the average BMD was calculated in lumbar vertebrae.

9.7 PPAR-γ COMPETITOR ASSAY[5]

To 2.5 nM Fluormone, a high affinity, 27 pM ligand binding domain of PPAR-γ (aa193-475) was added, and fluorescent tagged PPAR-γ ligand with 9 nM Kd value was formed. The measurement gave PPAR-c LBD/Fluormone complex a high fluorescence polarization value (mP). Upon adding competitive compound to this complex, Fluormone was displaced and the reaction mixture gave a lower fluorescence polarization value. Upon adding noncompetitive compound to this complex, Fluormone was not displaced and the reaction mixture gave an unchanged fluorescence polarization value.

9.8 HUMAN SERUM ESTROGEN LEVEL ASSAY[5]

Postmenopausal women, aged 45–86 years (mean 68 years), at least 1 year since last menstrual bleeding, were volunteers without diseases interfering with bone metabolism. Serum samples were collected from a random subset of women and the estrogen levels were measured. Estradiol was determined using RIA method with 4 pM sensitivity and 4% to 9% coefficient of variation (CV). Estrone was measured using sandwich ELISA with a 15 pM sensitivity and 10% CV. Sex hormone binding globulin (SHBG) was determined using an ELISA (IBL) with an 0.2 nM sensitivity and 9% CV. As the molar ratio of total estradiol to SHBG, the free estrogen index was calculated.

9.9 LUCIFERASE ACTIVITY ASSAY[6]

In DMEM (Gibco-BRL, Paisley, Scotland, UK) containing 8% FCS (Biological Industries, Kibbutz Beth Haemek, Israel), MG-63 cells (ATCC number CRL-1427) were plated at 60% to 80% confluence. On the next day using FuGene 6 Transfection reagent at 3/2 ratio (μL reagent/μg DNA), 1–2 μg of each construct and 0.5 μg of pSV-h-Galactosidase Control vector were co-transfected into MG-63 cells. At proper concentrations, test compounds were added 6 h after transfection. Twenty-four hours after transfection, in a TD-20/20 Luminometer, luciferase (Luciferase assay system, Turner Designs, Sunnyvale, CA, USA) and β-galactosidase (chemiluminescent β-gal reporter gene assay) activities were measured. For each construct or treatment, the independent transfection number (replicas) was not less than 3, while in each transfection an independent clone was used and tested in duplicate.

9.10 IL-1β AND TNF-α LEVEL ASSAY[7]

Patients receiving revision showed a loose implant with radiologic osteolysis and without infection sign. Before cutting into small pieces, tissue specimens from patients were thoroughly washed with PBS, at 37°C in α-MEM (Gibco, Carlsbad, CA) containing 1 mg/mL of collagenase (Sigma, St. Louis, MO) supplemented with 100 IU/mL penicillin-G (Sigma) and 50 μg/mL gentamicin sulfate (Sigma) digested for 30 minutes, incubated in Versene (Invitrogen, Carlsbad, CA) for 1 h, the isolated cells were resuspended in α-MEM containing 10% heat-inactivated FBS, and at 37°C in a humidified atmosphere containing 5% carbon dioxide incubated for 4 passages before subsequent experiments. Using commercially available high-sensitivity enzyme-linked immunosorbent assay kit (Quantikine, R&D Systems) according to the manufacturer's instructions, the levels of IL-1β and TNF-α presented in the media were determined. Each sample was at 1000 × *g* centrifuged for 10 min to discard the cell, the supernatants were collected, and frozen at −80°C until assayed.

9.11 ER BINDING AND RECEPTOR ACTIVITY ASSAYS[8]

1. *ER Binding Assay*: Using human cDNAs encoding ERα and Erβ as templates, receptor proteins were expressed *in vitro*. The proteins were produced with rabbit reticulocyte lysates supplied by TNT kit (rabbit reticulocyte lysates, Promega) coupling transcription and translation in a single reaction. Template amount for each reaction was empirically determined, while the expression in parallel reactions was monitored based on [^{35}S]methionine incorporated receptor, its gel electrophoresis and film exposure. The final volume of the binging reaction TEG buffer (10 mM pH 7.5 Tris, 1.5 mM EDTA, 10% glycerol) was 100 μL. For each binding, 5 μL *in vitro* transcribed translated receptor and 0.5 nM [^{3}H]estradiol (E_2) were used. All compounds were routinely tested as ethanol dilution ranging from 10^{-2} nM to 1 μM. The reaction mixtures were at 4°C incubated overnight and by adding 200 μL of dextran-coated charcoal, bound E_2 was quantified. After 4°C and 15 min rotation, the tubes were centrifuged for 10 min, 150 μL of supernatant was added to 5 mL of scintillation cocktail for determining cpms by liquid scintillation counting. As background controls, 5 μL of unprogrammed rabbit reticulocyte lysates were used for each experiment and the maximal counts were typically 10% to 15%. The maximum binding was obtained by competing bound E_2 with vehicle (ethanol), which was set to 100% (maximal E_2 binding), based on which percent inhibition was calculated. Using the Prism Software the data were plotted and Ki values were calculated.

2. *Receptor Activity*: In 12-well plates cells were typically seeded and allowed to grow until 60% to 80% confluence. Using a liposome-based delivery system, transfection of DNA into cells was carried out. During transfection experiments, the cells were maintained in phenol red free medium containing 10% fetal calf serum, which had been treated with dextran-coated charcoal to remove steroids. In standard conditions, in each well 100 ng of pCMX-hERα or pCMX-hERβ, 1000 ng of ERE-driven luciferase reporter plasmid, and 500 ng of pCMX-β-galactosidase was transfected for 24 h, the medium was removed, the cells were treated with 0.1% ethanol (vehicle), 10 nM E_2, or 10 nM E_2 containing increasing concentrations of test compound for 24–48 h, harvested by 4°C and 15 min lysing in 10 mM potassium phosphate buffer (pH 7.8) and 1% Triton X-100. Using standard methods, with 25 μL of cell lysate luciferase and β-galactosidase activity were determined. With the

corresponding β-galactosidase activity, luciferase values were normalized. Receptor activity was expressed as the percentage of the maximal E_2 stimulation without test compound and set to 100%.

9.12 FLUORESCENT ESTROGEN RECEPTOR ASSAY[9]

1. *Saturation and Binding Assays*: Under 400 mbar vacuum produced by the Multiscreen vacuum manifold, filters of Multiscreen MABV NOB filtration plate (Millipore) were prewetted with 250 μL of assay buffer (pH 7.4 PBS with 8 mM dithiothreitol, i.e., DTT, and 0.1% ovalbumin). A native fluorescent phytoestrogen, coumestrol, was used as the fluorescent ligand. For the saturation assay 20 μL of fluorescent ligand (hRec ER-α, final concentration 0.1–4 nM or hRec ER-β, 0.1–10 nM final concentration) in PBS was added in duplicate into the wells of the filtration plate. For determining total binding and nonspecific binding, 20 μL of assay buffer and 20 μL of PBS containing 10 μM 17-β-estradiol or test compound were added, respectively, to which 160 μL of receptor suspension (480 fmol receptor protein) was added, the plate was placed on an orbital micro-plate shaker (800 rpm), incubated at room temperature for 1 h, the receptor protein was precipitated by adding 50 μL of PBS containing 1% γ-globulin and 36% PEG 6000, incubated for 15 min, the supernatant was decanted, washed three times with 250 μL of assay buffer, 200 μL of 17-β-estradiol (E2) in PBS (10 μM) was added to dissociate the fluorescent ligand from the receptor protein, incubated for 1 h, and the supernatant was collected and 100 μL aliquots were directly injected into the HPLC system. During assays, a capmat was placed onto the microtiter plate to prevent evaporation.

2. *Binding Assays*: Under vacuum produced by the Multiscreen vacuum manifold, filters of Multiscreen MABV NOB filtration plate were prewetted with 250 μL of assay buffer (pH 7.4 PBS with 8 mM DTT and 0.1% ovalbumin), into the wells containing 20 μL of 17-β-estradiol or test compound in PBS (0.01–50 nM final concentration) and 20 μL of fluorescent ligand in PBS (1.2 nM, final concentration), 160 μL of receptor suspension was added, at room temperature incubated for 1 h, and the supernatant was collected and 100 μL aliquots were directly injected into the HPLC system. During assays, a capmat was placed onto the microtiter plate to prevent evaporation.

3. *HPLC System*: A Shimadzu SCL-10A system controller, a Shimadzu FCV 10 AL Low-pressure gradient flow, a Shimadzu LC 10 AD solvent delivery module, a Shimadzu CTO 10 AS column-oven, an automatic degasser, a 96-well-plate configured autoinjector, and a Shimadzu RF-10 Axl fluorescent detector were involved in the chromatographic system. The Shimadzu CLASS-VP software (version 6.10), a 125 mm × 4 mm i.d. column packed with 5 μM Lichrospher 100 RP-18 (VWR) and a potassium phosphate (50 mM, pH 6.8) MeOH buffer (35:65, v/v) at 0.8 mL/min flow rate were used for integrating peak areas, separating components, and eluting the fractions, respectively. Column temperature was a constant 30°C. The fluorescent detector was set excitation at 379 nm and emission at 436 nm. The limit of detection (LOD) for the fluorescent ligand was 5 pM (injection volume was 100 μL).

9.13 ELISA FOR URINARY HELICAL PEPTIDE[10]

Urinary levels of peptide α1 (I) 620–633 originating from the helical part of type I collagen and consisting of the 620–633 sequence of the α1 chain were determined by competitive ELISA, which used a mouse monoclonal antibody against the synthetic ^{620}Ala-Hyp-Gly-Asp-Arg-Gly-Glu-Hyp-Gly-Pro-Hyp-Pro-Ala^{633} sequence of the α1 chain of type I collagen. With the homologous peptides of type II and type III collagen, this antibody has no significant cross-reactivity. In assay, after adding 20 μL of standards (synthetic ^{620}Ala-Hyp-Gly-Asp-Arg-Gly-Glu-Hyp-Gly-Pro-Hyp-Pro-Ala^{633} sequence), controls, or unknown urine samples to microtiter wells coated with mouse monoclonal antibody, 150 μL of helical peptide alkaline phosphatase conjugate in 0.1 M sodium phosphate was added, the plates were at 4°C incubated overnight, and washed three times with PBS-Tween. During incubation, peptide α1 (I) 620–633 in the urine sample competed with peptide α1 (I) 620–633-alkaline phosphatase conjugate. The amount of antigen in urine was inversely proportional to the optical density measured at 405 nm.

Based on an immobilized synthetic peptide (Gly-Lys-Ala-His-βAsp-Gly-Gly-Arg, CrossLaps antigen) with an amino acid sequence specific for a part of the C-telopeptide of the α1 chain of type I collagen, urinary C-terminal cross-linking telopeptides of type I collagen was measured by an ELISA. The intra-assay and inter-assay CVs were less than 5% and 8%, respectively. Using a standard colorimetric method urinary helical peptide and urinary C-terminal cross-linking telopeptides of type I collagen data were corrected by the urinary creatinine (Cr) concentration measured.

9.14 URINE MIDMOLECULE OSTEOCALCIN ASSAY[11,12]

A shorter osteocalcin (Oc) peptide with sequence corresponding with 12–30 fragment of human Oc was used as calibrator and tracer to develop the radioimmunoassay. Peptide 12–30 (5–7 μg) was dissolved in 10 μL of 500 mM sodium phosphate (pH 7.5) and mixed with 1 mCi ^{125}I (specific activity 17.4 Ci/mg) to prepare the radioiodinated tracer via chloramine-T method leaded iodination reaction, which was separated from free iodine on a C18 Sep-Pak cartridge (Waters Associates, Milford, MA). In the separation, the cartridge was primed according to the manufacturer's instructions, the iodination mixture was loaded on the column, the column was washed with 10 mL of 0.1% TFA to remove unbound ^{125}I, and ^{125}I Oc peptide was eluted with 50% AcN in 0.1% TFA.

The diluent buffer, 50 mM sodium borate (pH 8.3) supplemented with 1% RIA grade BSA (Sigma, St. Louis, MO) and 0.05% sodium azide, was used for all RIA procedures. In this assay, 100 μL of tracer and 100 μL of primary antibody solution (1:70 dilution) were treated with either 100 μL of standard or 50 μL of unknown urine samples at 4°C overnight for equilibrium. Phase separation was accomplished by adding 100 μL of preimmune chicken serum (1:30 dilution), 400 μL of secondary antibody (goat anti-chicken IgG, 1:40 dilution), and 200 μL of PEG (8%, formula weight 8000). The samples were thoroughly mixed, at 40°C incubated for 3 h, and centrifuged to collect the precipitates. On gamma counter, the bound and total ^{125}I-labeled peptide was determined.

9.15 BMD AND OSTEOCALCIN ASSAY[13]

Patients were divided into group A (25 normal young women, 30 ± 2.5 years, with no evidence of abnormalities in serum calcium levels or skeletal, kidney, or liver function), group B (25 healthy premenopausal women, 42 ± 5 years, with normal laboratory values and no history of chronic illness or metabolic bone disease, BMD values within normal range), group C (25 postmenopausal women, 54 ± 4 years, with BMD measurements at the ultradistal radius within 2 standard deviations of the average value of subjects of matched age, after recruitment, all of them received 40 IU/d of carbocalcitonin intramuscularly every 2 months for 9 months), and group D (25 women, 50 ± 10 years, with chronic renal failure receiving maintenance hemodialysis for 4.5 ± 0.8 years). In the morning after an overnight fast, blood samples were collected from patients in groups A–C and patients in group D before and after dialysis. At the therapy end, an additional blood sample was collected from patients in group C. All blood samples were stored at –20°C until used. To eliminate interassay variation, the samples from each group were measured with all four assays. Using a computerized dual X-ray densitometer, BMD of the ultradistal forearm was measured at the same time only for groups B and C and at the treatment end for group C. Test osteocalcin was assayed according to the manufacturers' instructions with OSTK-PR, ELSA-osteo, ELSA-OST-Nat, and Osca.

9.16 *IN VIVO* ANTIOSTEOPOROSIS ASSAY ON MICE[14]

The procedure of this assay is summarized in Fig. 9.1. The assessments were performed based on a protocol reviewed and approved by the ethics committee, namely the welfare of the animals was maintained in accordance with the requirements of the animal welfare act and according to the guide for care and use of laboratory animals. The tested compound was dissolved in aqueous solution of 0.5% carboxymethylcellulose just before use and kept in an ice bath. ICR mice weighting 30.7 ± 3.1 g (7 weeks old) were anesthetized with sodium pentobarbital (40.0 mg/kg, ip). The mice of OVX groups were given abdominal ovariectomy by standard procedure, and the mice of the sham group were given ovariotomy only (Fig. 9.1). On the 5th day of surgical operation, the mice of the ovariotomy and sham groups were orally administrated 0.2 mL of aqueous solution of 0.5% carboxymethylcellulose, the mice of treatment groups were orally administered the solution of 110.3 nmol/kg estradiol, 110.3 nmol/kg of the test compounds in 0.2 mL of aqueous solution of 0.5% carboxymethylcellulose once a day. All of the mice were treated according to the corresponding procedure for 4 weeks. On the next day of the last administration, the mice were weighed, blood was drawn via eye orbit, anesthetized with sodium pentobarbital

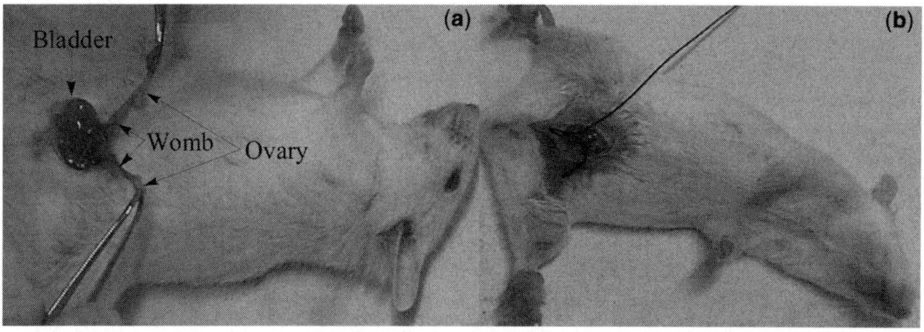

Figure 9.1 (a) Anatomic sites of bladder, womb and ovaries; (b) incision after removing ovaries.

(40.0 mg/kg, ip), and executed to remove the lungs, livers, spleens, uteri, and left femurs.

After 30 min standing, the blood was centrifuged at 3000 × *g* for 20 min and the serum was stored at −20°C before use. The calcium content of the serum was measured using the method of *o*-methylphenolphthalein complexing ketone. The phosphorous content of the serum was measured by the method of molybdenum bleu. The alkaline phosphatase content of the serum was measured by use of disodium phenylphosphate as the substrate.

The lungs, livers, spleens, and uteri were weighed. After completely removing the muscle, the lengths of left femurs were measured and then immersed in the solution of chloroform-methanol (2:1) for two times (each time 3 h). After defatting, the left femurs were heated at 120°C for 6 h, cooled, and weighed to record the dry weight. The femurs were incinerated in a muffle furnace at 800°C for 8 h, cooled, weighed to record the ash weight and calculate the rate of the ash weight to dry femur weight (namely the mineral content of the femur).

The ashes of the left femurs were dissolved in 0.5 mL of hydrochloric acid (6 N) and diluted to 5 mL with ultrapure water, from which 0.05 mL of the solution was drawn and diluted to 1 mL with ultrapure water before use. The calcium content of the aqueous solution was measured by the method of *o*-methylphenolphthalein complexing ketone. The phosphorous content of the aqueous solution was measured by the method of molybdenum blue.

REFERENCES AND NOTES

1. L. Qin, T. Han, Q. Zhang, D. Cao, H. Nian, K. Rahman, H. Zheng. Antiosteoporotic chemical constituents from Er-Xian decoction, a traditional Chinese herbal formula. J Ethnopharmacol 118 (**2008**) 271–279.
2. Y. Zhao, B. Zou, Z. Shi, Q. Wu, G. Chen. The effect of 3-hydroxybutyrate on the *in vitro* differentiation of murine osteoblast MC3T3-E1 and in vivo bone formation in ovariectomized rats. Biomaterials 28 (**2007**) 3063–3073. *Note*: The experiments were performed on 3-month-old female Wistar rats. After 7-day adaptation under 50 mg/kg sodium pentobarbital anesthesia, bilateral ovariectomy or a sham operation was performed.

On the dorsolateral body wall, a longitudinal incision was made inferior to the rib cage and the ovaries were exteriorized, ligated, and excised. Rats for sham surgical procedure had their ovaries exteriorized only and then replaced. At the end of each experiment, by the absence of ovarian tissue was verified to confirm the success of ovariectomy.

3. P. Pietschmann, M. Skalicky, M. Kneissel, M. Rauner, G. Hofbauer, D. Stupphann, A. Viidik. Bone structure and metabolism in a rodent model of male senile osteoporosis. Exp Gerontol 42 (**2007**) 1099–1108.
4. J.X. Li, J. Liu, C.C. He, Z.Y. Yu, Y. Du, S. Kadota, H. Seto. Triterpenoids from Cimicifugae rhizoma, a novel class of inhibitors on bone resorption and ovariectomy-induced bone loss. Maturitas 58 (**2007**) 59–69.
5. C.R. Hopkins, S.V. O'Neil, M.C. Laufersweiler, Y. Wang, M. Pokross, M. Mekel, A. Evdokimov, R. Walter, M. Kontoyianni, M.E. Petrey, G. Sabatakos, J.T. Roesgen, E. Richardson, T.P. Demuth Jr. Design and synthesis of novel N-sulfonyl-2-indole carboxamides as potent PPAR-c binding agents with potential application to the treatment of osteoporosis. Bioorg Med Chem Lett 16 (**2006**) 5659–5663.
6. N. García-Giralt, A. Enjuanes, M. Bustamante, L. Mellibovsky, X. Nogués, R. Carreras, A. Díez-Pérez, D. Grinberg, S. Balcells. In vitro functional assay of alleles and haplotypes of two COL1A1-promoter SNPs. Bone 36 (**2005**) 902–908.
7. Y. Qian, B. Zeng, X. Zhang, Y. Jiang. Substance P stimulates production of interleukin 1β and tumor necrosis factor α in fibroblasts from Hip periprosthetic membrane. J Arthroplasty 23 (**2008**) 581–585.
8. J.M. Schmidt, G.B. Tremblay, M.A. Plastina, F. Ma, S. Bhal, M. Feher, R. Dunn-Dufault, P.R. Redden. In vitro evaluation of the anti-estrogenic activity of hydroxyl substituted diphenylnaphthyl alkene ligands for the estrogen receptor. Bioorg Med Chem 13 (**2005**) 1819–1828.
9. T. de Boer, D. Otjens, A. Muntendam, E. Meulman, M. van Oostijen, K. Ensing. Development and validation of fluorescent receptor assays based on the human recombinant estrogen receptor subtypes alpha and beta. J Pharma Biomed Anal 34 (**2004**) 671–679.
10. P. Garnero, P.D. Delmas. An immunoassay for type I collagen α1 helicoidal peptide 620–633, a new marker of bone resorption in osteoporosis. Bone 32 (**2003**) 20–26.
11. A.K. Srivastava, S. Mohan, F.R. Singer, D.J. Baylink. A urine midmolecule osteocalcin assay shows higher discriminatory power than a serum midmolecule osteocalcin assay during short-term alendronate treatment of osteoporotic patients. Bone 31 (**2002**) 62–69.
12. P. Ravn, G. Neugebauer, C. Christiansen. Association between pharmacokinetics of oral ibandronate and clinical response in bone mass and bone turnover in women with postmenopausal osteoporosis. Bone 30 (**2002**) 320–324.
13. M. Cecchettin, S. Bellometti, G. Torri, L. Galzigna. Comparison of commercial osteoclacin assay kits in evaluating osteoporosis. Curr Ther Res 56 (**1995**) 163–167.
14. Y. Xiong, M. Zhao, C. Wang, H. Chang, S. Peng. Improved antiosteoporosis potency and reduced endometrial membrane hyperplasia during HRT with estrogen-RGD peptide conjugates. J Med Chem. 50 (**2007**) 3340–3353.

10

METHODS AND APPLICATIONS OF IMMUNOMODULATING ASSAYS

Shiqi Peng

The innate immune system is at the front line in the host's defense against microbial invasion, and immunomodulation is widely involved in the onset and progression, in the diagnosis and prevention, and in the treatment of numerous diseases. For instance, immediate hypersensitivity is the basis of acute allergic reactions caused by the activation of basophils and mast cells when an allergen interacts with membrane-bound IgE; a role for mast cells has been implicated in chronic inflammatory disorders including rheumatoid arthritis; an important role for CD83 in the immune response is known via selective expression and upregulation together with co-stimulatory molecules like CD80 and CD86; dendritic cells (DCs) are highly effective antigen presenting cells that have become popular in the field of cancer immunotherapy, and DC migration is a pivotal step in the initiation and modulation of the adaptive immune response to cancer; from human and animal studies, the evidence strongly suggests that immunosuppression and tumor cell growth are closely related; the vaccine has long been used in antituberculosis treatment and as a gold standard for the diagnosis of latent tuberculosis; a fundamental immunologic process has been discovered to be dependent upon carbohydrate sulfation; T cells may play a key role in alloreactive immune responses of acute rejection; and emerging evidence establishes

Pharmaceutical Bioassays: Methods and Applications. By Shiqi Peng and Ming Zhao

that the immune system employs carbohydrates as recognition determinants for various cellular interactions. All of these achievements in the immune system are supported by immunomodulating assays established in the past decade. In this chapter, 29 assays are presented: rat mast cell histamine-release assay,[1] rabbit aortic force assay,[1] dopaminergic cell death-based neural transplantation assay,[2] mast cell degranulation assay,[3] basophils assay as allergen,[4–6] RBL-2H3 cell desensitization assay,[7] migration assay of dendritic cell from PBMCs,[8] mouse EAE induction assay,[8] COSTIM assay for DC/T-cells,[9,10] cytokine assay for IL-6, IL-10, IL-12, and TNF-α of DCs,[11] ELISPOT assay for DC IFN-γ,[11] DC function assay for evaluating toll-like receptor function,[12] migration assay of dendritic cell from bone marrow of A/J mice,[13] lymphoid organ assay,[14–16] ELISA of IFN-γ from human myelomonocytic KG-1 cells,[17–19] human whole blood IFN-γ assays,[19,20] sheep whole blood IFN-γ assays,[21] ELISPOT assay for IFN-γ,[22,23] IFN-β RG assay,[24] anti-rHuEPO NAb assay,[25,26] chloramphenicol acetyltransferase assay,[27] chemotaxis assay,[28] fibroblast-populated microsphere assay,[29] fibroblast-populated concentric microsphere assay,[29] radial assay of chemotaxis,[30] antibody forming cell assay,[31] immunosuppressive assay,[32] cell-based ELISA,[33] and large animal lung transplantation assay.[34]

10.1 RAT MAST CELL HISTAMINE-RELEASE ASSAY[1]

Ice-cold PBS (10 mL, 137 mM NaCl, 2.68 mM KCl, 0.91 mM $CaCl_2$, 8.1 mM Na_2PO_4, 1.47 mM KH_2PO_4, 0.91 mM $MgCl_2$, 5.6 mM glucose, and 20.0 mM HEPES) was injected (ip) into male Wistar rats (200–250 g); 90–120 s after injection, PBS was collected, and peritoneal lavage was subsequently washed with 5 mL and 10 mL of PBS. The washing PBS were collected, combined with PBS, and centrifuged (200 × *g*, 5 min, 4°C). After washing twice with PBS, the pellet was resuspended in 10 mL of PBS. From the rat peritoneal lavage, mast cells were isolated.

Preincubation of 1.8 mL of mast cell suspensions was carried out at 37°C for 10 min, 0.1 mL of test compound was added to stimulate the mast cells, which were cooled in ice to terminate histamine release. After centrifugation (100 × *g*, 10 min, 4°C), to both supernatant and the pellet equal volume of 0.8 M $HClO_4$ and double volume of 0.4 M $HClO_4$ were added. A mixture of 125 μL of 5 M NaOH, 0.4 g NaCl, and 2.5 mL of *n*-butanol was added to 1 mL of solution of test compound, the mixture was centrifuged (200 × *g*, 1 min, room temperature) and separated. The upper organic phase was mixed with 2 mL of 0.1 M NaOH saturated with NaCl and centrifuged (200 × *g*, 1 min, room temperature) and this procedure was repeated, of which the upper organic phase was mixed with 2 mL of 0.1 M HCl and 7.6 mL of *n*-heptane. The lower aqueous phase (1 mL) was mixed with 0.1 mL of 10 M NaOH. By incubation with 0.1 mL of *o*-phthalaldehyde (OPT, 10 mg/mL methanol) at room temperature for 4 min OPT was conjugated with histamine, and by adding 0.6 mL of 3 M HCl this conjugation was terminated. The fluorescence of histamine-*o*-phthalaldehyde conjugate was assessed at 450 nm with emission excitation at 360 nm.

10.2 RABBIT AORTIC FORCE ASSAY[1]

Thoracic aorta of male Japanese White rabbits (3–3.5 kg) was cut into helical strips (approximately 4 mm wide and 20 mm long) and its endothelium was removed by gently rubbing the endothelial surface with cotton pellets. In a 1-mL organ bath containing the modified Krebs-Ringer-bicarbonate solution (120 mM NaCl, 4.8 mM KCl, 1.2 mM $CaCl_2$, 1.3 mM $MgSO_4$, 25.2 mM $NaHCO_3$, 1.2 mM KH_2PO_4, 5.8 mM glucose), the aortas were mounted and suspended. With a force-displacement transducer connected to a polygraph, the muscle tensions of the aortas were recorded isometrically. A passive tension of 1 g was initially applied and the aortas were equilibrated for 60 min, after which 60 mM NaCl in the modified Krebs-Ringer-bicarbonate solution was replaced by equimolar KCl for precontraction of the aortas. Reaching a steady level of the response, the test compound assay was started. Contractile response to histamine was normalized with that of high K^+.

10.3 DOPAMINERGIC CELL DEATH-BASED NEURAL TRANSPLANTATION ASSAY[2]

From pentobarbitone-anesthetized time-mated Wistar rats (200 mg/mL, dose 140 mg/kg) receiving abdominal laparotomy, uterine horns were collected in sterile Dulbecco's PBS. In aseptic conditions at room temperature, fetuses were collected in HBSS (w/o Ca^{2+} and Mg^{2+}). To verify fetal age, crown-rump lengths were measured. The ventral mesencephala (VM) having the developing substantia nigra and ventral tegmental area was dissected in HBSS at room temperature and stored in HBSS containing 0.05% deoxyribonuclease (DNase) on ice. At 37°C, pooled explants were incubated in 0.1% trypsin DNase in HBSS for 20 min and the excess trypsin was removed. At 37°C, the explants were rinsed briefly with 0.1% soybean trypsin inhibitor in HBSS/DNase and then rinsed three times in HBSS/DNase. Using 1000 μL and 100 μL Eppendorf pipette tips (20 strokes/tip), explants were dissociated by gentle trituration in HBSS/DNase (20 μL/explant). Using the trypan blue exclusion method, cell viability and density were estimated. Cells (3000 cells/mm^2) were plated into 4-well plates (2 cm^2/well; Nunc, coated with 100 μg/mL poly-D-lysine) containing 0.5 mL of culture medium warmed to 37°C and incubated at this temperature in a humidified 95% air/5% CO_2 atmosphere for 48 h. By removing 0.35 mL of medium from each well and adding 0.35 mL of 37°C fresh medium to each well, the medium was replenished, and by rinsing briefly with 37°C HBSS, cultures were stopped. Prior to subsequent immunocytochemistry, cultures were fixed with 4% paraformaldehyde in PBS at room temperature for 30 min.

After three-time rinse of fixed cultures with PBS, the cells were permeabilized for 30 min by use of 0.5% Triton X-100 in PBS. Using PBS containing 3% H_2O_2 and 10% methanol, endogenous peroxidase activity was quenched for 10 min. With 10% heat-inactivated normal goat serum in 0.1% Triton X-100 in PBS at room temperature, blocking nonspecific binding sites for 1 h background staining, the cells was reduced. Cells were incubated with the primary antibody (polyclonal rabbit anti-TH, diluted 1 : 1500 using 0.1% Triton X-100 in PBS) for 48 h at 4°C, rinsed with PBS, and incubated in the secondary antibody (biotinylated goat anti-rabbit IgG, 1 : 200 in PBS) at room temperature for 1 h. The cells were rinsed with PBS and incubated with an avidin-biotin horseradish peroxidase complex at room temperature for 1 h. After incubation with 0.05% 3,38-diaminobenzidine (DAB) in Tris-HCl-buffered saline (TBS) containing 0.01% H_2O_2 for 10 min, TH-immunoreactive neurons (TH1ve) were visualized.

Fixed cells were permeabilized and endogenous peroxidase activity was quenched. In the TUNEL reaction mixture, fixed cells were incubated at 37°C for 1 h, rinsed three times in PBS, incubated in the anti-fluorescein antibody-conjugated peroxidase (diluted 1:2 in TBS containing 1% normal goat serum) at 37°C for 45 min, rinsed three times in PBS, and incubated in TBS containing 0.05% DAB and 0.01% H_2O_2 for 10 min to visualize TUNEL-positive nuclei. As positive control, fixed and permeabilized cells were incubated in PBS containing 0.5 mg/mL DNase at room temperature for 15 min, rinsed and incubated in the TUNEL reaction mixture. As negative control, fixed and permeabilized cells were incubated in the label solution (without terminal transferase) instead of the TUNEL reaction mixture.

To confirm apoptosis and distinguish necrotic from apoptotic cells, by visualization of condensed and fragmented chromatin, 20 μL aliquot of the fluorescent DNA-binding dyes consisting of ethidium bromide (100 μg/mL) and acridine orange (100 μg/mL) dissolved in PBS was added to each well. On a Leica DM IRB fluorescence microscope with a standard fluorescein excitation filter, the cultures were viewed.

After permeabilization, all incubations of fixed cells were carried out in the dark. By use of the ApopTag In Situ Apoptosis Detection kit (Appligene Oncor, S7110 fluorescein kit), TUNEL labeling was performed. After rinsing with PBS, cells were incubated with a primary antibody to TH (monoclonal anti-TH, diluted 1:4000 in PBS containing 0.1% Triton X-100) at 4°C for 48 h, rinsed again with PBS, incubated in the secondary antibody (biotinylated anti-mouse IgG, made in horse, rat adsorbed, diluted 1:200 in PBS) at room temperature for 2 h, rinsed in PBS, and finally incubated in Texas red–Avidin D (diluted 1:200 in PBS) at room temperature for 2 h. After rinsing in PBS, cultures were examined: on a fluorescence microscope with standard fluorescein and rhodamine excitation filters.

10.4 MAST CELL DEGRANULATION ASSAY[3]

Isogenic female R/A Tor rats (weighing 250 g) were subcutaneously injected with albumin pool in the presence of adjuvant mixture of 5 mg of $Al_2(OH)_3$ and 5 mg of nigrosin and were anesthetized and bled by cardiac puncture after 15 days. The sera were pooled and tested by mast cell degranulation assay. Rats were euthanized with CO_2, and a peritoneal wash was performed with the injection of 20 mL of DMEM (Gibco) containing 12 U/mL of heparin. The abdomen was gently massaged for about 90 s. After carefully opening the peritoneal cavity, the fluid containing peritoneal cells was aspirated. The cells were incubated at 37°C in Petri plates for 30 min. Two thirds of the superior layer was aspirated and discarded. Mast cell-rich supernatant (1.8×10^5 mast cell/mL) was separated in aliquots of 100 μL in Eppendorf tubes. As a control, mast cells (100 μL) were incubated with Tor rat pre-immune serum and were initially activated at 37°C for 60 min using a 2S albumin polyclonal anti-rat IgE. Cells were sensitized with IgE and washed twice with DMEM. Each of these assays was carried without or with the 2S albumin pool (100 ng). After incubation with antibodies and allergens, cells (10 μL) were stained for 15 min with 10 μL of solution containing 0.1% toluidine blue, 10% formaldehyde, and 1% acetic acid, pH 2.8, to visualize degranulated mast cells. Granulated and degranulated mast cells were counted under a light microscope using the ×40 objective in a Neubauer chamber.

10.5 BASOPHILS ASSAY AS ALLERGEN[4–6]

1. *ELISA and Dot Immunoblot*: With 15 μg of *Dolichos biflorus* agglutinin (DBA), microtiter wells were at pH 9.6 and 4°C coated overnight, blocked with 3% gelatin, incubated with subjects' sera at 1:3 dilution and 4°C overnight, incubated with AP-conjugated murine monoclonal anti-human IgE at 1:1500 dilution and 37°C for 2 h, and color developed. Using 15 μg of DBA, which was applied as a spot on the nitrocellulose (NC) membrane, dot immunoblot was carried out with concanavalin A (con A) and ovalbumin (OVA) as reference controls and air-dried. The membrane was blocked, incubated with non-atopic or DBA-sensitized subject's serum (1:3 dilution) at 4°C overnight, and murine monoclonal anti-human IgE-AP conjugate (1:1500 dilution) was used as the secondary antibody. Using BCIP-NBT substrate (at 1:3 dilution, Bangalore Genei, Bangalore, India) the membrane was developed, having a positive blue spot against a white background.

2. *Isolation of Leukocytes Containing Basophils*: Using 6% dextran T 700 gradient (Hi-Media Laboratories, Mumbai, India) and from 10 mL of heparinized venous blood, leukocytes (buffy coat containing basophils) were isolated. After 4–5 isotonic PBS washes, the leukocyte layer was resuspended in Tris-CAM buffer. Using crystal violet, the isolated leukocytes were counted. Using Trypan blue dye exclusion, cell viability was determined.

3. *Isolation of Rat Peritoneal Exudate Cells (PEC)*: Male Wistar rats (4 week old, ca. 300 g) were injected with pH 7.4 Tyrode buffer containing 0.1% BSA. Five minutes later, the fluid containing PECs was collected. By treatment with 150 mM NH_4Cl buffer, the residual erythrocytes were removed and PECs were pelleted, washed, resuspended in Tris-CAM buffer, stained for mast cells using toluidine blue, and their viability was evaluated by Trypan blue dye exclusion.

4. *Histamine Release (HR) Assay*: Cells, test compounds (DBA or other proteins), and Tris-CAM buffer were mixed in polystyrene tubes at a final volume of 1 mL in an ice bath. The tube (containing ca. 2×10^6 cells/mL) was incubated at 37°C for 45 min. To obtain the total histamine content of cells (defined as Pc), perchloric acid (final concentration, 3%) was added to one set of samples or alternatively the tubes were reflexed at 100°C for 10 min. Tubes only containing cells and buffer were used as controls for nonspecific or spontaneous release (defined as Ps). After 45 min, the tubes were transferred to an ice bath to stop the reaction, and centrifuged

at 4°C and 275 × *g* for 20 min to get the supernatants for testing histamine content of test samples (defined as Pt). The histamine in the supernatant was extracted subsequently into *n*-butanol and HCl. After neutralization and *o*-phthalaldehyde derivatization, the reaction was arrested with phosphoric acid. The fluorescence intensity was measured at 360 nm and 450 nm. HR was represented as [(Pt – Ps)/(Pc – Ps)] × 100.

10.6 RBL-2H3 CELL DESENSITIZATION ASSAY[7]

1. *Degranulation-Based Desensitization*: RBL-2H3 cells grown in DMEM (Gibco-BRL, NY, USA) supplemented with 10% heat-inactivated FCS (Gibco-BRL, NY, USA), 2% L-Gln, and 0.2% combined antibiotics were detached, harvested from culture flasks at the confluent phase by 15 min treatment with 10 mM EDTA, resuspended in fresh DMEM, plated in 96-wells plates (1×10^5 cells/100 μL/well final concentration), and allowed to adhere at 37°C for 2 h in the presence of the relevant concentrations of monomeric IgE. Cells were washed three times with DMEM, incubated for 1–12 h with the desensitizing concentrations of either antigen or the IgE-specific mAb95.3, and stimulated with optimal doses of the clustering agents for 30 min to desensitize. After stimulation, 20 μL aliquots of supernatant were transferred into other 96-wells plate and at 37°C treated with 50 μL of solution of *p*-nitrophenyl-*N*-acetyl-D-glucosamine in 0.1 M citrate (1.3 mg/mL, pH 4.5) for 1 h. The reaction was terminated by adding 150 μL of glycine solution (0.2 M, pH 10.7). The color intensity of nitrophenol produced by the enzymatic reaction was read on a computer-assisted ELISA reader at 405 nm. The cells' secretory response was defined as net (i.e., minus basal secretion values) percentage of the cells' total enzyme activity content from lysis of control cells' sample with 1% Triton-100.
2. *Cytokines Secretion-Based Desensitization*: RBL-2H3 cells grown in DMEM supplemented with 10% heat-inactivated FCS, 2% L-Gln, and 0.2% combined antibiotics were detached from culture flasks with 10 mM EDTA, resuspended in fresh DMEM, plated in 96-well plates (final concentration, 1×10^5 cells/100 μL/well), and treated with the appropriate concentrations of IgE. Cells were incubated for 4 to 14 h with the determined subthreshold concentrations of antigen, washed three times with DMEM and stimulated with antigen, of which the concentration either ranged from 0.1 to 3000 ng/mL or was optimal, as determined separately for each IgE concentration, to desensitize. According to the manufacturer's instructions with the TNF-α and IL-4 specific kits, by ELISA the cells' TNF-α or IL-4 basal secretion and the secretory response after desensitization treatment or optimal stimulation were determined.

10.7 MIGRATION ASSAY OF DENDRITIC CELL FROM PBMCs[8]

Human peripheral blood mononuclear cells (PBMCs) isolated from leukapheresis by sedimentation in Lymphoprep were seeded into cell culture flasks at 37°C in 5% CO_2, incubated for 1 h in RPMI 1640 (BioWhittaker, Verviers, Belgium) containing 1% L-Gln, 1% penicillin/streptomycin (Sigma-Aldrich), 1% HEPES, and 1% heat-inactivated (56°C, 30 min) AB human plasma. The nonadherent cell fraction (NAF) was harvested, while the adherent cells were incubated for another 24 h. Cells were fed with GM-CSF (800 U/mL, Amgen GmbH, Munich, Germany) and IL-4 (500 U/mL) for 2 days, then were fed again with GM-CSF (400 U/mL) plus IL-4 (500 U/mL, Strathmann, Hamburg, Germany) for additional 2 days. On day 5, cells were matured with a maturation cocktail consisted of GM-CSF (40 U/mL), IL-4 (200 U/mL), IL-1β (1 ng/mL, Sigma-Aldrich), IL-6 (1000 U/mL, Strathmann) PGE2 (0.5 μg/mL, Cayman Chemicals, Ann Arbor, MI, USA), and TNF-α (1.25 ng/mL, Boehringer Ingelheim, Vienna, Austria).

Transwell inserts (Costar, London, UK) with a pore size of 5 mm and 24-well plates were preincubated with 100 μL of migration medium in 24-well plates, each well containing 600 μL of same medium. A total of 2×10^5 test compound treated DC and untreated DC (control) were resuspended in migration medium (RPMI 1640 supplemented with 500 U/mL GM-CSF, 250 U/mL IL-4, 1% autologous serum, and glutamine) and seeded in the upper compartment. To analyze migration toward the chemokine gradient, CCL19 (100 ng/mL) was added to the lower wells, while for migration against a CCL19-gradient the chemokine (100 ng/mL) was added to the upper well. DCs were migrated for 120 min, harvested from the lower chamber, and collected by brief centrifugation. On removal of supernatant, the cells were lysed with 25 μL of PBS and 5 μL of 1% Triton X-100 (Roche Diagnostics, Mannheim, Germany). β-Glucuronidase activity in the lysates was tested with *p*-nitrophenyl-β-D-glucuronide (Sigma) according to the manufacturer's instructions, and the resultant absorbance was measured on a Wallac Reader (Wallac, Turku, Finland) at 405 nm, and the number of migrated cells was calculated using a separate standard curve for each assay.

10.8 MOUSE EAE INDUCTION ASSAY[8]

To induce encephalomyelitis (EAE) on day 0, female C57Bl/6 mice were immunized subcutaneously (sc) with MOG peptide 35-55 (100 μg/mouse, Sigma Genosys) in 50 μL of H_2O emulsified in 50 μL of CFA enriched with 10 μg/mL *Mycobacterium tuberculosis* (H37Ra, Difco/PD PharMingen) in each flank. On days 0 and 2, 200 ng pertussis toxin (Pt) was also administered intraperitoneally (ip). EAE paralysis of mice was defined as (0) no disease, (1) tail weakness, (2) paraparesis, (3) paraplegia, (4) paraplegia with forelimb weakness, and (5) moribund or dead animals. Test compound MCS-18 was properly administered. The statistical significance of differences in clinical index between groups was analyzed using Student's *t*-test. Significance was accepted only when $p < 0.05$.

10.9 COSTIM ASSAY FOR DC/T-CELLS[9,10]

COSTIM assay selectively measured co-stimulatory activity, or functional potency of dendritic cells (DCs), in which T-cells stimulated with a suboptimal amount of anti-CD3 antibody were unable to proliferate unless a source of co-stimulation (DCs) was added to the culture. Before being used either for DC culture or for preparing T-cells, human peripheral blood mononuclear cells (PBMCs) isolated from freshly leukapheresed blood were centrifuged and cryopreserved in 90% autologous serum and 10% DMSO. By negative depletion and using anti-HLA-DR monoclonal antibody-conjugated paramagnetic beads (Dynal, Lake Success, NY), from allogeneic PBMCs enriched T-cells were prepared. The cells with potential co-stimulatory function were removed from the PBMCs; as suspension each batch of which was tested for the presence of any remaining B-cells, monocytes, and 80% to 90% T-cells. Using cell-specific paramagnetic bead preparations and by biomagnetic separation, the stimulators B-cells and T-cells were purified and the purity of each cell type was greater than 90%, as determined by flow cytometry.

The cryopreserved PBMCs were thawed in warm AIM-V medium (Life Technologies, Gaithersburg, MD), washed with PBS, and resuspended in Opti-MEM medium (Invitrogen Corp., Carlsbad, CA) supplemented with 1% heat-inactivated autologous plasma to prepare 5×10^6 to 1×10^7 cells/mL suspension. The suspension of 1×10^9 cells was transferred into T-75 culture flasks and cultured for 1 h, the nonadherent cells were resuspended, aspirated out, and stringently washed with cold PBS to remove loosely adherent cells. To each flask, 1.5×10^9 Opti-MEM medium containing 5% heat-inactivated autologous plasma, 500 U/mL rhGM-CSF, and 500 U/ml rhIL-4 was added, and the adherent cells were incubated for 6 days. These DCs were then treated either with Bacillus Calmette-Guerin (BCG) alone or BCG plus IFN-γ for 24 h.

In the COSTIM assay, allogeneic T-cells were thawed in warm AIM-V culture media, washed with PBS, and resuspended in PBS at 1×10^5 DCs or 1×10^6 T-cells/mL. To each of triplicate wells of a U-bottom 96-well plate, 100 μL of 1×10^4 DCs and 100 μL of 1×10^5 allogeneic T-cells were successively added, and with or without 0.005 μg/mL anti-CD3 monoclonal antibody, which was incubated at 37°C for 44 h in humidified atmospheric air containing 5% CO_2. To each well, 0.5 μCi tritiated (^{3}H)-thymidine in 50 μL of AIM-V was added, cultured for the last 18 h, the cells were harvested, and the incorporated radioactivity was quantified. Before 1 h of adding T-cells and anti-CD3, the sterile, azide-free, and low-endotoxin IgG1 monoclonal antibodies specific for CD54, CD80, CD86, and an isotype control were added to the DCs at 1 μg/well to observe the co-stimulatory molecule block. In the mixed lymphocyte reaction (MLR), the cryopreserved DCs (stimulators) and T-cells (responders) were thawed in warm AIM-V media, washed with PBS, and resuspended in PBS at 1×10^5 DCs or 1×10^6 T-cells/mL. To each well of a U-bottom 96-well plate, 100 μL of 1×10^4 DCs and 100 μL of 1×10^5 allogeneic T-cells were successively added, and incubated at 37°C for 6 days in humidified atmospheric air containing 5% CO_2. To each well, 0.5 μCi tritiated (^{3}H)-thymidine in 50 μL of AIM-V was added, cultured for the last 18 h, the cells were harvested, and the incorporated radioactivity was quantified.

10.10 CYTOKINE ASSAY FOR IL-6, IL-10, IL-12, AND TNF-α OF DCs[11]

After washing out of femurs, bone marrow cells were resuspended at 2×10^5 cells/mL in RPMI 1640 supplemented with 10 μM β-mercaptoethanol, 1%

penicillin/streptomycin, 10 mM HEPES, 2 mM L-Gln, and either 10% fetal calf serum (FCS, <1 ng/mL LPS as per LAL assay) or 2% autologous, freshly prepared normal mouse serum (NMS), and incubated in 65-mm-diameter low-adherence Petri dishes containing 200 U/mL rmGM-CSF (PeproTech, CA) at 37°C and 5% CO_2. On day 4, fresh medium was added. On day 8, DCs were either pulsed overnight with indicated doses of mycobacterial sonicate or left untreated. On day 9, DCs were in PBS washed twice to remove the soluble and passively adsorbed antigen, counted, and their viability was evaluated (always 490% by Trypan blue exclusion). DCs were resuspended in either PBS or in culture medium.at appropriate concentrations for *in vivo* and *in vitro* tests, respectively. Antigen-pulsed and control DCs were resuspended in PBS and injected in indicated numbers in either 0.05 mL (in the trachea or hind foot pads) or 0.2 mL (subcutaneously in the dorsum) for immunization. To evaluate the protective effect, DCs were injected subcutaneously 2 or 3 times, at 2-week intervals between doses.

DCs obtained from 8-day cultures were further cultured at 5×10^5/mL in supplemented RPMI-1640 medium containing either 5% FCS or 2% NMS, and stimulated with 10 μg/mL H37Rv sonicate or left unstimulated. Using ELISA kits OptEIA mouse TNF-α Set (sensitivity, 31 pg/mL), OptEIA mouse IL-12 Set (31 pg/mL), OptEIA mouse IL-10 Set (63 pg/mL), and OptEIA mouse IL-6 Set (15.6 pg/mL) and according to the manufacturer's instructions, IL-6, IL-10, IL-12, and TNF-α in the 48-h culture supernatants were measured.

10.11 ELISPOT ASSAY FOR DC IFN-γ[11]

After washing out of femurs, bone marrow cells were resuspended at 2×10^5 cells/mL in RPMI 1640 supplemented with 10 μM β-mercaptoethanol, 1% penicillin/streptomycin, 10 mM HEPES, 2 mM L-Gln, and either 10% FCS or 2% autologous, freshly prepared NMS, and incubated in 65-mm-diameter low-adherence Petri dishes containing 200 U/mL rmGM-CSF at 37°C and 5% CO_2. On day 4, fresh medium was added. On day 8, DCs were either pulsed overnight with indicated doses of mycobacterial sonicate or left untreated. On day 9, DCs were in PBS washed twice to remove the soluble and passively adsorbed antigen, counted, and their viability was evaluated (always 490% by Trypan blue exclusion). DCs were resuspended in either PBS or in culture medium at appropriate concentrations for *in vivo* and *in vitro* tests, respectively. Antigen-pulsed and control DCs were resuspended in PBS and injected in indicated numbers in either 0.05 mL (in the trachea or hind foot pads) or 0.2 mL (subcutaneously in the dorsum) for immunization. To evaluate the protective effect, DCs were injected subcutaneously 2 or 3 times, at 2-week intervals between doses.

With 2×10^6 cells/mouse of sonicate loaded DC, mice were vaccinated 3 times at 2-week intervals of doses. After the last vaccination, single cell suspensions were obtained individually from lungs and lymph nodes of three vaccinated and three naive control mice. Sterile filter Millipore plates were coated with rat antibody against murine IFN-γ, washed, and blocked with RPMI containing 10% FCS, to which cells were added at 1×10^6 cells/well with 4 doubling dilutions and cultured for 48 h with medium alone or 10 μg/mL sonicate. The sites of cytokine production were detected using biotin-labeled rat antibody against murine IFN-γ. Spots were counted on ELISPOT Bioreader 4000 Pro-X, calculated, and normalized for the bulk individual samples, and the results were expressed as the mean ± SD per organ.

10.12 DC FUNCTION ASSAY FOR EVALUATING TOLL-LIKE RECEPTOR FUNCTION[12]

Ten milliliters to 50 mL whole blood (WB) from healthy volunteer blood donors was drawn directly into 10-mL tubes containing 143 USP units of freeze-dried sodium heparin, gently rocked at room temperature, and used generally within 1 h. In stimulation experiments, WB was aliquoted at 180 μL/tube into prelabeled 5-mL round-bottom disposable 12 × 75 mm Falcon polystyrene test tubes (BD Labware Catalog No. 2058) and stimulated with ligand. Prior to being diluted, all ligands were thawed from −80°C immediately and held at 4°C. All dilutions were prepared in R5 medium (5% pooled human AB serum, 1% HEPES buffer, 100 U/mL penicillin, and 100 μg/mL streptomycin in RPMI 1640). In titration experiments, dilutions such as P3C (0.001, 0.01, 0.1, 1, 10, 100, and 200 μg/mL), Poly-I:C (0.1, 1, 10, 100, and 200 μg/mL), LPS (0.0001, 0.001, 0.01, 0.1, 1, and 10 μg/mL), CpG 2395 DNA, and CpG 2137 DNA (inactive control, 25, 50, 100, and 200 μg/mL) and resiquimod (0.001, 0.01, 0.1, 1, 10, and 100 μM) were involved. Dilutions were at a concentration 10-fold greater than the desired final concentration so that adding 20 μL of diluted ligand would bring up the final volume in the test tube to 200 μL. Unstimulated test tubes received 20 μL of R5 medium alone. After adding stimuli, tubes were incubated at 37°C in a 5% CO_2 humidified atmosphere at a 5-degree slant for 1, 2, 3, or 5 h or overnight. To block cytokine secretion, Brefeldin A (BFA, 10 μg/mL final concentration) was added to the 1, 2, 3, or 5 h incubation for the last 0.5, 1, 2, or 3 h, respectively, while to the overnight incubation BFA (10 μg/mL final concentration) was added for 3 h of the 16th to 18th h. After incubation, by adding either (1) 2 μL of HLA-DR-PerCP, 2 μL of Lin-1-FITC, 2 μL of CD11c-APC, and 2 μL of PE-conjugated maturation marker per tube or (2) 2 μL of HLA-DR-PerCP, 2 μL of Lin-1-FITC and 2 μL of CD123-PE per tube, surface staining was performed in WB. At room temperature, tubes were vortexed and incubated for 20 min. In the case using (2), after permeabilization of the cells, Alexa-fluor-647-conjugated intracellular cytokine antibodies were added to the left channel. By adding 2 mL of FACS Lysing Solution (BD Biosciences, San Jose, CA) to each tube, repeating vortexing and incubating samples in the dark at room temperature for 10 min, a gentle lysis of red blood cells and fixation of leukocytes was performed. After this step, cells were centrifuged at 1400 × *g* and 4°C for 5 min, washed twice by resuspending in 2 mL of wash buffer (WB, 1% bovine serum albumin, 0.02% sodium azide in PBS), and centrifugation at 1400 × *g*. By inverting tubes onto a stack of paper towels, samples were decanted and allowed approximately 50 μL of fluid to remain in each tube. In the case using (1), samples for maturation markers were subjected to a third wash in WB, and the pellets were resuspended in 200 μL of paraformaldehyde

(PFA, 1%, prepared daily from a 10% stock, Electron Microscopy Sciences, Hatfield, PA) and stored in the dark at 4°C prior to flow cytometric analysis. Samples for intracellular cytokine expression were fixed in 200 μL of PFA (4%) in the dark for 15 min at 4°C, washed in 1 mL of WB, vortexed, and centrifuged at 1400 × *g* and 4°C for 5 min. By inverting tubes onto a stack of paper towels, samples were decanted and allowed approximately 50 μL of fluid to remain in each tube. By washing samples in 1 mL of permeabilization buffer (PB, 1% bovine serum albumin, 1% saponin in PBS), vortexing, centrifuging at 1400 × *g* and 4°C for 5 min and decanting as above, cell permeabilization was performed. In PB, anti-human TNF-α Alexa-fluor-647 and anti-human IFN-α Alexa-fluor-647 antibodies were diluted 1:15 and 1:640, respectively. In the case of TNF-α, antibody was added at 30 μL per tube, and samples were incubated for 30 min in the dark at room temperature. In the case of INF-α, antibody was added at 30 μL per tube and samples were incubated for 30 min in the dark at 4°C. Samples were washed twice in 1 mL of PB and once in 1 mL of WB. Cell pellets were resuspended in 200 μL of PFA (1%) and stored in the dark at 4°C prior to flow cytometric analysis.

To isolate PBMCs, heparin-anticoagulated WB was subjected to standard density-gradient centrifugation. To cryopreserve PBMCs intended for later use, freshly isolated PBMCs were resuspended at 5.0×10^7 cells/mL in R10 medium (10% pooled human AB serum, 100 IU/mL penicillin, and 100 μg/mL streptomycin in RPMI 1640) and mixed with an equal volume of freezing medium (20% DMSO/80% heat-inactivated FCS, sterile-filtered). Cells were aliquoted into cryotubes, which were transferred into "Mr. Frosty" containers for gradual temperature reduction to −80°C overnight. On the next day, prior to stimulation the tubes were taken out of "Mr. Frosty" containers, transferred to a liquid nitrogen cryopreservation tank, and frozen PBMCs were first warmed in a 37°C water bath until almost thawed, to which a total of 1 mL of warm RPMI was added dropwise. Cells were resuspended gently, transferred into a 15 mL conical tube, and washed three times by centrifugation (200 × *g*, 4°C, 10 min) in RPMI 1640. For stimulation experiments, freshly isolated or freshly thawed PBMCs were resuspended in R10, counted, and aliquoted into prelabeled 5 mL round-bottom disposable 12 × 75 mm Falcon polystyrene test tubes in 1 mL volume at 1.5×10^6 cells per tube and processed. Tubes containing isolated PBMCs receiving either TLR ligand resiquimod (10 μM) or R5 medium alone were incubated at 37°C in a 5% CO_2 humidified atmosphere at a 5-degree slant to incubate for another 2 h. For maturation marker staining, a total of 5-h incubation was performed without BFA. All cultures were treated with 2 mM EDTA at 37°C for 10 min to reduce clumping and to detach the cells from tube walls. The cells were washed twice in 1 mL of FACS, centrifuged at 1400 × *g*, and 4°C for 5 min and stained.

10.13 MIGRATION ASSAY OF DENDRITIC CELL FROM BONE MARROW OF A/J MICE[13]

Dendritic cell progenitor cells generated from the bone marrow of A/J mice were co-cultured with or without TBJ cells in 6-well plates (4×10^6 cells/mL) using clear polyester membrane Transwell inserts (24-mm diameter, 0.4-μm pore size,

Costar; Corning Incorporated, Corning, NY, via Fisher Scientific, Pittsburgh, PA) in complete RPMI medium supplemented with 20 ng/mL granulocyte-macrophage colony-stimulating factor and IL-4. Dendritic cell progenitor cells and 2.6 mL of complete RPMI were placed in the lower chamber, washed, fed with fresh complete RPMI containing cytokines granulocyte-macrophage colony-stimulating factor and IL-4 (Sigma-Aldrich, St. Louis, MO) at 2-day intervals, maintaining a lower well volume of 2.6 mL, harvested from the lower wells on day 7, pooled into one group per condition, washed, and plated in 60-mm Petri culture dishes. On day 8, PGE2 (100 ng/mL, Sigma, St. Louis, MO) was added to the DC culture to stimulate DC maturation for 24 h. Using 24-well, 6.5-mm diameter, 5.0-μm pore size, polycarbonate membrane Transwell plates and recombinant mouse MIP-3β/CCL19 (1, 5, 100 ng/mL), the chemotaxis assay was conducted. Chemokine in complete RPMI or medium alone (600 μL) serving as a control was placed in the lower well, DCs (5×10^5 cells) in 100 μL of complete RPMI were added to the upper wells and allowed to migrate at 37°C in 10% carbon dioxide for 5 h. TBJ cells (5×10^5 cells) and 1.5 mL of complete RPMI medium or medium alone as a control were added to the upper chamber of the Transwell system, the medium was checked and replenished at 2-day intervals to maintain an upper chamber volume of 1.5 mL. Migrated cells were recovered from each lower well and counted using a hemocytometer. To eliminate lymphocyte contamination, the counts of cells larger than 12 μm were verified on a Coulter Counter (Coulter Corporation, Miami, FL). Each experiment was performed in duplicate, and the percentage of migrating cells was calculated from the ratio of migrating cells to starting cells.

10.14 LYMPHOID ORGAN ASSAY[14–16]

After a 4-week acclimation period, male Wistar rats (4 weeks old, 225–250 g, housed in polypropylene cages in a environment controlled room maintained at 22°C, 55% relative humidity, and a 12/12 h light/dark cycle) were allocated to six groups (n = 15 to 20 rats each). The untreated group was used as control (maintained on basal diet and sacrificed at week 4 and 30), during week 1 and 2 the DMBDD, DMBDD/PB, and DMBDD/2-acetylaminofluorene (2-AAF) groups were sequentially treated with initiators DEN (*N*-nitrosodiethylamine, 100 mg/kg, ip), MNU (*N*-methyl-*N*-nitrosourea, 20 mg/kg, ip, four times, two doses per week) and *N*-butyl-*N*-(4-hydroxybutyl)nitrosamine (BBN, 0.05% in drinking water during 2 weeks), during week 3 and 4 DHPN and DMH groups were sequentially treated with dihydroxy-di-n-propylnitrosamine (DHPN, 0.1% in drinking water during 2 weeks) and 1,2-dimethylhydrazine (DMH, 40 mg/kg, sc, four times, two doses per week). At the end of week 4, some rats of the DMBDD group were killed and the remainder was maintained on basal diet until week 30. After the initiation, the DMBDD/PB and DMBDD/2-AAF groups were supplied with phenobarbital (PB, 0.05%) and 2-acetylaminofluorene (2-AAF, 0.01%) in the diet for 25 weeks, respectively. From week 6 until week 30, two noninitiated groups received PB or 2-AAF in the diet. At week 4 or 30, all rats were sacrificed under pentobarbital (45 mg/kg) anesthesia. The liver, kidneys, spleen, thymus, mesenteric lymph nodes, and bone marrow removed from all rats were fixed in buffered formalin for 48 h for tissue processing and histologic analysis. Only at week 30, the lung, small and large intestine, and Zymbal's gland were examined. The removed liver, kidneys, spleen, and thymus were weighed immediately. The spleen was cut in two halves for evaluating cytokines and histologic analysis, respectively. All removed organs were embedded in paraffin and stained with hematoxylin and eosin for histologic analysis. The suspensions of spleen cells dispersed in a Petri dish containing RPMI-1640 culture medium were centrifuged, resuspended in RPMI-1640 supplemented with 20 mg/mL gentamicin, 2 mM glutamine, and 10% inactivated fetal calf serum, and washed twice by centrifugation at 1500 × *g* for 10 min. To aliquots of 2×10^6 cells/mL (500 μL/well) in 24-well flat-bottom microtiter plates, RPMI-1640 (500 μL/well) or concanavalin A (Con A 2.5 μg/mL, 500 μL/well, Sigma Co.) or *Staphylococcus aureus* Cowan's strain 1 (SAC-1:5000, 500 μL/well, Calbiochem, San Diego, CA, USA) was added and the plates were incubated at 37°C for 72 h in humidified atmospheric air containing 5% CO_2. After incubation, the collected supernatants were stocked at −70°C for quantification of cytokines. *In vitro* stimulated CONA samples were used to quantify IL-2, IFN-γ, IL-10, and TGF-β1. *In vitro* stimulated SAC samples were used to quantify TNF-α and IL-12. Using ELISA kits IL-2, IL-12, TNF-α, IFN-γ, and IL-10 levels were measured. Samples were acidified by 1 M HCl and measured by Quantikine anti-human TGF-h1 kit (R&D Systems, Minneapolis, MN,

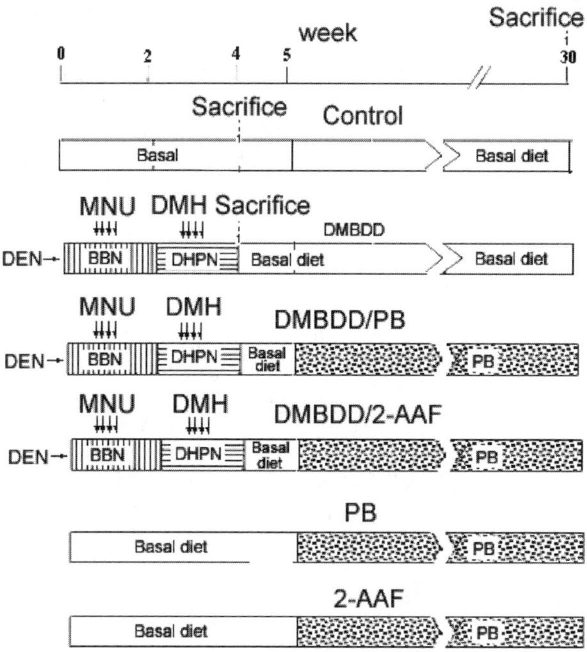

Figure 10.1 Design of the assay.

USA) for detecting the TGF-h1 immunoreactive form. The assay design is summarized in Fig. 10.1.

10.15 ELISA OF IFN-γ FROM HUMAN MYELOMONOCYTIC KG-1 CELLS[17–19]

Human myelomonocytic KG-1 cells (ATCCCCL246) were incubated with RPMI-1640 supplemented with 10% FCS, 100 μg/mL penicillin, and 100 μg/mL streptomycin at 37°C in air containing 5% CO_2. By expression of the corresponding cDNA in *E. coli*, HuIL-18 and MuIL-18 were prepared and purified to homogeneity. Using Pfu DNA polymerase (at 95°C for 45 s; at 72°C for 3.5 min; 10 cycles and at 95°C for 45 s; at 68°C for 3.5 min; 35 cycles) MuIL-18R cDNA (1.7 kbp, MuIL-1 Rrp, GenBank, accession number U43673) were amplified from murine liver RNA by RT-PCR and cloned into pCRScript Cam SK (+) to synthesize 5′-AGAGG-AACCACCCACAACGATCCT-3′ and 5′-TGAATAGGCACACGCATGACCTCT-3′. With the EF-1 promoter of pEF-BOS vector, the dihydrofolate reductase unit of pSV2dhfr (ATCC 37146) and the backbone of pRc/CMV vector pREF-XN were constructed. IL-18R cDNA was ligated into XhoI/NotI sites of the vector to form MuIL-18R expression vector pRcEFM18R, of which the structure is represented as Fig. 10.2.

KG-1 cells (1×10^7) were washed twice with RPMI-1640 and transfected with 50 mg of pRcEFM18R by electroporation (260 V, 960 μF, Gene Pulser, BioRad Labs, Hercules, CA). In the presence of 400 μg/mL G-418, the transformed cells were selected and cloned. The suspension of 2×10^6 cells, on which the receptor binding of ^{125}I-labeled MuIL-18 or HuIL-18 has been examined, in RPMI-1640 containing 0.1% NaN_3 and 100 mM HEPES (pH 7.2) was incubated at 4°C for 1 h with approximately 4 ng of ^{125}I-labeled MuIL-18 or HuIL-18. After separating unbound IL-18, the cell-bound ^{125}I count was determined. Subtracting the nonspecific binding from 3 μg of unlabeled cognate ligand, the specific binding of IL-18 was obtained. To prepare mice serum containing endogenous MuIL-18, C57BL/6 mice were treated with 500 μg of heat-killed *Propionibacterium acnes* for 1 week and challenged

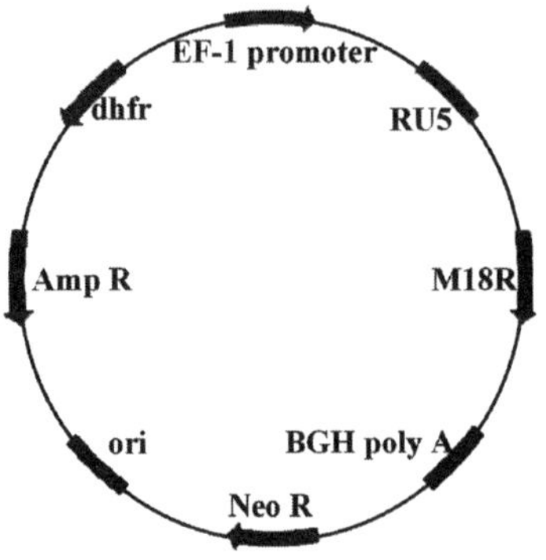

Figure 10.2 Structure of pRcEFM18R, a MuIL-18R expression vector.

with 1 μg of lipopolysaccharide for 2 h to induce endotoxic shock. From the heart under proper anesthesia, blood samples were taken to prepare sera. The MuIL-18R-expressing KG-1 cells were washed and resuspended at 5×10^5 cells/mL with RPMI-1640 for 2 days. The cells were adjusted to 1×10^6 cells/mL with RPMI-1640 containing 10% FCS, to which the indicated amounts of MuIL-18, HuIL-18, or serum sample were added. One day later, the culture supernatants were recovered and the quantitative analysis of the produced IFN-γ was performed with ELISA. The culture supernatant was incubated with mAb-IFN-γ-15. The bound IFN-γ was further incubated with mAb-IFN-γ-6 and detected with hydrogen peroxide and *o*-phenylene-diamine.

10.16 HUMAN WHOLE BLOOD IFN-γ ASSAYS[19,20]

The first and second stages of QuantiFERON-TB-Gold In Tube (QFT) assay involved incubation of whole blood with antigens and the measurement of IFN-γ production in harvested plasma by ELISA, respectively. Prior to tuberculin skin test (TST) administration, venous blood was directly collected into three tubes containing 1 mL of heparin, which contained only heparin as negative control, mitogen as positive control, and overlapping peptides representing the entire sequences of ESAT-6 and CFP-10 and another peptide from a portion of the TB antigen TB7.7 (Rv2654), respectively. Within 2–6 h of blood draw, the tubes were incubated at 37°C for exactly 24 h and centrifuged. Plasma was harvested and frozen at −70°C to perform ELISA (on average, ELISA was performed within 4–6 weeks of blood collection). Using ELISA the IFN-γ response was quantified. For tuberculosis-specific antigens and mitogen, IFN-γ values (IU/mL) were corrected for background by subtracting the respective negative control value. According to the manufacturers' instructions and the previous studies, the cutoff value for a positive test was IFN-γ $\geq$0.35 IU/mL. The QFT ELISA was not designed to estimate the absolute IFN-γ values when they exceed 10 IU/mL and IFN-γ values $>$10 IU/mL were reported as 10 IU/mL.

10.17 SHEEP WHOLE BLOOD IFN-γ ASSAYS[21]

Three ewes were obtained from a confirmed *Corynebacterium pseudotuberculosis* infected Suffolk cross flock and had external lesions suggestive of recent infection with *Corynebacterium pseudotuberculosis*. Using 10-mL glass Vacutainer tubes containing heparin, blood samples were collected, maintained at room temperature, and processed within 3 h. A closed mixed breed flock monitored as negative for 15 years for *Corynebacterium pseudotuberculosis* was used as a source of samples from known-negative sheep. Using 10-mL glass Vacutainer tubes containing heparin, blood samples were collected, maintained at room temperature, and processed within 3 h. Flock infected with *Corynebacterium pseudotuberculosis* was used as the naturally infected flock for field validation of the IFN-γ enzyme immunoassay (EIA). At 1-month intervals, blood was collected for 5 months, kept at room temperature, and processed within 8 h. A group of 20 naturally infected sheep were penned, fed and watered separately from the main infected flock, but shared a common barn and airspace. At 1-month intervals, blood was collected for 5 months, kept at room temperature, and processed within 8 h. Selected sheep were euthanized with barbiturate injected intravenously and a detailed postmortem examination was performed.

Corynebacterium pseudotuberculosis strain 301/82 isolated from a case of caseous lymphadenitis (CLA) was at 37°C grown in 500 mL of brain heart infusion broth (BHIB) for 72 h with shaking at 100 rpm. By centrifugation at 10,000 × *g* and 4°C for 30 min, the culture was harvested and the pellet was resuspended in 10 mL of BPS (pH 7.2). To the suspension, 37% (w/w) formaldehyde (0.1% final concentration) was added, the formalinized culture was incubated at 37°C for 72 h, checked for sterility, washed twice with sterile PBS, and resuspended in PBS. The OD of the suspension was measured at 540 nm and the concentration of each dilution was determined using the formula $\log x = (7.85492 + \log y)/0.9183$, with x representing concentration in the undiluted sample and y representing OD. To calculate the final concentration of bacterial cells per milliliter of suspension, the value obtained was multiplied by the dilution factor. The bacterial suspension was diluted with sterile PBS and stored at 4°C.

To optimize the IFN-γ assay, various concentrations of the formalin-inactivated antigen against two different volumes of heparinized blood collected from known-negative, naturally infected flock or vaccinated infected sheep were tested. Within 8 h of collection, triplicate 1 mL aliquots without antigen (100 μL PBS), with pokeweed mitogen (PWM, 3.75 μg/well, Sigma-Aldrich Incorporated, St. Louis, MO, USA), or with varying concentrations of formalin-inactivated antigen (100 μL) were incubated in wells of 24-well tissue culture plates at 37°C for 24 h in 5% CO_2 and centrifuged at 1400 rpm for 10 min. Plasma was harvested, stored at −20°C, and assayed for IFN-γ concentration using a bovine IFN-γ EIA according to the manufacturer's instructions.

10.18 ELISPOT ASSAY FOR IFN-γ[22,23]

Among 45 patients who received living donor kidney transplantations, ELISPOT tests were performed twice, then immunosuppression was initiated, which consisted of a combination of prednisone, mycophenolate mofetil, and cyclosporine or tacrolimus. The patients having flow cytometry- or cytotoxicity-based positive cross-match were excluded. With ELISA PRA (LAT-M, One Lambda Inc., Canoga Park, CA), panel-reactive antibody (PRA) test was performed. Delayed graft function (DGF) was defined as during the first week after transplantation needing dialysis. Acute rejection episode (ARE) was defined as creatinine level increase that was not attributable to other reasons, with a subsequent return to baseline after antirejection treatment. The simplified Modification of Diet in Renal Disease (MDRD) study group formula was used to calculate glomerular filtration rate (GFR) at 6 months.

Recipient peripheral blood lymphocytes (PBLs, 2×10^5 cells/well) in 100 μL of culture medium were placed in 96-well ELISPOT plates, which were precoated at 4°C overnight with mouse monoclonal anti-human IFN-γ capture antibody, stimulated with medium alone (negative control), phytohemagglutinin (1 μg/mL medium, positive control, Sigma-Aldrich, St. Louis, MO), and donor or third-party-cell non-depleted PBLs (2×10^5 cells/well). Third-party stimulator cells chosen from 2–4 normal persons with degrees of human leukocyte antigen (HLA) mismatching were comparable with that of the actual donor. Stimulator cells were used after 3000 rad (30 Gy) irradiation. After 37°C overnight incubation, at room temperature a biotinylated detection antibody was added for 2 h and then anti-streptavidin horseradish peroxidase was added for 45 min. The plates were developed with aminoethylcarbazole (10 mg/mL *N,N*-dimethylformamide), prepared in 0.1 M sodium acetate buffer (pH 5.0) mixed with H_2O_2. With AID ELISPOT analyzer (Autoimmun Diagnostika, Strassberg, Germany), the resulting spots were counted. From the total number of spots in wells containing mixed responder and stimulator PBLs, the spots in the negative control wells were subtracted, and the results were represented as the mean number of IFN-γ per 200,000 recipient PBLs based on duplicate or triplicate measurements in two individual assays.

10.19 IFN-β RG ASSAY[24]

Vero Mx-Luc cells (kidney of African green monkey, Rockville, MD, USA) were grown in 96-well plates (4×10^4 cells/well) with culture medium containing 20 pg/mL phenylbenzothiazole and known amounts of interferon-β (IFN-β). Samples and standard were tested at three serial 1/2 dilutions in duplicates. The plates were incubated at 37°C in 5% CO_2 for 21 h, washed once with PBS, and at room temperature with 40 μL of cell lysis buffer supplemented with 8 U/mL aprotinin lysed for 15 min. The 96-well microplate was transferred into the Dynatech luminometer ML 3000, into each well 150 μL of luciferase assay reagent was added, and the light emitted was measured for 5 s. The potencies of unknown samples were calculated versus the reference preparation following the statistical requirements of a 3×3 balanced parallel line assay. This assay was performed with three dose levels for standard and sample, with two replicates for each dose.

10.20 ANTI-rHuEPO NAb ASSAY[25,26]

32D-EPOR cells were incubated at 37°C in humidified atmospheric air containing 5% CO_2 with RPMI 1640 supplemented with 15% heat-inactivated FBS, 2 mM L-glutamine, and penicillin/streptomycin mixture (1%, v/v). The erythropoietin (EPO, EPOGEN®, Amgen) dependent cells were incubated in RPMI 1640 supplemented with 10 U/mL rHuEPO. Via two to three subcultures a week, cell densities were maintained between 3×10^4 and 1×10^6 cells/mL. Up to 30 days after thawing, the old cells in the cryopreserved cells were discarded and a vial of frozen cells was thawed and expanded in culture.

In the cell proliferation assay, 32D-EPOR cells were incubated overnight, harvested, and washed twice with RPMI 1640 lacking rHuEPO by centrifugation (200 to 300 $\times$ *g*). The supernatant was discarded, the cell pellet was resuspended in RPMI 1640 and centrifuged. The formed cell pellet was resuspended in RPMI 1640 and adjusted to 5×10^5 cells/mL. Cells were incubated at 37°C (humidified atmospheric air containing 5% CO_2, without rHuEPO) for 16–24 h, centrifuged, resuspended in fresh RPMI 1640, and counted. Staged 32D-EPOR cells (100 μL, 2×10^4) and prepared testing sample (100 μL) were incubated at 37°C for 44 h in humidified atmospheric air containing 5% CO_2. The solution of 2 μCi [methyl-^{3}H] thymidine diluted in 50 μL of RPMI 1640 were added to each well and incubated for 4 h. The contents of the plate were harvested, 25 μL of scintillation fluid was added, and the cells were counted.

All controls were prepared in a mixture of 5% human serum and 15% pooled rat serum. The background control (N) consisted of cells only, which was prepared by mixing 40 μL of pooled human serum with 240 μL of RPMI 1640 and 120 μL of pooled rat serum, the maximum growth control (M) consisted of cells and 1 ng/mL rHuEPO, which was prepared by mixing 40 μL of pooled human serum with 120 μL of rHuEPO at 6.67 ng/mL, 120 μL of RPMI 1640, and 120 μL of pooled rat serum, and the neutralizing antibody positive control (P) consisted of cells, 1 ng/mL rHuEPO, and 500 ng/mL the positive control antibody, which was prepared by mixing with 120 μL of RPMI 1640 and 120 μL of pooled rat serum, were included in anti-rHuEPO NAb assay controls. Samples were prepared by mixing 40 μL of individual donor serum with 120 μL of rHuEPO at 6.67 ng/mL, 120 μL of RPMI 1640, and 120 μL of pooled rat serum. Before addition to the cells, all controls and samples were preincubated at room temperature for at least 30 min.

10.21 CHLORAMPHENICOL ACETYLTRANSFERASE ASSAY[27]

Cell monolayers were rinsed several times with calcium and magnesium free PBS and lysed in either 0.5% Nonidet P-40, 20 mM HEPES-KOH (pH 7.5), 120 mM KCl, 5 mM $MgCl_2$, or a cell lysis buffer (5x cell lysis buffer). For standard assays, with a rubber policeman cells were scraped from 6-cm dishes, by centrifugation at 5000 $\times$ *g* for 2 min nuclei were removed and supernatants were heated at 70°C for 10 min to inactivate enzymes capable of hydrolyzing ace@-CoA (Sigma, St. Louis, MO). In the 96-well assay, monolayers were rinsed with PBS, at 0°C lysed on the dish in 60 μL of lysis buffer for 10 min, and the supernatants were transferred to a fresh 96-well dish using a multichannel pipettor. The wells were sealed with strips of Scotch tape, and the dish was placed in an oven at 70°C for 10 min with a preheated glass plate on top, which was then cooled on ice and from each well aliquots (50 μL) were removed for assays. Assays were carried out by pipetting 50 μL of sample to a 5 mL scintillation vial and adding 200 μL of cocktail containing 100 mM Tris-HCl (pH 8.0), 0.8 mM chloramphenicol, 0.1 μCi [^{3}H]acetyl-CoA (Amersham, Arlington Heights, IL) per reaction, and the reaction mixtures were overlaid with 5 mL of aqueous immiscible scintillation cocktail. After 60–120 min, the vials were counted on a Beckman liquid scintillation spectrometer at approximately 50%

efficiency. Controls consisted of lysis buffer as a blank and varying amounts of authentic chloramphenicol acetyltransferase as a calibration standard. In some assays, extracts were incubated for 1 or 2 h without adding scintillant, and reactions were stopped by adding 25 μL of iodoacetic acid (50 mg/mL). Scintillant was added and before counting the vials left for 15 min to allow diffusion of reaction product into the organic phase. Chloramphenicol acetyltransferase was also quantitated with an ELISA according to the manufacturer's instructions.

10.22 CHEMOTAXIS ASSAY[28]

Using general methods, Ts16 fetuses were generated, identified, and confirmed. By the crown-rump length of normal littermate fetuses, gestational age was determined. Cortices were removed, cleaned of meninges, and dissociated into single cell suspensions with a papain digest to prepare acute cell dissociates from fetuses at embryonic day 17 (E17).

In the presence of attractants, the cells were migrated onto the underside of the porous membrane, fixed, stained, and counted. The directed migration of cells toward regions of higher concentrations of a chemical attractant was defined as chemotaxis that was distinguished from random or spontaneous migration (media alone) in each experiment. By counting 5–10 fields of stained cells for each experimental condition, the average number of migrated cells per square millimeter was calculated. For a minimum of three different litter groups containing Ts16 fetuses and their normal littermates, each experimental condition was replicated in three to five trials.

10.23 FIBROBLAST-POPULATED MICROSPHERE ASSAY[29]

Rat dermal fibroblasts (RDFs) prepared using a primary explant technique were kept in liquid nitrogen after slow freezing. Cell lines were cultured by thawing an aliquot of cells and centrifuging at 4°C and 1000 rpm for 10 min. The pellet of cells was resuspended in DMEM (Grand Island, NY) supplemented with pen-strep (1% v/v) and L-glutamine (1% v/v), the cells were plated on 60 × 15 mm Petri dishes using 5 mL of DMEM with 10% FBS (HyClone Laboratories, Logan, UT), 1% pen-strep, 1% Fungizone, and 1% L-glutamine, and kept in a humidified CO_2 incubator at 37°C. The cells were passed with trypsin/EDTA (Sigma, St. Louis, MO) at a 1:4 dilution once a week, harvested with 0.5% trypsin/EDTA, washed twice with Medium 199 with Hanks' salts (M199) supplemented with 10% FBS, 1% pen-strep, and 1% L-glutamine, diluted to a standard concentration of 2×10^5 cells/mL, and passed or harvested at 60% to 90% confluence.

To the bottom of a 10 × 2 × 5 cm glass well, which contained a 1-cm layer of sg 1.1 "heavy" silicone oil, an approximately 3-cm-thick layer of sg 0.96 "light" silicone oil was added and separated into two distinct layers. The silicone oil wells were placed in a 37°C incubator. In a well of a 96-well plate, 100 μL of fibrinogen solution (3.33 mg/mL final concentration), 25 μL of thrombin solution (0.038 U/mL final concentration), and 25 μL of rat dermal fibroblast (RDF, 3.33×10^4 cells/mL) suspension were mixed. Into the glass well containing the 37°C silicone oils, an 80 μL aliquot of the RDF-thrombin-fibrinogen solution was immediately pipetted by placing the pipette tip just below the surface of the light oil, slowly ejecting the contents, and the tip quickly withdrawn. Care was taken to avoid introducing air bubbles into the forming fibroblast-populated microsphere (FPM) manufactured using general technique. The FPM slowly settled to the interface between the two silicone oils having its spherical shape retained. In the 37°C silicone oil, FPMs were incubated for 15 min to permit fibrin gelation, and the diameter of the resulting FPMs was approximately 5.2 mm. FPMs were removed, washed thoroughly in US-M199, and incubated with complete medium (M199, 10% FBS, 1% pen-strep, 1% L-glutamine, and 1% Fungizone), transferred to a new 24-well plate with complete medium, and incubated at 37°C for the duration of the assay.

The plate was placed on the stage of an Olympus IX-70 inverted light microscope and the images of the equatorial plane of each microsphere were captured with a Photometrics SenSys cooled CCD camera (Tucson, AZ) using a × 4 objective. Using a motorized microscope stage, a series of images was captured to form a mosaic of the entire FPM. After FPM preparation and at subsequent time points, the mosaic image of the microsphere was modified in an image manipulation program to perform diameter measurements immediately. The surface was marked with white dots. With image analysis software, the images were analyzed and the positions of the dots were reported. By fitting the data via least-squares regression to the equation for a circle, the diameter and center of the FPM shell and core were estimated to yield the diameters and centers of the best-fit circles. Compaction degree ρ was calculated as the percentage decrease of diameter with respect to its initial value.

10.24 FIBROBLAST-POPULATED CONCENTRIC MICROSPHERE ASSAY[29]

To the bottom of a 10 × 2 × 5 cm glass well, which contained a 1-cm layer of sg 1.1 "heavy" silicone oil, an approximately 3-cm-thick layer of sg 0.96 "light" silicone oil was added and separated into two distinct layers. A small fibrin sphere was created as the core of the fibroblast-populated concentric microsphere (FPCM, see Fig. 10.3). Around the core a fibrin gel shell was cast. The distinct core and shell regions of the FPCM made it possible to have spatially varying cell concentrations of one or more cell types. The Base FPCM consisted of a fibroblast-seeded shell and an acellular

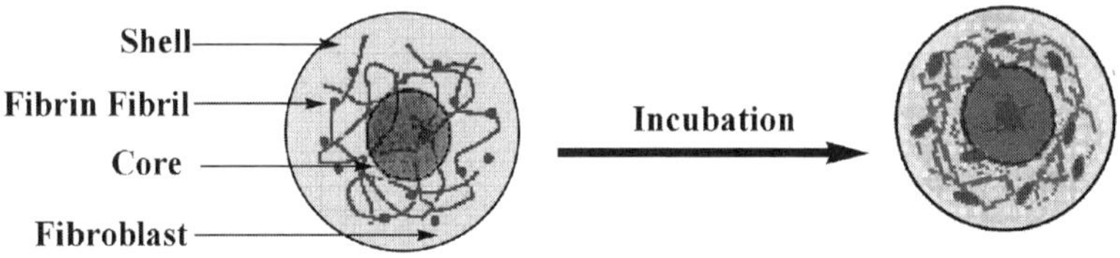

Figure 10.3 Schematic of FPCM fibrin gel co-culture assay.

core. Twenty microliters of fibrinogen solution was mixed with 5 μL of thrombin solution and 5 μL of US-M199. Into a well containing the layered silicone oils, 15 μL aliquot of this thrombin-fibrinogen solution was immediately pipetted. The FPCM core slowly settled to the interface between the two silicone oils having its spherical shape retained. In the 37°C silicone oil, FPCMs were incubated for 10 min to permit fibrin gelation. One hundred microliters of the fibrinogen solution was mixed with 25 μL of thrombin solution and 25 μL of rat dermal fibroblast (RDF) suspension. To form the shell of the Base FPCM, a 65 μL aliquot of this RDF-thrombin-fibrinogen solution was pipetted to a core in the silicone oil. A 0.05-mm-diameter glass rod was prewetted in US-M199 and used to draw the core into the gelling shell solution. A 22-gauge copper wire loop was connected to the negative terminal of a 1000-V power supply and used to position the core in the center of the Base FPCM via electrostatic repulsion. After incubation in the 37°C silicone oil for 15 min, the Base FPCMs were incubated, removed, washed with US-M199, transferred to a 24-well plate with complete medium, and incubated in air at 37°C for the course of the assay. The diameters of initial FPCM and core were approximately 5.2 mm and 2.9 mm, respectively.

Except both the core and the shell contained RDFs, the control FPCM was identical to the Base FPCM. By mixing 20 μL of fibrinogen solution with 5 μL of thrombin solution and 5 μL of RDF suspension, the forming solution was used to create the core. The rest fabrication procedure was as detailed for the Base FPCM. The fibroblast distribution of Inverse FPCM was the inverse of that of the Base FPCM. RDFs were seeded in the core formed from a solution prepared by mixing 20 μL of fibrinogen solution with 5 μL of thrombin solution and 5 μL of RDF suspension. The shell was formed from a solution prepared by mixing 100 μL of fibrinogen solution with 25 μL of thrombin solution and 25 μL of US-M199.

As an alternative, allowing FPCMs to float and compact freely during incubation, an embedding method was practiced to prevent compaction. By mixing 1.0% low melting agarose with 0.5 mg/mL fibrinogen in complete medium, a solution was obtained. Into a Millicell plate insert, a 350 μL aliquot of the embedding solution was pipetted. After positioning FPCM in the Millicell, 5 μL of thrombin was added, the embedded FPCM was incubated at 37°C for 20 min until fibrin gelation was complete and placed in a cold room (4°C) for 1 h until agarose gelation was complete. The Millicell inserts were placed in the wells of a 24-well plate contained 1.5 mL of complete medium and incubated at 37°C for the duration of the assay.

The plate was placed on the stage of an Olympus IX-70 inverted light microscope (Melville, NY) and the images of the equatorial plane of each microsphere were captured with a Photometrics SenSys cooled CCD camera (Tucson, AZ) using a × 4 objective. Using a motorized microscope stage, a series of images was captured to form a mosaic of the entire FPCM. After FPCM preparation and at subsequent time points, the mosaic image of the microsphere was modified in an image manipulation program to perform diameter measurements immediately. The surface was marked with white dots and smaller dots were also placed around the core for FPCMs. With image analysis software, the images were analyzed and the positions of the dots were reported. By fitting the data via least-squares regression to the equation for a circle, the diameter and center of the FPCM shell and core were estimated to yield the diameters and centers of the best-fit circles. Compaction degree ρ was calculated as the percentage decrease of diameter with respect to its initial value.

FPCM core concentricity χ with respect to the shell was determined from the initial time point image according to $\chi = 1 - [(\Delta x^2 - \Delta y^2 - \Delta z^2)/(R_{shell} - R_{core})]$, with χ ranging from 0 (core touching the FPCM surface) to 1 (concentric), Δx and Δy were calculated from the difference in location between the core and the shell centers as determined from the diameter measurements, while Δz was determined by recording the difference in z position between the plane of focus for the FPCM shell equatorial plane and that of the core.

10.25 RADIAL ASSAY OF CHEMOTAXIS[30]

From stock sorocarp cultures of *Dictyostelium discoideum* grown on agar plates, spores from strain v12 were harvested and heat-shocked at 45°C for 30 min. The suspension of the spores was mixed with a full loop of *E. coli* B/r, the 200 μL aliquots in new SM-agar plates were cultured at 22°C for 24 h, the cells were harvested and maintained vegetatively by shaking (200 rpm in Lpp medium, approximately 10^6 cells/mL), or were starved by washing four times by centrifugation (3000 × *g*, 30 s). The cells were shaken in 15 mM Tris-HCl (pH 7.0) for 2 h or 4 h, the latter was for routine chemotaxis assays. Chemotaxis of *Dictyostelium discoideum* amoebae was performed on thin agar plates. To each Petri plate, 1 mL of 1.0% agarose in 15 mM Tris-HCl (pH 7.0) was added and agitated to allow even spreading. The chemoattractants cAMP or folic acid was added to the agar with or without agonists prior to pouring plates. Chemotactically competent cells were centrifuged to form a viscous suspension and spotted on the agar plates. Each plate containing 6 aliquots of cells was covered, incubated at 22°C for 3 h, uncovered, and placed on a heater to fix cells and desiccate the agar. As a measure of chemotactic efficiency, the initial spot diameters were subtracted from the diameters of visible rings or "halos" formed by outwardly migrating cells. During a period of 60 min with a playback time of 3 min, the individual halos on the agar plates were monitored. Proteins (20 μg/lane) were separated from whole cell lysates by polyacrylamide gel electrophoresis in 12% SDS-PAG and transferred to nitrocellulose membranes in a mini-trans-blot cell using a buffer containing 25 mM Tris, 192 mM glycine (pH 8.3), and 20% methanol. Transfer was with 100 V for 40–60 min employing a frozen cooling unit. Using standard curves produced by running prestained Rainbow markers in parallel lanes, relative molecular weights were estimated. Nitrocellulose blots were blocked in 4% BSA at 4°C for 16 h, incubated for 1 h in 1/500 monoclonal anti-phosphotyrosine PT-66, and diluted by a 1/1000 dilution of peroxidase conjugated goat anti-mouse IgG at room temperature for 1 h to detect the proteins containing phosphorylated tyrosine residues. After each incubation, blots were washed three times for 5 min, and immunoreactive bands were visualized using the chromogenic substrate diaminobenzidine HCl and 0.025% peroxide.

10.26 ANTIBODY FORMING CELL ASSAY[31]

Four days prior to the assay, the mice were sensitized intraperitoneally with 0.2 mL of EBSS containing 5×10^8 sheep red blood cells (SRBCs). Single splenocyte suspensions were prepared in 3 mL of EBSS. Into melted 0.5% agar in EBSS, 0.05% DEAE-dextran was added and maintained at 47°C throughout the assay. Into 12 mm × 75 mm heated glass tubes, 350 μL of melted agar was dispensed, to which 25 μL of indicator SRBC, 100 μL of spleen cell suspension, and 25 μL of guinea pig complement were added. Before use, SRBCs were washed with EBSS at least three times. From the tube a 150 μL aliquot was immediately pipetted onto a 100 mm × 20 mm Petri dish, and the agar solution was covered with a 22 mm × 40 mm microscopic cover glass. At room temperature, the Petri dishes were placed for several minutes to solidify the agar and incubated in a humidified 37°C incubator for 3 h to form hemolytic plaques. The number of antibody-forming cells (AFCs) was counted and expressed as AFCs/10^6 spleen cells or AFCs $\times 10^3$ per spleen. On the day of necropsy, the body and organ weights of the animals used in the antibody response were determined.

10.27 IMMUNOSUPPRESSIVE ASSAY[32]

BALB/c mice (6–8 weeks old) were sacrificed, and spleens were removed aseptically. After removing cell debris and clumps, a single cell suspension was prepared. With ammonium chloride buffer, erythrocytes were depleted. After washing three times with PBS containing 2% FBS, lymphocytes were resuspended in RPMI 1640 (Gibco-BRL, Life Technologies) at the indicated concentration. From BALB/c mice (male, 7–9 weeks old), fresh spleen cells were obtained, cultured (5×10^5 spleen cells/well) in 96-well flat plates with 200 μL of RPMI 1640 containing 10% FBS, 100 U/mL penicillin, and 100 μg/mL streptomycin in 5% CO_2 at 37°C for 48 h in the presence or absence of various concentrations of test compounds. At the final 5 h culture, to each well 18 μL of MTT (5 mg/mL) was added. To each well, 90 μL of lysis buffer (10% SDS, 50% DMF, pH 7.2) was added for 6–7 h, and the absorbance values were recorded on microplate reader at 570 nm. Cell mortality was determined according to Cytotoxicity% = $(\text{Sample}_{OD570} - \text{Background}_{OD570})/(\text{Control}_{OD570} - \text{Background}_{OD570}) \times 100$.

Fresh spleen cells (5×10^5) from BALB/C mice (male, 7–9 weeks old) were cultured in 96-well flat plates with 200 μL of RPMI 1640 containing 10% FBS, 100 U/mL penicillin, and 100 μg/mL streptomycin in 5% CO_2 at 37°C for 48 h. The cultures were unstimulated or stimulated with 5 μg/mL concanavalin A (ConA, Sigma, St. Louis, MO) or 10 μg/mL lipopolysaccharide (LPS) for 48 h to induce T-cell or B-cell proliferative responses. To cultures, test compounds with indicated concentrations were added. Proliferation was assessed in terms of uptake of [^{3}H]thymidine (Shanghai Institute of Atomic Energy, China) during 8 h pulsing with 20 kBq [^{3}H]thymidine/well, and the cells were harvested onto glass fiber filters by a Basic 96 harvester. On a liquid scintillation counter, the incorporated radioactivity was counted.

10.28 CELL-BASED ELISA[33]

Male inbred C57Blr6 mice (6–8 weeks old) were injected (iv) with test compounds or same volume of sterile PBS, sacrificed 24 h later, and the spleen was removed through a stainless steel mesh and suspended in RPMI 1640. After washing twice with RPMI 1640, red blood cells were lysed with Tris-buffered ammonium chloride (17 mM Tris Cl–0.73% NH_4Cl, pH 7.65), and erythrocytes-free spleen cells were washed and suspended in complete RPMI (RPMI 1640 supplemented with 10% fetal bovine serum, streptomycin 10 μg/mL, and penicillin 10^4 U/mL).

In 6-well tissue culture plates, 4 mL aliquots of the spleen cells suspension (1×10^6 cells/mL) were distributed, stimulated with test compounds, and incubated at 37°C with controlled atmosphere of 5% CO_2 for 24 h. The supernatants (SN) were harvested, filtered through 0.22-μm pore size membranes, and stored at –20°C for macrophage activation or cytokine assay.

The mice were injected with 1 mL of 3% (w/v) thioglycolate broth (Difco, USA), 3 days later peritoneal macrophages were harvested, plated onto 96-well culture plates (2×10^5 cells/mL in complete RPMI) and allowed to adhere for 1 h. By washing with warm medium, nonadherent cells were removed and in the presence or in the absence of 20 ng/mL LPS (*S. tifosa* lipopolysaccharide W0901, Difco, USA), adherent cells (more than 98% macro-phages) were incubated with different volumes of test compound stimulated spleen cell supernatants in 250 μL of RPMI complete medium. As negative controls, the spleen cell supernatants were in the same conditions incubated without lectin. As positive controls for NO production, mouse IFN-γ (10 U/mL) stimulated macrophages were incubated with or without LPS. In some experiments, before macrophage stimulation, polymyxin (50 μg/mL) was added to the supernatants. In some experiments, during the incubation anti-mIFNγ (R46A2) or anti-mTNFα (XT22.1, Department of Immunology, UNICAMP, São Paulo, Brazil) was also added to macrophage cultures. After 24 h at 37°C and 5% CO_2, as a metabolite of NO, the amount of NO in the macrophage supernatants was assayed.

The amount of IFN-γ in test compound stimulated spleen cells was evaluated by cell-based ELISA. The amount of TNF-α content in the supernatants of lectin stimulated spleen cells was assayed using a cytotoxicity assay with the TNF-α sensitive cell line WEHI 164.

In proliferation experiments, spleen cells from mice receiving (iv 24, 48, or 72 h before) PBS or test compounds were collected, and spleen mononuclear cells were isolated through Ficoll-Hypaque gradient. Mononuclear cells were washed twice and cell pellet was suspended in RPMI complete medium. Triplicate cultures in 200 μL of spleen cell suspensions (2×10^5 cells/well) from PBS or test compound injected mice were at 37°C and 5% CO_2 incubated in the presence or absence of ConA or test compounds or with phorbol 12-myristate 13-acetate (PMA, 50 μg/mL) plus ionomycin (2.0 μg/mL) for 72 h. Individual cultures were pulsed with 1 μCi of [^{3}H]thymidine (Amersham, UK) for the last 8 h, cells were harvested, and [^{3}H]thymidine incorporation was evaluated by liquid scintillation counting on a Beckman LS6000 counter.

Pooled spleen cells (1×10^6 cells/well) from PBS or test compound injected mice were plated into 96-well microtiter round-bottom plates and centrifuged at $400 \times g$ for 1 min. To avoid unspecific background staining, samples were suspended and incubated for 30 min at 4°C in 50 μL PBS containing 0.1% sodium azide, 10% mouse serum (PBS-S), and Fcγ_{II}R block (CD16/CD32) monoclonal antibodies (MoAb), incubated with FITC (CD3, CD4, CD8, VLA-4, L-selectin, or CD69) and PE (CD3 or CD122) antibodies (San Diego, CA, USA), diluted in PBS-S at 4°C for 30 min, washed, and suspended in PBS/0.1% azide for analysis in a FACScalibur (Becton and Dickinson, USA). Based on the FSCrSSC light scattering and staining with CD3, lymphocyte gating was performed.

10.29 LARGE ANIMAL LUNG TRANSPLANTATION ASSAY[34]

Donors and recipients were from randomized weight-matched pairs of outbred pigs, using which lung transplantations were performed. After flushing with 1000 mL of LPD solution (Perfadex, Vitrolife AB, Kungsbacka, Sweden), 1°C/24 h storage, and 1 h reperfusion, the right contralateral lung was excluded from perfusion and ventilation. Two groups were observed. Before inflow occlusion, into the pulmonary artery (PA) of the pigs in group I ($n = 5$), sildenafil (0.7 mg/kg) was injected. In addition, LPD containing 0.7 mg sildenafil/L was used as flush solution, and during backtable preparation the isolated left lung was perfused retrograde with 500 mL of this flush solution. The recipients continuously received iv infusion with 0.7 mg/kg sildenafil over 5 h, starting at reperfusion. Before flushing, into the PA of the pigs in group II, 8 mg/kg prostaglandin E1 (PGE1; Prostin VR Pediatric, The Upjohn Company, Kalamazoo, MI, USA) was injected, as in clinical use. LPD was employed for *in vivo* and backtable perfusion. Preservation time of both groups I and II donor lungs was 24 h.

One hour after reperfusion, the right PA (including the branch to the upper lobe) and the right main bronchus of the transplanted lung were ligated to assess isolated allograft function for the following 5 h after reperfusion. During the assessment period, to maintain anesthesia, 1.5% isoflurane was used. FiO_2 was 1.0 tidal volume 5500 mL at a respiratory rate of 20/min and a PEEP of 5 mm H_2O. Systemic arterial, PA, central venous, and left atrial pressures were continuously recorded, and arterial and mixed venous blood was collected for gas analysis at 60-min intervals. For tissue myeloperoxidase (MPO) and thiobarbituric acid-reactive substance (TBARS) assay, upper lobe allograft samples were collected and snap frozen at −80°C in liquid nitrogen. Time of tissue harvest was at 2–3 h after reperfusion for the control pigs and at 5 h after reperfusion for the treated recipients.

Extravascular lung water was measured as direct assessment of reperfusion edema. Via the external carotid artery, a fiberoptic catheter was advanced into the descending aorta. The indicator bolus consisting of indocyanine green served as an intravascular marker and ice-cold 5% glucose as a thermal intravascular and extravascular indicator. With a temperature controlled injector, the bolus was injected via the external jugular vein. With the thermistor tipped fiberoptic catheter, the dilution curves for dye and temperature were recorded simultaneously in the descending aorta. Based on the measurement of the mean transit times for thermal and dye indicators and of the decay time, volumes calculated from the indicator dilution curves, thoracic intravascular and extravascular fluid volumes, were determined. A lung water computer was used to determine the mean transit time for the thermal indicator and for the dye indicator and to calculate total thermal volume (ITTV), intrathoracic blood volume (ITBV), and extravascular thermal volume (ETV), and ETV = ITTV − ITBV.

Donor and recipient lung samples were frozen and stored at −80°C immediately. Myeloperoxidase levels were measured using a myeloperoxidase assay kit according to the manufacturer's instructions. The change in optical density units per milligram of tissue protein per minute was considered as enzyme activity. Thiobarbituric acid reactive substance levels in lung tissue were measured using the NWLSS kit according to the manufacturer's instructions.

REFERENCES AND NOTES

1. S. Saito, M. Tanaka, K. Matsunaga, Y. Li, Y. Ohizumi. The combination of rat mast cell and rabbit aortic smooth muscle is the simple bioassay for the screening of anti-allergic ingredient from methanolic extract of corydalis tuber. Biol Pharm Bull 27 (**2004**) 1270–1274.
2. R.L. Branton, D.J. Clarke. Apoptosis in primary cultures of E14 rat ventral mesencephala: Time course of dopaminergic cell death and implications for neural transplantation. Exp Neurol 160 (**1999**) 88–98.
3. S.P. Felix, R.O. Mayerhoffer, R.A. Damatta, M.A. Verícimo, V.V. Nascimento, O.L.T. Machado. Mapping IgE-binding epitopes of Ric c 1 and Ric c 3, allergens from Ricinus communis, by mast cell degranulation assay. Peptides 29 (**2008**) 497–504. *Note*: The cells were incubated with antibodies and allergens and centrifuged at 170 × *g* for 10 min. Aliquot (20 μL) of the supernatant was removed, the remaining cell suspensions were sonicated for 30 s, both of which were used for the histamine content determination. The solutions containing histamine were separated by ionic chromatography on an Amino-Na column (Shimadzu) equilibrated with 0.2 M NaOH. Histamine was detected and quantified by fluorescent isoindole, a post-column-*o*-phthaldialdehyde derivative that formed in the presence of 2-mercaptoethanol and under alkaline conditions. In the detection of histamine, a fluorescence detector was used at 360 nm excitation wavelength and 450 nm emission wavelength. Using one pmol to one nmol of histamine, a standard curve was prepared. The release of histamine was calculated as a percentage of the total histamine content of the sample.
4. S.N. Pramod, T.P. Krishnakantha, Y.P. Venkatesh. Effect of horse gram lectin (Dolichos biflorus agglutinin) on degranulation of mast cells and basophils of atopic subjects: Identification as an allergen. Int Immunopharmacol 6 (**2006**) 1714–1722.
5. J. Vendelina, C. Laitinen, P.J. Vainio, E. Nissinen, T. Mäki, K.K. Eklun. Novel sulfhydryl-reactive compounds orazipone and OR-1958 inhibit cytokine production and histamine release in rat and human mast cells. Int Immunopharmacol 5 (**2005**) 177–184.
6. S.H. Hong, H.J. Jeong, H.M. Kim. Inhibitory effects of Xanthii fructus extract on mast cell-mediated allergic reaction in murine model. J Ethnopharmacol 88 (**2003**) 229–234.

7. A.E. Barbu, I. Pecht. Desensitization of mast cells' secretory response to an immunoreceptor stimulus. Immunol Lett 100 (**2005**) 78–87.

8. B. Horstmann, E. Zinser, N. Turza, F. Kerek, A. Steinkasserer. MCS-18, a novel natural product isolated from Helleborus purpurascens, inhibits dendritic cell activation and prevents autoimmunity as shown in vivo using the EAE model. Immunobiology 212 (**2008**) 839–853.

9. G. Shankar, M.S. Fourrier, M.A. Grevenkamp, A. Patricia. Validation of the COSTIM bioassay for dendritic cell potency. J Pharm Biomed Anal 36 (**2004**) 285–294.

10. K. Dheda, A. Pooran, M. Pai, R.F. Miller, K. Lesley, H.L. Booth, G.M. Scott, A.N. Akbar, A. Zumla, G.A. Rook. Interpretation of Mycobacterium tuberculosis antigen-specific IFN-g release assays (T-SPOT.TB) and factors that may modulate test results. J Infection 55 (**2007**) 169–173.

11. E. Rubakova, S. Petrovskaya, A. Pichugin, V. Khlebnikov, D. McMurray, E. Kondratieva, I. Baturina, T. Kondratieva. A. Apt, Specificity and efficacy of dendritic cell-based vaccination against tuberculosis with complex mycobacterial antigens in a mouse model. Tuberculosis 87 (**2007**) 134–144.

12. J.A. Ida, N. Shrestha, S. Desai, S. Pahwa, W.A. Hanekom, P.A.J. Haslett. A whole blood assay to assess peripheral blood dendritic cell function in response to Toll-like receptor stimulation. J Immunol Methods 310 (**2006**) 86–99.

13. S.R. Walker, P.D. Ogagan, D. DeAlmeida, A.M. Aboka, E.M. Barksdale Jr. Neuroblastoma impairs chemokine-mediated dendritic cell migration in vitro. J Pediatr Surg 41 (**2006**) 260–265.

14. A.L.T. Spinardi-Barbisan, R. Kaneno, L.F. Barbisan, J.L.V. de Camargo, M.A.M. Rodrigues. Chemically induced immunotoxicity in a medium-term multiorgan bioassay for carcinogenesis with Wistar rats. Toxicol Appl Pharmacol 194 (**2004**) 132–140.

15. C.F. Kuper, J.H. Harlerman, H.B. Richter-Reichelm, J.G. Vos. Histopathologic approaches to detect changes indicative of immunotoxicity. Toxicol Pathol 28 (**2000**) 454–466.

16. K.G Bowman, S. Hemmerich, S. Bhakta, M.S Singer, A. Bistrup, S.D Rosen, C.R. Bertozzil. Identification of an N-acetylglucosamine-6-o-sulfotransferase activity specific to lymphoid tissue: an enzyme with a possible role in lymphocyte homing. Chem Biol 5 (**1998**) 447–460. *Note*: (1) Microsomes were isolated from fresh porcine tissues. In brief, by intravenous administration of ketamine HCl, 8-month-old female Yucatan micropig was euthanized. The tissues were harvested, snap frozen in liquid nitrogen, placed in a motor driven tissue grinder, crushed to powder in liquid nitrogen, and suspended in a 10-fold volume of SKTM buffer. In a 10 mL glass homogenizer with a drill driven conico-cylindrical Teflon pestle (five passes at 800 rpm), the tissue suspensions were homogenized, centrifuged at 680 $\times$ g for 10 min to collect the supernatant, the nuclear/mitochondrial pellet was discarded, the supernatant was centrifuged at 10,000 $\times$ g for 10 min to collect the new supernatant and discard the lysosomal pellet. The supernatant was centrifuged at 100,000 $\times$ g for 1 h to collect the supernatant and the microsomal pellet for analysis and as an enzyme source, respectively. By resuspension in five volumes of 150 mM Tris-HCl [pH 8.0 supplemented with 1 $\times$ protease inhibitor cocktails (PIC)-I and PIC-II] and centrifugation at 100,000 $\times$ g for 1 h the microsomal pellet was washed. Using a handheld Microfuge tube homogenizer, the pellet was dissolved in lysis buffer (50 mM HEPES pH 7.5, 2.5% Triton X-100 supplemented with 1 $\times$ PIC-I and PIC-II). By recentrifugation for 1 h at 100, 000 $\times$ g, lysates were cleared, and to each cleared lysate 1/4 volume of 50% glycerol was added. Using the Pierce BCA protein assay, the

total protein concentrations in the lysates and cytosolic supernatants were determined. (2) From porcine lymph nodes, cellular fractionation was prepared. In brief, on a stainless steel screen, 4.2 g of fresh porcine peripheral lymph nodes were diced and teased, and by flushing with RPMI through the screen, lymphocytes were washed. In RPMI with collagenase I and DNAse I, the remaining stromal elements were digested for 15 min to degrade the extracellular matrix components. On the screen, the digested stroma were re-filtered to remove additional lymphocytes. After 1 h digestion with collagenase I and DNAse I, the liberated stromal cells were flushed with buffer (PBS, 1% BSA, 5 mM EDTA) through the screen. By immunomagnetic separation and with mAb MECA-79 as HEC-specific marker from the stromal cells, high endothelial cells (HECs) were purified. The stromal cells were at 4°C with MECA-79 incubated for 15 min, washed by centrifugation, resuspended in buffer, at 4°C with a biotinylated secondary antibody (anti-Rat IgM) incubated for 15 min, washed, and resuspended. After 15 min incubation with streptavidin magnetic beads, the labeled cells were applied to a WHM separation column in a magnetic field. HEC-depleted stromal cells were with buffer washed through the column, and by removing the column from the field and washing with buffer, magnetic HEC were obtained.

17. M. Taniguchi, K. Nagaoka, S. Ushio, Y. Nukada, T. Okura, T. Mori, H. Yamauchi, T. Ohta, H. Ikegami. Establishment of the cells useful for murine interleukin-18 bioassay by introducing murine interleukin-18 receptor cDNA into human myelomonocytic KG-1 cells. J Immunol Methods 217 (**1998**) 97–102.
18. P. Parnet, K.E. Garka, T.P. Bonner, S.K. Dower, J.E. Sims. IL-1 Rrp is a novel receptor-like molecule similar to the type I interleukin-1 receptor and its homologues T1 r ST2 and IL-1R Acp. J Biol Chem 271 (**1996**) 3967.
19. S.Y. Eum, Y.J. Lee, H.K. Kwak, J.H. Min, S.H. Hwang, L.E. Via, C.E. Barry III, S.N. Cho. Evaluation of the diagnostic utility of a whole-blood interferon-γ assay for determining the risk of exposure to mycobacterium tuberculosis in bacille calmette-guerin (BCG)-vaccinated individuals. Diagn Microbiol Infect Dis 61 (**2008**) 181–186. *Note*: (1) The detection limit of this assay was 31 pg/mL, and a positive response was defined as ≥62 pg/mL. (2) QuantiFERON-TB Gold In-Tube test kit was used according to the manufacturer's instructions. Blood was directly collected into two tubes containing 1 mL of heparin. As negative control, one tube contained heparin alone, the other tube contained overlapping peptides representing the entire sequences of CFP-10 and ESAT-6 and another peptide representing a portion of TB7.7. According to the manufacturer's instructions, the tubes were incubated for 20 to 24 h, and plasma was removed and frozen until used for ELISA. By subtracting the value of negative control, the IFN-γ values were obtained, and according to the manufacturer's instructions, the cutoff value was 0.35 IU/mL.
20. S. Dogra, P. Narang, D.K. Mendiratta, P. Chaturvedi, A.L. Reingold, J.M. Colford Jr, L.W. Riley, M. Pai. Comparison of a whole blood interferon-g assay with tuberculin skin testing for the detection of tuberculosis infection in hospitalized children in rural India. J Infection 54 (**2007**) 267–276.
21. V. Sunil, P.I. Menzies, P.E. Shewen, J.F. Prescott. Performance of a whole blood interferon-gamma assay for detection and eradication of caseous lymphadenitis in sheep. Vet Microbiol 128 (**2008**) 288–297.
22. S.H. Kim, E.J. Oh, M.J. Kim, Y.J. Park, K. Han, H.J. Yang, J.Y. Kim, B.S. Choi, C.W. Yang, Y.S. Kim, B.K. Bang. Pretransplant donor-specific interferon-γ ELISPOT assay predicts acute rejection episodes in renal transplant recipients. Transplant Proc 39 (**2007**) 3057–3060.

23. G. Bellisola, G. Tridente, F. Nacchia, F. Fior, L. Boschiero. Monitoring of cellular immunity by interferon-gamma enzyme-linked immunosorbent spot assay in kidney allograft recipients: Preliminary results of a longitudinal study. Transplant Proc 38 (**2006**) 1014–1017.

24. U. Canosi, M. Mascia, L. Gazza, O. Serlupi-Crescenzi, S. Donini, F. Antonetti, G. Galli. A highly precise reporter gene bioassay for type I interferon. J Immunol Methods 199 (**1996**) 69–76.

25. X. Wei, S.J. Swanson, S. Gupta. Development and validation of a cell-based bioassay for the detection of neutralizing antibodies against recombinant human erythropoietin in clinical studies. J Immunol Methods 293 (**2004**) 115–126.

26. S. Mason, S. La, D. Mytych, S. Swanson, J. Ferbas. Validation of the BIACORE 3000 platform for detection of antibodies against erythropoietic agents in human serum samples. Curr Med Res Opin 19 (**2003**) 651–659.

27. J.A. Lewis. A sensitive biological assay for interferons. J Immunol Methods 185 (**1995**) 9–17.

28. T.N. Behar, C.A. Colton. Redox regulation of neuronal migration in a down syndrome model. Free Radic Biol Med 35 (**2003**) 566–575.

29. B.A. Bromberek, P.A.J. Enever, D.I. Shreiber, M.D. Caldwell, R.T. Tranquillo. Macrophages influence a competition of contact guidance and chemotaxis for fibroblast alignment in a fibrin gel coculture assay. Exp Cell Res 275 (**2002**) 230–242.

30. D.D. Browning, T. The, D.H. O'Day. Comparative analysis of chemotaxis in Dictyostelium using a radial bioassay method: protein tyrosine kinase activity is required for chemotaxis to folate but not to camp. Cell Signal 7 (**1995**) 481–489.

31. J.K. Lee, J.A. Byun, S.H. Park, H.S. Kim, J.H. Park, J.H. Eom, H.Y. Oh. Evaluation of the potential immunotoxicity of 3-monochloro-1,2-propanediol in Balb/c mice I. Effect on antibody forming cell, mitogen-stimulated lymphocyte proliferation, splenic subset, and natural killer cell activity. Toxicology 204 (**2004**) 1–11.

32. Z. Yang, J. Wang, Y. Zhou, J. Zho, Y. Lia. Synthesis and immunosuppressive activity of new artemisinin derivatives. Part 2: 2-[12(b or a)-dihydroartemisinoxymethyl- (or 10-ethyl)]phenoxyl propionic acids and esters. Bioorg Med Chem 14 (**2006**) 8043–8049.

33. J.E. Lima, A.L.F. Sampaio, M. das G.M.O. Henriques, C. Barja-Fidalgo. Lymphocyte activation and cytokine production by Pisum sativum agglutinin (PSA) in vivo and in vitro. Immunopharmacology 41 (**1999**) 147–155.

34. S. Korom, S. Hillinger, M. Cardell, W. Zhai, Q. Tan, A. Dutly, B. Leskosek, W. Weder. Sildenafil extends survival and graft function in a large animal lung transplantation model. Eur J Cardiothorac Surg 29 (**2006**) 288–293.

11

METHODS AND APPLICATIONS OF ANTI-INFLAMMATORY ASSAYS

Ming Zhao

In the development of serious diseases, such as cancer, rheumatoid arthritis, sprains, bronchitis, muscle pains, chronic inflammatory bowel disease, persistent asthma, and liver fibrosis, inflammation is frequently involved. In the development of inflammatory events, chemicals such as glucocorticoids (GCs) and mometasone furoate (MF); endogenous factors such as tumor necrosis factor alpha (TNF-α); enzymes and proteins such as copper and zinc-superoxide dismutase (SOD1), proinflammatory peptide substance P (SP), RGD peptides, interleukin-4 (IL-4), IL-10, interferon-γ (IFN-γ), COX-1, COX-2, 5-LOX, macrophage inflammatory protein (MIP)-1R, glucocorticoid regulated protein CD163, FK506 binding protein 51 (FKBP51), and monocyte chemoattractant protein-1 (MCP-1); pleiotropic proinflammatory cytokines such as interleukin-1 (IL-1); adhesions such as fibrous adhesions and cell adhesions resulting from the receptor, such as $\alpha_{II}\beta_{III}$, human glucocorticoid receptor, and chemokine receptors (CCR5); inflammatory mediators such as lipopolysaccharide (LPS), reactive oxygen species (ROS), nitric oxide (NO), and prostaglandin E2 (PGE2); as well as proinflammatory cells such as CD4$^+$ T cells, CD8$^+$ T cells, and NK cells are well known to have an important role. Based on these correlations, numerous assays are reported in the literature, used for inflammatory mechanism

Pharmaceutical Bioassays: Methods and Applications. By Shiqi Peng and Ming Zhao

research and developed to screen anti-inflammatory agents. In this chapter, 17 assays are described: adhesion formation assay,[1] ligand complex-based adhesion assay,[2,3] human umbilical vein endothelial cell assay,[4,5] pleurisy mouse assay,[6–11] proliferation of PBMC assay,[5,12–15] COX-1, COX-2, and 5-LOX assay with [1-^{14}C]arachidonic acid,[16] COX-1 and COX-2 assay with human whole blood,[17] *o*-hydroxyleukotriene B4 assay,[18–20] leukocyte rolling and adherence assay,[21] CCR5 receptor binding assay,[22] tissue binding affinity assay,[23–25] G93A-SOD1 transgenic mouse assay,[26–28] MCP-1-induced ERK1 and ERK2 phosphorylation assay,[29,30] LPS- and IL-6-induced ERK1 and ERK2 phosphorylation assay,[31] ELA4.NOB-1/CTLL cell assay,[32] FK506 binding protein 51 (FKBP51) mRNA assay,[33–35] and xylene-induced ear edema assay[36].

11.1 ADHESION FORMATION ASSAY[1]

For assessment of neurokinin-1 receptor antagonist (NK-1RA) on peritoneal adhesion formation, a laparotomy was performed through a midline incision, and four ischemic buttons spaced 1 cm apart were created on both sides of the parietal peritoneum by grasping 5 mm of peritoneum with a hemostat and ligating the base of the segment with 4-0 silk suture. To assess the effects of NK-1RA on adhesion formation, peritoneal adhesions were induced in male Wistar rats (200–250 g) that were randomized to groups receiving specific nonpeptide NK-1RA, test compounds, or vehicle. In the initial assay, the rats in the group received 0.2 mL ip injections of 25 mg/kg NK-1RA twice a day for 2 days. At the time of surgery, 1 mL of test compound (0.75 mg/mL) was given as a peritoneal lavage, and the rats then received ip injections for 7 days. Control rats were similarly injected/lavaged with sterile vehicle. This experiment was repeated with 10 mg/kg NK-1RA per day. At day 7, all the rats were sacrificed, and the adhesions were quantified in a blinded fashion. Each rat received a percentage adhesion score based on the number of ischemic buttons with attached adhesions.

For assessment of NK-1RA on peritoneal tPA and plasminogen activator inhibitor (PAI-1) expression and activity, the temporal expression pattern of tPA and PAI-l mRNA in peritoneal tissue collected from an 0.5-cm radius of the ischemic buttons was determined by RT-PCR analysis at day 0, 1, 3, and 7 after surgery. Based on the results the effects of NK-1RA administration on tPA and PAI-1 mRNA and protein, levels were determined at postoperative day 1 in peritoneal tissue and fluid by RT-PCR analysis and bioassay, respectively. The rats received 5.0 mg/kg NK-1RA or vehicle per day. Control samples were collected from six nonoperated rats. All samples were frozen immediately in liquid nitrogen and stored at −80°C until used.

Total RNA was isolated from 50 mg of peritoneal tissue with the SV Total RNA Isolation System (Promega), and RT-PCR was conducted with Gene-Amp RNA PCR System (Applied Biosystems). To amplify tPA and PAI-1, 28 cycles of 95°C, 60°C, and 72°C for 30 s each were used. In the amplification of tPA, the primer sets of 5′-TCTGACTTCGTCTGCCAGTG-3′ (sense) and 5′-GAGGCCTTGGATGTGGTAAA-3′ (antisense) were used. In the amplification of PAI-1, the primer sets of 5′-ATCAACGACTGGGTGGAGAG-3′ (sense) and 5′-AGCC-TGGTCATGTTGCTCTT-3′ (antisense) were used. PCR products (15 μL) were subjected to electrophoresis on 2% agarose gels containing 0.03 μg/mL ethidium bromide, and the quantitative level of the transcript was determined on scanned photographs of gels. Levels of mRNA expression were normalized to GAPDH, a constitutively expressed gene that does not vary among treatment groups.

Using the corresponding kits, total levels of tPA and PAI-1 in peritoneal fluid samples were measured. From 12 vehicles administered, 12 NK-1RA administered [(3R,4S,5S,6S)-6-diphenylmethyl-5-(5-isopropyl-2-methoxybenzlyamino)-1-azabicyclo[2,2,2]octane-3-carboxylic acid, 10.0 mg/kg per day], and 12 nonoperated control rats, peritoneal fluid was collected in 5 mM citrate and 0.1 M acetate for assessment of fibrinolytic activity caused by tPA activation of plasminogen (Athens Research and Technology, Athens, GA). Peritoneal fluid samples (1.0 μL) were run on 10% SDS polyacrylamide gels containing 0.1% gelatin and 0.002% plasminogen. After electrophoresis, the gels were washed twice in 2% Triton X-100 and incubated overnight at 37°C in 0.1 M glycine (pH 8.3). The gels were stained with a 0.25% Coomassie blue solution, and PA activity was visualized as clear bands produced by plasmin lysis of gelatin. Determining the contribution of serine proteases, 10 mM serine protease inhibitor PMSF was added to the developing buffer. In the identification of the zones of lysis corresponding with tPA activity, tPA and/or urokinase-type plasminogen activator (uPA) were immunoprecipitated from peritoneal fluid samples. To the mixture of 10 μL of peritoneal fluid and 10 μL of buffer containing 40 mM phosphate, 1 M NaCl, 0.2% SDS, 2% Igepal CA-630, and 1% deoxycholate (pH 7.5), 1 μg tPA and/or uPA antibody (American Diagnostica, Greenwich, CT) was added and incubated at 4°C overnight. To each sample, 10 μL of UltraLink protein A/G slurry (50%) was added and incubated overnight at 4°C. Samples were centrifuged at 4°C and 16,000 × *g* for 1 min, and the supernatant (50%) was analyzed by zymography for comparison with human recombinant tPA and uPA standards.

11.2 LIGAND COMPLEX-BASED ADHESION ASSAY[2,3]

In a 1.5-mL Teflon tube, FITC or PE labeled anti-human Fcγ specific IgG $F(ab')_2$ fragments (in a 1/6.25 dilution, Jackson ImmunoResearch) and ICAM-1-Fc (200 μg/mL, University of Bonn, Germany) were incubated at 4°C overnight in PBS to generate ICAM-1-Fc-$F(ab')_2$ complexes (scICAM-1). In the case of whole blood, the samples were incubated with 1/5 dilution scICAM-1 and 2/5 PBS containing test compounds or solvent alone. For phorbol-12-myristate-13-acetate (PMA) stimulation and the corresponding control, PBS, supplemented with 5 mM Mg^{2+} with the naturally occurring Mg^{2+} of the blood sample together to reach a final concentration of 2 mM, was used. All added compounds were Ca^{2+} free, and due to the Ca^{2+} content of the blood sample, Ca^{2+} had a final concentration of about 1 mM. After the indicated times, 10 volumes of FACS lysing solution (BD/Pharmingen) supplemented with 2 mM Mg^{2+}/1 mM Ca^{2+} in the case of PMA stimulation or 10 mM Mg^{2+}/1 mM EGTA in the case of Mg^{2+}/EGTA stimulation were added, vortexed, and incubated for 5 min at 37°C. To the whole blood, FACS-lysing solution

(BD/Pharmingen) was added to simultaneously lyse red blood cells and fix leukocytes. By adding ice-cold FACS buffer, the fixation was stopped. The samples were centrifuged, cells were stained for subpopulations, and analyzed by flow cytometry for $CD45^+$ events of peripheral blood stem cells and other leukocyte subsets within the lymphocyte gate.

After erythrocyte lysis and the washing steps, whole blood leukocytes were used for the blocking experiments. Cells were resuspended in PBS saturated with the respective blocking antibody, incubated at room temperature for 30 min, spun down and resuspended in PBS with the respective cations, 1/4 dilution of blocking antibodies (20 μg/mL), 1/4 dilution of scICAM-1 and 1/2 of dilution PBS containing test compounds. Prior to fixation with 10 volumes of FACS-lysing solution containing the respective cations, samples were incubated for 30 min and stained for flow cytometric analysis (FACS Calibur).

11.3 HUMAN UMBILICAL VEIN ENDOTHELIAL CELL ASSAY[4,5]

1. Expression of surface adhesion molecules, vascular cell adhesion molecules (VCAM-1 and ICAM-1), was assessed with ELISA (absorbance units, AU) on cells grown in 96-well plates. To block nonspecific binding sites, human umbilical vein endothelial cells (HUVECs) were washed three times, fixed in 2% paraformaldehyde for 15 min, and incubated at 37°C for 1 h with PBS (PAA, Austria) containing 3% BSA (PBSA, Sigma). At 37°C, the monolayers were incubated with primary monoclonal antibody (5 μg/mL in PBSA) or appropriate control isotype (5 μg/mL in PBSA) for 2 h, and the cells were washed and incubated with secondary goat anti-mouse IgG-peroxidase conjugate antibody (diluted 1/10,000 in PBSA) for 1 h to perform ELISA. Between each incubation step, cells were washed three times with PBS containing 1% BSA, and the integrity of the monolayer was monitored by phase-contrast microscopy. In the final step, *o*-phenylenediamine dihydrochloride was added as peroxidase substrate, and color development was stopped with HCl (3 M). By absorbance determination at 490 nm (Titertek Instruments Inc.), ICAM-1 and VCAM-1 expression was quantified. The absorbance of the control isotype was subtracted from the readings.

2. MTT assay-produced formazan was used as an index of cell viability responses to experimental interventions. The plates were incubated at 37°C in 5%CO_2 environment for 2 h, to the wells fresh medium containing a fixed volume of reconstituted MTT (0.5 mg/mL final concentration) was added. Absorbance was measured by spectrophotometry at 570 nm.

3. By modified alkaline single cell gel electrophoresis (comet), DNA damage in response to exogenous hydrogen peroxide was assessed, which was based on the fact that in high pH electrophoresis, the broken DNA loops freely moved toward the anode and result in comet-like tail-to-head structures whose relative fluorescence intensity is proportional to DNA damage. Slides were spread with 1% normal melting agarose in Ca^{2+} and Mg^{2+} free PBS and left to solidify until used. After hydrogen peroxide exposure and test compound treatment, HUVECs were centrifuged,

resuspended in 0.5% low melting agarose, and incubated at 37°C. Onto the top of the first agarose layer, cell suspension was rapidly dropped, covered with a coverslip, and allowed to solidify at 4°C. After removal of coverslip, to the slide a final layer of 0.5% low melting agarose was added. By immersion in a coupling jar containing freshly prepared cold lysing buffer (2.5 M NaCl, 100 mM Na_2 EDTA, 10 mM Tris, 10% DMSO, and 1% Triton X-100, pH 10.0) at 4°C for 1 h, the slides were lysed, placed in a horizontal gel electrophoresis chamber filled with freshly prepared cold alkaline buffer (300 mM NaOH, 1 mM Na_2EDTA, pH >13), and left in this solution for 20 min to allow DNA unwinding. DNA fragments were separated by electrophoresis for 20 min at 25 V (300 mA), the slides were washed with neutralizing buffer (0.4 M Tris, pH 7.5) and stained with ethidium bromide (20 μg/mL). On a fluorescence microscope with ×200 magnification, coded slides were viewed. For each experiment, 100 randomly selected cells (50 cells from each replicate slide) were scored and analyzed with Comet Image Analysis System (version 5.5, Kinetic Imaging Ltd.).

11.4 PLEURISY MOUSE ASSAY[6–11]

Adult Swiss mice (18–22 g, both sexes, aged 2 months) were kept in a room at constant temperature (25 ± 2°C), with alternating 12-h periods of light and dark and fed on food and water *ad libitum*. By a single intrapleural injection of 0.1 mL of carrageenan (Cg, 1%), mice were induced to suffer from pleurisy. After 4 h, the mice were sacrificed with an overdose of ether, their thoraxes were opened, and pleural cavities were washed with 1.0 mL of sterile PBS (pH 7.6, consisting of 137 mmol of NaCl, 2.7 mmol of KCl, and 10 mmol of phosphate buffer salts) containing heparin (20 IU/mL). Pleural fluid samples were collected for further determination of leukocytes, myeloperoxidase (MPO), nitric oxide (NO), and cytokines (TNF-α, IL-1β) levels. After diluting the pleural fluid with Türk solution (1:200), total leukocyte counts were performed in a Neubauer chamber. The cytospin preparations of pleural wash were stained with May–Grünwald Giemsa, and under oil immersion objective the differential leukocytes count was performed.

1. MPO levels in the pleurisy induced by Cg were determined according to the general methods by means of colorimetric measurements (absorbance at 450 nm) in an ELISA plate reader (Organon Tecknica, Roseland, NJ, USA). Results were expressed as mU/mL. Samples of pleural cavity fluid from control and test compound-treated mice were collected and immediately processed to analyze the MPO levels.

2. Nitric oxide levels in the pleurisy induced by Cg were determined as its breakdown products nitrite (NO_2^-) and nitrate (NO_3^-) according to the general methods. Samples of pleural cavity fluid from control and test compound-treated mice were collected and stored at −20°C to determine the levels of nitrate/nitrite.

3. In the determination of TNF-α and IL-1β levels in samples obtained from pleurisy induced by Cg, monoclonal specific USA antibody (Bioscience Pharmigen, USA) and Japan antibody (Immuno Biological Laboratories Co. Ltd.,

Japan) were used. According to the manufacturers' instructions, TNF-α and IL-1β concentrations were measured by means of colorimetric measurement at 450 nm on an ELISA plate reader by interpolation from a standard curve. Samples of pleural cavity fluid from control and test compound-treated mice were collected and immediately used to determine the cytokines.

11.5 PROLIFERATION OF PBMC ASSAY[1,5,12–15]

Healthy male volunteers (mean age 27 years) were chosen for this assay, and 60 mL of heparinized peripheral blood was collected. The peripheral blood mononuclear cells (PBMCs) were isolated by the Ficoll-Paque (specific gravity 1.077) gradient-density method. To remove the plasma, peripheral blood was centrifuged at 4°C and 2000 × *g* for 10 min. Blood cells were diluted with PBS and centrifuged in a Ficoll-Paque discontinuous gradient at 1500 × *g* for 30 min. To remove red blood cells, the PBMC layers were collected and washed with cold distilled water and ×10 Hanks' buffered saline solution and resuspended to a concentration of 2×10^6 cells/mL in RPMI-1640 medium supplemented with 2% FCS, 100 U/mL penicillin, and 100 μg/mL streptomycin. To test the lymphoproliferation, 100 μL of PBMC suspension was deposited into a 96-well flat-bottomed plate with or without 5 μg/mL phytohemagglutinin (PHA, Life Technologies, Grand Island, NY). Cyclosporin A was used as a positive control. After addition of various concentrations of the test compounds, the plates were incubated at 37°C for 3 days in humidified atmospheric air containing 5% CO_2 followed by addition of tritiated thymidine. After 16 h incubation, the cells were harvested on glass fiber filters using an automatic harvester and measured with a scintillation counting. To determine cell viability, approximately 2×10^5 T cells with or without PHA were cultured with 0.1% DMSO or various concentrations of the test compound for 3 days. The viable cell numbers were counted using a microscope with a hemocytometer after staining by Trypan blue, and the percentage of viable cells was calculated. To analyze cell cycle, 1 mL of PBMC suspension was placed in a 6-well flat-bottomed plate with or without 5 μg/mL PHA. After addition of 25 μg/mL test compounds, the plates were incubated at 37°C for 3 days in humidified atmospheric air containing 5% CO_2. The cells were harvested by means of centrifugation, washed with PBS, and fixed in 70% ethanol at −20°C for 30 min. DNA was stained with 4 μg/mL propidium iodide containing 100 μg/mL ribonuclease A. Flow cytometric analysis (Beckman Coulter, USA) was conducted.

After incubation with PHA alone or in combination with varying concentrations of test compounds for 3 days, PBMC (2×10^5 cells/well) supernatants were collected and assayed for IL-2, IL-4, IL-10, and IFN-γ concentrations by means of enzyme immunoassays.

To extract the total cellular RNA from PBMCs, after stimulation with or without PHA and co-culture with 25 μg/mL test compounds for 18 h, the collected cells (5×10^6) were subjected to lysis with RNA-Beek. After centrifugation, supernatants were extracted with a phenol-chloroform mixture. The extracted RNA was precipitated with isopropanol. Total cellular RNA was obtained as pellets by centrifugation and redissolved in diethyl pyrocarbonate (DEPC)-treated H_2O. The concentration of the extracted RNA was measured using the optical density at 260 nm.

In the synthesis of the first-strand cDNA, 1 μg aliquots of RNA were reverse-transcribed using the Advantage RT-for-PCR kit (BD Biosciences, Palo Alto, CA) according to manufacturer's instructions. In brief, a mixture of 1 μg RNA, 12.5 μL DEPC-treated H_2O and 20 μM oligodeoxythymidine 18 [Oligo(dT)18] was heated at 70°C for 10 min, and then quick-chilled on ice, to which 6.5 μL concentrated synthetic buffer (50 mM Tris-HCl, pH 8.3, 75 mM KCl, 3 mM $MgCl_2$, 0.5 mM deoxynucleotides triphosphates, and 0.5 U ribonuclease inhibitor) and 200 U moloney murine leukemia virus reverse transcriptase was added. The mixture of 1 μg of RNA in 12.5 μL of DEPC-treated H_2O and 20 μM oligodeoxythymidine 18 [oligo(dT)18] was heated at 72°C for 2 min, then quick-chilled on ice. The reaction mixture was incubated at 42°C for 1 h and then at 94°C for 5 min. To the reaction mixture, 80 μL of DEPC-treated H_2O was added and stored at −20°C for PCR use. To 10 μL of first-strand cDNA 0.75 μM primers, 4 U Taq polymerase, 10 μL of reaction buffer (2 mM Tri-HCl, pH 8.0, 0.01 mM EDTA, 0.1 mM dithiothreitol, 0.1% Triton X-100, 5% glycerol, and 1.5 mM $MgCl_2$), and 25 μL of water were added and the total volume was 50 μL. According to the air thermocycler, namely denaturing temperature of 94°C for 1 min, annealing temperature of 60°C for 1 min, and elongation temperature of 72°C for 80 min for the first 25 cycles, and finally 72°C for 10 min, PCR was carried out. After the reaction, the amplified product was taken out of the tubes and run on 1.8% agarose gel.

11.6 COX-1, COX-2, AND 5-LOX ASSAY WITH [1-^{14}C]ARACHIDONIC ACID[16]

For cyclooxygenase (COX)-1 and COX-2 assay, the dilution of test compound was incubated with COX-1 or COX-2 according to the standard method, and the enzyme reaction was initiated by adding [^{14}C] arachidonic acid. From the incubation medium prostaglandin E_2 and prostaglandin D_2 (PED_2) were extracted to measure radioactivity. Indomethacin was used as a positive control. For 5-lipoxygenase (5-LOX) assay, the inhibitory activity of the test compound was evaluated. The test compound was incubated with 5-LOX (46 μg of protein) at 24°C for 10 min and then the enzyme reaction was initiated by addition of [^{14}C]-arachidonic acid (50 nCi). After 5 min, 4 M formic acid was added to terminate the enzyme reaction. Indomethacin was used as a positive control.

11.7 COX-1 AND COX-2 ASSAY WITH HUMAN WHOLE BLOOD[17]

By gentle shaking on deep-well plates, 480 μL of fresh blood (0.1% Li-Heparin Liquemin 25000) from human volunteers (female, via venous puncture) with no pharmacologic therapy in the last 2 weeks preceding sampling was mixed with 1 μL of test compounds or controls (in DMSO). For measuring COX-2-induced

PGE2 synthesis, LPS (1 μg/mL end-concentration/well, total volume of 20 μL PBS) stimulation of blood samples was performed. On an orbital shaker at 37°C in humidified atmosphere, controls were adjusted to the same volume with PBS and mixed for 1 min. After 24 h incubation, blood samples were centrifuged (2250 × *g* or 3700 rpm, 10 min, 4°C) to harvest plasma supernatant on ice. According to the manufacturer's instructions, PGE2 levels were analyzed by EIA. For analyzing COX-1, fresh blood was collected into Vacutainers containing no anticoagulants, allowed to clot for 1 h (37°C, humidified atmosphere), and 2 μL of diclofenac (25 mM, equivalent to 100 mM end concentration/well) was added and mixed for 1 min on an orbital shaker to stop TxB2 release. Plates were centrifuged (2250 × *g* or 3700 rpm, 10 min, 4°C), serum supernatant was harvested on ice and analyzed with a TxB2 EIA (R&D Systems, Germany). IC_{50} values were calculated with GraphPad Prism (version 4.02, GraphPad Software Inc., San Diego, CA) from 4 to 10 independent female donors on at least 4 different days.

11.8 *o*-HYDROXYLEUKOTRIENE B4 ASSAY[18–20]

Test compounds were assayed at 50 mM final concentration and in at least three independent experiments, and each experiment was performed in duplicate. IC_{50} values were calculated by semilogarithmic presentation of dose versus inhibitory activity and logarithmic regression analysis. In a 96-well plate format, purified prostaglandin H synthase (PGHS)-1 from ram seminal vesicles (Cayman Chemical Company, Ann Arbor, MI, USA) was used for COX-1 assay and purified PGHS-2 from sheep placental cotyledons (Cayman Chemical Company, Ann Arbor, MI, USA) was used for COX-2 assay. The concentration of the main arachidonic acid metabolite in this reaction, PGE2, was determined using a competitive PGE2 EIA kit (Assay Designs Inc., Ann Arbor, MI, USA) and with indomethacin (IC_{50} COX-1 0.9 mM) and NS-398 (IC_{50} COX-2 2.6 mM) as positive controls. In the assay for inhibition of leukotriene formation, polymorphonuclear leukocytes with 5-LOX activity were isolated from venous human blood based on sedimentation rates and lysis tolerance. With 0.4% Trypan blue solution, cell viability was checked. The cell suspension (4500 cells/mL) was incubated with test compounds, $CaCl_2$, calcimycin A23187, and arachidonic acid in a shaking water bath at 37°C for 10 min, stopped by adding 10% HCO_2H and then centrifuged. The samples were diluted, and the concentration of *o*-hydroxyleukotriene B4 (LTB4) formed during incubation was determined by means of a competitive LTB4 EIA kit. Zileuton (IC_{50} 5.0 mM) was used as positive control.

11.9 LEUKOCYTE ROLLING AND ADHERENCE ASSAY[21]

The umbilical vein was incubated with 0.1% dispase. After washing, HUVECs were cultured in endothelial basal medium with endothelial cell growth supplement at 37°C in a humidified atmosphere containing 5% CO_2, while the second passages were cultured in endothelial basal medium only. Cells were grown in 6-well culture plates (1×10^5 cells/well) to confluence and used for stimulation. Bovine aortas were incubated with a solution containing 0.25% collagenase for 10 min, flushed with PBS to collect detached bovine aortic endothelial cells (BAECs), which were cultured in medium containing RPMI 1640 and M199 (1 : 1), penicillin (100 IU/mL) and streptomycin (100 μg/mL) in 20% FCS.

BAECs or HUVECs were preincubated for 4 h with medium alone (control) or with medium containing 10 ng/mL TNF-α/IL-1β (Sigma, München, Germany) in the presence/absence of test compounds dissolved in DMSO. After incubation, BAECs or HUVECs were washed with PBS and on their monolayer a flat-polished flow chamber was placed. Mono Mac 6 (MM6) cells were suspended at 1×10^6 cells/mL in RPMI medium plus 1% FCS and perfused through the flow chamber at a constant shear stress of 1 dyne/cm^2 for 4 min. Using an inverted phase-contrast microscope connected to a video camera and a video recorder, rolling and adherence of MM6 cells in each well in four separate randomly chosen fields within the first minute were visualized and counted.

11.10 CCR5 RECEPTOR BINDING ASSAY[22]

In the determination of chemokine receptor (CCR5) binding activity, a 96-well scintillation proximity assay (SPA) format and CHO cells overexpressing the human CCR5 receptor were used. From the CHO cells, macrophage inflammatory protein (MIP)-1α and membranes were obtained; after ^{125}I-labeling the former was converted into [^{125}I] human MIP-1α. The solution of test compound, 12.5% aqueous DMSO, 12 μg membranes, 0.17 nM [^{125}I] MIP-1α, 0.25 mg Wheat Germ Agglutinin–scintillation proximity assay (SPA) beads, and assay buffer (50 mM HEPES, 1 mM $CaCl_2$, 1 mM $MgCl_2$, 1% BSA, and a protease inhibitor cocktail) was incubated at room temperature for 5 h with shaking. The beads were settled for 2 h, and the total binding was measured with the radioactivity test. In the presence of 1 μM recombinant human MIP-1α, the nonspecific binding was defined. The IC_{50} of 2.7 nM of human MIP-1α was used as a reference.

11.11 TISSUE BINDING AFFINITY ASSAY[23–25]

In hydrophobic Teflon bags, the test compound and blood monocytes isolated from pooled buffy coats at a density of 2×10^8 cells/mL McCoy's 5a medium supplemented with 20% fetal calf serum were cultured for 2 days. Monocytes were washed with cold PBS (pH 7.4) and incubated with BSA (1%) at 4°C for 30 min. Monocytes were washed and incubated with monoclonal antibody anti CD163 (5–10 mg/mL, Bachem, Weil am Rhein, Germany) at 4°C for 45 min and subsequently washed with PBS and incubated with fluorescein-isothiocyanat (FITC)-labeled secondary antibody goat-antimouse IgG1 (Dianova, Hamburg, Germany) in 1% BSA at 4°C for 30 min. At the last 3 min of the incubation, propidium iodide was added to determine cell viability and exclusion of dead cells using fluorescence activated cell sorter (FACS) analysis. The parameters were 488 nm excitation wavelength, 250 mW, and logarithmic amplification.

Human lung cancer free tissue from patients with bronchial carcinomas and who received no glucocorticoids for the last 4 weeks prior to surgery was immediately washed with Krebs-Ringer-HEPES buffer (118 mM NaCl, 4.84 mM KCl, 1.2 mM KH_2PO_4, 2.43 mM $MgSO_4$, 2.44 mM $CaCl_2 \cdot 2H_2O$, and 10 mM HEPES, pH 7.4) and sliced into pieces of 1 mm^3 or deeply frozen immediately in liquid nitrogen and stored at $-70°C$.

The frozen human lung tissue was pulverized and homogenized in three aliquots buffer solution A (10 mM Tris, 10 mM $NaMoO_4$, 30 mM NaCl, 10% glycerol, 4 mM DTT, 5 mM dichlorvos, and 1 mM Complete) with an Ultra Turrax mixer in an ice bath and then centrifuged at 4°C and $105{,}000 \times g$ for 1 h to prepare glucocorticoid receptors (30–60 fmol/mg cytosol). The receptor binding experiments were performed according to the standard method, the radiochemical purity of the labeled glucocorticoid was determined by HPLC and TLC, and scintillation counting was performed with a Rackbeta 1214 LKB using Emulsifier-Safe.

11.12 G93A-SOD1 TRANSGENIC MOUSE ASSAY[26–28]

EOC-20 cells grown in 75 cm^2 cell culture flasks with DMEM supplemented with 10% FCS and 20% L292 fibroblast-conditioned medium were transferred into 24-well cell culture plates and treated with the solution of test compound in DMSO or DMSO alone (1% final volume) for 30 min and challenged with TNF-α (Schützdeller, Tübingen, Germany) for 24 h. The culture medium was removed and assayed for determining the IC_{50} of test compound that suppresses TNF-α-induced NO_2^- by 50% via the determination of NO_2. The medium was removed, and DMEM lacking phenol red indicator was added. To each well, MTS reagent [3-(4,5-dimethylthiazol-2-yl)-5-(3-carboxymethonyphenol)-2-(4-sulfophenyl)-2H-tetrazolium, inner salt, Sigma, Deisenhofen, Germany] was added and incubated at 37°C for 30–45 min. The aliquots of media were removed and evaluated spectrophotometrically at 540 nm. The MTS solution prepared for the blank was incubated without cells. Viability was formally calculated as the ratio of OD_{540} in test compound–treated wells, relative to the same variable measured in wells treated with approximately 25 μM (typically 20 ng/mL) TNF-α alone, after subtraction of the blank.

Transgenic mice expressing high copy numbers of human mutant G93A-SOD1 were maintained in the hemizygous state by mating G93A males with B6SJL-TGN females. The mice were fed *ad libitum* standard AIN93G diets (Dyets, Inc., Bethlehem, PA, USA) or the same diets formulated with nordihydroguaiaretic acid at 2500 ppm with drug administration started on day 90 to demonstrate motor weakness and fine limb tremors. The feeding study used 16 mice per group. The group receiving nordihydroguaiaretic acid (NDGA, Aldrich, Milwaukee, WI, USA) consisted of 5 male and 11 female mice. The control group consisted of 4 male and 12 female mice. Mice were placed on a horizontal rod rotating 1 rpm every 10 s until they fell from the rod. Rotarod tests were conducted at 90 days and 100 days of age, and subsequently at 5-day intervals. Mice were sacrificed when no longer able to right themselves within 10 s of being placed on their sides.

Using TRI reagent (Sigma, Milwaukee, WI, USA), total RNA was isolated from spinal cords of non-transgenic control and G93A+ transgenic mice. In the presence of avian myeloblastosis virus (AMV) RT (Roche, Indianapolis, IN, USA) using oligo(dT)15 to prime the reaction, 5 μg of RNA samples were reverse transcribed. On completion, each reaction was diluted to a final volume of 50 μL with TE buffer (10 mM Tris, 1 mM EDTA, pH 8.0). PCR amplification of a 309-bp 5LOX gene product was accomplished with Taq DNA polymerase (Roche, Indianapolis, IN, USA), utilizing the buffer supplied, and final concentrations of 1.5 mM $MgCl_2$, 0.2 mM dNTP, and 0.3 μM primer (5′-GGCACCGACGACTACATCTAC-3′, forward and 5′-CAATTTTGCACGTCCATCCC-3′, reverse), of which the final volumes were 50 μL. A 353-bp PCR product of β-actin primers (5′-CGGCCAGGTCATCACTATTG-3′, forward and 5′-ACTCCTGCTTGCTGATCCAC-3′, reverse) was used as normalization controls. The optimal cycling conditions of PCR amplification of a 309-bp 5LOX gene product were at 94°C for 2 min for 1 cycle, at 94°C, 56°C, and 72°C each for 1 min for 27 cycles, and at 72°C for 7 min for 1 cycle. Conditions for β-actin primers were the same except that an annealing temperature of 54°C was used, and 24 cycles were performed. Samples (25 μL) from each reaction were electrophoresed in 2% agarose/TBE [tris/borate/EDTA buffer (0.09 M Tris, 0.09 M borate, 0.002 M EDTA)] gels for 1.5 h, stained with ethidium bromide, and photographed with a NucleoVision imaging system (Nucleotech, Westport, CT, USA). The positive control for 5-LOX Western blots was SL-29 fibroblast lysate (Transduction Laboratories). On 4% to 20% gradient polyacrylamide gels, the electrophoresis was performed and bands were visualized with chemiluminescence detection reagents. In 50 mM Tris-HCl (pH 6.8), 0.3 M NaCl, 1% β-mercaptoethanol, 1 mM phenylmethyl-sulfonyl fluoride, and 5 μM leupeptin, spinal cord samples were homogenized. Homogenates were centrifuged at 4°C and 30,000 × *g* for 5 min. Supernatants were refluxed for 10 min and centrifuged for 30 min at 30,000 × *g* and 4°C. The final supernatants were dialyzed overnight against 50 mM Tris-HCl. From the heat-soluble fraction of total spinal cord lysate, cleaved microtubule-associated tau protein (C-tau) was measured by sandwich ELISA, using affinity purified monoclonal antibody 12B2. In immunohistochemical experiments, the mice were terminally anesthetized and perfused transcardially with PBS (pH 7.4), followed by 4% paraformaldehyde in 0.1 M phosphate buffer. The spinal cord was removed, and the lumbar L5 region was processed for paraffin embedding. Serial cross sections of the L5 spinal cord region were cut at 5 μM thickness for immunostain with antibodies to glial fibrillary acidic protein (GFAP, dilution 5 lg/mL, Chemicon). Negative immunohistochemical controls were treated in the same way without primary antibody. GFAP-labeled sections were counterstained with hematoxylin.

11.13 MCP-1-INDUCED ERK1 AND ERK2 PHOSPHORYLATION ASSAY[29,30]

By reverse transcriptase PCR from phorbol myristate acetate (PMA, Calbiochem)-treated THP-1 cells, cDNA encoding CCR2B was isolated and subcloned into the expression vector pXMT3-neo (a derivative of pXMT), which contained an adenovirus late promoter and neo- and DHFR selectable markers. After transfection with pXMT3-neo-CCR2B, selection with G418, amplification with methotrexate, and clone of CHO DUK-X cells, the resultant CHO-CCR2B cell line (Gibco/Invitrogen) was cultured with modified MEMα (without ribonucleosides and deoxyribonucleosides, Gibco/Invitrogen) containing Glutamax-I, 100 IU/mL penicillin, 100 μg/mL streptomycin, 10% v/v dialyzed FBS, and 80 nM methotrexate. Parental CHO cells (Gibco/Invitrogen) were cultured with original α-MEM containing 100 IU/mL penicillin, 100 μg/mL streptomycin, and 10% v/v FBS. Cells were split twice per week by incubation for 5 min with enzyme-free cell dissociation buffer and reseeded at a density of 1×10^4 to 2×10^4 cells/cm^2. Prior to an assay, cells were seeded in methotrexate at 2×10^4 cells/well in flat-bottomed 96-well tissue culture plates and allowed to attach for 6 h. The cells were washed with 250 μL/well PBS and then incubated overnight in methotrexate containing endotoxin-free BSA (0.1%, w/v). After preincubation of 60 μL of monocyte chemoattractant protein-1 (MCP-1, 20 nM, R&D Systems) with 60 μL 0.4–1.8 μg/mL mAb at 37°C for 30 min, 50 μL MCP-1-mAb solution was transferred to the cells. The cells were preincubated at 37°C for 45 min with test compound diluted in 50 μL of serum-free medium. After addition of 50 μL of 20 nM MCP-1, the cells were incubated at 37°C for 5 min, and the medium was replaced by 100 μL/well methanol equilibrated at 20°C. Plates were incubated at 20°C for 20 min and washed three times with PBS containing Triton X-100 (0.1%, w/v, Gibco/Invitrogen). The cells were incubated with H_2O_2 (0.6%, v/v) in washing buffer for 20 min and washed as above, by which the activity of endogenous peroxidase was quenched. By adding 250 μL/well assay buffer (washing buffer containing fraction V BSA, 5%, w/v) and incubation at room temperature for 1 h, nonspecific binding sites were blocked, which was replaced by 100 μL/well monoclonal anti-phospho-ERK antibody (1 μg/mL in assay buffer, Gibco/Invitrogen). After 37°C and 1 h incubation and three washes with washing buffer, to the wells 100 μL/well goat anti-mouse-IgG coupled with horseradish peroxidase (0.5 μg/mL in assay buffer) was added. After 37°C and 1 h incubation and six washes with washing buffer, to the wells 100 μL/well tetramethylbenzidine was added. After color development for 10–30 min, the reaction was stopped by adding 50 μL 2 M H_2SO_4, and plates were read at 450 nm on a Spectramax Plus microtiter plate reader (Molecular Devices).

11.14 LPS- AND IL-6-INDUCED ERK1 AND ERK2 PHOSPHORYLATION ASSAY[31]

H187 and H188 (IL-6-dependent clones of hybridomas 187g3 and 188ND9) growing exponentially in the presence of interleukin-6 (IL-6) were washed three times in RPMI 1640 (Sigma Chemical Co.) and left to starve for 5 h in RPMI 1640 medium without serum and growth factors. After adding purified lipopolysaccharide (LPS, 10 μg/mL) and IL-6 (100 U/mL), cells were incubated for different time periods, washed once in PBS, and lysed in 1 mL of ice-cold lysis buffer (PBS with 0.5% sodium deoxycholate,

1% Nonidet P-40, and 0.1% SDS) containing freshly added test compounds and 1 mM sodium orthovanadate. After 30 min incubation on ice, cell lysates were centrifuged at 10,000 × *g* and 4°C for 10 min to collect supernatant of total cell lysates. The mixture of equal volumes (containing equal protein amounts, measured by the Bradford method) of the cell lysates and sample buffer was resolved by 10% SDS–polyacrylamide gel electrophoresis (PAGE). The proteins were transferred to PVDF membranes (Millipore) and incubated with the phospho-specific antibodies according to the manufacturer's instructions. After incubation with secondary peroxidase-conjugated antibody, proteins were visualized with Chemiluminescence Luminol Reagent (Santa Cruz Biotechnology).

11.15 ELA4.NOB-1/CTLL CELL ASSAY[32]

The ELA4.NOB-1 cells and cytotoxic T lymphocyte line (CTLL) maintained in RPMI 1640 containing (Gibco BRL, Uxbridge, UK) 2 mM glutamine, 40 μg/mL gentamicin, and 100 IU/mL penicillin (Sigma Chemical Co., Poole, UK) were transferred into a tissue culture medium (TCM) containing 5% heat-inactivated FCS (Sigma Chemical Co., Poole, UK), 10% FCS, and recombinant human L-2 (100 U/mL), respectively. Cell lines were centrifuged at room temperature and 400 × *g* for 10 min, washed twice by resuspension and centrifugation in fresh TCM supplemented with 10% FCS. By enumeration using Trypan blue (0.4%, w/v) in saline, ELA4.NOB-1 and CTLL cells were adjusted to concentrations of 1×10^6 and 4×10^4 viable cells/mL, respectively. To 96-well round-bottom microtiter plates containing 100 μL of cytokine (control) or test compound, 50 μL aliquots of each cell suspension were placed. After incubation in air containing 5% CO_2 at 37°C for 24 h, the cells were treated with 50 μL of 2 μCi/mL tritiated thymidine, incubated for another 24 h, and harvested onto printed filtermats, which were dried at 30°C for 1 h and heat sealed into sample bags containing 4.5 mL of β-scintillant. Radioactivity (cpm) was measured using a 'Microbeta plus' liquid scintillation counter (WALLAC).

To each well containing 100 μL aliquots of the standard dilutions, 50 μL of ELA4.NOB-1 and 50 μL of CTLL cell suspensions were added. Using human recombinant IL-2 (0.025–1.6 U/mL) or test compound, the integrity of CTLL responses in the absence of ELA4.NOB-1 cells was measured. Human recombinant IL-1ra (0.1–100 ng/mL), ultrapure natural human TGF-β1 (0.01–10 ng/mL, Genzyme, Cambridge, MA, USA) and test compound were tested on ELA4.NOB-1 and CTLL co-cultures in a similar way, generating standard dose-response curves for these cytokines. In the tests of the effects of IL-1ra (Serotec, Oxford, UK) and TGF-β1 on IL-1-induced activity of the co-culture bioassay and on IL-2-induced CTLL proliferation, serial dilutions of IL-1β (0.78–50 pg/mL), IL-2 (0.025–1.6 U/mL), IL-1ra (1–100 ng/mL), TGF-β1 (0.01–10 ng/mL), and test compound were prepared in TCM supplemented with 10% FCS. Aliquots (50 μL) of each dilution of IL-1ra were dispensed into 96-well microtiter plates, and 50 μL of each dilution of IL-1β was pipetted into the plate. Suspension (50 μL) of ELA4.NOB-1 and CTLL cells were aliquoted into each well, incubated, and treated with [^{3}H]thymidine as described above. Similar assays were set up for IL-1ra and IL-2, for TGF-β1 and IL-1β, and for TGF-β1 and IL-2.

11.16 FK506 BINDING PROTEIN 51 (FKBP51) mRNA ASSAY[33–35]

Blood of nine healthy controls without a history of glucocorticoid (GC) medication (mean 37 years old) and one GC hyposensitive patient harboring one nonfunctional GC receptor allele resulting in a net functional GC receptor expression of approximately 50% was collected in sodium heparin tubes by venipuncture within a 2-week period between 0900 and 1000 h and centrifuged to isolate PBMCs, which were washed twice with RPMI 1640 and resuspended in assay medium consisting of RPMI 1640 (Gibco BRL, Uxbridge, UK) supplemented with 4 mM L-Gln, 100 IU/mL penicillin, 100 μg/mL streptomycin (Sigma Chemical Co., Poole, UK), and 10% dextran-coated charcoal steroid-stripped FCS at a concentration of 0.5×10^6 cells/mL. One-milliliter aliquots of the PBMC suspension were added onto 24-well plates and incubated overnight at 37°C in humidified atmospheric air containing 5% CO_2. To the medium, dexamethasone (DXM, Sigma-Aldrich Chemie, Steinheim, Germany) or test compound was added and incubated for another 24 h, after which RNA was isolated.

For activation of PBMCs, 0.1-mL aliquots of 0.5×10^5 cells were added onto a 96-well round-bottom plate and incubated for 96 h with tetanus toxoid in a concentration of 1.5 μg/mL. Seventy-two hours later, DXM was added. Ninety-six hours later, RNA was isolated. In a simultaneous control experiment, 78 h later 1 μCi [^{3}H]thymidine was added, the cells were harvested, and incorporated radioactivity was determined.

For purification of PBMC subsets, cells were isolated, washed twice, resuspended in PBS to a concentration of 1×10^7 cells/mL, and incubated at room temperature for 30 min with phycoerythrin labeled primary antibodies (mouse-anti-human CD20, CD4, CD8, or CD14, Becton Dickinson, San Jose, CA). After two washes with PBS, PBMCs were resuspended in PBS to a concentration of 1×10^8 cells/mL, incubated at 4°C for 30 min, and labeled with goat-anti-mouse IgG. The PBMCs were washed and resuspended in 0.5 mL of PBS. The labeled and unlabeled PBMCs were separated, resuspended in assay medium, divided in 0.1-mL aliquots onto a 96-well round-bottom plate, and incubated overnight. To the wells, DXM was added and incubated for another 24 h to isolate RNA.

For performing real-time PCR on the LightCycler (Roche Applied Science), after the isolation of RNA by using TriPure reagent (Roche Applied Science, Mannheim, Germany) and reverse transcription by using MMLV-RT RNase H Minus, the synthesized cDNA was diluted 20–40 times in ddH_2O. The relative expression level of the housekeeping gene β_2-microglobulin (β_2m) was determined in each sample to calculate the relative expression level of the genes of interest. Real-time PCR was performed in 10 μL of medium containing 5.0 μL of diluted cDNA (0.5 pmol/μL each primer), 10% either DNA master SYBR-green I (Roche Applied Science, Mannheim, Germany) solution (for the FKBP51 and β_2m PCR) or DNA hotstart mix (for the FKBP12, FKBP13, FKBP22, FKBP25, FKBP52, and Cyp40 PCR), and an optimal concentration of $MgCl_2$ (4 mM β_2m PCR, 3 mM FKBP51 PCR). The mixture was denatured at 95°C for 30 s and 10 min for DNA Master I kit and Fast Start kit, respectively, and subjected to up to 40 amplification cycles: denaturing 15 s, 95°C, annealing 10 s, 56°C for β2 m or 67°C for FKBP51, and elongation 30 s, 72°C. A single measurement of fluorescence was taken after each elongation step at 82°C.

11.17 XYLENE-INDUCED EAR EDEMA ASSAY[36]

Male Kunming mice (about 25 g) were randomly divided into three groups of 12 mice, namely the test group, vehicle control group, and positive control group. The mice in vehicle control group were orally administered a suspension of aspirin in carboxymethyl cellulose (CMC) at a dosage of 20 mg/kg, and a concentration of 0.3 mg/mL, whereas the mice in the test group were orally administered a suspension of test compound in CMC at a dosage of 20 mg/kg, 4.0 mg/kg, 0.8 mg/kg, and a concentration of 2.0 mg/mL, 0.4 mg/mL, 0.08 mg/mL. Thirty minutes later, 0.03 mL of xylene was applied to both the anterior and posterior surfaces of the right ear. The left ear was considered as control. Two hours after xylene application, the mice were sacrificed, and both ears were removed. Using a cork borer with a diameter of 7 mm, several circular sections were taken and weighed. The increase in weight caused by the irritant was measured through subtracting the weight of the untreated left ear section from that of the treated right ear section.

REFERENCES AND NOTES

1. K.L. Reed, A.B. Fruin, A.C. Gower, A.F. Stucchi, S.E. Leeman, J.M. Becker. A neurokinin 1 receptor antagonist decreases postoperative peritoneal adhesion formation and increases peritoneal fibrinolytic activity. Proc Natl Acad Sci USA 101 (**2004**) 9115–9120.
2. M.H. Konstandin, G.H. Wabnitz, H. Aksoy, H. Kirchgessner, T.J. Dengler, Y. Samstag. A sensitive assay for the quantification of integrin-mediated adhesiveness of human stem cells and leukocyte subpopulationsin whole blood. J Immunol Methods 327 (**2007**) 30–39.
3. M.H. Konstandin, U. Sester, M. Klemke, T. Weschenfelder, G.H. Wabnitz, Y. Samstag. A novel flow-cytometrybased assay for quantification of affinity and avidity changes of integrins. J Immunol Methods 310 (**2006**) 67–78.
4. S. Cianchetti, A.D. Fiorentino, R. Colognato, R.D. Stefano, F. Franzoni, R. Pedrinelli. Anti-inflammatory and anti-oxidant properties of telmisartan in cultured human umbilical vein endothelial cells. Atherosclerosis 198 (**2008**) 22–28.
5. S.K. Yeap, N.B. Alitheen, A.M. Ali, A.R. Omar, A.R. Raha, A.A. Suraini, A.H. Muhajir. Effect of rhaphidophora korthalsii methanol extract on human peripheral blood mononuclear cell (PBMC) proliferation and cytolytic activity toward HepG2. J Ethnopharmacol 114 (**2007**) 406–411.
6. A.B. Montanher, S.M. Zucolotto, E.P. Schenkel, T.S. Fröde. Evidence of anti-inflammatory effects of passiflora edulis in an inflammation model. J Ethnopharmacol 109 (**2007**) 281–288.

7. T.S. Rao, J.L. Currie, A.F. Shaffer, P.C. Isakson. Comparative evaluation of arachidonic acid (aa)- and tetradecanoylphorbol acetate (tpa)-induced dermal inflammation. Inflammation 17 (**1993**) 723–741.

8. T.S. Fröde, Y.S. Medeiros. Myeloperoxidase and adenosine-deaminase levels in the pleural fluid leakage induced by carrageenan in the mouse model of pleurisy. Mediators Inflamm 10 (**2001**) 223–227.

9. M. Di Rosa, A. Lalenti, A. Ianaro, L. Sautenbin. Interaction between nitric oxide and cyclooxygenase pathway. Prostaglandins Leukot Essent Fatty Acids 54 (**1996**) 229–238.

10. T.S.F. Saleh, J.B. Calixto, Y.S. Medeiros. Effects of anti-inflammatory drugs upon nitrate and myeloperoxidase levels in the mouse pleurisy induced by carrageenan. Peptides 20 (**1999**) 949–956.

11. H.J. Lee, E.A. Hyun, W.J. Yoon, B.H. Kim, M.H. Rhee, H.K. Kang, J.Y. Cho, E.S. Yoo. In vitro anti-inflammatory and anti-oxidative effects of Cinnamomum camphora extracts. J Ethnopharmacol 103 (**2006**) 208–216. *Note*: (1) In RPMI 1640 supplemented with 100 IU/mL penicillin, 100 μg/mL streptomycin, and 10% fetal bovine serum, RAW264.7 and U937 cells were maintained and grown at 37°C and 5% CO_2 in humidified air. Test compounds were dissolved in RPMI 1640 before assay. Their inhibitory effect on TNF-α, IL-1β, and IL-6 productions from the LPS-treated RAW264.7 cells were determined generally. Supernatants were harvested and assayed for TNF-α, IL-1β, and IL-6 by ELISA. (2) After 18 h preincubation, RAW264.7 cells (1×10^6 cells/mL) were incubated with each test compound, LPS (1 μg/mL) and IFN-γ (50 U/mL) for 24 h. By adding 100 μL of Griess reagent (1% sulfanilamide and 0.1% *N*-[1-naphthyl]-ethylenediaminedihydrochloride in 5% phosphoric acid) to 100 μL of medium samples, the nitrite in culture supernatants was measured. (3) After 18 h preincubation with or without LPS (1 μg/mL), RAW264.7 cells (1×10^6 cells/mL) were incubated with each test compound for 24 h and PGE2 in the culture supernatants was measured by ELISA kit. (4) A conventional MTT assay was used to evaluate the cytotoxic effect of test compound. Three hours prior to culture termination, 10 μL of MTT solution (10 mg/mL in a phosphate-buffered saline, pH 7.4) were added, the cells were continuously cultured until termination, and the incubation was stopped by the adding 15% sodium dodecyl sulfate into each well to dissolve formazan. The OD was measured at 570 nm ($OD_{570-630}$) on a Spectramax 250 microplate reader. (5) According to the manufacturer's instructions, total RNA from the LPS-treated RAW264.7 cells was prepared by adding TRIzol Reagent and its solution was stored at −70°C until used. Using MuLV reverse transcriptase semiquantitative RT reactions were carried out. The total RNA (1 μg) was incubated with oligo(dT)15 at 70°C for 5 min, mixed with 5x first-strand buffer, 10 mM dNTP, and 0.1 M DTT and further incubated at 37°C for 5 min, and after adding MuLV reverse transcriptase (2 U) incubated at 37°C for another 60 min. The reactions were terminated at 70°C for 10 min and by adding RNase H, the total RNA was depleted. Under the incubation conditions (a 45-s denaturation time at 94°C, an annealing time of 45 s at 55°C to 60°C, an extension time of 60 s at 72°C and final extension of 7 min at 72°C at the end of 30 cycles), the PCR reaction was carried out with the incubation mixture [2 μL cDNA, 4 μM 5′ and 3′ primers, a 10x buffer (10 mM Tris-HCl, pH 8.3, 50 mM KCl, 0.1% Triton X-100), 250 μM dNTP, 25 mM $MgCl_2$, and 1 unit of Taq polymerase]. The primers used in this experiment were TNF-α (forward) 5′-TTGACCTCAGCGCTGAGTTG-3′, (reverse) 5′-CCTGTAGCCCACGTCGTAGC-3′; IL-1β (forward) 5′-CAGGATGAGGACATGAGCACC-3′, (reverse) 5′-CTCTGCAGACTCAAACTCCAC-3′; IL-6 (forward) 5′-GTACTCCAGAAGACCAGAGG-3′, (reverse) 5′-TGCTGGTGACAACCACGGCC-3′; iNOS (forward)

5′-CCCTTCCGAAGTTTCTGGCAGCAGC-3′, (reverse) 5′-GGCTGTCAGAGCCTC-GTGGCTTTGG-3′; COX-2 (forward) 5′-CACTACATCCTGACCCACTT-3′, (reverse) 5′-ATGCTCCTGCTTGAGTATGT-3′; β-Actin (forward) 5′-GTGGGCCGCCCTAGG-CACCAG-3′, (reverse) 5′-GGAGGAAGAGGATGCGGCAGT-3′.

12. Y.C. Kuo, Y.L. Huang, C.C. Chen, Y.Z. Lin, K.A. Chung, W.J. Tsai. Cell cycle progression and cytokine genes expression of human blood mononuclear cells modulated by Agaricus blazei. J Lab Clin Med 140 (**2002**) 176–187.
13. Y.C. Kuo, H.C. Meng, W.J. Tsai. Regulation of cell proliferation, inflammatory cytokines production and calcium mobilization in primary human T lymphocytes by emodin from Polygonum hypoleucum Ohwi. Inflamm Res 50 (**2001**) 73–82.
14. Y.C. Kuo, N.S. Yang, C.J. Chou, L.C. Lin, W.J. Tsai. Regulation of cell proliferation, gene expression, production of cytokines, and cell cycle progression in primary human T lymphocytes by piperlactam S isolated from Piper kadsura. Mol Pharmacol 58 (**2000**) 1057–1066.
15. S.H. Shin, M.K. Ye, H.S. Kim, H.S. Kang. The effects of nano-silver on the proliferation and cytokine expression by peripheral blood mononuclear cells. Int Immunopharmacol 7 (**2007**) 1813–1818.
16. R.W. Li, D.N. Leach, S.P. Myers, G.D. Lin, G.J. Leach, P.G. Waterman. A new anti-inflammatory glucoside from Ficus racemosa L. Planta Med 70 (**2004**) 421–426.
17. G.F. Sud'inaa, M.A. Pushkareva, P. Shephard, T. Klein. Cyclooxygenase (COX) and 5-lipoxygenase (5-LOX) selectivity of COX inhibitors. Prostaglandins Leukot Essent Fatty Acids 78 (**2008**) 99–108.
18. S.L. Crockett, E.M. Wenzig, O. Kunert, R. Bauer. Anti-inflammatory phloroglucinol derivatives from Hypericum empetrifolium. Phytochem Lett 1 (**2008**) 37–43.
19. M. Knödler, J. Conrad, E.M. Wenzig, R. Bauer, M. Lacorn, U. Beifuss, R. Carle, A. Schieber. Anti-inflammatory 5-(11′Z-heptadecenyl)- and 5-(8′Z,11′Z-heptadecadienyl)-resorcinols from mango (Mangifera indica L.) peels. Phytochemistry 69 (**2008**) 988–993.
20. B.S. Zweifel, M.M. Hardy, G.D. Anderson, D.R. Dufield, R.A. Pufahl, J.L. Masferrer. A rat air pouch model for evaluating the efficacy and selectivity of 5-lipoxygenase inhibitors. Eur J Pharmacol 584 (**2008**) 166–174.
21. H. Ulbrich, O. Soehnlein, X. Xie, E.E. Eriksson, L. Lindbom, W. Albrecht, S. Laufer, G. Dannhardt. Licofelone, a novel 5-LOX/COX-inhibitor, attenuates leukocyte rolling and adhesion on endothelium under flow. Biochem Pharmacol 70 (**2005**) 30–36.
22. K. Yoganathan, C. Rossant, R.P. Glover, S. Cao, J.J. Vittal, S. Ng, Y. Huang, A.D. Buss, M.S. Butler. Inhibition of the human chemokine receptor CCR5 by variecolin and variecolol and isolation of four new variecolin analogues, emericolins A-D, from Emericella aurantiobrunnea. J Nat Prod 67 (**2004**) 1681–1684.
23. A. Valotis, K. Neukam, O. Elert, P. Hogger. Human receptor kinetics, tissue binding affinity, and stability of mometasone furoate. J Pharm Sci 93 (**2004**) 1337–1350.
24. O.H. Lowry, N.J. Rosebrough, A.L. Farr, R.J. Randell. Protein measurement with the Folin phenol reagent. J Biol Chem 193 (**1951**) 265.
25. P. Högger, P. Rohdewald. Binding kinetics of fluticasone propionate to the human glucocorticoid receptor. Steroids 59 (**1994**) 597–602.
26. M. West, M. Mhatre, A. Ceballos, R.A. Floyd, P. Grammas, S.P. Gabbita, L. Hamdheydari, T. Mai, S. Mou, Q.N. Pye, C. Stewart, S. West, K.S. Williamson, F. Zemlan, K. Hensley. The arachidonic acid 5-lipoxygenase inhibitor nordihydroguaiaretic acid inhibits

tumor necrosis factor alpha activation of microglia and extends survival of G93A-SOD1 transgenic mice. J Neurochem 91 (**2004**) 133–143.

27. M. Marzinzig, A.K. Nussler, J. Stadler, E. Marzinzig, W.P. Barthlen, N.C. Nussler, H.G. Beger, S.M. Jr. Morris, U.B. Bruckner. Improved methods to measure end products of nitric oxide in biological fluids: nitrite, nitrate and S-nitrosothiols. Nitric Oxide Biol Chem 1 (**1997**) 177–189.
28. C. Guégan, S. Przedborski. Programmed cell death in amyotrophic lateral sclerosis. J Clin Invest 111 (**2003**) 153–161.
29. J.H. Terra, I. Montaño, A. Schilb, T.A. Millwarda. Development of a microplate bioassay for monocyte chemoattractant protein-1 based on activation of p44/42 mitogen-activated protein kinase. Anal Biochem 327 (**2004**) 119–125.
30. C. Binkert, J. Landwehr, J.L. Mary, J. Schwander, G. Heinrich. Cloning, sequence analysis and expression of a cDNA encoding a novel insulin-like growth factor binding protein (IGFBP-2). EMBO J 8 (**1989**) 2497–2502.
31. I. Iankov, G. Atanasova, M. Praskova, S. Kalenderova, D. Petrov, V. Mitev. Ivan Mitova. Bacterial lipopolysaccharide induces proliferation of IL-6-dependent plasmacytoma cells by MAPK pathway activation. Immunobiology 208 (**2004**) 445–454.
32. P. Ashwood, R. Harvey, T. Verjee, R. Wolstencroft, R.P.H. Thompson, J.J. Powell. Competition between IL-1, IL-1ra and TGF-β1 modulates the response of the ELA4.NOB-1/CTLL bioassay: implications for clinical investigations. Inflamm Res 53 (**2004**) 60–65.
33. H. Vermeer, B.I. Hendriks-Stegemana, D. van Suylekoma, Ger.T. Rijkers, S.C. van Buul-Offers, M. Jansen. An in vitro bioassay to determine individual sensitivity to glucocorticoids: induction of FKBP51 mRNA in peripheral blood mononuclear cells. Mol Cell Endocrinol 218 (**2004**) 49–55.
34. G. Blackhurst, P.K. McElroy, R. Fraser, R.L. Swan, J.M. Connell. Seasonal variation in glucocorticoid receptor binding characteristics in human mononuclear leukocytes. Clin Endocrinol (Oxf.) 55 (**2001**) 683–686.
35. H. Vermeer, B.I. Hendriks-Stegeman, B. van der Burg, S.C. van Buul-Offers, M. Jansen. Glucocorticoid-induced increase in lymphocytic FKBP51 mRNA expression: a potential marker for glucocorticoid sensitivity, potency and bioavailability. J Clin Endocrinol Metab 88 (**2003**) 277–284.
36. L. Bi, Y. Zhang, M. Zhao, C. Wang, P. Chan, J.B.H. Tok, S. Peng. Novel synthesis and anti-inflammatory activities of 2,5-disubstituted-dioxacycloalkanes. Bioorg Med Chem 13 (**2005**) 5640–5646.

12

METHODS AND APPLICATIONS OF ANTIOXIDANT ACTIVITY ASSAYS

Shiqi Peng

Living cells are continuously exposed to a variety of challenges exerting oxidative stress, which could stem from endogenous sources through normal physiologic processes such as mitochondrial respiration and hemoglobin oxidation and which could result from exogenous sources such as exposure to pollutants, ionizing radiation, and other extreme factors. Oxidative stress is either associated with the generation of ROS or leads to the generation of ROS, including free radicals. The amount of ROS depends not only on generation rate but also on the antioxidant defense system of the human body, particularly of blood. ROS are deeply immersed in the pathophysiology of diseases such as cancer, heart disease, atherosclerosis, aging, diabetes mellitus, and renal, inflammatory, infectious, and neurologic diseases. In the past decade, it became well known that there are different mechanisms of antioxidant action in biological fluids, including decreasing the level of active products from oxygen reduction, removing the transition metals (Fe, Cu) by their binding with proteins, and eliminating ROS by interaction with antioxidants. In parallel with the mechanisms, a large variety of corresponding assays exist. In this chapter, 28 assays are presented: blood and plasma total antioxidant capacity (TAC) assay,[1] ferric reducing-antioxidant power (FRAP) assay,[2] human LDL oxidation assay,[1–6] DPPH radical cation scavenging assay,[2,3,5,6] $ABTS^{+}$ radical cation scavenging assay,[3,5] lipid peroxidation assay

Pharmaceutical Bioassays: Methods and Applications. By Shiqi Peng and Ming Zhao

using rat brain tissue,[4,5] flow-through chemiluminescence (FTCL) assay,[6] superoxide radical scavenging assay,[7] deoxyribose assay for hydroxyl radical scavenging activity,[7] DNA nicking assay for hydroxyl radical scavenging activity,[7] oxidative lag-time assay,[8] TBARS and electrophoresis assay,[8] reporter and electrophoretic mobility shift assay (EMSA),[9] $[Ca^{2+}]_{cyt}$ assay,[10] quantitative real-time PCR assay,[10] FTIR-based assay for antioxidation activity of ionol and piperidone,[11] HPLC assay for antioxidation potential of polyphenol,[12] ROS production assay,[13] rabbit LDL oxidation assay,[14,15] ROS scavenging assays,[16–18] DNA damage assay,[18] egg yolk thiobarbituric acid reactive substances (TBARS) assay,[19] mouse catalase (CAT) assay,[19] rat tissue thiobarbituric acid reactive substances (TBARS) assay,[19,20] antioxidant activity assay for β-carotene/linoleic acid system,[20] rat brain tissue NO assay,[21] rat brain antioxidative enzyme assay,[22–25] and rat brain hippocampi protein oxidation assay.[26]

12.1 BLOOD AND PLASMA TOTAL ANTIOXIDANT CAPACITY (TAC) ASSAY[1]

Healthy blood donors (21–54 years of age) were on a normal diet and used as controls for determining the reference interval of total antioxidant capacity (TAC). Male and female patients with chronic renal failure received hemodialysis three times per week 3 h each for either less than 1 year or more than 1 year. Before and after dialysis or 24 and 48 h after session of hemodialysis, venous and arterial blood was collected in glass tubes containing heparin as anticoagulant. For preparing plasma, blood was at 3000 rpm centrifuged for 5 min, whereas for preparing serum, blood was placed into pure drawn tubes and the clots were rimmed with a wooden applicator stick and then centrifuged at 2000 rpm for 20 min, which was immediately analyzed for low density lipoproteins (LDL) and TAC.

To determine TAC, bromine was electrogenerated from aqueous 0.2 M KBr in 0.1 M H_2SO_4 using a P-5827M potentiostat at a current density of 5 mA/cm^2. With two polarized platinum electrodes ($\Delta E = 300$ mV), amperometric titration end-points were measured. A smooth platinum plate with a surface area of 1 cm^2 was used as working electrode, and a platinum coil separated from the anodic compartment with a semipermeable diaphragm was used as auxiliary electrode. In 50.0 mL of cell containing 20.0 mL of supporting electrolyte, coulometric titration was performed. The generating circuit was switched on, the indicator current reached a certain value, 20 μL of blood or plasma was added to the cell, a timer was simultaneously started, the titration end-point was detected by the indicator current reaching initial value, the timer was stopped, and the generating circuit was turned off. TAC was expressed in units of electric quantity (coulombs), which was spent for titration on 1000 mL of plasma or blood. All assays were carried out at temperature 25 ± 2°C.

With lyophilic albumin of human serum and 10% human serum albumin solution as standards, coulometric titration was performed by electrogenerated bromine and iodine, which were carried out from 0.1 M KI in tartrate buffer solution with pH 3.56 and 0.2 M KBr in 0.5 M H_2SO_4.

12.2 FERRIC REDUCING-ANTIOXIDANT POWER (FRAP) ASSAY[2]

The principle of FRAP assay was based on reducing Fe^{3+}-2,4,6-tripyridyl-s-triazine (TPTZ, Sigma, St. Louis, MO, USA) complex to the ferrous form (Fe^{2+}). The antioxidant activity of individual compound was measured by monitoring the absorbance change at 593 nm. For this assay, 0.3 M acetate buffer (pH 3.6) was prepared by dissolving 3.1 g of $C_2H_3O_2Na \cdot 3H_2O$ and 16 mL of acetic acid in 1000 mL of double-distilled water, whereas TPTZ solution was prepared by dissolving 10 mol of TPTZ in 1000 mL of 40 mM HCl, and 20 mM ferric solution was prepared using $FeCl_3 \cdot 6H_2O$. By freshly mixing acetate buffer, TPTZ, and ferric solutions at a ratio of 10/1/1, the final working FRAP reagent was prepared, 500 μL of which was mixed with 480 μL of double-distilled water and warmed to 37°C in a water bath for use. The reagent blank reading was recorded at 593 nm, and 20 μL of test compound solution was added. When the reading was constant, the absorbance was taken at 10 min. The difference between the absorbances of test compound and the blank reading was calculated and expressed as mM of ferric reduced to ferrous form.

12.3 HUMAN LDL OXIDATION ASSAY[1–6]

Fresh blood was collected, 2.7 mM EDTA (Fluka, Buchs, Switzerland) and 7.7 mM NaN_3 were immediately added to protect the lipoprotein from oxidative modification, LDL was isolated, and the protein content of isolated LDL was determined. The stock LDL fraction (5 mg protein/mL) was dialyzed in the dark with 100 volumes of the degassed dialysis solution (pH 7.4) containing 0.01 M sodium phosphate, 9 mg/mL NaCl, 10 μM EDTA, and 7.7 mM NaN_3 for 24 h, for which the dialysis solution was changed at least four times, and the dialyzed LDL was diluted to 250 μg protein/mL with 0.01 M SPB (pH 7.4). As control group, 0.4 mL of LDL (250 mg/L) was mixed with 50 μL of $CuSO_4$ solution (50 μM) and 50 μL of sodium phosphate buffer (0.01 M, pH 7.4) and at 37°C incubated for up to 20 h. As treating group 0.4 mL of LDL (250 mg protein/L) was with 12 μM individual test compound preincubated for 5 min, 50 μL of $CuSO_4$ solution (50 μM) was added to initiate the oxidation, and incubated at 37°C for up to 20 h. The oxidation was stopped by adding 25 μL of EDTA (27 mM) and cooling at 4°C. By measuring the production of thiobarbituric acid-reactive substances (TBARS), the degree of LDL oxidation was monitored. Briefly, in 0.1 N HCl solution, LDL-incubated tubes were immediately treated with 2 mL of thiobarbituric acid (0.67%) and trichloroacetic acid (15%), heated at 9°C for 1 h, cooled on ice, centrifuged at 1000 × *g* for 20 min, and TBARS were at 532 nm determined for absorbance and expressed as nmol malondialdehyde (MDA)/mg LDL protein. The calibration was performed using a MDA standard solution prepared from tetramethoxylpropane.

12.4 DPPH RADICAL CATION SCAVENGING ASSAY[2,3,5,6]

The mixture of 0.5 mL of methanol containing individual test compound and 2.5 mL of methanol containing 75 μM 2,2-diphenyl-1-picrylhydrazyl (DPPH•, a stable free radical that has a typical absorbance at 517 nm) was at room temperature in dark maintained for 90 min, and the free radical scavenging activity was tested via measuring the absorbance at 517 nm and calculated as Scavenging activity $\% = [A_a - (A_b - A_c)/A_a] \times 100$, where A_a, A_b, and A_c are the absorbance of the incubation DPPH$^\bullet$ solution without test compound, the absorbance of the incubation DPPH$^\bullet$ solution with test compound, and the absorbance of the blank solution without DPPH$^\bullet$, respectively.

12.5 ABTS$^+$ RADICAL CATION SCAVENGING ASSAY[3,5]

By mixing 5 mL of 7 mM 2,2′-azinobis-(3-ethylbenzthiazoline-6-sulfonic acid; ABTS) with 88 μL of $K_2S_2O_8$, ABTS reagent (140 mM) was prepared, which was kept at room temperature in the dark for 16 h to allow the completion of 2,2′-azino-bis(3-ethylbenzothiazoline-6-sulphonic acid) diammonium salt radical (ABTS$^{\bullet+}$) generation. Using a spectrophotometer at 734 nm, the absorbance of ABTS reagent was with 95% ethanol adjusted to 0.70 ± 0.05. To determine the scavenging activity, 1 mL of ABTS reagent was mixed with 10 μL of test compound or negative control (ethanol), kept for 6 min, and the absorbance was measured at 734 nm using ethanol as blank. Based on the equation Inhibition percentage $\% = [1 - (A_{test\ compd}/A_{control})] \times 100\%$, the inhibition percentage of test compound was calculated. $A_{test\ compd}$ and $A_{control}$ represent the absorbances of test compound and ethanol at 734 nm, respectively. A dose-response curve of inhibition percentage against different concentrations of Trolox standard ranging from 0.5 nM to 2 mM was prepared. The antioxidant activity of test compounds was expressed as Trolox equivalent antioxidant capacity (TEAC), which represented the concentration (mM) of Trolox, having the same activity as 1 mg of test compounds.

12.6 LIPID PEROXIDATION ASSAY USING RAT BRAIN TISSUE[4,5]

Brains of SD rats were dissected, homogenized with a Polytron in ice-cold Tris-HCl buffer (20 mM, pH 7.4), the produced 1:2 (w/v) brain tissue homogenate was centrifuged at $3000 \times g$ for 10 min, a 0.1-mL aliquot of the supernatant was incubated with 0.2 mL of test compound solution, 0.1 mL of $FeSO_4$ (10 μM, AnalaR), and 0.1 mL of ascorbic acid (0.1 mM, Sigma Chemical Co., Poole, UK) at 37°C for 1 h, after adding 0.5 mL of trichloroacetic acid (TCA, 28%, w/v, Univar) and 0.38 mL of thiobarbituric acid (TBA, 2%, w/v, Sigma), the mixture was heated at 80°C for 20 min to stop the reaction, centrifuged at $3000 \times g$ for 10 min to remove the precipitated protein, and the color intensity of the malondialdehyde (MDA)-TBA complex in the supernatant was measured at 532 nm using a spectrophotometer (Spectronic Genesys, Milton Roy Co., Ivyland, PA, USA) for its absorbance. Butylated hydroxyanisole (0.2 mg/mL) served as a positive control. Inhibition ratio $= [(A - B)/A] \times 100\%$ was used to calculate the inhibition ratio (%), where A and B were the absorbance of the control and the test compound, respectively.

12.7 FLOW-THROUGH CHEMILUMINESCENCE (FTCL) ASSAY[6]

FTCL assay used the emission of light resulted from a particular chemical reaction. With the catalyst cytochrome C, a hydroperoxide such as *tert*-butylhydroperoxide (*t*-BHPO, Sigma Chemical Co., Poole, UK) was decomposed to radical products such as hydroxyl radical, alkoxyl radical, and peroxyl radical, which reacted with chemiluminescent reagent luminol and led to emission of light. According to the radical quenching activity of the particular antioxidant, adding an antioxidant to the system reduced the intensity of the light produced. Assays were carried out, and IC_{50} values were calculated.

12.8 SUPEROXIDE RADICAL SCAVENGING ASSAY[7]

In 3.0 mL of 16 mM Tris-HCl (pH 8.0) containing 78 mM β-nicotinamide adenine dinucleotide (NADH, Sigma-Aldrich Chemical Co., USA), 50 μM nitroblue tetrazolium (NBT, Sigma-Aldrich Chemical Co., USA), and 10 μM phenazin methosulfate (PMS, Sigma-Aldrich Chemical Co., USA), superoxide radical was generated. After reacting with test compound at varying concentrations, the color reaction of test superoxide radical with NBT was monitored at 560 nm using a Hitachi U-3000 spectrophometer. Deionized water and ascorbic acid served as the blank and positive controls, respectively. Superoxide radical scavenging activity was calculated based on Superoxide radical scavenging activity (%) $= [(A_{blank} - A_{test\ compd})/A_{blank}] \times 100$.

12.9 DEOXYRIBOSE ASSAY FOR HYDROXYL RADICAL SCAVENGING ACTIVITY[7]

Using deoxyribose assay, non-site-specific hydroxyl radical scavenging activity of test compound was measured. To 1.0 mL of potassium phosphate buffer (20 mM, pH 7.4) containing 2.8 mM 2-deoxy-ribose, 104 μM EDTA, 100 μM $FeCl_3$, 100 μM ascorbate, and 1 mM hydrogen peroxide, test compound was added at varying concentrations, the mixture was incubated at 37°C for 1 h, an equal volume of 10% trichloroacetic acid containing 0.5% thiobarbituric acid was added, and the solution was refluxed at 100°C for 15 min. Deionized water and reduced glutathione served as a blank and positive controls. The absorbance was at 532 nm measured. Based on the equation Superoxide radical scavenging activity (%) $= [(A_{blank} - A_{test\ compd})/A_{blank}] \times 100$, superoxide radical scavenging activity was calculated. This procedure was also used to measure site-specific hydroxyl radical scavenging activity except EDTA was changed to an equal volume of buffer.

12.10 DNA NICKING ASSAY FOR HYDROXYL RADICAL SCAVENGING ACTIVITY[7]

To assess the non-site-specific hydroxyl radical scavenging activity, a suitable mixture of test compound, 10 μg/mL pBR 322 plasmid DNA, 13 mM potassium phosphate buffer (pH 7.2, containing 5.7 mM EDTA), 3 mM $FeCl_2$, and 5 mM H_2O_2 was incubated at 37°C for 1 h, while the molar ratio of $FeCl_2$/EDTA was kept at 0.53. To the incubated solution, 2 μL of loading buffer (50% sucrose and 0.25% bromophenol blue in H_2O) was added, and using Tris-Acetate-EDTA (TAE, 40 mM Tris-acetate, and 1 mM EDTA, pH 7.4) as running buffer, 10 μL afforded mixture was subjected to electrophoresis (agarose gel, 0.7%). With ethidium bromide, the properly developed gel was immediately stained and visualized under a UV trans-illuminator.

12.11 OXIDATIVE LAG-TIME ASSAY[8]

Based on the fact that conjugated dienes at 234 nm having absorbance may form via LDL oxidation, the oxidative lag-time was assayed and the antioxidant capacity of the test compounds was determined by continuously monitoring conjugated diene formation. In assay, a BioTek Powerwave 200 plate reader (Laguna Hills, CA) with Costar low-UV absorbance 96-well plates (Corning, NY) was used to continuously monitor a series of LDL samples at one time. In 200 μL of samples containing 100 μL of LDL-protein/mL, 3.00 μM Cu^+ and estrogenic compounds (EC) in PBS, the metal-initiated oxidation was examined. At 37°C oxidation was carried out, the absorbance changes at 234 nm were monitored for 240 min, and the absorbance of PBS cells was set to zero as absorbance reference for the spectrophotometer. The time from reaction initiation to achievement of Vmax was considered as the lag-time and arbitrarily determined using the maximum velocity (Vmax) function of the CD4 software package. Samples showing no oxidation (no change in absorbance) or not reaching Vmax during 240-min monitor period were arbitrarily assigned a 240-min lag-time.

12.12 TBARS AND ELECTROPHORESIS ASSAY[8]

In 24-well tissue plates, 250 μL of dialyzed LDL (100 μg LDL-protein/mL) in PBS was at 37°C incubated with full exposure to air to oxidize. Metal-promoted oxidation was carried out with 10 μM cupric sulfate via 95 min co-incubation. To evaluate relative antioxidant activities, wells were prepared without and with test compounds diluted in PBS, incubated for 95 min, with TBARS test LDL oxidative status was determined, and with the Beckman–Pergamon agarose lipogel system (Brea, CA) relative electrophoretic mobility was assessed. Using tetramethoxy propane (8 ng/100 μL incubate, TMPP) as a standard, TBARS test measured malondialdehyde expressed as percentage of control (full oxidation). Gel electrophoresis measured the migration distances of treated LDL relative to LDL that were maximally oxidized in the absence of estrogenic compounds. LDL oxidized in the absence of test compounds was used to represent 100% REM (relative electrophoretic mobility), and treatment migrations were expressed as a fraction of this distance.

12.13 REPORTER AND ELECTROPHORETIC MOBILITY SHIFT ASSAY (EMSA)[9]

1. *Reporter Assay*: HepG2 and Caco2 cells maintained at 37°C under an atmosphere of 95% air and 5% CO_2 in DMEM supplemented with 10% FBS were seeded in 6-well plates 1 day before the transfection, which was carried out using Fugene 6 reagent according to the manufacturer's instructions. β-Galactosidase expression plasmid pSV-β-GAL was co-introduced into the cells to normalize the transfection efficiency. After the transfection, cells were treated with test compound for 24 h, harvested, and luciferase activity was measured on a Micro Lumat LB96P Luminometer (Berthold Japan Co. Ltd., Tokyo) with a luciferase assay system (Promega Corp.). With a β-galactosidase enzyme assay system (Promega Corp.), β-galactosidase activity was measured spectrophotometrically, and luciferase activity was normalized to β-galactosidase activity for each sample. The luciferase reporter constructs used in this assay is summarized in Fig. 12.1.

2. *EMSA*: Fluorescence-labeled oligonucleotides containing human GSTP1 ARE/AP1 (5′-CCGGCGCCGTGACTCAGCACTGGGGCG-3′ and 5′-CGCCCCA-GTGCTGAGTCACGGCGCCGG-3′) were prepared. HepG2 nuclear extracts and labeled probe were incubated at room temperature for 30 min, resolved in a polyacrylamide gel, and analyzed with a FLA-5100 fluoroimage analyzer (Fuji Film, Tokyo, Japan). Prior to the addition of labeled probe, unlabeled probe or anti-Nrf2 antibody was incubated with nuclear extracts for 15 min.

12.14 $[Ca^{2+}]_{cyt}$ ASSAY[10]

The apoaequorin-expressing BY-2 cell suspension was incubated with 1 μM coelenterazine (Dojindo Laboratories, Kumamoto, Japan) for at least 6 h. With a

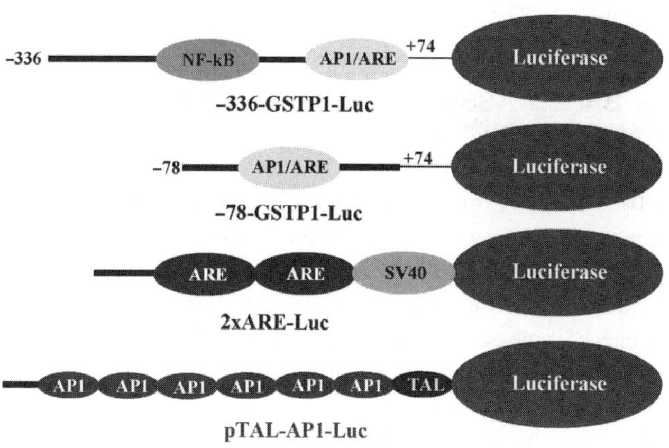

Figure 12.1 Constructs of the luciferase reporter.

luminometer (Lumicounter 2500, Microtech Nition, Funabashi, Japan) equipped with an A/D converter (MacLab, AD Instruments, Castle Hill, Australia), the aequorin luminescence reflecting $[Ca^{2+}]_{cyt}$ was measured and the data were analyzed using the program Chart v. 3.6.8 (AD Instruments). To cylindrical plastic cuvette, 200 μL of cell samples was transferred, the cuvette was placed in the luminometer, rotated at 3-s intervals 17 times alternately clockwise and counterclockwise to stir and supply air to the cells during luminescence measurement, a 20% volume of 1 M $CaCl_2$/20% ethanol solution was added, the total luminescence intensities were analyzed, and the aequorin luminescence curves were normalized to the background luminescence of each cell line.

12.15 QUANTITATIVE REAL-TIME PCR ASSAY[10]

According to manufacturer's instructions, Plant RNeasy Extraction Kit (Qiagen, Hilden, Germany) was used to isolate the total RNA of the cells for quantitative real-time PCR analyses. Using Superscript Choice system (Invitrogen, Carlsbad, CA, USA) from 1 μg of total RNA, the first-strand cDNA was synthesized, for which 1 μg of total RNA was mixed with 2 μL of oligo(dT) primer (0.5 μg/μL), at 70°C heat-denatured for 10 min, cooled on ice, 4 μL of first-strand buffer, 2 μL of dithiothreitol (0.1 M), 10 μL of dNTP (10 mM), and 1 μL of Super Script III (200 U/μL) were added and at 42°C incubated for 60 min. After reverse transcription, the reaction mixture was at 70°C incubated for 15 min and cooled on ice. For RNA degradation, 1 μL of RNaseH was added and incubated at 37°C for 20 min. With nuclease-free water, synthesized cDNA was diluted to 100 ng/μL and stored at −20°C until used. Using an ABI PRISM 7900HT Sequence Detection System (Applied Biosystems, Foster City, CA), real-time quantitative PCR was performed. TaqMan Universal PCR master mix, 900 nM forward and reverse primers, 200 nM TaqMan probe, and 100 ng template cDNA were involved in the PCR mixture. The PCR conditions consisted of 50°C for 2 min, 95°C for 10 min, and 40 cycles of amplification at 95°C for 15 s and 60°C for 1 min. All quantifications were normalized to 10,000 molecules of actin.

12.16 FT-IR-BASED ASSAY FOR ANTIOXIDATION ACTIVITY OF IONOL AND PIPERIDONE[11]

On an FT-IR 1600 Perkin-Elmer interferometer, typically 256 FT-IR spectra (KBr or thin films) were recorded from 4400 to 450 cm^{-1} with resolution 4 cm^{-1} and signal averaged for each sample. All spectra were normalized, and each spectrum was analyzed by second derivatization, namely Fourier self-deconvolution (FSD) and curve fitting. Before curve fitting, spectra were subjected to FSD using a 6 cm^{-1} bandwidth and a 5 smooth factor. The spectra were subjected to a line shape analysis with GRAMS software and SPECTRUM for Windows. Using the numbers and the bands determined by the FSD and with derivative spectra as starting parameters,

Gaussian curve fitting was performed. The fitting procedure consisted of iteratively adjusting the heights and half-width of the component bands with a 5.5 cm^{-1} peak full width at half-height (FWHH). After the completion of the iterative process, the band area of the peaks between 1680 and 1770 cm^{-1} was determined and expressed as a percentage of the total peak area in this range. The criteria judging the resulting fit were (1) minimization of the reduced χ^2 value, (2) no systematic deviations in the difference between fitted and measured spectra, and (3) convergence of the solution.

12.17 HPLC ASSAY FOR ANTIOXIDATION POTENTIAL OF POLYPHENOL[12]

The concentration of aqueous solutions of authentic standards was 15 μg/mL. A Waters 1525 Binary Pump, 717-plus autosampler (fitted with 10 μL sample loop), and C18 column (Waters Symmetry, 5 mm, 30 × 0.4 cm) were used for HPLC analyses. Mobile phase A (methanol/formic acid/water, 20/0.3/79.7, v/v/v) and B (ethanol/formic acid, 99.7/0.3, v/v) were used for gradient elution separating polyphenols. The gradient protocol for 35 min in a linear gradient mode was 0 mL of B/100 mL for 10 min, 10 mL of B/100 mL for 15 min, 30 mL of B/100 mL. For all elutions, the flow rate was 1.0 mL/min, and both mobile phase flasks were degassed by continuously using an on-line degasser. The separated polyphenols were detected at 280 nm (Waters 2487 Variable Wavelength detector), and using Waters Breeze chromatography software, the chromatogram was processed. Using the individual standard curves for each polyphenol, quantitative analysis was carried out. Recovery studies were carried out on specific amounts of standards. Samples were analyzed in triplicate before and after the addition of standards.

12.18 ROS PRODUCTION ASSAY[13]

SKBR3 cell line was incubator cultured in RPMI 1640 medium supplemented with 10% (v/v) FSC, 0.1 mg/mL of streptomycin, and 100 U/mL of penicillin at 37°C in 5% CO_2. Serial dilutions of test compound (50 μL) in medium or medium alone were added into each of 96-well plates, cells were plated at a density of 2×10^5 cells/well, incubated for 24–48 h, detached with trypsin-EDTA, washed once with PBS, suspended in 0.5 mL of PBS containing 10 μM 2′,7′-dichlorodihydrofluorescein diacetate (DCFH-DA, Sigma, St. Louis, MO) at 37°C for 30 min, with 4 mM H_2O_2 (as inducer for ROS production) incubated for 30 min, and ROS production of cells was evaluated on luminescence spectrophotometer (Perkin-Elmer, MA).

12.19 RABBIT LDL OXIDATION ASSAY[14,15]

Male and female homozygous KHC rabbits (1.75–2.40 kg) at the age of 3 months were housed individually under controlled environmental conditions (21°C to 25°C,

45% to 65% relative humidity, 12-h dark/light cycle lighting, more than 10 cycles/h ventilation). The control rabbits received commercially available standard diets. The diets for rabbits in the treatment group were prepared in Oriental Yeast by mixing test compound into the standard diet at 1%. The mixed diets were stored at $-20°C$ before feeding. Rabbits received 100 g of respective diets per day and tap water *ad libitum*, and their food consumption was measured daily. All rabbits were clinically observed; body weights, plasma lipid levels, and LDL oxidation were measured 1 month after the initiation of the study and at a monthly intervals. At the end of month 6, rabbits were euthanized by administering an overdose of pentobarbital. Under EDTA-3K treatment (1 mg/mL final concentration), from the auricular artery of each rabbit blood was collected and centrifuged to obtain plasma, of which a portion was ultracentrifuged to separate very-low-density lipoprotein (VLDL; $d < 1.020$ g/mL), LDL ($d = 1.019$ to 1.063 g/mL), and high-density lipoprotein (HDL; $d > 1.064$ g/mL).

Using COBAS FARA II (Roche, Switzerland) and commercially available kits (Wako Pure Chemical, Osaka, Japan), total cholesterol, triglyceride, and phospholipid levels in the plasma and in each lipoprotein fraction were measured enzymatically. Using fluorophotometer (model 650-10S; Hitachi, Tokyo, Japan) and a commercially available kit (Wako Pure Chemical, Osaka, Japan), the concentration of TBARS in the plasma was measured. In a bicinchonic acid assay using bovine serum albumin as standard, the LDL concentration was determined. As oxidant, 2,2′-azobis(4-methoxy)-2,4-dimethylvaleronitrile (Sigma-Aldrich, St. Louis, MO, USA) was added to LDL sample (70 μg/mL final concentration) and incubated at 37°C, using a Beckman Model DU-640 spectrophotometer the formation of conjugated dienes was monitored at 234 nm to obtain the lag time, the time the amount of conjugated dienes began to increase.

12.20 ROS SCAVENGING ASSAYS[16–18]

1. *$ONOO^-$ Scavenging Activity and Level Assays*: In $ONOO^-$ scavenging activity assay, a mixture of test compound, 10 μM 3-morpholinosydnomine, and 5 mM dihydrorhodamine 123 (DHR 123) was kept at 37°C for 20 min, and the fluorescence of rhodamine 123 (the reduced form of DHR) was determined with a microplate fluorescence reader at 485 nm excitation and 535 nm emission. DL-Penicillamine served as positive control. In $ONOO^-$ level assay, to rhodamine buffer (pH 7.4) containing 6.25 μM DHR 123 and 125 μM diethylenetriaminepentaacetic acid samples were added, incubated at 37°C for 5 min, and the absorbance of rhodamine 123 was measured with a microplate fluorescence reader (Tecan SPECTRAFluor, Tecan UK, Goring-on-Thames, UK) at 500 nm.

2. *$^{\bullet}O_2$ Scavenging Activity and Level Assays*: In $^{\bullet}O_2$ scavenging activity assay, the mixture of samples, 200 μL of freshly prepared 125 μM EDTA, 62 μM nitro blue tetrazolium (NBT), and 98 μM β-NADH in 50 mM PBS (pH 7.4) was treated with 25 μL of freshly prepared 33 μM phenazine methosulfate in 50 mM phosphate buffer (pH 7.4) pipetted in microplate wells for 5 min, and using a microplate reader,

the absorbance at 550 nm as an index of NBT reduction was estimated. L-(+)-Ascorbic acid served as a positive control. In $^{\bullet}O_2$ level assay, culture supernatant was mixed with 50 mM PBS, 125 μM EDTA, 62 μM NBT, and 98 μM NADH, treated with 33 μM 5-methylphenazium methyl sulfate for 5 min, and the absorbance was measured (on Tecan SPECTRAFluor, Tecan UK, Goring-on-Thames, UK) at 540 nm as an index of NBT reduction.

3. *NO• Scavenging Activity and Level Assays*: In NO• scavenging activity assay, 5 mM sodium nitroprusside (SNP) in PBS (pH 7.4) was mixed with various concentrations of samples, at 25°C incubated for 150 min, the reaction solution was transferred to 96-well plates for 5 min, and the amount of NO• generated from SNP was assayed by estimating nitrite accumulation, of which the absorbance was determined at 550 nm as an index of produced nitrite. Curcumin served as a positive control. In NO• level assay, 100 μL of culture supernatant was allowed to react with 100 μL of Griess reagent, the reaction mixture was at room temperature incubated for 5 min, and using a microplate reader, the optical density of the samples was at 540 nm determined.

4. *•OH Scavenging Activity Assays*: In •OH scavenging activity assay, 5,5-dimethyl-1-pyrroline *N*-oxide (DMPO) served as a spin trapping reagent, and Mn^{2+} was used as an external standard for calculating relative ESR signal intensity. A mixture of 20 μL of DMPO (1/10 diluted with DW, v/v), 38 mL of $FeSO_4 \cdot 7H_2O$ (0.2 mM), and 37 μL of diethylenetriaminepentaacetic acid (1 mM) was treated with 30 μL of sample solution (dissolved with distilled water or acetone) and 75 μL of H_2O_2 (1 mM), the reaction mixture was transferred to a capillary tube and placed in the cavity of the ESR spectrometer for 5 min, the ESR signal was measured to determine samples' •OH inhibition. With the ratio of the peak height of the second signal from the DMPO-•OH spin adduct to the signal of Mn^{2+} samples' OH inhibition was evaluated and then compared with the control ratio. Maltol and thiourea served as reference compounds, and the ESR spectra were recorded on a JES-TE100 ESR spectrometer (JEOL, Tokyo, Japan). The experimental magnetic field, power, modulation frequency, amplitude, and sweep time were 336.5 ± 5 mT, 1 mW, 9.41 GHz, 1×200, and 4 min, respectively. Following the equation •OH scavenging activity $= [(1-H)/H_0] \times 100\%$, •OH scavenging ability was calculated, in which H and H_0 were relative peak height of radical signals with and without sample, respectively.

12.21 DNA DAMAGE ASSAY[18]

To assay the protective effects of test compound on hydroxyl radical-induced DNA damage, 4 μL of H_2O_2 (30%) was added to an Eppendorf tube at a total volume of 12 μL containing 0.5 μL of PBR 322 DNA in 3 μL of phosphate buffer (50 mM, pH 7.4), 3 μL of $FeSO_4$ (2 mM), and 2 μL of test compound at various concentrations, incubated at 37°C for 30 min, the mixture was subjected to 0.8% agarose gel electrophoresis, and DNA bands (supercoiled, linear and open circular) were stained with ethidium bromide.

12.22 EGG YOLK TBARS ASSAY[19]

The egg yolk (1 g) was mixed with hexane/isopropanol (3/2), filtered, at 60°C, dried in a rotary evaporator, 0.2 g of obtained phospholipid was diluted with water to 10 mL, centrifuged, and used as a homogenates. The homogenates were at 1400 × *g* centrifuged for 10 min, pellet was discarded, and a low-speed supernatant was used for the assay. The homogenates (100 μL) were treated at 37°C with or without 50 μL of various freshly prepared oxidants (iron and nitroprusside), different concentrations of test compounds, and an appropriate volume of deionized water at a total volume of 300 μL, incubated for 1 h, the color reaction was carried out by adding 600 μL of acetic acid (pH 3.4) and 600 μL of thiobarbituric acid (TBA, Sigma, St. Louis, MO) and serial dilutions of 0.03 mM standard MDA (Sigma, St. Louis, MO), the reaction mixtures were incubated at 97°C for 1 h, after cooling the tubes with tap water, 2 mL of *n*-butanol was finally added, the mixture was centrifuged, the supernatant was taken, and the absorbance was read at 532 nm in a spectrophotometer.

12.23 MOUSE CATALASE (CAT) ASSAY[19]

Male albino mice weighing 24–30 g were divided into group 1 (control), which received only distilled water, group 2 (paracetamol control), which received only 250 mg/kg acetaminophen, and group 3 (paracetamol + test compound), which received a single dose of acetaminophen (250 mg/kg) + test compound (indicated dose). Test compound was administered 3 h after administering acetaminophen. The treatments were continued for 7 days, and on day 8, all mice were anesthetized, killed by heart puncture, blood samples were collected, while livers were quickly removed, placed on ice, homogenized in seven volumes of NaCl (0.9%), and the homogenates were centrifuged at 4000 × *g* for 10 min to yield a low-speed supernatant fraction for CAT assay. In CAT assay, 10 μL of liver supernatant was added to a quartz cuvette, the reaction was started by the adding freshly prepared 30 mM H_2O_2 in phosphate buffer (50 mM, pH 7.0), H_2O_2 decomposition rate was measured spectrophotometrically at 240 nm during 120 s, and CAT activity was expressed as 1 mol H_2O_2 mg liver^{-1} min^{-1}.

12.24 RAT TISSUE TBARS ASSAY[19,20]

Wistar rats (200–250 g) were killed after ether anesthesia, liver, kidney, and brain were quickly removed and placed on ice, of which 1 g was homogenized in cold 100 mM Tris-buffer (1/10 w/v, pH 7.4) with 10 up-and-down strokes at approximately 1200 rev/min in a Teflon glass homogenizer. The homogenates were at 1400 × *g* centrifuged for 10 min, pellet was discarded, and a low-speed supernatant was used for the assay. The homogenates (100 μL) were incubated at 37°C with or without 50 μL of various freshly prepared oxidants (iron and nitroprusside), different concentrations of test compounds, and an appropriate volume of deionized water at a total volume of 300 μL for 1 h, the color reaction was carried out by adding 8.1% SDS (200, 500, and 500 μL), acetic acid (pH 3.4), 0.6% TBA (Sigma, St. Louis, MO), and serial dilutions of 0.03 mM standard MDA, the reaction mixtures were incubated at 97°C for 1 h, and after cooling the tubes the absorbance was read at a wavelength of 532 nm in a spectrophotometer.

12.25 ANTIOXIDANT ACTIVITY ASSAY FOR β-CAROTENE/LINOLEIC ACID SYSTEM[20]

The antioxidant activity was assayed based on the coupled oxidation of β-carotene/linoleic acid system. According to its ability in preventing β-carotene/linoleic acid system oxidation, antioxidant compound was comparatively evaluated and monitored by bleaching β-carotene at 470 nm. The ability of different concentrations of antioxidant compound preventing oxidation of the β-carotene/linoleic acid system was compared with a control (no antioxidants added) and to reference antioxidant. An initial reading of absorbance at 470 nm (Spectrophotometer-Coleman, SP-395-UV, São Paulo, Brazil) was taken immediately for each tube and subsequently monitored at 15-min intervals for 2 h. During measurements, the cuvettes were kept in a water bath at 50°C. Based on $AA = [(As - Ac)/(Ao - Ac)] \times 100$, the antioxidant activity (AA) was calculated as percentage inhibition relative to the control, where As and Ac are absorbance of antioxidant compound and control, respectively, and Ao is absorbance at 470 nm of antioxidant compound at the start of the test.

12.26 RAT BRAIN TISSUE NO ASSAY[21]

Male Wistar rats weighing 260–290 g were killed by decapitation, and brains were rapidly removed to prepare homogenates. To prevent undesirable changes in the examined brain tissue that could happen *ex vivo* during the postmortem period, homogenization and centrifugation were carried out extremely fast at 4°C, and crude tissue homogenates containing membranes and all cellular components were kept on ice until further processing. The level of nitrites (the final products of NO metabolism) was determined based on the reaction of nitrites, sulfanilamide and *N*-(1-naphthyl)-ethylenediamine dihydrochloride (Sigma, Steinheim, Germany) and the yielding reddish-violet diazo dye, of which the absorbance was measured in the visible range at 540 nm. To determine the nitrite level in the brain samples, 100 μL of homogenate was added to 400 μL of redistilled water, the mixture was placed in a 100°C water bath for 15 min to stop all enzymatic processes, after cooling to each sample, 30 μL of Carrez I (0.36 M $K_4[Fe(CN)_6] \cdot 3H_2O$) and 30 μL of Carrez II (1 M $ZnSO_4 \cdot 7H_2O$) were added, by adding 4 μL of NaOH (10 M) the samples were alkalized to pH 8 and centrifuged at 10,000 × *g*, 75 μL of supernatant and 75 μL of redistilled water were placed in microplate wells, at room temperature incubated for 30 min, and the absorbance at 540 nm was measured. In blank samples, redistilled water was used instead of supernatant. To each well, by introducing

50 μL of solution of sulfanilamide in 2.5% H_3PO_4 (1%) and 50 μL of solution of *N*-(1-naphthyl)ethylenediamine dihydrochloride in 2.5% H_3PO_4 (0.1%) and in the dark mixing with the examined samples for 15 min, the colored reaction was developed, absorbance was measured at 540 nm, the results were calculated according to standard curves obtained for solutions of sodium nitrite (6–600 μM), and the absorbance change before and after incubation with sulfanilamide and *N*-(1-naphthyl)ethylenediamine dihydrochloride. As a measure of NO in brain tissue, concentrations of nitrites were calculated and expressed in nmol per mg of protein.

12.27 RAT BRAIN ANTIOXIDATIVE ENZYME ASSAY[22–25]

Male Wistar rats (250–280 g) housed in metal cages with free access to drinking water and standard pellet diet received once daily 0.1 mL/100 g of olive oil (control groups) and antioxidant compound at indicated dose, were sacrificed on day 14 and day 28 of the exposure, and brain samples were removed and washed with ice-cold 0.9% NaCl solution containing 0.16 mg/mL heparin to determine the activities of SOD, CAT, glutathione peroxidase (GPx), and reductase (GR) as well as reduced glutathione level (GSH).

To measure SOD activity, 10% brain homogenates prepared in 0.25 M saccharose were at 4°C and 8500 × *g* centrifuged for 10 min, and using BIOXYTECH SOD-525 Assay kit (OXIS International, Inc., Portland, OR, USA), the supernatant was assayed. To measure CAT activity, 10% brain homogenates prepared in phosphate buffer were at 4°C and 9000 × *g* centrifuged for 15 min, and activity was determined accordingly. To measure brain activity of GPx, the brain samples were prepared by homogenization in 8 volumes of cold buffer (50 mM Tris-HCl, pH 7.5, containing 5 mM EDTA and 1 mM β-mercaptoethanol), and at 4°C and 8500 × *g* centrifuged for 10 min. To measure brain activity of GR, BIOXYTECH GR-340 Assay kit (OXIS International, Inc., Portland, OR, USA) in a supernatant of tissue homogenate obtained by mincing the tissue in cold buffer (50 mM potassium phosphate pH 7.5, containing 1 mM EDTA) was at 4°C and 8500 × *g* centrifuged for 10 min. To measure brain GSH content with BIOXYTECH GSH-400 Assay kit (OXIS International, Inc., Portland, OR, USA), brain tissues were minced in ice-cold metaphosphate acid solution, centrifuged at 3000 × *g* for 10 min, and the assay was performed in a clear supernatant.

12.28 RAT BRAIN HIPPOCAMPI PROTEIN OXIDATION ASSAY[26]

At the treatment end, Wistar rats of each group were killed by cervical dislocation, brains were quickly taken out, cooled in a deep freezer, hippocampi were rapidly dissected out on ice plate, and the left and right hippocampi of the brain of one rat were pooled to make one sample of the tissue. Tissue samples were homogenized in 50 mM Tris (pH 7.4) with a Potter-Elevehijam type homogenizer fitted with Teflon plunger, the homogenate was with Tris (pH 7.4) diluted (1 : 10), at 6000 rpm centrifuged for

5 min, the resulting pellet consisting of nuclear and cellular material was discarded, the supernatant containing mitochondria, synaptosomes, microsomes, and cytosol was at 25,000 rpm further ultracentrifuged for 25 min to form mitochondrial pellets, and the resulting supernatant was used as such as cytosolic fraction.

Aliquots of cytosolic fraction were used for measuring protein oxidation by estimating the protein carbonyl levels. Protein carbonyl content in the samples was determined by measuring di-nitrophenylhydrazine (DNPH) adducts at 375 nm on a Shimadzu UV-160A spectrophotometer. Using a molar extinction coefficient (e) of 22,000/M · cm, carbonyl contents were calculated and data were expressed as nanomole carbonyl per milligram of soluble extracted protein.

REFERENCES AND NOTES

1. G.K. Ziyatdinova, H.C. Budnikov, V.I. Pogorel'tzev, T.S. Ganeev. The application of coulometry for total antioxidant capacity determination of human blood. Talanta 68 (**2006**) 800–805.
2. C.H. Lee, L. Yang, J.Z. Xu, S.Y.V. Yeung, Y. Huang, Z.Y. Chen. Relative antioxidant activity of soybean isoflavones and their glycosides. Food Chem 90 (**2005**) 735–741.
3. K.M. Lo, P.C.K. Cheung. Antioxidant activity of extracts from the fruiting bodies of Agrocybe aegerita var. alba. Food Chem 89 (**2005**) 533–539.
4. L.M. Cheung, P.C.K. Cheung. Mushroom extracts with antioxidant activity against lipid peroxidation. Food Chem 89 (**2005**) 403–409.
5. H.J.D. Dorman, A. Peltoketo, R. Hiltunen, M.J. Tikkanen. Characterization of the antioxidant properties of de-odourised aqueous extracts from selected Lamiaceae herbs. Food Chem 83 (**2003**) 255–262.
6. D.A. Gray, M.J. Clarke, C. Baux, J.P. Bunting, A.M. Salter. Antioxidant activity of oat extracts added to human LDL particles and in free radical trapping assays. J Cereal Sci 36 (**2002**) 209–218.
7. C. Sun, J.W. Wang, L. Fang, X.D. Gao, R.X. Tan. Free radical scavenging and antioxidant activities of EPS2, an exopolysaccharide produced by a marine filamentous fungus Keissleriella sp. YS 4108. Life Sci 75 (**2004**) 1063–1073.
8. T. Wilson, H. March, W.J. Banz, Y. Hou, S. Adler, C.Y. Meyers, T.A. Winters, M.A. Maher. Antioxidant effects of phyto-and synthetic-estrogens on cupric ion-induced oxidation of human low-density lipoproteins in vitro. Life Sci 70 (**2002**) 2287–2297.
9. T. Nishinaka, Y. Ichijo, M. Ito, M. Kimura, M. Katsuyama, K. Iwata, T. Miura, T. Terada, C. Yabe-Nishimura. Curcumin activates human glutathione S-transferase P1 expression through antioxidant response element. Toxicol Lett 170 (**2007**) 238–247.
10. Y. Kadota, T. Furuichi, T. Sano, H. Kaya, W. Gunji, Y. Murakami, S. Muto, S. Hasezawa, K. Kuchitsu. Cell-cycle-dependent regulation of oxidative stress responses and Ca^{2+} permeable channels NtTPC1A/B in tobacco BY-2 cells. Biochem Biophys Res Commun 336 (**2005**) 1259–1267.
11. A. Georgiev, I. Karamancheva, L. Topalova. Evaluation of the antioxidation activity of ionol and piperidone towards transformer oil using FT-IR spectroscopy. J Mol Structure 797 (**2006**) 25–33.

12. O. Hassan, L.S. Fan. The anti-oxidation potential of polyphenol extract from cocoa leaves on mechanically deboned chicken meat (MDCM). LWT 38 (**2005**) 315–321.

13. P. Moongkarndi, N. Kosem, S. Kaslungka, O. Luanratana, N. Pongpan, N. Neungton. Antiproliferation, antioxidation and induction of apoptosis by Garcinia mangostana (mangosteen) on SKBR3 human breast cancer cell line. J Ethnopharmacol 90 (**2004**) 161–166.

14. T. Kurosawaa, F. Itoh, A. Nozaki, Y. Nakano, S. Katsuda, N. Osakabe, H. Tsubone, K. Kondo, H. Itakura. Suppressive effects of cacao liquor polyphenols (CLP) on LDL oxidation and the development of atherosclerosis in Kurosawa and Kusanagi-hypercholesterolemic rabbits. Atherosclerosis 179 (**2005**) 237–246.

15. O.T. Iijima, H. Takedau, T. Matsumiya. Effects of san'o-shashin-to on the antioxidative mechanism in spontaneous familial hypercholesterolaemic rabbits. Pharmacol Res 41 (**2000**) 137–141.

16. Y.J. Kim, K.S. Kang, T. Yokozawa. The anti-melanogenic effect of pycnogenol by its antioxidative actions. Food Chem Toxicol 46 (**2008**) 2466–2471.

17. D. Štajner, B.M. Popović, J. Čanadanović-Brunet, P. Boža. Free radical scavenging activity of three Equisetum species from Fruška gora mountain. Fitoterapia 77 (**2006**) 601–604.

18. Z.J. Qian, W.K. Jung, H.G. Byun, S.K. Kim. Protective effect of an antioxidative peptide purified from gastrointestinal digests of oyster, Crassostrea gigas against free radical induced DNA damage. Bioresource Technol 99 (**2008**) 3365–3371. *Note*: In ESR spectroscopy–based superoxide anion radical scavenging activity assay, superoxide anion radicals were generated in UV-irradiated riboflavin/EDTA system containing 0.3 mM riboflavin, 1.6 mM EDTA, 800 mM DMPO, and indicated concentrations of test compound. This mixture was under UV lamp at 365 nm irradiated for 1 min and transferred to 100 μL quartz capillary tube of the ESR spectrometer for measurement. The experimental magnetic field, power, modulation frequency, amplitude, and sweep time were 336.5 ± 5 mT, 1 mW, 9.41 GHz, 1×200, and 4 min, respectively. Following the equation $^{\bullet}O_2$ scavenging activity $= [(1 - H)/H_0] \times 100\%$, $^{\bullet}O_2$ scavenging ability was calculated, in which H and H_0 were relative peak height of radical signals with and without sample, respectively.

19. S.M. Sabir, J.B.T. Rocha. Water-extractable phytochemicals from Phyllanthus niruri exhibit distinct in vitro antioxidant and in vivo hepatoprotective activity against paracetamol-induced liver damage in mice. Food Chem 111 (**2008**) 845–851.

20. M.A. Nascimento, A.K. Silva, L.C.B. França, E.L.J. Quignard, J.A. López, M.G. Almeida. Turnera ulmifolia L. (Turneraceae): preliminary study of its antioxidant activity. Bioresource Technol 97 (**2006**) 1387–1391.

21. A. Bilska, M. Dubiel, M. Sokołowska-Jeżewicz, E. Lorenc-Koc, L. Włodek. Alpha-lipioic acid and differently affects the reserpine-induced oxidative etress in the striatriatus and prefrontal cortex of rat brain. Neuroscience 146 (**2007**) 1758–1771

22. A. Lukaszewicz-Hussain. Subchronic intoxication with chlorfenvinphos, an organophosphate insecticide, affects rat brain antioxidative enzymes and glutathione level. Food Chem Toxicol 46 (**2008**) 82–86.

23. A.D. Muthuswamy, K. Vedagiri, M. Ganesan, P. Chinnakannu. Oxidative stress-mediated macromolecular damage and dwindle in antioxidant status in aged rat brain regions: role of l-carnitine and dl-a-lipoic acid. Clin Chim Acta 368 (**2006**) 84–92.

24. P. Montilla, I. Espejo, M.C. Muñoz, I. Bujalance, J.R. Muñoz-Castañeda, I. Tunez. Protective effect of red wine on oxidative stress and antioxidant enzyme activities in the brain and kidney induced by feeding high cholesterol in rats. Clin Nutr 25 (**2006**) 146–153.
25. E. Skrzydlewska, A. Augustyniak, K. Michalak, R. Farbiszewski. Green tea supplementation in rats of different ages mitigates ethanol-induced changes in brain antioxidant abilities. Alcohol 37 (**2005**) 89–98.
26. A. Jyoti, D. Sharma. Neuroprotective role of Bacopa monniera extract against aluminium-induced oxidative stress in the hippocampus of rat brain. NeuroToxicology 27 (**2006**) 451–457.

13

METHODS AND APPLICATIONS OF ANALGESIC ASSAYS

Ming Zhao

Pain is one of the most prevalent conditions limiting productivity and diminishes quality of life. For instance, visceral pain is very frequently accompanied by secondary muscle hyperalgesia, an early phenomenon accentuated in extent by the repetition of the visceral episodes and by being long-lasting in the referred area. Peripheral nerve injury usually causes neuropathic pain, which can be manifested as allodynia (pain response to non-noxious stimulus), hyperalgesia (increased response to noxious stimulus), and spontaneous pain. Functional abdominal pain, particularly prevalent in women, is characterized by pain and discomfort in the abdominal and pelvic regions in the absence of clear or demonstrable pathology of any component of the abdominal cavity including the viscera and its associated nerves. Bone pain, a major clinical problem in both osteolytic and osteoblastic metastatic bone disease resulting from structural damage of the bone, manifests as periosteal irritation and nerve entrapment, is usually localized in a compromised area, and has been described as a deep, boring sensation that aches and burns and is accompanied by episodes of stabbing discomfort. Cancer pain, a kind of common clinical pathologic pain syndrome, decreases the cancer patient's quality of life greatly. Approximately 30% to 50% of patients with cancer will experience moderate to severe pain, and 75% to 95% of patients with advanced-stage or metastatic cancer experience substantial pain that leads to substantial deterioration in quality of life. To explore the mechanism of different types of pain and to evaluate the potency of different analgesics, a series of assays have

Pharmaceutical Bioassays: Methods and Applications. By Shiqi Peng and Ming Zhao

been established. In this chapter, 17 assays are described: algogenic activity assay for rat ureteral stone,[1] cold pressor-based assay for acute pain of healthy volunteers,[2] heat-based assay for acute pain of healthy volunteers,[2] electrical stimulation-based assay for acute pain of healthy volunteers,[2,3] radiant heat tail-flick assay for mice,[4,5] pain behaviors/responses assays in rats,[6–8] hot plate assay in rats,[8,9] plantar assay in rats,[9] hot plate assay in mice,[10,11] paw and tail formalin assays in mice,[8,12] rat assays for bone cancer pain,[13] mouse assay for hind paw cancer pain,[14] visceral pain assay,[15] canine nociceptive thermal escape assay in dog,[16] carrageenan assay in rats,[17] electrophysiologic assay for mice with tumor-evoked hyperalgesia,[18] and mouse assay for bone cancer pain.[19–25]

13.1 ALGOGENIC ACTIVITY ASSAY FOR RAT URETERAL STONE[1]

Female SD rats (180–220 g) were housed in single Plexiglass cages with water and food *ad libitum*, kept on a 12-h light-dark cycle (artificial light: 08:00–20:00 h), anesthetized with pentobarbital (50 mg/kg, ip), received stone implantation in the left ureter (via injection of dental cement) or sham surgery (injection of isotonic saline), were videotaped using a telecamera and videotape time-lapse system with ultrared light for filming during the dark phase to record ureteral crises, and were killed 2 (nine stone and two sham) or 4 (16 stone and two sham) days after surgery [period of peak hyperalgesia of left oblique muscle (OE)]. Based on the videotape, post-operative ureteral crisis number was calculated for each stone-implanted rat with (1) hump-backed position, (2) licking lower abdomen and/or flank, (3) contraction of ipsilateral oblique musculature with inward moving hindlimb, (4) body stretching, (5) squashing lower abdomen against the floor, and (6) supine position with ipsilateral hindlimb adducted and compressed against abdomen as criteria. Each criterion required at least three movements and a minimum total duration of 2 min, whereas a lesser number of movements (1 or 2) and/or duration of the sequence (<2 min) was regarded as a subcrisis. During assay for qualitatively testing bilateral OE sensitivity (verification of presence or absence of vocalization upon light digital pressure), daily recordings were interrupted for a few minutes. No instrumental assessment of the hyperalgesia was used to avoid any strong manipulation of the OE interfering with the evaluation of contraction indices. Visceral crises number was taken as an indirect index of hyperalgesia degree, the two parameters being linearly and directly correlated.

13.2 COLD PRESSOR-BASED ASSAY FOR ACUTE PAIN OF HEALTHY VOLUNTEERS[2]

In the assay, the nondominant hand was immersed in ice-saturated water, the temperature of the ice-saturated water was measured and ranged from 1.89 ± 0.39°C to 2.08 ± 0.16°C across all assays. On a 0% to 100% electronic visual analog scale, perceived pain intensity was continuously rated with the dominant hand and stored electronically for analyzing peak pain intensity, area under the intensity time curve, and mean pain intensity. If a subject exhibited 50% pain (i.e., midpoint between no pain and maximum pain) for 120 s, the area under curve (AUC) value was defined as 50% $\times$ 120 s, or 6000% s. If pain was intolerable and had elapsed before 2 min, the subject could withdraw, in which case pain intensity was considered to be maximal until the end of the 2-min period (for calculation of the AUC).

13.3 HEAT-BASED ASSAY FOR ACUTE PAIN OF HEALTHY VOLUNTEERS[2]

A computer-driven Peltier element was used to assess pain detection level and pain tolerance. On the left forearm, a 32 $\times$ 32 mm thermode was applied and its temperature was from 30°C continuously increased to a maximum of 56°C at a rate of 1.4°C/s. Because thresholds for heat tolerance could exceed 53°C in some individuals, a relatively high cutoff temperature of 56°C was chosen to avoid a ceiling effect and an underestimation of a potential analgesic effect. As a safety precaution, a temperature rise rate of 1.4°C/s was used to avoid skin damage occurring with lower rise rates and consequently longer stimulus durations. In this assay course, the maximal individual value for heat tolerance without medication was 53.2°C and 55.0°C under medication [with 25 μg/h transdermal fentanyl (TDF) at 24 h]. In one series of trials, subjects were asked to indicate heat pain threshold; in another trial they were asked to indicate when heat became intolerable. On pressing the switch or ending the assay, the thermode temperature returned to baseline settings (30°C). Four measurements were taken for each run, and the mean of measurements 2, 3, and 4 was involved in the assays.

13.4 ELECTRICAL STIMULATION-BASED ASSAY FOR ACUTE PAIN OF HEALTHY VOLUNTEERS[2,3]

A commercially available constant current stimulator triggered by a train/delay generator was used in the electrical stimulation assay. Onto the lateral upper nondominant arm and delivering 3-s stimuli, two electrodes (3.6 cm apart) were applied to deliver totally 30 stimuli, of which the parameters were 1 ms width, 10 Hz for 3 s, 64 mA constant current until maximum tolerated value, 8 stimulus train frequency, and approximately 15-s interstimulus interval (ISI). An ascending staircase design between 1.2 and 2.4 mA was used for starting and randomly varied stimuli to determine pain threshold. The stimulus intensity was increased by 30% and 15% increments until the pain threshold and the maximally tolerated pain were reached, respectively. Once the pain threshold and the maximally tolerated electrical stimulus had identical value, a series of stimuli (stimuli of 8 trains) spaced evenly between 30% less than the pain threshold and the maximally tolerated stimulus were randomly administered, and the subject rated the pain intensity on a visual analog scale, which was determined for each subject individually and used throughout the assay. Parameters describing the slope of regression of stimulus intensity versus visual analog scale were derived by treatment group and by time-point (not baseline corrected) and used for comparing the relation of stimulus intensity (mA) versus visual analog scale (mm). Using the simple linear model $y = a \times (x - b)$, a line was fitted to the data, where a, b, x, and y were the electrical current, the visual analog scale score, the slope of the relation, and the x-intercept, respectively.

13.5 RADIANT HEAT TAIL-FLICK ASSAY FOR MICE[4,5]

Male ICR BR mice (20–25 g) were used, and the assay was performed on the portion of the tail immersed in the treatment solution, as the analgesic actions of test compounds administered in this manner were restricted to the exposed part of the tail (proximal regions were not affected). Using a tail-flick apparatus, the tail-flick latency was recorded for an individual mouse, and the latency twice its baseline latency or greater was defined quantitatively as analgesia. Because thermal stimulus was usually applied to the superficial surface of the skin and the thermal energy absorbed in the skin was concentrated in the dermal/epidermal junction where nerve terminals were located, this assay was especially useful in examining topical analgesics. Baseline latencies typically ranged from 2.5 to 3.5 s, with a maximum cutoff latency of 10 s to minimize tissue damage in analgesic mice. Because analgesia was assessed quantitatively, group comparisons were performed with Fisher's exact test, and ED_{50} values were determined with GraphPad Software.

13.6 PAIN BEHAVIORS/RESPONSES ASSAYS IN RATS[6–8]

1. *Formalin Assay*: Before 50 μL of formaldehyde solution in saline (0.5%) was subcutaneously injected into the plantar surface of the rat's left hindpaw, the rats were initially acclimated to the acryl cages (Muromachi Kikai, Tokyo, Japan) for 15 min. By measuring the time spent in licking/biting the injected paw at 5 min intervals with a stopwatch, nociceptive behaviors of the rats were quantified. Biphasic changes occurred to the time spent in nociceptive behaviors. The first reaction between 0 and 10 min after formalin injection was defined as the early phase, and the second reaction between 10 and 45 min after the first reaction was defined as the late phase. Test compounds were given orally 30 min before formalin injection (about 1 h before the peak of the late phase). At the end of the assay, formalin-injected rats were killed with CO_2.

2. *Chronic Constrictive Injury Model-Based Neuropathic Pain Assay*: The rats were anesthetized with sodium pentobarbital (45 mg/kg, ip); by blunt dissection at the level of midthigh, the left common sciatic nerve was exposed, 4 loose ligatures (3-0 chromic gut, Matsuda Ika Kogyo Co., Ltd., Tokyo, Japan) were placed around the nerve, and the muscle and skin were sutured. As a sham operation, the right sciatic nerve was isolated in the same way without ligation. To minimize discomfort and painful mechanical stimulation, the rats were housed in plastic cages having their floors covered with soft bedding. The neuropathic rats were used for the behavioral assay 14–15 days after the operation, when about 80% rats developed distinct neuropathic behaviors (allodynia or hyperalgesia).

3. *Behavioral Assay in the Neuropathic Rats*: In the plastic cages with mesh bottoms, the neuropathic rats were individually placed, and with a set of von Frey filaments (Stoelting Company, WI, USA) ranging from 0.69 g to 28.84 g, the withdrawal thresholds of the rats to mechanical stimuli were measured. In a period of 3 s, each filament was in ascending order vertically applied to the midplantar skin and at the thresholds the rats quickly flicked their paws, otherwise the force of the thickest filament (28.84 g) was assigned as the withdrawal threshold.

4. *Heat Hyperalgesia Assay*: In transparent plastic chambers with a glass floor, the neuropathic rats were individually placed, acclimated for 15 min, and through the glass the radiant heat source (Plantar test No. 7370; Ugo Basile, VA, Italy) was aimed onto the midplantar area of both paws. Throughout the assay, the heat stimuli intensity was kept constant. At $\geq$3-min intervals, each paw was tested twice. As the withdrawal latencies, the latencies from initial heat activation to paw withdrawal were recorded to the nearest 0.1 s. To avoid tissue damage, 30-s cutoff latency was used.

13.7 HOT PLATE ASSAY IN RATS[8,9]

Female Wistar rats (140–180 g) and a computer-controlled increasing temperature were used for this assay. A rat was placed onto the plate, plate temperature was linearly increased from 30°C to a temperature at which a rat showed nocifensive behavior confined to either hindpaw (licking or lifting), which was regarded as the noxious heat threshold. After conditioning and control measurements, test compound (10–200 μg/kg) or its solvent was administered intraperitoneally, and 30 min later, threshold was determined. The heat thresholds before and after test compound or its solvent administration were compared using the Student's *t*-test for paired samples.

13.8 PLANTAR ASSAY IN RATS[9]

Male SD rats (300–350 g) were placed individually in a clear plastic chamber (Ugo Basile Plantar test apparatus) and left to acclimatize for 5 min before assay. Light from an 8 V, 50 W halogen bulb (64607 Osram) was through the base of the plastic box delivered to the plantar surface of one of the rat's hindpaws, of which the beam was about 12 mm in diameter. The time for the rat withdrawing its left hindpaw was recorded. Rats not responding were removed after 30 s (cutoff time). Two radiant heat stimuli were applied and separated by 1- to 2-min intervals, and the mean of the two measures were taken. To reach a basal withdrawal latency of 8–10 s and to reduce the variability, the intensity was selected. After predrug trial, rats were given vehicle or drug by oral, intraperitoneal, or subcutaneous routes and reassessed 30 (ip and sc) or 60 (po) min later.

13.9 HOT PLATE ASSAY IN MICE[10,11]

Three-month-old male albino Swiss mice (30–40 g) were housed at controlled temperature (22 $\pm$ 2°C) with a 12-h light/dark cycle, received standard lab chow and tap water *ad libitum*, and were habituated to the experimental room for at least 2 h before hot plate assay. In the assays, the hot plate apparatus (Socrel, model-DS 37) was maintained at $50 \pm 0.1^{\circ}C$, mice were placed into an acrylic cylinder (20 cm in diameter) on the heated surface, and the time (s) between placement and licking hindpaws or jumping (whichever occurred first) was recorded as the response latency. To prevent tissue damage, a 90-s cutoff was used. To minimize novelty-induced antinociception, 24 h before the assay all mice were habituated to the assay procedure including intraperitoneal injection of vehicle and measurement of latency. Mice presenting training latencies higher than 30 s were excluded. On the day of the assay, the mice were intraperitoneally injected with vehicle (rice oil or saline, 10 mL/kg), dipyrone (1.5 mmol/kg, which was included in all experiments as an internal standard), or test compounds (0.1–1.0 mmol/kg) and subjected to the hot plate assay 15, 30, and 60 min thereafter. The dose of dipyrone was selected in pilot experiments.

13.10 PAW AND TAIL FORMALIN ASSAYS IN MICE[8,12]

1. *Paw Formalin Assay*: Before assays, mice had been individually exposed to the observation chamber for 45 min. For formalin injection, a 30-gauge needle was used, and into the plantar surface of the right hindpaw, 20 μL of formalin (5%) was administered subcutaneously. The mice were placed in a 20 × 30 cm clear Plexiglas cylinder for observation. Over 40–60 min using 5-min period, by determining the time the mouse spent for licking the injected paw, the pain behavior was quantified. After the formalin injection, two phases of spontaneous licking behavior were observed with phase I corresponding with the interval from 0 to 10 min, which was thought to be due to a direct chemonociceptive effect of formalin, and phase II corresponding with the interval from 15 to 40 min, which was mainly mediated by inflammatory reactions. Incomplete formalin injection and excessive bleeding from the injection site were used as exclusion criteria. Time response data were expressed as mean ± SEM of 5-min bins over 40 min. All test compounds were given 30 min before formalin administration, and saline was used as control. In the dose-response analysis, the observed data in phase I and II were considered separately. In each case, the cumulative licking response was calculated for each mouse, and the dose-response curve was represented with percentage of control. Using linear regression analysis, ED_{50} values and its 95% confidence intervals were calculated.

2. *Tail Formalin Assay*: Using a Hamilton syringe with a 30-gauge needle, 20 μL of formalin (5% to 20%) was injected intradermally into the dorsal surface of the tail, and the mouse was put back into a chamber to start the observation. The only behavior suggestive of pain was licking the injection site as well as the tail base. Using a digital camera, the time the mouse spent licking the whole tail was recorded for 1 h and expressed as AUC without observing the two classic phases. In the time course and dose-response assays, the observation period was restricted to 40 min.

13.11 RAT ASSAYS FOR BONE CANCER PAIN[13]

1. *Mechanical Allodynia Assay*: As the hindpaw withdrawal response to von Frey filament stimulation, mechanical allodynia was measured. Using von Frey filaments with logarithmically incremental stiffness, which delivers force ranging from 0.132 g to 20.9 g to the hind paw of female SD rats (150–180 g) bearing bone tumor, mechanical stimulation was carried out. In the assay, a L15 × W10.7 × H13.8 cm^3 perspex box was situated on a wire mesh platform, the rat was placed in this box, allowed to settle for 5–10 min, and each von Frey filament was applied to the plantar surface of rat hindpaw in ascending order beginning with 0.132 g filament. As a single trial within 2–3 s, six to eight filaments were applied at $\geq$2-min intervals, and five trials were administered for each von Frey filament. When a hindpaw withdrawal response occurred, a trial was suspended immediately. The equation Response frequency % = (number of paw withdrawals/five trials) × 100 was used to express the amount of paw withdrawals in the five trials.

2. *Hindpaw Weight Assay*: Using a Dual Channel Weight Averager (Churchill Electronic Services Ltd.), hindlimb weight bearing of female SD rats (150–180 g) suffering from bone tumor was measured. The rats were placed in a special Perspex chamber and each hindpaw rested on a separate transducer pad. The averager was set, the load on the transducer was recorded over 3 s, and the displayed two numbers represented the distribution of the rat's body weight on each paw. For each rat, two readings from each paw were taken, averaged, and presented as weight-bearing difference (contralateral reading minus ipsilateral reading) or as percentage difference between pretreatment and posttreatment.

3. *Mechanical Hyperalgesia Assay*: Using a paw pressure apparatus equipped with a wedge-shaped probe, the nociceptive pressure threshold was determined on both hindpaws of female SD rats (150–180 g) suffering from bone tumor. Paw withdrawal, struggling, or vocalization was taken as the end point. Both ipsilateral and contralateral paw readings were taken and presented as paw withdrawal threshold.

13.12 MOUSE ASSAY FOR HINDPAW CANCER PAIN[14]

Male C57BL/6 mice (6 weeks old) were housed in plastic cages under controlled conditions (24 ± 0.5°C, 6 a.m. to 6 p.m. alternate light-dark cycles, free access to food and water). B16-BL6 melanoma cells were cultured in EMEM containing 5% fetal bovine serum, 2% penicillin/streptomycin, released from plastic by exposure to a mixture of 0.25% trypsin and 0.02% EDTA, collected by centrifugation for 3 min at 1200 rpm, the resulting pellet was washed twice with 10 mL of PBS, recentrifuged for 3 min at 1200 rpm, the pellet was suspended in PBS, counted using a hemocytometer, diluted, and kept on ice until injected. B16-BL6 melanoma cells (2×10^5) or heat-killed cells were injected subcutaneously into the plantar region of the unilateral hindpaw of the mouse in a volume of 20 μL and equivolume of PBS into the contralateral hindpaw. To determine melanoma volume, the varicose perimeter of the paw *in situ* was examined by wrapping silk thread three times around the perimeter of the paw.

Mice with melanoma cells inoculation were randomly divided into electroacupuncture, sham (treatment without electrical current), and model (nontreatment) groups. During electroacupuncture treatment, with a specially designed holder, mouse trunk was kept motionless while the head and four limbs maintained freedom of movement. Electroacupuncture was administered with two stainless steel needles (0.18 mm in diameter). When restraining the mice into the holders and inserting the acupuncture needles to the acupoint, the mice were anesthetized by inhaling ether. The two needles (one in ST36 and the other in BL60) were connected with the output terminals of an electroacupuncture apparatus.

Alternating trains of dense sparse frequencies were 100 Hz for 1.05 s, 4 Hz for 2.85 s, bidirectional asymmetric pulse with 0.6-ms pulse width. The stimulation intensity was approximately 1 mA. For each train, the stimulation lasted for 30 min. At treatment end, the immediate analgesic action of electroacupuncture was tested for 60 min, and the paw withdrawal latency and paw licking response latency were recorded. Electroacupuncture was applied at 24-h intervals. The mice in the sham group received same manipulation.

13.13 VISCERAL PAIN ASSAY[15]

Adult female C57/BL6 mice (10–28 weeks old and 20–30 g body weight at the start of the assay) were used. The assay was developed based on intracolonic instillation of capsaicin. To avoid the stimulation of somatic areas by contact with the irritant chemicals, Vaseline was applied in the perianal area, and 0.05 mL of capsaicin (0.1%, w/v) in a mixture of 10% ethanol, 10% Tween-80, and 80% saline or 0.9% saline was administered by introducing a fine cannula with round tip (external diameter 0.61 mm, 4 cm long) into the colon via the anus. The spontaneous behavior of the mice to this stimulus was directly observed for 20 min. Licking the abdomen, stretching the abdomen, squashing the lower abdomen against the wire mesh, and abdominal retractions were defined as pain-related behaviors.

13.14 CANINE NOCICEPTIVE THERMAL ESCAPE ASSAY IN DOG[16]

Male and female beagle dogs (11.0–15.8 kg, 12–21 months old) deemed healthy based on physical exam, complete blood count, and serum chemistry analysis were housed in an AALAC-approved vivarium and provided food and water *ad libitum*. Food was withheld for 12 h prior to all assays, but water made available. Prior to entering assays, dogs underwent an extensive adaptation period (2 weeks minimum) and were acclimatized to the testing paradigm and additional handling.

Briefly, a cart (Intermetro Industries, Welks-Barre, PA, USA) accommodated a fabric canine restraint sling over a wire grill. On the grill, a 45.7 × 76.2 cm,

0.3-cm-thick glass plate was secured. Beneath the grill was a box with shelf holding the paired radiant lamps, device timer, and intensity control. The lamps, timer, and intensity control device were components of a rodent thermal nociceptive testing device. With an arbitrary scale of 0–100 to paired, calibrated, focused (5 mm aperture diameter) projection bulb lamps, the intensity control box delivered variable amperage. Using a foot pedal released by the researcher when a dog moved the hindpaw being tested, a silent relay circuit between the lamp, control box, and timer was independently activated for each lamp. The lamps fixed 28 cm apart in an aluminum frame allowed the bulbs to lie directly under the glass plate at one end of the cart and project vertically through the glass. The spacing and orientation was set to accommodate the dogs' natural stance and hindpaw position, thus the light beam was naturally focused on the anterior one-third of the metatarsal paw pad when dogs stood in the restraint sling.

In a closed, normally lighted room with care taken to avoid extraneous auditory and handling stimulation, all behavioral assays were performed between 08:00 and 18:00 h. Using an implantable thermocouple microprobe, which was regularly calibrated against a mercury thermometer in constant-temperature water baths between 34°C and 60°C and secured to the glass surface, a thermometer attached to a microprocessor and placed at the center of the lamp beam measured rate-of-rise of the uppermost glass surface temperature during lamp activation. To prevent light scatter, the thermocouple microprobe was covered with dark fabric. Left and right lamps were individually activated, and temperatures were recorded at 5-s intervals between 5 and 25 s after activation of the lamp. To simulate the normal testing interval employed during dog assays, trials with each lamp were repeated three times with approximately 20- to 40-s delays between trials.

Prior to assay, dogs were acclimated to the environment and the restraint sling. During training and baseline testing, to discourage the dog from turning or moving in the sling, one researcher stood at its head, while another researcher stood at its left side with one hand placed gently on the anterior tibial or metatarsal region of each hindlimb. The lamp under the paw to be tested was actuated using a foot pedal once the dog was quietly standing with one hindpaw centered over each lamp. The stimulus was terminated when the dog purposefully moved the tested hindpaw. In every trial, each hindpaw was tested twice. The order of testing between left and right paws was randomly trained, and each train was at least 20 s. Thus, in any given observation, for both hind paws the response latency was assessed twice. Since for a given dog the left and right paw variation was not significant, in order to give a representative latency a mean, which was defined as the "response latency" of that dog at that time point, four tests (two left, two right) were calculated. To alert the dog to the testing situation without alerting it to which paw was being tested or when testing had started or was completed, additional environmental modifications, including consistent background noise (fan), regular but unpatterned speaking to the dog, and positive verbal reinforcement were used. Prior to all assays with dogs completing total trials, left and right hindpaw baseline latencies were recorded. Linear regression curves were fit for all left and right paw trials and the slopes were compared. For left, right, and population withdrawal latency, mean and standard deviations were calculated, and significance was set at $p < 0.05$.

13.15 CARRAGEENAN ASSAY IN RATS[17]

Male SD rats weighing 250–300 g were examined for the level of thermal hyperalgesia induced by plantar carrageenan injection (carrageenan assay). Under halothane anesthesia, a suspension of 2 μg of lambda carrageenan (Sigma, St. Louis, MO) in 0.1 mL of normal saline was injected subcutaneously via a 24-gauge needle into the plantar surface of the right hindpaw of the rats. Using von Frey filaments with logarithmically incremental stiffnesses (0.41, 0.70, 1.20, 2.00, 3.63, 5.50, 8.50, and 15.1 g, Stoelting, Wood Dale, IL), mechanical thresholds were measured to calculate the 50% probability thresholds for mechanical paw withdrawal. Starting from a 2.00 g probe, filaments were applied to the plantar surface of a hindpaw for 6–8 s either in a stepwise ascending order after negative withdrawal responses or descending order after positive withdrawal responses, until six consecutive responses were noted. When continuous negative responses were observed, 15.00 g was assigned to the limits of the stimulus set; when continuous positive responses were observed, 0.25 g was assigned to the limits of the stimulus set. For measuring the thermal nociceptive threshold, the rats were placed in a clear plastic cage (10 × 20 × 24 cm) placed on an elevated floor of clear glass (2 mm thick). Inside a movable holder beneath the glass floor, a radiant heat source (eye projector halogen lamp JRC-12-V-100W, Iwasaki Electric, Tokyo, Japan) with an aperture diameter of 5 mm was placed. The halogen lamp beneath the floor was positioned to focus on the plantar surface of the carrageenan-injected paw that contacted the glass plate. The interval between the application of the light beam and the brisk hindpaw withdrawal response was measured to the nearest 0.1 s and designated as response latency. In the absence of a response within 20 s, the trial was terminated and the lamp was removed, and the response was assigned as a 20-s latency.

13.16 ELECTROPHYSIOLOGIC ASSAY FOR MICE WITH TUMOR-EVOKED HYPERALGESIA[18]

Male C3H/HeNCr mice (>5 weeks old, weighing 20–30 g) used in the assay were housed four per box, had free access to food and water, and were maintained on a 12-h light/dark schedule.

NCTC 2472 fibrosarcoma cells were grown to confluency in 75 cm^2 flasks in NCTC 135 medium (pH 7.4) containing 10% horse serum and prepared for implantation by creating a cell suspension with trypsin. Cells were counted, pelleted, resuspended in PBS, and implanted into the hindpaw. For implantation, mice were briefly anesthetized with halothane (2% to 3%); using a 0.3-mL insulin syringe with a 29.5-gauge needle, fibrosarcoma cells (2×10^5 cells/10 μL) were injected into and around the calcaneus bone of the left hindpaw. After implantation of fibrosarcoma cells, none of the mice showed signs of motor dysfunction.

On days 10–21 after implantation of fibrosarcoma cells, mice with tumor-evoked hyperalgesia and control mice were anesthetized with acepromazine maleate (20 mg/kg, ip) and sodium pentobarbital (48 mg/kg, ip). As needed to maintain areflexia, additional doses of sodium pentobarbital (15 mg/kg) were added. On removal of hair in the skin overlying the thoracic and lumbar parts of the vertebral column, an incision was made and by a laminectomy the lumbar enlargement was exposed. In a spinal frame, mice were secured, and the spinal cord was continually bathed in warm (37°C) mineral oil. With stainless steel microelectrodes (10 mΩ, Frederick Haer and Co., Brunswick, ME) lowered into the spinal cord, using an electronic micromanipulator (Burleigh, Fisher, NY) in 3-μm steps, extracellular recordings from dorsal neurons at the L3-L5 spinal levels with receptive fields (RFs) located on the plantar surface of the hindpaw were obtained. Neurons recorded at depths of 20–200 μm and at depths greater than 200 μm from the spinal cord surface were considered to be located in superficial laminae and in intermediate to deep laminae, respectively. Single neurons with easily discriminated action potentials were assayed only.

Electrophysiologic activity of dorsal horn neurons was amplified, audio-monitored, and displayed on a storage oscilloscope. Using a customized data acquisition program, neuronal activity, discriminated impulses, and stimulus temperatures were collected for off-line analyses.

13.17 MOUSE ASSAY FOR BONE CANCER PAIN[19–25]

1. *Bone Cancer Pain Model*: C3H/He male mice (weighing 25–30 g) were housed in boxes of temperature- and humidity-controlled environment, maintained on a 12-h light/dark cycle with free access to chow and water, and adapted to the laboratory conditions for at least 1 week prior to surgery. A mouse skin incision was made, and the patellar ligaments were cut to expose the condyles of the distal femur. At the level of the intercondylar notch and the intramedullary canal of the femur, a 23-gauge needle was inserted and a cavity was created for injection of the cells. Using a syringe (Becton Dickinson and Company, Franklin Lakes, NJ), approximately 2.5×10^5 NCTC clone 2472 fibrosarcoma cells derived originally from a connective tissue tumor in a C3H mouse in 20 μL of medium were injected unilaterally into the intramedullary cavity of the femur, the injection site was sealed with dental acrylic to prevent leakage of cells outside the bone, and by stitching the skin of the knee, the surgical procedure was ended. Sham control mice were subjected to operation procedure and housed in the same room and conditions with the exception that the medium was injected without any cell loading.

2. *Spontaneous Lifting*: Before the assay, mice were habituated to the laboratory room for at least 30 min, and at room temperature spontaneous lifting was carried out. In the assay, mice were placed and habituated in a transparent acrylic cylinder with 20 cm diameter and on the glass plate surface, observed for 4 min for spontaneous lifting behavior of the left hindpaw, and data were expressed as percentage withdrawal time over total session time. All mice in nonoperated and sham-operated groups gave 0% as normal value.

3. *Limb-Use Impairment on Rotarod*: After spontaneous lifting, assay mice were immediately placed on a mouse rotarod (ENV-575M®, Med Associates Inc., GA,

USA) at a speed of 16 rpm for a 2-min rotation to score the limb-use during the forced ambulation, following 4 = normal, 3 = limping, 2 = partial non-use of left hindpaw, 1 = substantial non-use of left hindpaw, and 0 = non-use of left hindpaw. The sedative effects of test compounds were also examined by determining mouse ability to support its own body weight on rotarod at a speed of 16 rpm. Mice that fell from the rotarod within 1 min were defined as positive for sedation/motor impairment.

4. *Mechanical Hypersensitivity*: In a quiet room, 30 min prior to assay, mice were placed in acrylic cages with wire grid floors. A hand-held force transducer (electronic anesthesiometer, Senselab® Somedic, Horby, Sweden) was applied manually to the midplantar hindpaw with a gradual increase in pressure, and a hindpaw flexion reflex was evoked. Mouse successively removing and flinching its paw was considered as the assay's end point, after which the intensity of the pressure was automatically recorded. Mice were assayed before and after treatments, and the values for this response were obtained by averaging three measurements. Effects of test compounds were expressed as withdrawal threshold ratio (%) and calculated using the equation Withdrawal threshold ratio (%) = [(Left paw posttreating threshold − Left paw pretreating threshold)/(Right paw pretreating threshold − Left paw pretreating threshold)] × 100. Once the left paw threshold increased to normal values, withdrawal threshold ratio was 100%.

REFERENCES AND NOTES

1. M.A. Giamberardino, G. Affaitati, R. Lerza, G. Fanó, S. Fulle, S. Belia, D. Lapenna, L. Vecchiet. Evaluation of indices of skeletal muscle contraction in areas of referred hyperalgesia from an artificial ureteric stone in rats. Neurosci Lett 338 (**2003**) 213–216.
2. M. Koltzenburg, R. Pokorny, U.E. Gasser, U. Richarz. Differential sensitivity of three experimental pain models in detecting the analgesic effects of transdermal fentanyl and buprenorphine. Pain 126 (**2006**) 165–174.
3. C. Shanahan, A.R. Ward, V.J. Robertson. Comparison of the analgesic efficacy of interferential therapy and transcutaneous electrical nerve stimulation. Physiotherapy 92 (**2006**) 247–253.
4. Y. Kolesnikov, D. Sõritsa. Analgesic synergy between topical opioids and topical non-steroidal anti-inflammatory drugs in the mouse model of thermal pain. Eur J Pharmacol 579 (**2008**) 126–133.
5. A. Wesolowska, S. Young, M. Dukat. MD-354 potentiates the antinociceptive effect of clonidine in the mouse tail-flick but not hot-plate assay. Eur J Pharmacol 495 (**2004**) 129–136.
6. Y. Akada, S. Ogawa, K. Amano, Y. Fukudome, F. Yamasaki, M. Itoh, I. Yamamoto. Potent analgesic effects of a putative sodium channel blocker M58373 on formalin-induced and neuropathic pain in rats. Eur J Pharmacol 536 (**2006**) 248–255.
7. U. Herzberg, E. Eliav, G.J. Bennett, I.J. Kopin. The analgesic effects of R(+)-WIN 55,212-2 mesylate, a high affinity cannabinoid agonist, in a rat model of neuropathic pain. Neurosci Lett 221 (**1997**) 157–160.
8. J. Szolcsányi, K. Bflcskei, A. Szabó, E. Pintér, G. Pethő, K. Elekes, R. Bfrzsei, R. Almási, T. Szfts, G. Kéri, Z. Helyes. Analgesic effect of TT-232, a heptapeptide somatostatin

analogue, in acute pain models of the rat and the mouse and in streptozotocin-induced diabetic mechanical allodynia. Eur J Pharmacol 498 (**2004**) 103–109.

9. P. Girard, Y. Pansart, M.C. Coppe, J.M. Gilardin. Nefopam reduces thermal hypersensitivity in acute and postoperative pain models in the rat. Pharmacol Res 44 (**2001**) 540–545.
10. J. Milano, S.M. Oliveira, M.F. Rossato, P.D. Sauzem, P. Machado, P. Beck, N. Zanatta, M.A.P. Martins, C.F. Mello, M.A. Rubin, J. Ferreira, H.G. Bonacorso. Antinociceptive effect of novel trihalomethyl-substituted pyrazoline methyl esters in formalin and hot-plate tests in mice. Eur J Pharmacol 581 (**2008**) 86–96.
11. A.K. Baker, V.L.H. Hoffmann, T.F. Meert. Dextromethorphan and ketamine potentiate the antinociceptive effectsof m- but not d- or k-opioid agonists in a mouse model of acute pain. Pharmacol Biochem Behav 74 (**2002**) 73–86.
12. Y. Kolesnikov, M. Cristea, G. Oksman, A. Torosjan, R. Wilson. Evaluation of the tail formalin test in mice as a new model to assess local analgesic effects. Brain Res 1029 (**2004**) 217–223.
13. S.J. Medhurst, K. Walker, M. Bowes, B.L. Kidd, M. Glatt, M. Muller, M. Hattenberger, J. Vaxelaire, T. O'Reilly, G. Wotherspoon, J. Winter, J. Green, L. Urban. A rat model of bone cancer pain. Pain 96 (**2002**) 129–140.
14. Q.L. Mao-Ying, K.M. Cui, Q. Liu, Z.Q. Dong, W. Wang, J. Wang, H. Sha, G.C. Wu, Y.Q. Wang. Stage-dependent analgesia of electro-acupuncture in a mouse model of cutaneous cancer pain. Eur J Pain 10 (**2006**) 689–694.
15. R. Sanoja, F. Cervero. Estrogen-dependent abdominal hyperalgesia induced by ovariectomy in adult mice: a model of functional abdominal pain. Pain 118 (**2005**) 243–253.
16. K. Wegner, K.A. Horais, N.A. Tozier, M.L. Rathbun, Y. Shtaerman, T.L. Yaksh. Development of a canine nociceptive thermal escape model. J Neurosci Methods 168 (**2008**) 88–97.
17. T. Yamamoto, O. Saito, K. Shono, S. Tanabe. Anti-hyperalgesic effects of intrathecally administered neuropeptide W-23, and neuropeptide B, in tests of inflammatory pain in rats. Brain Res 1045 (**2005**) 97–106.
18. S.G. Khasabov, D.T. Hamamoto, C. Harding-Rose, D.A. Simone. Tumor-evoked hyperalgesia and sensitization of nociceptive dorsal horn neurons in a murine model of cancer pain. Brain Res 1180 (**2007**) 7–19.
19. M.E. Mouedden, T.F. Meert. Pharmacological evaluation of opioid and non-opioid analgesics in a murine bone cancer model of pain. Pharmacol Biochem Behav 86 (**2007**) 458–467.
20. M.A. Sevcik, J.R. Ghilardi, K.G. Halvorson, T.H. Lindsay, K. Kubota, P.W. Mantyh. Analgesic efficacy of bradykinin B1 antagonists in a murine bone cancer pain model. J Pain 6 (**2005**) 771–775.
21. M.E. Mouedden, T.F. Meert. Evaluation of pain-related behavior, bone destruction and effectiveness of fentanyl, sufentanil, and morphine in a murine model of cancer pain. Pharmacol Biochem Behav 82 (**2005**) 109–119.
22. H.C. Park, J. Seong, J.H. An, J. Kim, U.J. Kim, B.W. Lee. Alteration of cancer pain-related signals by radiation: proteomic analysis in an animal model with cancer bone invasion. Int J Radiat Oncol Biol Phys 61 (**2005**) 1523–1534. *Note*: Bone cancer pain model: Male C3H/HeJ mice (8–10 weeks old) maintained at the specific-pathogen-free colony with 22°C temperature, 55% humidity and water/diet supplied *ad libitum* were used for establishing model. Syngeneic murine hepatocarcinoma cells (HCa-1, 5×10^5) were injected into the

periosteal membrane of mouse's hind foot dorsum, mice were regularly examined for tumor formation and growth, killed at regular intervals, their tumor-bearing hindpaws were fixed overnight at 4°C in 4% zinc-buffered formalin in 0.1 M PBS, decalcified in 10% EDTA (pH 7.4) for 2 weeks, and embedded in paraffin. The paraffin blocks were sectioned at 7 μm thickness, stained with hematoxylin and eosin and under light microscopy examined for invasion of cancer cells to the bone.

23. C. Zhao, P.W. Wacnik, J.M. Tall, D.C. Johns, G.L. Wilcox, R.A. Meyer, S.N. Raja. Analgesic effects of a soy-containing diet in three murine bone cancer pain models. J Pain 5 (**2004**) 104–110.
24. L.J. Kehl, D.T. Hamamoto, P.W. Wacnik, D.L. Croft, B.D. Norsted, G.L. Wilcox, D.A. Simone. A cannabinoid agonist differentially attenuates deep tissue hyperalgesia in animal models of cancer and inflammatory muscle pain. Pain 103 (**2003**) 175–186. *Note*: Cannabinoids produced catalepsy, a state of decreasing responsiveness to external stimuli associated with a waxy rigidity of the extremities. It was possible that increasing forelimb muscle rigidity would result in increased grip force interpreted mistakenly as reversal of hyperalgesia. To distinguish the antihyperalgesic and cataleptic effects of test compounds on grip force, the effect of catalepsy on grip force in mice receiving no hyperalgesic stimuli was evaluated using the bar assay, for which each mouse was placed with its forelimbs on a plastic bar (1 cm diameter) positioned 5 cm above and parallel to a countertop with its hindpaws resting on the countertop. The total time each mouse spent with its forelimbs on the bar in 3 consecutive 60-s trials was recorded. To evaluate the effect of catalepsy on grip force, just prior to grip force measurements at each time point, bar assays were collected. Before and at 30 min after injection of test compounds (30 mg/kg, ip) or vehicle, mice were assayed.
25. L. Menéndez, A. Lastra, M.F. Fresno, S. Llames, Á. Meana, A. Hidalgo, A. Baamonde. Initial thermal heat hypoalgesia and delayed hyperalgesia in a murine model of bone cancer pain. Brain Res 969 (**2003**) 102–109.

14

METHODS AND APPLICATIONS OF EPILEPSY ASSAYS

Ming Zhao, Shiqi Peng, and Guohui Cui

Epilepsy affects up to 0.5% to 1% of the population and is characterized by recurrent unprovoked seizures. Despite adequate pharmacologic treatment, more than 30% of the epilepsy patients have uncontrolled seizures or unacceptable medication-related side effects, therefore have "refractory epilepsy." As an alterative strategy, epilepsy surgery is a curative treatment option that aims at removing the ictal onset zone, but is invasive and believed to be responsible for seizure occurrence. If a patient's ictal onset zone is not well circumscribed or localized in functional brain tissue, few treatment options are left. A continuous impetus to investigate novel forms of treatment is provided due to the inability to adequately treat all patients with refractory epilepsy, and development of epilepsy assays has thus been continuous. In this chapter, 17 assays are presented: handling-induced seizure susceptibility (HISS) assay,[1,2] mouse locomotor activity assay,[2] mouse diathesis-stress assay,[3] social interaction assay,[4] spatial learning ability assay for recurrent seizure rats,[5] timed pentylenetetrazol (PTZ) infusion assay for mice,[6] PTZ seizure assay for mice,[6–8] maximal electroconvulsions threshold assay for mice,[7,9,10] maximal electroshock seizure (MES) assay,[11,12] 6 Hz psychomotor seizure assay for mice,[13] subcutaneous bicuculline and picrotoxin assay for mice,[13] *N*-methyl-D-aspartate (NMDA)-induced convulsions assay for mice,[13] audiogenic seizure (AGS) assay,[13] kindled rat assay for focal seizures,[13] cobalt/homocysteine assay for status epilepticus of rats,[13]

Pharmaceutical Bioassays: Methods and Applications. By Shiqi Peng and Ming Zhao

PTZ-induced kindling assay,[14,15] and increasing-current electroshock seizure (ICES) assay for mice.[16]

14.1 HISS ASSAY[1,2]

Seizure-susceptible El/Suz (El) and seizure-resistant ddY/hydrocephalus II (ddY) mice were fed standard chow and water *ad libitum* and housed in a controlled environment with a reverse light cycle (lights off at 1000 h and on at 2200 h) at 22°C and 48% humidity. All assay procedures were conducted between 1200 and 1800 h during the nocturnal phase of the light cycle, when mice were naturally awake and active.

1. *Transfer Cage Apparatus*: In standard cages a circular hole was cut into one cage wall and a tubular insert held by a threaded cap inserted. When the threaded insert of the home cage adjoined the threaded insert of the new cage, dams/litters relocated spontaneously from one cage to another, which was completed weekly at the same time (between 1000 and 1400 h). Regular husbandry was performed neither in the parental generation of mice beginning with the first day of mating and continuing to the time of weaning nor in the generation of offspring throughout their lifetimes with the exception of postnatal day (PND) 1 culling, PND 5/6 testing, and PND 21 cage allocation, and mice were not handled by humans until completing the handling-induced seizure susceptibility (HISS) assay on PNDs 80–90. Avoiding human animal contact had the effect of increasing latency to first seizure in El mice for up to 60 days beyond the seizure onset latency that characterized handled El mice.

2. *Mating, Rearing, and Weaning Procedures*: A total of 37 group-housed breeder male and female El (19 pairs) and ddY (18 pairs) strains were first raised to reproductive maturity (approximately 90 days of age), for which mice of each strain were assigned randomly to male–female mating pairs in either mother-alone condition (ddY = 6, El = 7) or biparental condition (ddY = El = 6). To assess the gestational condition of a female, dams were observed in the cage for a protruding abdomen and enlarged nipples and not disturbed. For mother-alone mating pairs, 16–18 days after conception, males were removed and females were moved to a cage transfer apparatus for the remainder of gestation, parturition, and weaning. For biparental mating pairs, males remained with the female for the duration of gestation, parturition, and weaning. On the day of pregnancy detection (16–18 days postconception), all mice were given a nesting pad. Cages were checked daily for pregnancies and litters, and the birthday was assigned as PND 0. On PND 1, pups were removed from the cage, briefly weighed, and sexed. Litters were culled to eight pups, and those consisting of fewer than four pups were excluded from the assay. On PND 21, pups were weaned with body weight and gender recorded. Male and female pups separately housed were grouped no more than five to a cage, and each treatment group involved offspring from no fewer than six litters to result in treatment groups matched for litter size, gender, and number of mice housed per cage.

3. *Undisturbed and Pup Retrieval Assays*: In the home cage using an infrared camcorder, both parental and nonparental behaviors of adult mice were videotaped. On PND 3/4, observations began with a 20-min, undisturbed videotaping without handling or moving the cage or the mice. On PND 5/6, by removing two pups from the home nest and placing them in an opposite corner of the cage, a 20-min retrieval assay was performed. The pups having negative retrieval were replaced in the original nest location. It was necessary to differentiate videotaping male and female caregivers, which was achieved by using a nontoxic black Sharpie permanent marker marking the male with black stripes on both sides of his body. On the day prior to a videotaping session, the male was marked to decrease the amount of disturbance on the day of testing, which produced no seizures during assay course. For parental behaviors such as nest building, nursing (maternal), crouching over pups, sniffing pups, licking/mouthing pups, and resting with pups, a total of 120 10-s intervals were scored. Component behaviors such as "resting with pups" and "interaction with pups" were not strictly speaking parental behaviors and quantified to provide a comprehensive accounting of the frequency of social interaction in the litter. Scored nonparental behaviors were self-grooming, resting alone, eating/drinking, running in circles, walking, rearing, and climbing. Total parental and nonparental scores equaled the cumulative number of a total of 120 intervals constituting each 20-min test session, in which adults engaged in parental or nonparental behaviors, respectively.

4. *HISS Assay*: Beginning on PND 80, HISS assay was performed in El strain mice, which included two trials separated by a 30-min rest period. During each trial, each mouse was picked up by its tail, held 10 to 15 cm above the bedding of its home cage for 30 s, placed in a cage containing fresh bedding for 2 min, held by the tail for another 15 s (10–15 cm above the bedding of the cage), and then returned to the home cage. Even if the mouse experienced a seizure in trial 1, 30 min later trial 2 was performed.

The assay procedure was repeated after 10 days (on PND 90), and mice were undisturbed over this interval. On PND 90, a subset of mice (without father, $n = 15$; with father, $n = 12$) were HISS assayed for the first time. From each litter, equal numbers of El mice were randomly selected for the two separate PND 80 and PND 90 HISS I assaying groups so that litter bias could not impact the results.

14.2 MOUSE LOCOMOTOR ACTIVITY ASSAY[2]

Because older mice of the El strain were more susceptible to seizures, seizure-susceptible El mice were at least 6 months old at the time testing. Prior to assay, mice had free access to drinking water and were fed chow *ad libitum*, of which the composition was 59.5% carbohydrate, 22% protein, 5% fat, 11% fiber, and 2.5% added minerals.

The locomotor apparatus consisted of photocell grids, each measuring 25 cm × 48 cm and positioned around the perimeter of a standard mouse housing cage. Two centimeters above the bedding at 16-cm intervals, four infrared photocell beams were located along the long axis, and eight beams were located across the width. Mice continually consumed their respective experimental diets and were tested in the locomotor apparatus for 24 h after HISS assay.

14.3 MOUSE DIATHESIS-STRESS ASSAY[3]

Male and female seizure-prone El mice were used with ddY mice as control, among them pregnant females were monitored daily, and the birthday was assigned as postnatal day 0. On postnatal day 21, pups were weaned and housed with same-sex littermates in modified shoebox caging. The modified transfer cage apparatus allowed the current explicit handling assays to be conducted in mice having no prior history of human handling between weaning and the time of testing. For the second assay, 120- to 150-day-old adult El and ddY mice in mixed gender groups were selected from the general colony and as such had a history of weekly handling for cage changing husbandry. Seizure induction was monitored as a consequence of tail suspension handling, and only El strain mice with a known history of seizure susceptibility were used. For both assays, mice were supplied with chow and water *ad libitum*, housed in a reverse light–dark cycle colony (lights off 1000, lights on 2200) at 21°C to 23°C, and SaniChip bedding was used in all cages.

One third of the mice were perfused, which experienced neither tail suspension nor foot-shock (unhandled control group) and were not exposed to mice in the other two experimental groups. Another third of the mice were suspended for 30 s by the tail approximately 10–15 cm above the floor of the home cage, and 15 min later this assay was repeated. The remaining one third of mice were exposed to two episodes of foot shock (1/s over 90 s at 0.4 mA) separated by a 2-min interval. One hour after the initial tail suspension or foot-shock exposure, mice were transported to a procedure room separated from the mouse colony and sacrificed. Because the assays were performed 10–20 days prior to the postnatal days 80–90 onset of seizure susceptibility in El mice, no mouse of either strain experienced a seizure as a result of tail suspension or foot-shock exposure.

14.4 SOCIAL INTERACTION ASSAY[4]

Experimental El and ddY mice and nonexperimental stimulus mice were composed of male and female mice in equal 50/50 ratios and were at least 120 days old at the time of the social interaction test. Twenty-four hours prior to data collection, the stimulus mice were housed singly, and to facilitate environmental acclimation on cage transfer, some of the soiled bedding from the home cage was transferred into each new cage.

Fifteen minutes before the start of social interaction testing, all experimental and stimulus mice were moved to a testing room separate from the colony. In the home cage of a stimulus mouse of the same gender and strain, the experimental mouse was gently placed. Using a video camera, the experimental mouse's social behavior, specifically olfactory investigation of the home cage mouse, was recorded during a 5-min social interaction assay. Before assay, using a nontoxic marker, stimulus mice were marked with black stripes to distinguish them from the experimental subjects. The mice were in random order assayed, while social behavior was scored with treatment-blind observers. Using separate naive groups of mice, the social interaction phenotyping experiment was replicated twice and the results were pooled.

14.5 SPATIAL LEARNING ABILITY ASSAY FOR RECURRENT SEIZURE RATS[5]

Male SD healthy rats (180–220 g) were kept at 22 $\pm$ 1°C on a 12-h light/dark schedule with free access to standard laboratory food and water and randomly divided into kainic acid (KA) and saline-treated (control) groups. The rats in the KA group were injected with KA freshly dissolved in NS (10 mg/kg, 5 mg/mL) subcutaneously (sc), whereas rats in control group were injected with NS alone. According to the Racine scale, seizure stages of the rats were behaviorally rated as stage 1, 2, 3, 4, and 5 for chewing, head nodding, unilateral forelimb clonus, rearing with bilateral forelimb clonus, and rearing with bilateral forelimb clonus and falling back, respectively. During 21 days after injection, KA-treated rat behavior was monitored for 10 h each day. Spontaneous recurrent seizures (SRS) corresponded with stage 3–5, and the rats rated stage 3–5 were regarded as the KA-treated group with SRS, otherwise regarded as KA-treated group without SRS.

Morris water maze consisted of a monitor with the video camera set in the ceiling, a computerized tracking system (DMS-2), and a circular metal tank (1.20 m in diameter, 0.50 m in height) filled with water at 25 $\pm$ 1°C. The maze was divided it into four equal quadrants, and around its edge four start positions with red mark were located equidistantly. During training and assay, 0.5 cm below the surface of the water, a platform (0.12 m in diameter, 0.24 m in height) was submerged. To exclude the effect of other factors on spatial learning memory, skimmed milk was put the in the water. Throughout all assays, distal visual cues outside of the pool and the location of the platform remained constant. After the final treatment with KA, all rats were trained for 3 consecutive days in the Morris water task. The rat was placed on the platform for 20 s and then put into the water facing the maze's wall in four quadrants consecutively at the same time each day. If the rat reached the platform in 60 s, it was kept on the platform for 20 s. If the rat failed to find the platform within 60 s, it was guided and placed on the platform for 20 s as repetition training. After 3 days of training, each rat was put into the water from the middle point of the quadrant facing the wall randomly every day, the latency and swim path of the rat were monitored by a video and DMS-2 for 4 consecutive days to measure its spatial learning ability. On day 5, the hidden platform was removed, the rat was placed into the water from the middle point of the edge in the target quadrant, and the latency, swim path, and the time spent in the training and opposite quadrants in 60 s were recorded to measure its spatial learning ability.

14.6 TIMED PTZ INFUSION ASSAY FOR MICE[6]

Male CD-1 mice (5–6 weeks old, weighing 25–30 g) were housed in polypropylene cages at 22 $\pm$ 2°C and relative humidity of 60% to 70%, with automatically controlled 12/12 h light/dark cycle, lights on at 7:00 a.m., and provided with the pelleted standard diet and with drinking water in water bottles. All assays were conducted between 9:00 a.m. and 4:00 p.m.

All test compounds and PTZ solutions were freshly prepared before administration. PTZ was dissolved in heparinized sterile 0.9% saline to prepare 10 mg/mL and 8.5 mg/mL solutions for iv and sc administrations, respectively. Test compounds were dissolved in water or suspended in 0.5% carboxymethyl cellulose sodium (CMC). Before drug administrations, mice were deprived of food for 5 h and provided water *ad libitum*. Test compounds were administered by oral route 60 min before administration of PTZ using oral gavage in a volume of 0.01 mL/g. Dose selection was based on literature reports and preliminary assays such that the selected dose of a test compound would cause suppression of seizures in either the MES or sc PTZ assay without producing apparent motor toxicity.

A butterfly cannula (needle size 27 gauge, $\frac{3}{4}$ in.) was attached to a 5-mL syringe prefilled with heparinized PTZ solution. For infusion, rat was restrained, and the needle was inserted into its tail vein. By appearance of blood in the cannula, the accuracy of needle placement in the vein was confirmed. By a special tape, the needle was secured to the tail. The transparent Perspex box keeping the rat had holes for ventilation. The syringe was held in the adjustable motor driven infusion pump, thus the mouse could move freely in the box without strain on the attached cannula with no severe struggling. PTZ was infused at 0.5 mL/min constant rate, during which mice were observed for the onset of different types of seizures. After longer than 5 s, time latencies from start of infusion to appearance of first clonus (characterized by the rapid involuntary rhythmic contraction and relaxation of limbs), tonic forelimb and/or hindlimb extension (characterized by stretching of limbs) was recorded. Infusion was stopped at the appearance of tonic extensor in each animal.

14.7 TIMED PTZ SEIZURE ASSAY FOR MICE[6–8]

Adult male Swiss mice (22–26 g) were kept in colony cages with free access to food and tap water in standardized housing conditions (natural light–dark cycle, 21 $\pm$ 1°C, relative humidity of 55 $\pm$ 5%), and after 7-day adaptation to laboratory conditions randomly assigned to assay groups consisting of eight mice. Each mouse was used only once, and all assays were performed between 9:00 a.m. and 2:00 p.m.

The PTZ-induced seizures in rodents were considered to be an assay of myoclonic convulsions in humans. By subcutaneous administration of 100 mg/kg PTZ, clonic convulsions were induced in mice, which was its CD97 (a dose necessary to induce clonic seizures in 97% of animals tested). After PTZ injection, mice were separately placed into 25 $\times$ 15 $\times$ 10 cm transparent Plexiglas cages and observed for 30 min for the occurrence of clonic seizures. Clonus of the whole body lasting over 3 s with an accompanying loss of righting reflex was defined as clonic seizure. The anticonvulsant activity of test compound alone was assayed at a series of doses against PTZ-induced clonic seizures in mice. The convulsing mouse number out of the totally assayed mouse number was noted for each treatment condition. The ED_{50} (in mg/kg) of each test compound alone was evaluated as its median effective dose protecting 50% of mice against PTZ-induced convulsions.

14.8 MAXIMAL ELECTROCONVULSIONS THRESHOLD ASSAY FOR MICE[7,9,10]

Adult male Swiss mice (22–26 g) were kept in colony cages with free access to food and tap water and in standardized housing conditions (natural light–dark cycle, temperature of 23 ± 1°C, relative humidity of 55 ± 5%). After 7-day adaptation to laboratory conditions, mice were randomly assigned to experimental groups, and all assays were performed between 0800 and 1500 h. By alternating current (0.2-s stimulus duration, 50 Hz, maximum stimulation voltage of 500 V) delivered via ear clip electrodes and a Rodent Shocker Generator, electroconvulsions were produced. The electrical system provided constant current stimulation, and the changes in impedance resulted in no alterations of current intensity. The tonic hindlimb extension (i.e., the hindlimbs of mice outstretched 180 degrees to the plane of the body axis) was used as the criterion of the occurrence of seizure activity. To evaluate maximal electroconvulsions threshold, at least four groups of mice each with eight mice were challenged with electroshocks of various intensities to yield 10–30%, 30–50%, 50–70%, and 70–90% of mice with seizures and construct a current intensity response relationship curve using a log-probit method, from which a median current strength (CS_{50} in mA) was calculated. CS_{50} value was defined as the current intensity required to induce tonic hindlimb extension in 50% of the mice challenged. Again, after administration of a single dose of test compound, the mice in four groups were subjected to electroconvulsions (each group with a constant current intensity). The maximal electroconvulsions threshold was recorded for six different doses of test compound. The percentage of increase in CS_{50} values for mice injected with increasing doses of test compound over the control (vehicle treated animals) was calculated.

14.9 MES ASSAY[11,12]

Adult male Albino Swiss mice (22–26 g) were kept in colony cages with free access to food and tap water in standardized housing conditions (12-h light–dark cycle, 21 ± 1°C). After 7-day adaptation to laboratory conditions, the mice were randomly assigned to experimental groups, and each mouse was assayed only once. All tests were performed between 9:00 a.m. and 2:00 p.m.

The protective activity of test compounds was assayed as their ED_{50} values (mg/kg) against MES-induced seizures (fixed current intensity of 25 mA, 50 Hz, 500 V, 0.2-s stimulus duration). The mice were administered different doses of test compound to obtain a variable percentage of protection against MES and construct a dose-effect curve for each test compound administered alone. ED_{50} value with 95% confidence limits was defined as the dose of test compound required to protect half of the mice tested against MES.

14.10 6-Hz PSYCHOMOTOR SEIZURE ASSAY FOR MICE[13]

This assay was based on the fact that psychomotor seizures produced by low-frequency and long-duration electrical stimulation of the brain in mice resembled the seizures seen in patients with partial epilepsy, and the understanding that the ability to protect against these seizures could be an indicator that a drug or test compound, particularly one that has been found to be inactive in the MES assay and the subcutaneous pentylenetetrazol assay, may be effective for therapy of pharmacoresistant partial seizures.

Groups of eight male albino CF No. 1 mice (18–25 g) were maintained in temperature- and humidity-controlled AAALAC-approved quarters with a 12-h on/12-h off light cycle, permitted access to food and water *ad libitum*, and pretreated intraperitoneally with lacosamide. At 30 min after pretreatment, mice were stimulated with sufficient current (32 mA at 6 Hz for 3 s delivered through corneal electrodes) to induce a psychomotor seizure. These seizures were typically defined as a minimal clonic phase followed by stereotyped automatistic behaviors, which were similar to the aura of human patients with partial seizures. Mice not displaying this behavior were assigned as protected.

14.11 SUBCUTANEOUS BICUCULLINE AND PICROTOXIN ASSAY FOR MICE[13]

GABAA receptor (GABAAR) antagonist bicuculline (BIC) and chloride-channel blocker picrotoxin (PIC) potentially induced convulsant seizures and were used to further characterize the anticonvulsant profile of lacosamide. Groups of eight male albino CF No. 1 mice (18–25 g) were maintained in temperature- and humidity-controlled AAALAC-approved quarters with a 12-h on/12-h off light cycle, permitted access to food and water *ad libitum*, and pretreated intraperitoneally with lacosamide. At 60 min after pretreatment, BIC and PIC dissolved in 0.9% saline were injected subcutaneously in a volume of 0.01 mL/g body weight in the mice at doses of 2.7 mg/kg and 2.5 mg/kg for BIC and PIC, respectively. Mice that received BIC and PIC were observed for at least 30 min and 45–60 min, respectively. Mice not displaying a 3-s clonic episode during these observation periods were assigned as protected.

14.12 NMDA-INDUCED CONVULSIONS ASSAY FOR MICE[13]

This assay was based on the fact that in the CNS NMDA receptor, activation contributed to many aspects of neuronal signaling and excitability; NMDA receptor antagonists blocked or delayed seizure activity in rats and mice, and thus NMDA receptor activation has a role in epileptogenesis. In NMDA seizure assay, intracerebroventricular (icv) injection of NMDA was involved. Groups of 12 male Rj : NMRI mice (21–27 g) were maintained in temperature- and humidity-controlled AAALAC-approved quarters with a 12-h on/12-h off light cycle, permitted access to food and water *ad libitum*, 30 min prior to receiving either NMDA (6 μg/mouse) or saline administered icv, received either saline vehicle, a dose of lacosamide, or the NMDA antagonist MK-801 maleate (0.125, 0.25, or 0.5 mg/kg, ip). On the basis of pilot experiments, which established a threshold for clear but not excessive convulsive activity, the dose of NMDA was chosen. In all assays, mice were treated blindly, and mice received both ip and icv injections to ensure that all mice were treated similarly. When NMDA injection was not indicated, the mice received saline icv.

14.13 AGS ASSAY[13]

The ability to protect genetically bred mice against audiogenic seizures was considered as a feature of many AEDs active in the CNS. This assay was used primarily to further differentiate the anticonvulsant profile of a drug. Groups of eight male and female Frings AGS-susceptible mice (18–25 g) were maintained in temperature- and humidity-controlled AAALAC-approved quarters with a 12-h on/12-h off light cycle, permitted access to food and water *ad libitum*, individually placed into a Plexiglass cylinder (diameter 15 cm, height 18 cm) fitted with an autotransducer, and exposed to a sound stimulus of 110 dB (11 kHz) delivered for 20 s at 30 min after receiving drug or test compounds, and were characterized by Wild running followed by generalized tonic-clonic seizures (GTCS), with loss of righting reflex accompanied by forelimb and hindlimb extension defined as the characteristics of sound-induced seizures. Mice not displaying hindlimb tonic extension were considered protected.

14.14 KINDLED RAT ASSAY FOR FOCAL SEIZURES[13]

As a suitable system for evaluating the antiepileptic (versus anticonvulsant) potential of a drug or a test compound, in the kindled seizure assay seizures were stimulated by an initially subconvulsive electrical or chemical stimulus that culminated in a generalized seizure. Reducing seizure score from 5 (maximum) to 3 without any effect on after-discharge duration was considered as indicative of a drug that may be useful against secondary generalized seizures but not against focal seizures. Greater decreases in seizure score (e.g., 5 to <1) with reductions in after-discharge were considered potential for efficacy against focal seizures. Groups of eight adult male SD rats (275–300 g) were maintained in temperature- and humidity-controlled AAALAC-approved quarters with a 12-h on/12-h off light cycle, permitted access to food and water *ad libitum*, and under ketamine-xylazine anesthesia into the ventral hippocampus of rats, a bipolar electrode was stereotaxically placed (−3.6, ML −4.9, VD −5.0 from dura, incisor bar +5.0). After a 1 week recovery period, using a stimulus consisting of a 50-Hz and 10-s train of 1-ms biphasic 200-μA pulses delivered at 30-min intervals for 6 h (12 stimulations per day) on alternating days for a total of 60 stimulations, rats were kindled to stage 5 behavioral seizure. After a 1-week stimulus-free period, assay was started. On each day of a trial, rats received 2–3 suprathreshold stimulations delivered at 30-min intervals prior to lacosamide treatment to assess the stability of the behavioral seizure stage and after-discharge duration.

Fifteen minutes prior to seizure induction, lacosamide was administered intraperitoneally at 7 mg/kg, 13 mg/kg, 19 mg/kg, and 25 mg/kg. After 15 min, each rat was stimulated at 30-min intervals for 3–4 h. The mean seizure score and after-discharge duration were calculated. Rats were used for multiple assays, and between two assays there would be four or five drug-free and stimulus-free days as rat recovery period.

14.15 COBALT/HOMOCYSTEINE ASSAY FOR STATUS EPILEPTICUS OF RATS[13]

Focal motor seizures that will undergo secondary generalization can be induced in rats via brain lesions produced with cobalt, and suppression of these seizures in rats is considered to be useful to identify the drugs or test compounds having clinical efficacy for the acute treatment of status epilepticus. Groups of 10 adult male SD rats (275–300 g) were maintained in temperature- and humidity-controlled AAALAC-approved quarters with a 12-h on/12-h off light cycle, permitted access to food and water *ad libitum*, and via surgical application of cobalt powder (25 mg) on the left frontal cortical surface of these rats, epileptogenic lesions of the motor cortex were created. In a square grid (5.0 mm) with the left frontal electrode over the cobalt lesions, four epidural electrodes were arranged, and the EEG was monitored daily beginning 4–5 days after surgery. Focal cobalt was epileptogenic and on average 7–9 days after cobalt exposure caused seizures. As soon as focal motor seizures occurred and accompanied by appropriate epileptiform EEG activity, 5.5 mmol/kg homocysteine thiolactone was administered intraperitoneally to induce status epilepticus. Two independent assays were performed in different groups of rats.

1. *Treatment of Status Epilepticus*: Solutions of five different concentrations (1.25 mg/mL, 2.5 mg/mL, 5.0 mg/mL, 10.0 mg/mL, and 12.5 mg/mL) of lacosamide in NS with 0.5% benzyl alcohol as a bacteriostat were prepared and stored in vials labeled in a blinded manner and assigned by a predetermined random schedule. Immediately after the second generalized tonic-clonic seizures following homocysteine injection, treatment was given at 8.0 mL/kg body weight ip, rats were observed for 30 min, and all evidence of seizure activity was documented.

2. *Diazepam as Adjunct Therapy in Status Epilepticus*: The synergism efficacy of low-dose lacosamide in combination with low-dose diazepam was assayed. In one assay, 5 min after receiving 0, 1.25, 2.5, 5, or 10 mg/kg lacosamide, diazepam (0.75 mg/mL in 40% propylene glycol, 10% ethanol, 50% normal saline) was administered intraperitoneally at a dose of 0.75 mg/kg, and each rat was observed for 120 min or until it had two additional GTCS. In a second assay, 12.5 mg/mL solution of lacosamide in NS containing 0.5% benzylalcohol was prepared and each rat received 12.5 mg/kg lacosamide immediately after the second GTCS following homocysteine injection, whereas diazepam was prepared in concentrations of 0.25 mg/kg, 0.5 mg/kg, and 1.0 mg/kg in 40% propylene glycol, 10% ethanol, and 50% NS and stored in vials labeled in a blinded manner, and treatment was administered at 1.0 mL/kg body weight in random order. Rats were observed

either for 90 min or until the additional GTCS occurred, during which EEG was continuous recorded.

14.16 PTZ-INDUCED KINDLING ASSAY[14,15]

Experimental A2AR WT and A2AR KO mice (20–40), the first generation resulting from the syngenic cross between founders, were housed by groups of 20 in Makrolon cages (38 × 24 × 18 cm) with free access to food and water, and were kept in a ventilated room at 21 ± 1°C under a 12-h light/12-h dark cycle for at least 1 week of habituation to the facilities before assays. Prior to assay, the mice were isolated in individual test cages (27 × 13 × 13 cm) for 30 min. All assays were performed between 9 a.m. and 7 p.m.

In the assay, 3 days before the first administration of PTZ, mice received test compound orally in the drinking water. Test compound administration was maintained throughout assay course. For PTZ-induced kindling, initially subconvulsive doses of PTZ were injected intraperitoneally daily 30 mg/kg for 3 days, 35 mg/kg for 3 days, 40 mg/kg for 14 days, and 50 mg/kg on day 21. The development of kindling was directly proportional to the dose, the repeated administration of subconvulsant doses of PTZ induced a gradual increase in CNS excitability, and fully kindled mice developed seizures in response to a dose of PTZ that was initially subconvulsant. After each injection, the behavior was observed at 5-min intervals for a total period of 30 min every day. During each period of observation, each mouse that did not experience spontaneous convulsion was picked up by the tail, and if no seizure occurred was gently handled. Seizure score was assessed every 5 min with scale 0, 1, 2, 3, 4, and 5 for no seizure after tail lift or handling, no seizure after tail lift but facial clonus after handling, clonic seizures after handling, clonic seizures after tail lift, spontaneous clonic seizures, and spontaneous tonic-clonic seizures with loss of righting reflex, respectively. Mice experiencing a seizure scored 5 were given an undisturbed 10 min and scored 5 twice. Mice that died as a result of seizure activity within the observation period were scored 30 on the day of death. The occurrence of spontaneous clonic motor seizures on three consecutive days was defined as a kindled state.

14.17 ICES ASSAY FOR MICE[16]

Groups of 7 or 10 mice were used in assays where electroshock was applied via ear electrodes (forceps style) using an electric stimulator. The electroshock consisted of a single train of pulses (square wave, 5 ms, 20 Hz), and the intensity was increased linearly (initial current of 5 mA, increment of 0.1 mA/0.1 s). The current producing tonic hindlimb extension was recorded and assigned as the seizure threshold current (STC) for each mouse. If a 30-mA current produced no tonic hindlimb extension, electroshock was terminated and this cutoff current was used in the analysis. Using the Kruskal–Wallis H-test, the statistical significance of differences between control and treatment groups was determined, multiple comparisons were performed by the method of Dunnett, and differences were considered significant at the 5% level. Using the percentage changes in STC against each control group, dose-response curves for anti convulsant and proconvulsant compounds were constructed.

REFERENCES AND NOTES

1. L.L. Orefice, S.C. Heinrichs. Paternal care paradoxically increases offspring seizure susceptibility in the El mouse model of epilepsy. Epilepsy Behav 12 (**2008**) 234–241.
2. A. Richman, S.C. Heinrichs. Seizure prophylaxis in an animal model of epilepsy by dietary fluoxetine supplementation. Epilepsy Res 74 (**2007**) 19–27.
3. P.A. Forcelli, L.L. Orefice, S.C. Heinrichs. Neural, endocrine and electroencephalographic hyperreactivity to human contact: a diathesis-stress model of seizure susceptibility in El mice. Brain Res 1144 (**2007**) 248–256.
4. L.H. Turner, C.E. Lim, S.C. Heinrichs. Antisocial and seizure susceptibility phenotypes in an animal model of epilepsy are normalized by impairment of brain corticotropin-releasing factor. Epilepsy Behav 10 (**2007**) 8–15.
5. S. Yin, Z. Guan, Y. Tang, J. Zhao, J. Hong, W. Zhang. Abnormal expression of epilepsy-related gene ERG1/NSF in the spontaneous recurrent seizure rats with spatial learning memory deficits induced by kainic acid. Brain Res 1053 (**2005**) 195–202.
6. S.N. Mandhane, K. Aavula, T. Rajamannar. Timed pentylenetetrazol infusion test: a comparative analysis with s.c.PTZ and MES models of anticonvulsant screening in mice. Seizure 16 (**2007**) 636–644. *Note*: Mice were administered 85 mg/kg PTZ subcutaneously into a loose fold of skin of the neck between two shoulder blades and over the course of 60 min observed for appearance of clonus and tonic extensor. During this period, mice not displaying seizures were assigned an arbitrary cutoff time latency of 30 min for calculating mean onset latency for clonus. In the assays designed to determine intraindividual and interindividual variability in the intravenous and subcutaneous PTZ-induced seizure responses, PTZ was administered at 48 h intervals.
7. J.J. Luszczki, S.J. Czuczwar. Isobolographic characterization of interactions between vigabatrin and tiagabine in two experimental models of epilepsy. Prog Neuropsychopharmacol Biol Psychiatry 31 (**2007**) 529–538.
8. A. Cuadrado, J.A. Armijo. Beneficial interaction between vigabatrin and valproate against seizures induced by pentylenetetrazole in mice. Pharmacol Res 51 (**2005**) 489–496. *Note*: Neurotoxicity was assessed by the rotarod assay. Before assay, mice were placed on a 3-cm rod rotating at 6 rpm for a 10-min training session and a 5-min training session. After anticonvulsant administration, mice were again assayed on the rotarod. The end point of a minimal neurotoxicity assessment was the inability of the mice to maintain their equilibrium for 1 min in each of three trials.
9. J.J. Luszczki, K. Glowniak, S.J. Czuczwar. Time course and dose response relationships of imperatorin in the mouse maximal electroshock seizure threshold model. Neurosci Res 59 (**2007**) 18–22.
10. W. Löscher, M. Fiedler. The role of technical, biological, and pharmacological factors in the laboratory evaluation of anticonvulsant drugs. VII. Seasonal influences on anticonvulsant drug actions in mouse models of generalized seizures. Epilepsy Res 38 (**2000**) 231–248.

11. J.J. Luszczkia, K.M. Sawicka, J. Kozinska, K.K. Borowicz, S.J. Czuczwar. Furosemide potentiates the anticonvulsant action of valproate in the mouse maximal electroshock seizure model. Epilepsy Res 76 (**2007**) 66–72.

12. J.J. Luszczki, P. Czuczwar, A. Cioczek-Czuczwar, S.J. Czuczwar. Arachidonyl-2′-chloroethylamide, a highly selective cannabinoid CB1 receptor agonist, enhances the anticonvulsant action of valproate in the mouse maximal electroshock-induced seizure model. Eur J Pharmacol 547 (**2006**) 65–74.

13. T. Stőhr, H.J. Kupferberg, J.P. Stables, D. Choi, R.H. Harris, H. Kohn, N. Walton, H.S. White. Lacosamide, a novel anti-convulsant drug, shows efficacy with a wide safety margin in rodent models for epilepsy. Epilepsy Res 74 (**2007**) 147–154.

14. M. El Yacoubi, C. Ledent, M. Parmentier, J. Costentin, J.M. Vaugeois. Evidence for the involvement of the adenosine A2A receptor in the lowered susceptibility to pentylenetetrazol-induced seizures produced in mice by long-term treatment with caffeine. Neuropharmacology 55 (**2008**) 35–40.

15. C. Höcht, A. Lazarowski, N.N. Gonzalez, J. Auzmendi, J.A.W. Opezzo, G.F. Bramuglia, C.A. Taira, E. Girardi. Nimodipine restores the altered hippocampal phenytoin pharmacokinetics in a refractory epileptic model. Neurosci Lett 413 (**2007**) 168–172. *Note*: Male Wistar rats (280–330 g) were injected with a solution of 3-mercaptopropionic acid (MP; 45 mg/kg), a convulsant drug, for 10 days to induce repeated seizures. Five minutes to 10 min after injection, MP resulted in onset of seizure episodes characterized by excitation with sudden running fits and seizures. Rats receiving saline solution were used as control group.

16. Y. Kitano, C. Usui, K. Takasuna, M. Hirohashi, M. Nomura. Increasing-current electroshock seizure test: a new method for assessment of anti- and pro-convulsant activities of drugs in mice. JPM 35 (**1996**) 25–29.

15

METHODS AND APPLICATIONS OF DIABETES ASSAYS

Ming Zhao, Shiqi Peng, and Guohui Cui

Diabetic nephropathy characterized by a progressive accumulation of extracellular matrix components in the glomerular mesangium and tubular interstitium is one of the major complications associated with type 2 diabetes and is becoming the most frequent single cause of end-stage renal disease in many developed countries. The underlying mechanisms of diabetic nephropathy evolution are extremely complex, and several growth factors or metabolic products such as TGF-β, IGF-I, platelet-derived growth factor, angiotensin II, and advanced glycation end products have been known as contributing factors involved in the progression of diabetic glomerulopathy. Although both type 1 and type 2 diabetes mellitus (DM) can lead to diabetic nephropathy, type 1 DM is an autoimmune disease characterized by the destruction of the insulin-producing β-cells of the islets of Langerhans, and type 2 DM is a multifactorial disease characterized by insulin resistance with a relative impairment in insulin secretion. Long-term hyperglycemia associates with both type 1 and type 2 DM, and the functional and structural lesions in the lens, retina, kidney, and peripheral nerve are slowly developed, but the mechanisms responsible for all the diverse pathologic changes in these tissues are not completely understood. Because the prognosis of diabetic patients is still poor, much effort has been made to treat the complication and to evaluate new therapy using appropriate assays. In this chapter, 10 assays are described: islet xenograft assay for diabetic mice,[1] assays for spontaneous diabetes and adoptive transfer of diabetes,[2] oral glucose tolerance assays for patients,[3]

Pharmaceutical Bioassays: Methods and Applications. By Shiqi Peng and Ming Zhao

injection glucose tolerance assays for rats,[4] renal cortical TGF-β1 protein assays for rats,[4] low-dose streptozotocin-treated heminephrectomized rat assay,[5–7] rat early diabetic nephropathy assay,[8] urinary endothelin-1 excretion assay for type 2 diabetes rats,[9] subtotally nephrectomized rat assay,[10,11] and type 2 diabetes mice assay.[11,12]

15.1 ISLET XENOGRAFT ASSAY FOR DIABETIC MICE[1]

Under the kidney capsule of streptozotocin-induced diabetic C57B1/6 mice or spontaneously diabetic non-obese diabetic (NOD) mice, 2000 porcine islets were transplanted, and graft survival was monitored using blood glucose (BG) measurement. Islet recipients were treated ip with 200 μg of anti-OX40L or 250 μg of CTLA4Ig 3 days a week starting on day 0 or 2 days a week starting on day X, respectively, 0.5×10^6 spleens and pancreatic lymph node cells from diabetic female NOD donors were adoptively transferred into NOD-scid mice. Mice were treated ip with 500 μg of anti-OX40L or 250 μg of CTLA4Ig alone or in combination 3 times a week for 2 weeks. Treg phenotyping was performed after staining CD4, CD25, and FoxP3.

15.2 ASSAYS FOR SPONTANEOUS DIABETES AND ADOPTIVE TRANSFER OF DIABETES[2]

Female NOD/MrKTacfBR (NOD), NOD.NON-Thy1^a (Bar Harbor, ME, USA) and NOD-scid mice were maintained in a pathogen-free animal facility, blood samples were obtained from mouse tail vein, and weekly blood glucose measurements were determined using blood glucose monitor Accu-chek Easy, and the incidence of spontaneous diabetes development in female NOD mice in this facility was 60%. C3.5 T-cell clone specific for a peptide representing residues 437–460 (VLGGGCALLRCIPALDSLTPANED) of hsp60 was generated. Briefly, at the base of the tail, an NOD mouse was immunized with hsp60 peptide 437–460 (200 μg in CFA), and 7 days later the draining lymph nodes and spleen were harvested. After several restimulation rounds, the generated T-cell line was cloned by limiting dilution. The specificity of C3.5 clone was determined using antigenic and irrelevant peptides related proliferation assays. C3.5 clone phenotype in terms of CD4/CD8 and TCR Vβ chain expression was determined with anti-Vβ-specific mAbs on flow cytometry. C3.5 clone cytokine secretion pattern was determined using ELISA for IFN-γ and IL-10 and assays for IL-2, IL-4, and GM-CSF.

1. *Spontaneous Diabetes*: With irradiated syngeneic APCs and 10 μg/mL antigen, C3.5 T cells were activated for 72 h, purified by centrifugation over Ficoll-Hypaque, washed three times in cold PBS, and 10^7 T cells in 100 μL of PBS were injected iv into young prediabetic 5-week-old NOD mice. As a control, a group of mice were injected with PBS alone. By weekly blood glucose measurements, the mice were monitored for developing spontaneous diabetes; mice on two consecutive weeks having a >200 mg/dL blood glucose were considered as diabetic ones, which

always exhibited increasing blood glucose level with eventual overt glycosuria and death.

2. *Adoptive Transfer of Diabetes*: Eight-week-old NOD mice were irradiated (750 rads) for 1 day, injected iv with either 10^7 activated C3.5 T cells, 10^7 pooled splenic cells from already diabetic mice (diabetogenic spleen) alone or with 10^7 activated C3.5 T cells in 100 μL of PBS, whereas the control mice were irradiated and received PBS alone, and mice were monitored for diabetes induction for 8 weeks. In some assays, 7-week-old NOD-scid or NOD.NON-Thy1a mice were injected with 10^7 diabetogenic spleen cells alone, 10^7 C3.5 cells, or 10^7 C3.5 cells alone, or 10^7 diabetogenic spleen cells together, and blood glucose was weekly measured for 6–10 weeks for development of diabetes.

15.3 ORAL GLUCOSE TOLERANCE ASSAYS FOR PATIENTS[3]

After overnight fast, 7 male and 5 female type 2 diabetic patients [53.6 $\pm$ 2.5 years old, 81.9 $\pm$ 3.6 kg (BW), 30.5 $\pm$ 1.1 kg/m^2 body mass index (BMI)] and 5 male and 5 female normal glucose tolerant subjects (39.6 $\pm$ 3.7 years old, 90.1 $\pm$ 3.9 kg BW, 31.1 $\pm$ 0.9 kg/m^2 BMI), which matched for BMI and were used as control subjects, underwent a double-tracer oral glucose tolerance assay consisting of a primed-constant infusion of [3-^{3}H]glucose (0.25 μCi/min) starting at time −120 and −180 min in normal and diabetic subjects, respectively, and an oral glucose load at time 0 of 75 g glucose diluted in water containing 75 μCi [1-^{14}C]glucose. Blood samples were collected starting at time −30 min until 240 min at 15-min intervals for determining plasma concentrations of unlabeled and labeled glucose and insulin. The first three samples ($t \leq 0$) were used for calculating basal concentrations (mean value) and coefficients of measurement variations at basal (100 times the standard deviation of the three values divided by their mean). Model fitting was performed on data from samples taken at positive times. Concentration (activity) measurements were expressed as follows: glucose (mg/dL), [3-^{3}H]glucose (dpm/mL), [1-^{14}C]glucose (dpm/mL), and plasma insulin (μU/mL).

15.4 INJECTION GLUCOSE TOLERANCE ASSAYS FOR RATS[4]

Ten-week-old male OLETF rats and age-matched control LETO rats were individually housed and maintained on a 12/12-h light-dark cycle in a temperature (23 $\pm$ 2°C) and humidity (55 $\pm$ 5%) controlled vivarium, fed standard rat chow and allowed free access to tap water, and randomly treated with either lithospermic acid B (LAB, 20 mg/kg) or PBS. LAB and PBS were given orally by gavage (2 mL) once daily for 28 weeks. During the assay, the effects of LAB concentrations (1–20 mg kg^{-1} day^{-1}) on albuminuria in OLETF rats were compared. Body weights were measured monthly, and 24-h urinary albumin excretion amounts were measured 12 and 26 weeks after LAB treatment. By tail-cuff photoplethysmography, systolic blood pressure in conscious prewarmed rats was measured monthly. For each measurement, the average of five pressure readings was recorded. At 38 weeks of age, rats were fasted overnight. At 8:00 a.m. of the following morning, rats were given an intraperitoneal injection of glucose (2 g/kg), and blood samples were obtained from the tail vein to determine blood glucose levels using a glucometer.

15.5 RENAL CORTICAL TGF-β1 PROTEIN ASSAYS FOR RATS[4]

Ten-week-old male OLETF rats and age-matched control LETO rats were individually housed and maintained on a 12/12-h light-dark cycle in a temperature (23 ± 2°C) and humidity (55 ± 5%) controlled vivarium, fed standard rat chow and allowed free access to tap water, and randomly treated with either lithospermic acid B (LAB, 20 mg/kg) or PBS. LAB and PBS were given orally by gavage (2 mL) once daily for 28 weeks. During the assay, the effects of LAB concentrations (1–20 mg kg^{-1} day^{-1}) on albuminuria in OLETF rats were compared. Body weights were measured monthly, and 24-h urinary albumin excretion amounts were measured 12 and 26 weeks after LAB treatment. By tail-cuff photoplethysmography, systolic blood pressure in conscious prewarmed rats was measured monthly. For each measurement, the average of five pressure readings was recorded. At 38 weeks of age, rats were sacrificed, kidney was homogenized, at room temperature treated with HCl (0.2 M final concentration) for 30 min and neutralized with equimolar NaOH to convert the TGF-β1 in the renal cortex homogenate into its active form. According to the manufacturer's instructions using TGF-β1 Emax Immuno-Assay System (Promega, Madison, WI), TGF-β1 amount was determined with a quantitative sandwich enzyme immunoassay.

15.6 LOW-DOSE STREPTOZOTOCIN-TREATED HEMINEPHRECTOMIZED RAT ASSAY[5–7]

Twenty male SD rats at the age of 8 weeks were anesthetized with diethyl ether, injected with streptozotocin (STZ, 40 mg/kg) into the right jugular vein, and 9 days later, blood glucose levels were checked to select 14 rats with 300–500 mg/dL blood glucose. After anesthesia with pentobarbital sodium 50 mg/kg ip, the skin was sterilized and the right kidney was surgically removed. Six rats receiving vehicle were similarly heminephrectomized. Two weeks after the nephrectomy, blood glucose levels were measured to select 14 rats with <200 mg/dL blood glucose and start assay (week 0). Seven rats (ST) were fed on normal chow and the other seven (ST + HF) on high-fat chow. Throughout the assay, all rats housed individually within metabolic cages in a temperature-controlled animal facility (24°C) with a 12-h light/dark cycle were provided with food and water *ad libitum*.

After allocation, food consumption was calculated, body weight was measured, and blood and 24-h urine samples were collected at 1-day, 1-week, and 5-week intervals, respectively, for 35 weeks. One week before each blood sampling, blood pressure was measured. With an automatic analyzer, blood samples were measured for hemoglobin A1c(HbA1c). By centrifugation, plasma was isolated to measure glucose (Glu), creatinine (Cr), urea nitrogen (BUN), total cholesterol (TC), HDL-cholesterol (HDL), and triglyceride (TG) levels with an automatic analyzer, and to measure the plasma insulin with an ELISA kit. At week 37, rats were anesthetized, sacrificed, and histologically analyzed. Urine samples were measured for creatinine (uCr), protein (uPro), and glucose (US), while the urinary albumin (UAE) level was measured with an ELISA kit.

15.7 RAT EARLY DIABETIC NEPHROPATHY ASSAY[8]

Male SD rats (250–400 g) were randomized into control, diabetic, and diabetic + pravastatin pretreated groups. Using a single intraperitoneal dose of STZ (65 mg/kg), diabetes mellitus was established. One week after STZ injection, blood glucose was measured from the tail vein, and if readings were more than 22 μmol/mL, the rats were deemed diabetic. By this technique, 90% of rats had diabetic induction. Rats in the control group received an intraperitoneal injection of a corresponding amount of normal saline (NS). Before harvesting, pravastatin was administered by gavage (0.4 mg/kg in dilution of 1 mL of NS) for 5 days, while control rats received a similar anesthetic and volume of NS by gavage for 5 days. Twelve days after initial intraperitoneal injection, all rats were harvested and maintained on a purified chow pellet diet and water *ad libitum* prior to the procedure.

Rats were anesthetized with inhalational halothane, kept supine for the duration of the experiment, and kept warm using an infrared heating lamp. Using a rectal temperature probe, core temperature was measured and maintained at 36.5°C to 37°C for the duration of the experiment. Using laser Doppler fluximetry, renal cortical blood flow was established. At 36.5°C to 37°C, rat core body temperatures were stabilized for 30 min, and through a laparotomy incision, probe was applied to the left kidney, which was held in a constant position for at least 120 s until a consistent waveform was visible on the monitor and an average value was derived. Tissue perfusion (Flux) was calculated according to the equation Flux = concentration of red cells × speed of red cells.

At the end of the experiment, via cardiac puncture blood was obtained for serum urea, creatinine, cholesterol, and triglycerides. By bladder puncture, the total urine volume per hour produced was measured, and a sample was analyzed for urinary creatinine and total urinary protein leakage. Glomerular filtration rate was calculated using the equation $Uc \times Vu/Pc$, where Uc, Vu, and Pc were urinary creatinine concentration, urine volume, and plasma creatinine concentration, respectively.

Snap-frozen renal samples were sectioned at 6 μm, stained with hematoxylin and eosin, and scored using a histologic scoring system. Evaluated parameters were interstitial edema, tubular edema, glomerular distortion, and inflammatory infiltrate. Scores 0, 1, and 2 were assigned as absent, present, and marked, respectively, with maximum 8 and minimum 0.

15.8 URINARY ENDOTHELIN-1 EXCRETION ASSAY FOR TYPE 2 DIABETES RATS[9]

Male OLETF and LETO rats were kept in a room under the specific pathogen-free environment with controlled temperature, humidity, and a 12-h light/dark cycle, housed two per cage and given free access to standard rat chow and water before and during the assay. After basal measurements, OLETF rats were randomly divided into vehicle (distilled water) alone group, enalapril maleate (5 mg/kg daily), 0.1 mg/kg (E)-*N*-[6-methoxy-5-(2-methoxyphenoxy)-2-(pyrimidin-2-yl)-pyrimidin-4-yl]-2-phenylethenesulfonamidate (YM598), and 1 mg/kg YM598 groups. YM598 was an orally active endothelin type A (ETA) receptor antagonist. More than 0.1 mg/kg of oral YM598 suppressed the pressor response of rats to exogenously administered endothelin-1 in a dose-dependent manner. Maximum antagonism was observed at 1 mg/kg YM598, and the effect lasted for 24 h. Enalapril maleate was dissolved in distilled water and administered once daily by gastric gavage. The treatment was initiated at 22 weeks of age, continued for the following 32 weeks, and at 22 weeks of age, diabetes mellitus was already established. LETO rats ($n = 10$) were used as a control and received vehicle alone. Agent dose was adjusted by body weight measured twice a week. Endothelin-1 was measured before and during the treatment period at 8-week intervals. To determine urinary excretions of endothelin-1, urine was collected for 4 h in metabolic cages after water loading (2% of body weight) by gastric gavage.

15.9 SUBTOTALLY NEPHRECTOMIZED RAT ASSAY[10,11]

After several days of adaptation, male Wistar rats (about 150 g) in one group underwent the resection of half of the left kidney and total excision of the right kidney at 10- to 12-day intervals under intraperitoneal 30 mg/kg sodium pentobarbital anesthesia and were injected intraperitoneally with 35 mg/kg streptozotocin in citrate buffer (10 mM, pH 4.5). The rats in the other group underwent sham treatments. On day 80 after the streptozotocin injection, urine was collected for 24 h, rats were killed by decapitation, blood was collected into a conical centrifuge tube, and the serum was separated immediately by centrifugation. The kidneys were removed quickly, fixed in Bouin's solution, embedded in paraffin, cut into semithin sections 2 μm in thickness, stained with hematoxylin and eosin, periodic acid–methenamine silver (PAM), periodic acid–Schiff (PAS), or phosphotungstic acid–hematoxylin (PTAH) and examined by light microscopy. Serum glucose, insulin, creatinine (Cr), triglyceride, and total cholesterol were determined using commercially available kits Glucose BTest, Insulin-EIA Test, Triglyceride E-Test, and Cholesterol E-Test.

15.10 TYPE 2 DIABETES MICE ASSAY[11,12]

Male ob/ob mice (about 55 g) and db/db mice (about 37 g) were maintained on a 12/12-h light/dark cycle, fed standard rodent diet *ad libitum* and had unlimited access to water. Acute assays were used for determining the magnitude of blood glucose changes in response to subcutaneous 5-aminoimidazole-4-carboxamide 1-β-ribofuranoside (AICAR) in mice and for determining appropriate doses for chronic dosing. db/db mice were given subcutaneous 0.25–0.5 mg/g AICAR in sterile saline, whereas ob/ob mice were given between 0.375 and 0.5 mg/g. From tail nick samples using a Glucometer Elite Monitor, blood glucose levels were measured periodically (for up to 300 min). For chronic assays, mice received subcutaneous AICAR at approximately 8.00 a.m. and 5.00 p.m. for 8 days. Approximately 18 h after the preceding dose of AICAR, blood glucose was measured each morning. On the eighth day of dosing, mice were fasted for 5 h, blood was collected from the retro-orbital sinus for measuring serum cholesterol and triglyceride concentrations, mice were killed with CO_2, and liver and gastrocnemius/plantaris muscle samples were rapidly removed and frozen in liquid N_2.

On ice using a motor-driven stirrer at 1200 rpm with a glass-on-glass homogenizer/tissue grinder, gastrocnemius/plantaris muscle samples were homogenized individually in 1 mL of 20 mM HEPES, 250 mM sucrose, 1 mM ethylenediamine tetraacetic acid (EDTA), pH 7.4 (HES), containing complete protease inhibitor cocktail until the suspension contained little particulate matter. After electrophoretic separation, proteins were transferred to 0.2-μm PVDF membranes in 25 mM Tris, 190 mM glycine, 20% methanol, 0.005% sodium dodecyl sulfate (SDS), blocked in 5% Carnation nonfat milk in TBS, and immunoblotted with rabbit polyclonal affinity-purified antibodies (each at 5 μg/mL, BioSource/Quality Controlled Biochemical) against Glut4 (491-509, acetyl-CEQEV KPSTELEYLGPDEND) or Glut1 (477–492, acetyl-CKTPEELFHPLGADSQV). In the detection, horseradish peroxidase–conjugated secondary antibodies were used, washed with TBS/0.3% Tween-20, and exposed to the reagents for enhanced chemiluminescence. Using a Molecular Dymanics Personal Densitometer SI and ImageQuant 5.0, the images were scanned and quantified, respectively.

Using a Tissue Tearor, aliquots of liver (150–200 mg) were homogenized in chloroform/methanol (2 : 1) and filtered through Whatman 0.45-μM filters, to the filtrate 1 mL of Milli-Q water was added, samples were vortexed vigorously, centrifuged at 4°C and 3000 rpm, the lower chloroform phase was transferred to a glass tube, solvent was evaporated under nitrogen, the residue remaining in the tube was suspended in 0.5% Triton X-100 by sonication, and this suspension was measured for triglycerides or cholesterol using sonicated 0.5% Triton X-100 as the diluent for the standard curves.

REFERENCES

1. M. Honkanen-Scott, J. Johnson, B. Hering, P. Bansal-Pakala. Blockade of OX40 signals enhance survival of xenoislet grafts in spontaneously diabetic NOD mice. Transplant Proc 40 (**2008**) 483–485.
2. M. Feili-Hariri, M.O. Frantz, P.A. Morel. Prevention of diabetes in the NOD mouse by a Th1 clone specific for a hsp60 peptide. J Autoimmun 14 (**2000**) 133–142.
3. K. Thomaseth, A. Pavan, R. Berria, L. Glass, R. DeFronzo, A. Gastaldelli. Model-based assessment of insulin sensitivity of glucose disposal and endogenous glucose production from double-tracer oral glucose tolerance test. Computer Methods Programs Biomed 89 (**2008**) 132–140.

4. E.S. Kang, G.T. Lee, B.S. Kim, C.H. Kim, G.H. Seo, S.J. Han, K.Y. Hur, C.W. Ahn, H. Ha, M. Jung, Y.S. Ahn, B.S. Cha, H.C. Lee. Lithospermic acid B ameliorates the development of diabetic nephropathy in OLETF rats. Eur J Pharmacol 579 (**2008**) 418–425.
5. M. Sugano, H. Yamato, T. Hayashi, H. Ochiai, J. Kakuchi, S. Goto, F. Nishijima, N. Iino, J.J. Kazama, T. Takeuchi, O. Mokuda, T. Ishikawa, R. Okazaki. High-fat diet in low-dose-streptozotocin-treated heminephre-ctomized rats induces all features of human type 2 diabetic nephropathy: a new rat model of diabetic nephropathy. Nutr Metab Cardiovasc Dis 16 (**2006**) 477–484.
6. M. Okumura, M. Imanishi, M. Okamura, M. Hosoi, N. Okada, Y. Konishi, T. Morikawa, K. Miura, T. Nakatani, S. Fujii. Role for thromboxane A2 from glomerular thrombi in nephropathy with type 2 diabetic rats. Life Sci 72 (**2003**) 2695–2705.
7. M. Wei, L. Ong, M.T. Smith, F.B. Ross, K. Schmid, A.J. Hoey, D. Burstow, L. Brown. The streptozotocin-diabetic rat as a model of the chronic complications of human diabetes. Heart Lung Circ 12 (**2003**) 44–50.
8. R.G. Casey, M. Joyce, G. Roche-Nagle, G. Chen, D. Bouchier-Hayes. Pravastatin modulates early diabetic nephropathy in an experimental model of diabetic renal disease. J Surg Res 123 (**2005**) 176–181.
9. K. Sugimoto, S. Tsuruoka, A. Fujimura. Renal protective effect of YM598, a selective endothelin ETA receptor antagonist, against diabetic nephropathy in OLETF rats. Eur J Pharmacol 450 (**2002**) 183–189.
10. T. Yokozawa, T. Nakagawa, K. Wakaki, F. Koizumi. Animal model of diabetic nephropathy. Exp Toxic Pathol 53 (**2001**) 359–363.
11. K.M. Elased, K.A. Gumaa, J.B. de Souza, H. Rahmoune, J.H.L. Playfair, T.W. Rademacher. Reversal of type 2 diabetes in mice by products of malaria parasites II. Role of inositol phosphoglycans (IPGs). Mol Genet Metab 73 (**2001**) 248–258.
12. A.E. Halseth, N.J. Ensor, T.A. White, S.A. Ross, E.A. Gulve. Acute and chronic treatment of ob/ob and db/db mice with AICAR decreases blood glucose concentrations. Biochem Biophys Res Commun 294 (**2002**) 798–805.

16

METHODS AND APPLICATIONS OF ASSAYS FOR TOXINS FROM MICROORGANISMS

Shiqi Peng

Some fusarium mycotoxins and the metabolites of filamentous fungi and molds occur naturally in a variety of feeds, foodstuffs, crops, vegetables, fruits, water, and plants. Ingestion of the contaminated items can produce health problems, such as reduced inhibiting growth and anorexia, inducing emesis, enhancing the proliferation of estrogen responsive tumor cells, promoting cervical cancer and causing premature initial breast development in humans. The common toxins verrucarin A, T2-toxin, roridin A, deoxynivalenol (DON), zearalenone (ZEN), fumonisin B1 (FB1), moniliformin (MON), microcystin-LR (MCLR), and the toxins from microorganisms are a threat to humans and animals worldwide. The lactose-utilizing yeast *Kluyveromyces marxianus* exhibits well-characterized β-galactosidase activity and has chromogenic substrates; the toxicants interfering with any one of the cellular functions required for induction and expression of the β-galactosidase gene are able to suppress β-galactosidase activity; DON and fumonisin B inhibit cell growth; the dehydrogenase involved in mitochondrial functions can be inhibited by the toxicants; in axenic culture, the mutant strain of the ciliate *Tetrahymena thermophila* that overproduces melanin precursor can be grown to high concentrations; and glutathione *S*-transferase (GST) and glutathione peroxidase (GPX) play essential roles in protecting plants from the hazardous effects of xenobiotics. These facts imply that detecting toxins

Pharmaceutical Bioassays: Methods and Applications. By Shiqi Peng and Ming Zhao

from microorganisms is important for the health of humans and animals, and numerous assays have been established. In this chapter, 21 assays are described: colorimetric yeast assay for trichothecene mycotoxins,[1] colorimetric cell proliferation assay,[2] yeast DEL assay,[3] NCCLS and EUCAST assays,[4] alcohol dehydrogenase-based colorimetric assays,[5] colorimetric assay for iron in yeast,[6] microplate redox assays of *Escherichia coli*,[7] ciliate *Tetrahymena thermophila* assay for trichothecene mycotoxins,[8,9] fluorescent dyes-based cell viability assay for Triton X-100 toxicity,[10] MTT assay for fusarium mycotoxins,[11,12] mortality and frass production assay for toxicity of bacterial strains,[13–17] cell toxicity assay,[18] dehydrogenase (Dhase) inhibition assay,[19] sediment toxicity assay,[20] toxicity assay of particle-associated arsenite and mercury,[21] genetic toxicity assay,[22] Microtox assays,[23] DNA piezoelectric biosensor assay,[24] protein phosphatase inhibition assay and enzyme-linked immunosorbent assay (ELISA) of microcystins,[25] *Lepidium sativum* assay for microcystin toxicity,[26–30] and antiproliferative assay for interleukin-4.[31]

16.1 COLORIMETRIC YEAST ASSAY FOR TRICHOTHECENE MYCOTOXINS[1]

Kluyveromyces marxianus GK1005 was routinely maintained and grown on 1% (w/v) yeast extract, 1% (w/v) bacteriological peptone, and 2% (w/v) glucose and solidified with 2% (w/v) agar when required. Cultures were prepared by adding a single-cell colony from an agar plate to 50 mL of liquid medium and incubated at 35°C and 200 rev/min in a rotary incubator for 16 h. Cell density was determined by measuring absorbance at 560 nm and calibrated by direct hemocytometer counts. One A_{560} unit corresponded to 1.1×10^9 cells. Mycotoxins-containing sample was dissolved in spectroscopic-grade methanol to make stock solutions (typically 0.1 mg/mL). Cetyl trimethyl ammonium bromide, polymyxin B sulfate, and polymyxin B nonapeptide were dissolved in water, filter sterilized, and kept no more than a day as stock solutions. 5-Bromo-4-chloro-3-indolyl-β-D-galactopyranoside (X-gal, Calbiochem Novabiochem, Beeston, Nottinghamshire, UK) was dissolved in dimethylformamide (DMF, 100 mg/mL) and stored at −20°C in the dark; before each assay it was immediately diluted in aqueous DMF (2 parts water/3 parts DMF) to make a working solution (20 mg/mL), excess of which was discarded after each experiment. Before use, *o*-nitrophenyl-β-D-galactopyranoside (ONPGal, Sigma, St. Louis, MO) was immediately dissolved in water (4 mg/mL) and discarded. Prior to use, MTT (Sigma, St. Louis, MO) was dissolved in PBS and filter sterilized to prepare a 5 mg/mL stock solution.

In the wells of a microtiter plate (sterile, flat-bottomed), 136 μL of growth medium consisting of 1% (w/v) yeast extract, 1% (w/v) bacteriological peptone, and 50 mM glucose (YPD-50) were mixed with polymyxin B sulfate to give final assay concentration of 15 μg/mL, to which 8 μL of stock solution of mycotoxin-containing sample or methanol (controls) and 16 μL of yeast inoculum were successively added to give an initial assay cell density of ca. 2×10^8 cells/mL. Blank wells contained 152 μL of medium and 8 μL of methanol. The plate was mixed, sealed with a plate sealer, and incubated at 35°C for the duration of the assay, during which cell density was regularly monitored at 560 nm. When the control (mycotoxin free) cultures reached stationary phase (after about 10 h, with an A_{560} of ca. 1.2), the cultures were assayed for β-galactosidase or mitochondrial (MTT-cleavage) activity.

To each well of the microtiter plate, 5 μL of SDS (0.1%, w/v) and 3 μL of chloroform were added to permeabilize the cells. For *in vivo* experiments and experiments examining the effects of different carbon sources, 1 μL of 100 mg X-gal/mL DMF was added to each well. For examining the effects of altering glucose concentration and inoculum cell density, 5 μL of 20 mg X-gal/mL DMF was added to each well. For methanol and ethanol toxicity experiments and the standardized bioassay, 8 μL of 20 mg X-gal/mL aqueous DMF was added. The contents in the wells were mixed, and the plates were incubated at 35°C in a plate shaker for a maximum of 30 min. Using a test filter at 666 nm and a reference filter at 560 nm, the plates were read. β-Galactosidase activity was expressed as product formation ($A_{666} - A_{560}$), as a function of cell density (A_{560}) as well.

After addition of 16 μL of MTT in PBS to each well and statical incubation at 35°C for 4 h, the medium was displaced by 200 μL of DMSO. Thorough mixing and dissolution, the MTT cleavage product was affected by repeated pipetting. Using a test filter at 560 nm and a reference filter at 666 nm, the plates were read.

Into 50 mL of solution consisting of 1% yeast extract, 1% bacteriological peptone, and 50 mM lactose, a single colony of *Kluyveromyces marxianus* GK1005 was inoculated and incubated at 35°C and 200 rev/min in an orbital shaker for 16 h. Into each of two universal bottles, 10 mL of samples of the culture were transferred, and 0.2 mL of chloroform and 0.1 mL of SDS (0.1% w/v) were added. To permeabilize the cells, to one bottle 20 μL of X-gal (100 mg/mL in DMF) was added, and to the other bottle 20 μL of DMF was added. In an orbital shaker both bottles were incubated at 35°C and 200 rev/min until indigo precipitates were clearly visible in the bottle containing X-gal. Using the solvent (DMF) control as a reference, the absorption spectrum of the X-gal-containing sample was then determined.

16.2 COLORIMETRIC CELL PROLIFERATION ASSAY[2]

Batches of medium (5 mL in a glass tube) were inoculated from fresh culture plates and at 25°C to 37°C incubated for 18 h. Culture mediums YM broth, tryptic soy broth plus yeast extract, and Man-Rogasa-Sharpe (MRS) medium were used for yeast, bacteria except lactic acid bacteria, and lactic acid bacteria, respectively. The current method mainly followed the reaction mechanism. First, the electron mediator quinone was reduced to hydroquinone by microbial cells. Consequently, the hydroquinone reduced tetrazolium salts extracellularly to formazans having absorbance maxima ranging from 438 to 550 nm and can be quantified spectrophotometrically. The cultivated microorganisms were diluted with the medium containing 0.25% yeast extract, 0.5% peptone, and 0.1% glucose (pH 7.0) to adjust the microbial cell density. To each well (96-well microtiter plate), 190 μL of microbial suspension and 10 μL of detection reagent consisting of an electron mediator and tetrazolium salt were added and incubated at 30°C or 37°C. The formation of formazan was measured on a microplate reader at 438, 550, 550, 460, 495, and 470 nm for 2-(4-iodophenyl)-3-(4-nitrophenyl)-5-(2,4-disulfophenyl)-2H-tetrazolium monosodium salt, 2-benzothiazolyl-3-(4-carboxy-2-methoxyphenyl)-5-[4-(2-sulfoethylcarbamoyl)phenyl]-2H-tetrazolium, 2,2′-dibenzothiazolyl-5,5′-bis[4-di(2-sulfoethyl)carbamoylphenyl]-3,3′-(3,3′-dimethoxy-4,4′-biphenylene)ditetrazolium disodium salt, 2-(2-methoxy-4-nitrophenyl)-3-(4-nitrophenyl)-5-(2,4-disulfophenyl)-2H-tetrazolium monosodium salt, 2-(4-nitrophenyl)-5-phenyl-3-[4-(4-sulfophenylazo)-2-sulfophenyl]-2H-tetrazolium monosodium salt, and 2,3-bis(2-methyloxy-4-nitro-5-sulfophenyl)-5-[(phenylamino)carbonyl]-2H-tetrazolium hydroxide (all from Dojindo, Kumamoto, Japan), respectively. After serial dilution, onto plastic plates (90 mm i.d.) microorganism cultures were plated with a standard agar containing 0.25% yeast extract, 0.5% peptone, 0.1% glucose, and 1.5% agar, incubated at 25°C to 37°C, and colonies were counted after 1 or 2 days.

Serial twofold dilutions of each antimicrobial agent in Mueller-Hinton broth were made within dilution schemes of 0.0625 to 64 μg/mL. Antimicrobial agents ciprofloxacin, cefotaxime, chloramphenicol, and gentamicin were used. With PBS, *Bacillus cereus* was adjusted to a turbidity equaled to that of 0.5 McFarland standard and diluted 10-fold. With antimicrobial agent solution, the suspension was further diluted to provide a final inoculum density of approximately 10^5 CFU/mL. After inoculating each well of a plate with 100 μL of inoculum, the plate was incubated at 35°C for 24 h and the MIC was read as the lowest concentration without visible growth for antimicrobial agent. In the susceptibility assays, the inoculum was prepared, and after 6 h of incubation at 35°C, to each well 5 μL of detection reagent was added. After 2 h incubation at 35°C, the produced formazans were measured at 460 nm with a microplate reader. MIC was read as the lowest concentration without absorbance change for antimicrobial agent.

16.3 YEAST DEL ASSAY[3]

For 384-well plate format of the yeast plate-based deletion (DEL) recombination assay, 1 μL of yeast (ca. 1×10^5 cells) was pipetted into eight microplate wells for each compound, among them four containing 70 μL of SC medium and four containing 70 μL of SC medium lacking histidine, each well was supplemented with 14 μL of [3-(4,5-dimethylthiazol-2-yl)-5-(3-carboxymethoxyphenyl)-2-(4-sulfophenyl)]-2H-tetrazolium (MTS, Sigma, St. Louis, MO) and 5 μL of compound. For 96-well plate format of DEL recombination assay, the above was consistent except media, MTS, and compound volumes were 100, 20, and 7 μL, respectively, whereas control wells were treated with water in lieu of compound. The outermost columns of the 96-well plate and outer two columns of each 384-well plate were excluded from the experiment lest edge evaporative effects alter the data. Plates were incubated at 30°C at normal atmosphere for yeast growth in the presence of the tested compound, and 490 nm absorbance was measured on a Molecular Devices SpectraMax M5 microplate reader 10–18 h later.

A mock experiment was performed to measure the sensitivity of 96-well plate format of the well-based DEL assay. Wells containing 1×10^5 RS112 yeast cells in

100 μL of SC medium lacking histidine were supplemented in six-plicate with 5 μL of RS112 His+ revertant cells ranging from 0 to 1000 cells. To each well, 20 μL of MTS was added, plates were incubated at 30°C, and 490 nm absorbance was read hourly between 12 and 24 h on a Molecular Devices SpectraMax M5 microplate reader.

16.4 NCCLS AND EUCAST ASSAYS[4]

Broth medium consisted of RPMI-1640 medium supplemented with glutamine and phenol red (Sigma-Aldrich, UK), without bicarbonate (10.4 g), 3-(*N*-morpholino)propanesulfonic acid (34.53 g in 400 mL of distilled water, Sigma-Aldrich, UK), and 1 M sodium hydroxide adjusting the pH to 7.0 at 25°C, and was diluted with additional water to a final volume of 500 mL, filter sterilized, and stored at 4°C until required. For the EUCAST assay, the medium was further supplemented with glucose (18 g) to achieve a final concentration of 2% glucose (w/v).

1. In South Africa National Committee for Clinical Laboratory Standards (NCCLS) assays of yeasts, *Candida parapsilosis*, *Candida krusei*, or *Candida albicans* cultured from frozen stocks and maintained at 37°C were subcultured once onto Sabouraud dextrose agar and at 37°C incubated for 24 h. By transferring several colonies to 5 mL of sterile distilled water, inocula were obtained. After mixing 15 s, the suspensions were diluted to match the turbidity of an 0.5 McFarland standard (i.e., OD = 0.12 to 0.15 at 530 nm, corresponding with 1×10^6 to 5×10^6 CFU/mL). By further dilutions in sterile distilled water, the working suspension (1×10^3 to 5×10^3 CFU/mL) was obtained. Assay's colorimetric readings employed resazurin as an indicator of cell growth, and the working suspension was supplemented with 0.1 mL of sterilized solution of resazurin (20 mg/mL in water). Stock solutions of test compounds (100 mg/mL) and itraconazole (1.6 mg/mL, positive control) in DMSO were prepared and further diluted (1 : 50) in broth for assays.

In flat-bottom 96-well clear plates containing broth medium (0.1 mL), in each well microdilution susceptibility was tested, and then 0.1 mL of sample solutions were serially diluted twofold in the plates with the broth, starting with the final concentration of 0.5 mg/mL for test compounds and 8 μg/mL for itraconazole. The working inoculum suspension (0.1 mL) was added to give a final inoculum concentration of 0.5×10^3 to 2.5×10^3 CFU/mL for the assay. Itraconazole was used as the standard antifungal drug. Sterility and growth controls in the presence of organic solvents employed in sample preparation were also included. The plates were incubated at 37°C for 48.

2. In EUCAST assay of yeasts, *Candida parapsilosis*, *Candida krusei*, or *Candida albicans* cultured from frozen stocks and maintained at 37°C were subcultured once onto Sabouraud dextrose agar and at 37°C incubated for 24 h. By transferring several colonies to 5 mL of sterile distilled water, inocula were obtained. After mixing 15 s, the suspensions were diluted to match the turbidity of an 0.5 McFarland standard (i.e., OD = 0.12 to 0.15 at 530 nm, corresponding with 1×10^6 to 5×10^6 CFU/mL). By further dilution in sterile distilled water, the working suspensions (1×10^5

to 5×10^5 CFU/mL) were obtained. The assay's colorimetric readings employed resazurin as an indicator of cell growth, and the working suspension was supplemented with 0.1 mL of sterilized solution of resazurin (20 mg/mL in water). Stock solutions of test compounds (100 mg/mL) and itraconazole (1.6 mg/mL, positive control) in DMSO were prepared and further diluted (1:50) in broth for assays.

In flat-bottom 96-well clear plates containing broth medium (0.1 mL), in each well microdilution susceptibility was tested, 0.1 mL of sample solutions were serially diluted twofold in the plates with the broth, starting with the final concentration of 0.5 mg/mL for test compounds and 8 μg/mL for itraconazole. The working inoculum suspension (0.1 mL) was added to give a final inoculum concentration of 0.5×10^5 to 2.5×10^5 and 0.5×10^3 to 2.5×10^3 CFU/mL for the assay. Itraconazole was used as the standard antifungal drug. Sterility and growth controls in the presence of organic solvents employed in sample preparation were also included. No inhibitory effects were observed in the presence of DMSO at the highest concentration used (0.5% v/v). The plates were incubated at 37°C for 24 h.

16.5 ALCOHOL DEHYDROGENASE-BASED COLORIMETRIC ASSAYS[5]

Using a 5-mL screw-capped test tube (flat bottom), 1 mL of sodium citrate buffer (2 mM, pH 6.2) containing 2 mM β-mercaptoethanol (Sigma-Aldrich, UK), and 1.5 g of glass beads (425–600 μm), and by six shaking periods of 30 s in a vortex, at intervals the cells (100 mg, dry weight) were disrupted and the tubes were kept in an ice bath during the resting periods. By centrifugation (10 min at 12,000 rpm in centrifuge Eppendorf 5415 R), the supernatant was separated and used for the alcohol dehydrogenase assays.

Enzymatic procedure for lactate dehydrogenase assay was modified for the current assay. Using blank cuvettes containing all the constituents of the reaction, except the ethanol (substrate of the enzymatic reaction), which was replaced by water, background reductions were eliminated from the assay. With 570 nm as test wavelength and 655 nm as reference wavelength, the developed coloration was measured and alcohol dehydrogenase activity was obtained by averaging two determinations. The optimum pH of alcohol dehydrogenase activity was defined by assaying the enzyme activity at 37°C and pH ranging from 4.0 to 10.0. By measuring the enzymatic activity in standard enzyme assay conditions after incubating the enzyme for 80 h at room temperature at pH ranging from 4.0 to 10.0, the pH stability of the enzyme was determined, in which 0.1 M acetate buffer (NaAc-HAc, Sigma, St. Louis, MO) was used for pH 4.0–5.0, 0.1 M phosphate buffer (NaH_2PO_4-Na_2HPO_4, Sigma, St. Louis, MO) was used for pH 6.0–7.0, and 0.1 M Tris buffer (Tris-HCl, Sigma, St. Louis, MO) was used for pH 8.0, 8.5, 9.0, and 10.0. By assaying enzyme activity at pH 8.5 in 0.1 M Tris-HCl buffer containing Triton X-100 1% at temperatures of 32°C, 37°C, 40°C, 45°C, 50°C, and 55°C, the optimum temperature of alcohol dehydrogenase activity was determined. By measuring the enzymatic activity in standard enzyme assay conditions after incubating the enzyme solution for 1 h at 0°C, 30°C, 40°C, 50°C, 60°C, and 70°C, the thermal stability of the enzyme was determined. Using production measures of the reduced derivative formazan of MTT (Sigma, St. Louis, MO), ethanol in samples of alcoholic beverages was measured. Using bovine serum albumin as the standard protein, total protein was assayed according to general method.

16.6 COLORIMETRIC ASSAY FOR IRON IN YEAST[6]

W303 and MML298 strains were grown on glucose-rich media, harvested during exponential growth, washed twice with ultrapure water, resuspended in 0.5 mL of nitric acid (3%), incubated at 98°C in 1.5 mL polypropylene tubes tightly capped for 16 h, and centrifuged in a tabletop centrifuge at 12,000 rpm for 5 min. After 5 min mixing, 400 μL of supernatant containing either standards or digested W303 and MML298 strains with 160 μL of sodium ascorbate (38 mg/mL), 320 μL of either bathophenanthrolinedisulfonic acid (Sigma, St. Louis, MO) or ferrozine (1.7 mg/mL, Sigma, St. Louis, MO), and 126 μL of ammonium acetate solution (saturated ammonium acetate diluted 1/3), the specific absorbance of the resulting iron-bathophenanthrolinedisulfonic acid complex was recorded at 535 nm or iron-ferrozine complex was recorded at 565 nm. The accuracy of both assays was improved by subtracting nonspecific absorbance recorded at 680 nm. Cell volumes were calculated in a Z2 particle count and size analyzer from Coulter (cell volume = number of cells × mean volume). To eliminate the contribution of contaminant iron, absorbance was recorded against blanks containing all the reagents.

16.7 MICROPLATE REDOX ASSAYS OF *E. coli*[7]

The mixture of 1 mL of 3TYG (30 g/L tryptone, 15 g/L yeast extract, and 3 g/L of glucose) and 0.25 mL of Alamar blue, 3-(4,5-dimethylthiazol-2-yl)-2-(4-sulfophenyl)-2H-tetrazolium salt, or 2-(2-methoxy-4-nitrophenyl)-3-(4-nitrophenyl)-5-(2,4-disulfophenyl)-2H-tetrazolium monosodium salt (Promega Co., WI) was defined as 3T/AB, 3T/MTS, or 3T/WST8, respectively. Overnight cultured *Escherichia coli* were diluted 10-fold to approximately 10^8 to 10^2 CFU/mL with PBS (pH 7.2), their 0.1 mL aliquots were placed in three wells in 96-well flat-bottom polystyrene microplates, while 0.1 mL of PBS was aliquoted as the reagent blank and 0.05 mL of 3T/AB, 3T/MTS, or 3T/WST8 was added. The 96-well plate was immediately placed in a model 550 microplate reader, and the optical density of the culture in each well of the 96-well plate was immediately assayed at 550–595 nm for Alamar blue and 490–655 nm for MTS and WST-8. The plate was incubated at 35°C to measure OD value at 30-min intervals for maximum. OD data from each well and corresponding times subtracted the OD of the blank reagent to give redox values (RV). The linear curves of RV versus time were analyzed by linear regression. Using the calculated slope and y intercept, the time achieving an OD of 50% of maximal value (T50%) was calculated for each dilution, plotted against the log CFU/well, and the relationship was analyzed by linear regression. By spread plate counts of suitable dilutions on heart infusion agar plates incubated at 35°C for 24 h, the number of CFU in each dilution was determined.

To each dilution (0.1 mL) of *Escherichia coli*, *Staphylococcus aureus*, and *Bacillus cereus* cultures in the three wells, 0.05 mL of 3T/AB containing 300 IU/mL penicillin and 0.3 mg/mL streptomycin was added. The plate was incubated at 35°C, and the OD was recorded at 15- or 30-min intervals for total 240 min.

Overnight cultured *Escherichia coli* and *Staphylococcus aureus* cultures were diluted 10-fold to approximately 10^8 to 10^2 CFU/mL with PBS (pH 7.2), and their 0.1 mL aliquots were placed in three wells in 96-well flat-bottom polystyrene microplates. The 18-h culture of *Staphylococcus aureus* was centrifuged at 1200 rpm for 5 min and washed with PBS before use. To the *Escherichia coli* dilutions, 0.05 mL of 3T/AB, of a solution of three times the concentration of AquaTest containing 5-bromo-4-chloro-3-indolyl-beta-D-galactopyranoside or of a solution of three times the concentration of PYP (phenol-red, yeast, peptone), medium containing 1.5% (w/v) lactose was added. In addition, 0.05 mL of 3T/AB, 3TYG containing 22.5% (w/v) NaCl, a solution of three times the concentration of PYP containing 1.5% mannitol (3P/Mat) or 3P/Mat containing 22.5% (w/v) NaCl was added to *Staphylococcus aureus* dilutions. The plate was incubated at 35°C, and the OD was recorded at 30-min intervals for maximum OD. The number of CFU in each dilution was determined as above.

16.8 CILIATE *Tetrahymena thermophila* ASSAY FOR TRICHOTHECENE MYCOTOXINS[8,9]

The phenotype characteristics of *Tetrahymena thermophila* strain BI3840 (University of California) were amicronucleated (amc), pigment producing (pig), resistant to 25 μg/mL of cycloheximide (cy-r), and mating type IV. The PP210 medium consisting of aqueous solution of proteose peptone (2%, w/v) was supplemented with 10 μM $FeCl_3$, 250 μg/mL streptomycin sulfate, and penicillin. Mycotoxins containing samples were dissolved in propylene glycol or acetonitrile to make stock solutions at 5 mg/mL. Patulin and T-2 were used as toxin.

The cells were grown axenically in PP210 medium for 24 h and maintained constantly at 28 ± 1°C. The exponential phase cellular suspensions were distributed in microtiter plates (100 μL cell culture/well). According to the required final concentration, different dilutions of the stock solutions of mycotoxins (Sigma, St. Louis, MO) containing samples in PP210 medium were added into the cultures (100 μL/well). To each 100 μL/well of the cell culture, 100 μL/well PP210 medium without mycotoxins sample was added to obtain one type of control and 100 μL/well PP210 medium plus the highest concentration of the corresponding solvent was added to obtain another type of control. In all cases, the microtiter plates were incubated at 28 ± 1°C for 48 h, and the wells with a dark-brown pigmentation were considered as the positive growth wells.

16.9 FLUORESCENT DYES-BASED CELL VIABILITY ASSAY FOR TRITON X-100 TOXICITY[10]

The fish cell lines RTL-W1 from the liver and RTgill-W1 from the gill of rainbow trout were cultured routinely in 75 cm^2 culture flasks at 20°C in Leibovitz's L-15 culture medium (Gibco BRL, Burlington, ON, Canada) supplemented with FBS (Sigma, St. Louis, MO, 10% for RTgill-W1 and 5% for RTL-W1) and 2% penicillin-streptomycin (100 μg/mL streptomycin, 100 IU/mL penicillin). Prior to exposure, cells were plated in 96-well tissue culture plates at a density of 5×10^4 cells per 200 μL of L-15 with FBS and grown to confluency. Mammalian cell line rat liver hepatoma cell line H4IIE cells were routinely cultured in 75 cm^2 vented culture flasks at 37°C in a humidified 5% CO_2–95% atmosphere with Dulbecco's medium supplemented with 10% FBS, 0.1% gentamicin sulfate (50 mg/mL), 2% L-glutamine, 2% MEM nonessential amino acids, 3% MEM amino acids, and 1.5% MEM vitamins. Human cell lines Caco-2 and HepG2 were routinely cultured in 75 cm^2 vented culture flasks at 37°C in a humidified 5% CO_2–95% atmosphere with Dulbecco's MEM modified with Earle's salts supplemented with 10% FBS, 0.1% gentamicin sulfate (50 mg/mL), 1% L-Gln, 2% MEM nonessential amino acids, and 0.2% sodium pyruvate (55 mg/mL). Prior to exposure, cells were plated in 96-well tissue culture plates at a density of 5×10^4 cells per 200 μL of L-15 with FBS and grown to confluency.

Approximately 2–3 mL of ciliate *Tetrahymena thermophila* cultured axenically in 10 mL of proteose peptone yeast extract medium (PPYE) was routinely transferred at 2- to 3-week intervals into 10 mL of fresh and sterile PPYE medium to keep a supply culture of *Tetrahymena thermophila* on hand. To prepare the testing cultures, 1 mL of stock culture was transferred into 10 mL of sterile PPYE and grown for 24 h. Into 50 mL of sterile PPYE in a 250-mL Erlenmeyer flask, 10 mL of culture was transferred, grown on orbital shaker at 50 rpm for 1–2 days, centrifuged, washed three times with spring water (pH 7.32, 235 ppm measured as calcium hardness and conductivity 840 μS/cm), centrifuged again, resuspended in 10 mL of spring water, counted using a Coulter Z2 particle counter and adjusted to 5×10^5 cells/mL (±10%) using spring water. By removing the culture medium from confluent cells, exposure was initiated. The cells were washed once with L-15/ex, into which 200 μL/well of the serial dilutions of Triton X-100 in L-15/ex was added, exposed at 20°C for the fish cell lines, and at 37°C for the mammalian cell lines. After 2 h exposure to Triton X-100, cell viability was assayed.

A stock solution of 1 mg/mL Triton X-100 in spring water was serially diluted with spring water to a range from 0 (control) to 400 mg/mL. *Tetrahymena thermophila* was exposed in microcentrifuge tubes, to which 1 mL of *Tetrahymena thermophila* in spring water (5×10^5 cells/mL) was added. After pelleting by centrifugation and removing supernatant by aspiration, the cells were resuspended in the microcentrifuge tube in 1 mL of the Triton X-100 exposure medium prepared in spring water and exposed for 2 h, spun down, and assayed for viability.

The medium with 100 μL of PI (10 μg/mL) in L-15/ex per well and the plates were incubated for 1 h at room temperature for the fish cells and at 37°C for the mammalian cells, after which fluorescence was measured. For all dyes, fluorescence was quantified as fluorescent units (FUs) with the CytoFluor Series 4000 microplate reader at the respective excitation and emission wavelengths of 530 and 595 nm for AB, 485 and 530 nm for CFDA-AM, 530 and 645 nm for NR, and 530 and 620 nm for PI.

Tetrahymena thermophila was exposed to Triton X-100 for 2 h, terminated by centrifugation and aspiration, and resuspended in working solutions of the fluorescent indicator dyes prepared as for the cell lines. In the microcentrifuge tubes, ciliates were at room temperature incubated in a combined AB and CFDA-AM solution for 30 min and in PI and in NR for 60 min. Into microculture wells of a 96-well plate, a 100 μL aliquot from each tube was transferred in six replicates, and fluorescence was quantified using the CytoFluor microwell plate reader.

16.10 MTT ASSAY FOR FUSARIUM MYCOTOXINS[11,12]

CHOK1, Caco-2, C5-O, and V79 cell lines were used at passage numbers between 30 and 50. HepG2 cell line was used at passage numbers between 80 and 100. Caco-2, C5-O, V79 cells (in DMEM, Life Technologies, Gibco BRL Products, Rockville, MD), CHO-K1 cells (in DMEM/F-12), and HepG2 cells (in MEM, Life Technologies, Gibco BRL Products, Rockville, MD) were grown as monolayers in 80 cm^2 culture flasks. The media were supplemented with 1.5 mg/mL sodium bicarbonate, 0.11 mg/mL sodium pyruvate, 1% nonessential amino acid, 25 mM HEPES, 100 units of penicillin/mL, 100 μg streptomycin/mL, 25 ng amphotericin B/mL, and 10% FBS (Life Technologies, Gibco BRL Products, Rockville, MD). The cell cultures were maintained in a humidified atmosphere of 5% CO_2 at 37°C. The cell lines were harvested when they reached 80% confluence to maintain exponential growth.

The cell monolayers in exponential growth were harvested using trypsin-EDTA (0.25% trypsin and 1 mM EDTA · 4Na) and after repeated pipetting the single cell suspensions were obtained. The single cell suspensions (cell densities ranging from 1×10^2 to 5×10^4 cells per 200 μL medium/well) were added to 96-well plates by serial dilution. The number of seeded cells was determined from the linear correlation between the number of seeded cells and the OD values. Caco-2, C5-O, HepG2 (at a cell density of 1×10^4 cells per 100 μL medium), CHO-K1, and V79 (at a cell density of 5×10^3 cells per 100 μL medium) were seeded to each well of the 96-well plates and incubated in a humidified atmosphere of 5% CO_2 at 37°C for 24 h. Mycotoxin-containing sample in 100 μL of medium was added from high to low concentrations to the wells and adjusted to final concentrations needed. MTT (Sigma, St. Louis, MO) was dissolved in PBS (5 mg/mL final concentration), filtered (0.22 μm filter) and stored in the amber vials at 4°C for a month. After 48 and 72 h incubation, 25 μL of MTT solution was added to each well of 96-well plates and incubated in a humidified atmosphere of 5% CO_2 at 37°C for 4 h. At the end of the incubation period, the media were discarded using a suction pump. The extraction buffer of 20% (w/v) SDS in a solution of 50% DMF in demineralized water (50:50, v/v) was prepared at pH 4.7 and filtered (0.22 μm filter). The extraction buffer in 100 μL of SDS (20%, w/v, Sigma, St. Louis, MO) was added into each well of the 96-well plates to solubilize formazan crystals. The culture plates were placed on an orbital shaker at 37°C overnight. The absorbance was measured at the test wavelength of 570 nm and the reference wavelength of 690 nm. The positive control contained an adjusted seeding cell number in log phase, of which the culture medium contained 0.1% ethanol.

16.11 MORTALITY AND FRASS PRODUCTION ASSAY FOR TOXICITY OF BACTERIAL STRAINS[13–17]

Bacillus thuringiensis subsp. *tenebrionis* and six unidentified strains of *Bacillus thuringiensis* labeled A30, A299, A311, A409, A410, and A429 (North American Technical Centre, Ontario, Canada) were grown on nutrient agar buffered with an equimolar concentration of KH_2PO_4 (50 mM, pH 7.0) at 30°C for 5 days. The bacterial culture consisted of vegetative cells, sporangia, spores, and crystals. These stages were lyophilized and stored at −20°C. The spore crystal suspensions of each strain with 1 : 1 in the spore : crystal ratio were used in the toxicity assays. The concentrations of extractable proteins were determined and adjusted to a given concentration of protein in the crystal spore mixture. Spore crystal mixtures were suspended in Triton X-100 solution (0.01%, v/v) and centrifuged at 22°C and 11,750 × *g* for 3 min to wash three times. To extract the proteins, the final pellet was resuspended in a putatively selective crystal solubilizing buffer (40.5 mM Na_2CO_3, 0.5 mM phenylmethylsulfonylfluoride, 0.1 mM dithiothreitol, pH 10.0) and solubilized at 42°C by incubating for 2 h and vortexing every 30 min. The suspension was at 22°C and 11,750 × *g* centrifuged for 3 min, and the supernatant was assayed for total protein five times per sample using a Bio-Rad protein assay kit (Bio-Rad, Ontario, Canada) with bovine serum albumin as the standard. The final protein solution was referred to as extractable protein. The laboratory cultures of plants were periodically supplemented with pest insects to maintain hybrid vigor. The colony was reared in an incubator at 24°C under light for 16 h and at 6°C in the darkness for 8 h. Abundant food was placed in plastic Petri dishes (100 mm diameter) with moist filter paper, to which 10 adults were added. Three replicate plates per treatment were placed in the incubators, and the number of frass pellets on filter paper was counted daily for 3 days. Adult mortality and modified frass production assay were used to measure the toxicity of the extractable proteins of the bacterial strains against plants. The modified frass production assay was regarded as a rapid method to determine the specificity of numerous bacterial strains toxins against pest insect. The optimum medium for insect feeding was offered to the insects as food cylinders (6 mm in length, 36 mm in diameter, 10 cylinders/plate) cut from foliage homogenate supplemented with agar. The cylinders were dipped into distilled water containing different concentrations of the extractable proteins, agitated for 5 s, air dried, and then coated with test spore crystal suspensions. The concentration of extractable proteins included 75, 150, 225, and 300 μg protein/mL with distilled water as a control. Starved adults were collected 3–8 days after eclosion and added to the diet in Petri dishes (15 × 100 mm diameter) containing moistened filter paper. Four replicates containing 15 adult insects (8 females and 7 males) per plate per dose were used. In the mortality assay, insects were fed at 25°C in darkness for 8 days. Mortality was monitored daily, and the median lethal concentration (LC_{50}) for each test strain was calculated by probit analysis. In the frass assay, insects were fed for 6 days. The total number of frass pellets was calculated.

16.12 CELL TOXICITY ASSAY[18]

1. *Assays for Cytotoxicity and Ethoxyresorufin-O-deethylase (EROD) Activity in Mouse Hepatoma Cell Line Hepa-1*: Mouse hepatoma cell line Hepa-1 retaining the CYP1A1 activity was grown as monolayer and well attached. When about 60% culture was confluent, by replacing the culture medium with a medium containing either serially diluted *Bacillus cereus* culture extract or the positive control substances, 2,3,7,8-tetrachlorodibenzo-p-dioxin (2,3,7,8-TCDD, State Institute of Agricultural Chemistry, Helsinki, Finland) or β-naphthoflavone (β-NF) or 2,4-dinitrophenol (Sigma, St. Louis, MO), the cells were exposed to the test extracts for EROD activity or the total protein content (TPC), and untreated cells were used as a negative control. After freezing of the cells at −70°C and thawing, 50 μL and 100 μL of sodium phosphate buffer were added into the test and reference wells, respectively. With sodium phosphate buffer, 0.4 mM ethoxyresorufin in DMSO was diluted to final concentration of 2 mM, and its fluorescence was measured at excitation/emission wavelengths of 530/590 nm, and a resorufin standard curve was measured for each assay, in which 2,3,7,8-TCDD (1.25 nM) and β-NF were used as positive controls. By measuring EROD activity of hepatic cells from a 2,3,7,8-TCDD exposed rat, the assay's functionality was verified, which served as a historical control. The cytotoxicity was measured with TPC as an end point, and EROD activity (pmol of resorufin formed per min per mg protein) was calculated per protein content of the cells.

2. *Single Cell Gel (SCG)/Comet Assay*: HepG2 cells stored in ampoules in liquid nitrogen were thawed quickly, washed with EMEM supplemented with 2 mM L-Gln, 1% nonessential L-amino acids, and 10% heat-inactivated FBS (Sigma, St. Louis, MO) and grown in an incubator at 37°C in a 5% CO_2 atmosphere. In the SCG assay, wells of a 96-well tissue culture plate were plated with 5×10^4 HepG2 cells in 0.2 mL of EMEM supplemented with 2 mM L-Gln, 1% nonessential L-amino acids, and 10% heat-inactivated FBS, incubated for 28 h, and treated with the *Bacillus cereus* extract (dissolved in methanol) or B[*a*]P (dissolved in DMSO, positive control chemical) or the solvent controls (2.5 μL/mL). The cells were exposed in complete EMEM supplemented with 0.5% FBS for 20 h, washed with Ca^{2+} and Mg^{2+} free PBS, and harvested with trypsin-EDTA solution. Olive tail moment (a measure of length between the head and the tail and the amount of DNA in the tail) was used as the metric to characterize the DNA damage in individual cells. Six wells per test concentration were established.

3. *Micronucleus Assay*: HepG2 cells stored in ampoules in liquid nitrogen were thawed quickly, washed with EMEM supplemented with 2 mM L-Gln, 1% nonessential L-amino acids, and 10% heat-inactivated FBS and grown in an incubator at 37°C in a 5% CO_2 atmosphere, incubated for 24 h, exposed to the test substances or the vehicle in 2 mL of complete EMEM supplemented with 0.5% FBS for 20 h, washed with PBS containing Ca^{2+} and Mg^{2+}, incubated in EMEM with cytochalasin B (4.5 μg/mL,

Sigma, St. Louis, MO) for 30 h, swelled, harvested after trypsinization, and fixed for microscopy. The fixed cells were spotted on glass slides, stained in acridine orange (62.5 mg/mL) in Sørensen buffer (pH 6.8) for 1 min, washed with Sørensen buffer (2×2 min, pH 6.8) and covered in Sørensen buffer (pH 6.8) for the analysis. With "the diameter of a micronucleus had to be less than 1/3 of that of the main nucleus, the micronucleus had to be surrounded by a clear membrane and without touching the main nucleus" as the criteria, micronuclei were scored in binucleated cells (BNCs) by a fluorescence microscope. To evaluate the cytotoxicity, the proportions of mononucleated cells, BNCs, and polynucleated cells were counted for 1000 cells, and cell proliferation (cytokinesis block proliferation index, CBPI) was calculated. In the assays, duplicates of 1.5×10^6 HepG2 cells in 3 mL of EMEM supplemented with 2 mM L-Gln, 1% nonessential L-amino acids, and 10% heat-inactivated FBS were established for each test concentration.

16.13 DHASE INHIBITION ASSAY[19]

1. *Dhase Inhibition Assay*: To avoid any spectrophotometric interference, sterile reagent solution (RS) was prepared by bringing up 10 mL of triphenyl tetrazolium chloride solution (1%) and 5 mL of bacterial suspension to a final volume of 100 mL of *Pseudomonas* Phage 3 medium diluted 10-fold. Sterile 10-mL centrifuge tubes were filled with 2 mL of RS, to which 100 μL of heavy metal solution at five serial concentrations (final concentrations 2.5 to 15 mg/L for $ZnSO_4 \cdot 7H_2O$, 1.5 to 24 mg/L for $NiSO_4 \cdot 6H_2O$, 0.31 to 5.0 mg/L for $CuSO_4 \cdot 5H_2O$, and 0.38 to 6.0 mg/L for $3CdSO_4 \cdot 8H_2O$) were added and with sterile ultrapure water as a negative control incubated at 30°C in the dark for 48 h; triphenyl formazan was extracted with acetone for 2 h, and triphenyl formazan concentration was determined spectrophotometrically at 482 nm.

2. *Phenol Dhase Inhibition Assay*: To avoid any spectrophotometric interference, sterile RS was prepared by bringing up 10 mL of triphenyl tetrazolium chloride solution (1%) and 5 mL of bacterial suspension to a final volume of 100 mL of *Pseudomonas* Phage 3 medium diluted 10-fold. Sterile 10-mL centrifuge tubes sealed with Teflon caps were filled with 2 mL of RS, to which 100 μL of phenol solution at five serial concentrations (final concentrations 25 to 400 mg/L, a range within the compound water solubility) were added and with sterile ultrapure water as a negative control incubated at 30°C in the dark for 48 h; triphenyl formazan was extracted with acetone for 2 h, and triphenyl formazan concentration was determined spectrophotometrically at 482 nm.

16.14 SEDIMENT TOXICITY ASSAY[20]

In the assays, three sediment samples, LP (pool where logs were stored), LF (brook through landfill area), KN (Kaskesniemi), and one reference sediment, REF (Höytiäinen), were used.

1. *Direct Contact Exposure Bioluminescence Test*: Sediment samples were diluted, mixed, pH adjusted to 6.5, and mixed with the bacterial suspension. Light output was measured on Bio-Orbit luminometer 1251 (Turku, Finland) at 0.05-s intervals for the first 30 s and 30 min after mixing. By comparing the initial peak luminescence at the beginning of the test to the luminescence at the end, percentage inhibition was observed. To find the dilution that caused a 50% decrease in light output (EC_{50}), a set of sediment dilutions (% v/v) was measured.

2. *Growth and Emergence Test*: *Chironomus riparius* was used as the test organism, and the tests were conducted with 12 replicates using individual exposure and first instar larvae. In a 10-day toxicity test, growth (dry weight), development (instar), and survival were used as end points, and in a 50-day emergence test, time to emergence and survival were used. The experiments were done at static conditions with 16 h light and 8 h dark, and the midges were fed with 0.12 mg of Tetramin fish food per day per organism. The temperature of growth and emergence experiment was 16.2 ± 0.6°C and 17.1 ± 1.0°C, respectively.

3. *Arsenite-Specific Biosensors and Concentration Measurements*: *Pseudomonas fluorescens* strain OS8 with arsenite sensor plasmid pTPT31 was grown with vigorous shaking at 28°C in King's medium B supplemented with tetracycline (50 mg/mL) and rifampicin (75 mg/mL), by centrifugation the bacteria were harvested during exponential growth phase at an OD_{600} of 0.5, and the cells were suspended in an equal volume of PBS (10 mM, pH 7.1). In Luria Berthani (LB) medium supplemented with kanamycin (30 mg/mL), *Escherichia coli* MC1061 (pTOO31) was at 37°C cultivated for 24 h, centrifuged, washed with minimal M9 medium, and resuspended into LB to a density of 3×10^6 cells/mL. An 0.5-g sediment sample and 5 mL of aqua sterilisata was vortexed for 10 min and allowed to settle for 15 min to have only larger particles sediment. For induction, 100 μL of bacterial suspension and 100 μL of appropriate dilution of the sample were mixed for 2 h, 100 μL of luciferin (2 mM D-luciferin in 0.1 M Na-citrate buffer, pH 5.0) was added, incubated for 20 min, and luminescence was measured using a LUMAC Biocounter M1500 (Lumac bv, Landgraaf, The Netherlands), a 1420 Victor2TM, or a portable 1253 Luminometer. The induction coefficient (I) was calculated from the equation RLU of sample/RLU of uninduced sample (the background light), where RLU was relative light units, photons measured in 10 s. To determine the dose-response curve, the bacterial strain was exposed to metal solutions from 1 pM to 10 mM of $NaAsO_2$ using triplicate determinations, and the samples were diluted to give I values corresponding with $NaAsO_2$ concentration range from 10 nM to 10 mM.

16.15 TOXICITY ASSAY OF PARTICLE-ASSOCIATED ARSENITE AND MERCURY[21]

Plasmids pTPT11 (mercury) and pTPT31 (arsenite) were used as reporter constructs with the heavy metal operon-promoters fused to the luciferase genes lucGR. The lucGR gene from pPP was cloned into vector pDN18-N containing the lacZ′ operator/T7 promoter and tetracycline resistance gene. Plasmid (pNEP01) was conjugated into the soil bacterium *Pseudomonas fluorescens* OS8, where the reporter was constitutively expressed. Plasmid-containing bacteria was grown on King's medium B agar plates at 28°C. Plates and liquid media were supplemented with 75 μg/mL rifampicin and 50 μg/mL tetracycline. The strain with plasmid pTPT11, pTPT31, and pNEP01 were referred to as P11, P31, and K15, respectively.

The humus (H) and clay (C) soil were spiked with either 100 μg/g mercuric chloride (which corresponded with $Hg^{2+} = 74$ μg/g) or sodium arsenite (which corresponded with $AsO_2^- = 82$ μg/g). Samples were incubated at room temperature in the dark for 10 weeks. The organic matter content was 49 and 8 g/kg, and pH was 4.0 and 4.6 in H and C soil, respectively. Soil conductivity was 279 μS/cm and 402 μS/cm for H and C, respectively. To reach concentration, 100 μg/g fresh weight at 20% water content aqueous $HgCl_2$ or $NaAsO_2$ were added to air-dried samples. From the incubation, four subsamples were taken, extracted with distilled water (1:5, soil:water), at room temperature shaked at 200 rpm for 1 h, and centrifuged (15 min, 1100 × *g*, the clear sample). Alternately, a sample was taken immediately after extraction (slurry). Six solid samples of the different contaminated areas were assayed.

Bacterial strains with plasmids were grown at 28°C with vigorous shaking, harvested during the exponential growth phase at OD_{600} of 0.5 (3.8×10^7 CFU/mL), by centrifugation, and suspended in an equal volume of PBS (10 mM, pH 7.1). After mixing, 50 μL of bacterial suspension with 50 μL of sample or heavy metal solution and 30 min (K15) or 2 h (P11, P31) incubation, 100 μL of luciferin (0.5 mM D-luciferin in 0.1 M Na-citrate buffer, pH 5.0) was added, incubated for 20 min, and luminescence was measured with a Wallac 1420 Victor2TM multilabel counter (Wallac Oy, Turku, Finland). The EC_{50} value was determined by K15 and expressed as v/v%. The induction coefficient (I) of P11 and P31 was calculated from the equation RLU of sample/RLU of uninduced sample (the background light).

16.16 GENETIC TOXICITY ASSAY[22]

Five strains of *Salmonella typhimurium* TA98, TA100, TA102, TA1535, and TA1537 were included. Strains TA98, TA100, TA1535, and TA1537 detected mutations at G-C sites and strain TA102 detected mutagens modifying A-T bp at target histidine genes. Confirmation of tester strain mutations was conducted at the source laboratory (American Type Culture Collection, ATCC, Rockville, MD, USA) with each shipment of tester strain and at MRI to confirm histidine requirement, rfa mutation, and uvrB mutation. Strains TA98, TA100, and TA102 have R-factor plasmid as confirmed by the presence of ampicillin resistance factor. Prior to the assay, L5178Y $tk^{+/-}$ mouse lymphoma cells (strain 3.7.2C, American Type Culture Collection, Rockville, MD, USA) were grown 10 h to a density of 1×10^9 to 2×10^9 cells/mL, as determined by absorbance greater than 1.2 at 660 nm. The mixture of polymethoxylated flavone (PMF) in a volume of 100 μL at five concentrations spanning four log doses was added in a test tube to each of five bacterial tester strains (0.1 mL), top agar (2 mL), and S9 fraction where indicated (0.5 mL), mixed, poured onto a minimal agar plate, incubated at 37°C for 48 ± 4 h, and revertant colonies were counted using an automated colony counter (LAI-HRCC). Triplicate plates were used for both negative and positive controls, and duplicate plates were used for PMF mixture. As separate positive control, 2-nitrofluorene (2-NF; 2 mg/plate), sodium azide (1.5 mg/plate), mitomycin C (MMC; 2 mg/plate), and 9-aminoacridine (9-AA; 100 mg/plate) were used for TA98, TA100 and TA1535, TA102, and TA1537, respectively, while S9, 2-aminoanthracene (2-AA, 10 mg/plate) was used as all strains' positive control.

Five days prior to sacrifice, SD rats (6–8 weeks old) were induced by a single ip injection of Aroclor 1254 (500 mg/kg). Aseptically removed livers were rinsed in cold sterile 0.15 M KCl buffer, homogenized, centrifuged at $9000 \times g$ for 20 min, and the supernatant (S9) was recovered and frozen at −70°C. Prior to receipt, protein concentration, sterility (no growth on agar plates or in liquid media), and biological activity of each batch were analyzed.

L5178Y $tk^{+/-}$ mouse lymphoma cells were maintained in RPMI-1640 containing 10% heat-inactivated horse serum, 0.1% Pluronic F68, and 0.22 mg/mL sodium pyruvate, incubated for growth at approximately 37°C in an atmosphere of approximately 5% CO_2 in air, and by periodically diluting the cultures with additional media their densities were maintained below 1.5×10^6 cells/mL. A 10,000-fold range of concentrations from 0.0005 mg/mL to 5 mg/mL were evaluated for cytotoxicity. For comparison, one solvent control culture in triplicate was included. Cultures were treated with PMF mixture for 4 h without or with S9. To evaluate the effect of the test article on cell growth, cell densities were determined on days 1 and 2 after treatment, and results were expressed as relative suspension growth (RSG).

In mutagenesis experiments, triplicate and duplicate cultures were assayed for negative and positive controls and test article, respectively. Five concentrations of test article spanning three log doses (0.0005–0.50 mg/mL) were evaluated. The cells (6×10^6 cells in 10 mL media) with or without the S9 mix were in a roller drum at 37°C exposed to 100 μL of PMF mixture for 4 h, centrifuged, and resuspended (twice) in 15 mL of fresh culture media at a concentration of 3×10^5 cells/mL in conical tubes, which were placed in a roller drum at 37°C for 48 h to permit expression of the tri-fluorothymidine (TFT)-resistant mutants. On each day, the cells were counted and diluted back to 3×10^5 cells/mL. At the end of incubation, cells were resuspended in 10 mL of culture medium, split into two sets, of which the first set was diluted 1:50 into 10 mL of fresh culture medium and diluted to 6 cells/mL in the final cloning medium containing 0.3% purified agar, while the second set was diluted 1:10 in 90 mL of cloning medium containing 0.3% purified agar. TFT (0.25 mL) was added to the second set of cells having a final concentration of 3×10^6 cells/mL. Prior to the automatic counting of colonies, both sets of cells were split (poured) into three dishes each (approximately 33 m per Petri dish) and incubated at approximately 37°C for 12 days. Triplicate cultures for negative and positive controls, EMS (0.5 μL/mL) without S9 and 3-methylcholanthrene (MCA; 5 mg/mL) with S9, and duplicate cultures for test article were assayed.

16.17 MICROTOX ASSAYS[23]

The bacterial strain *Arthrobacter* sp. N2 was precultured and cultured for 24 h on Trypcase soya broth. Cells were placed in 500 mL conical flasks containing 100 mL of medium, incubated at 27°C or 30°C and 200 rpm on a rotatory shaker, harvested under sterile conditions by filtration on a sintered glass filter or by centrifugation at 8000 rpm for 15 min, washed twice with Knapp buffer, 1000 mL of which contained 1 g of KH_2PO_4, 1 g of K_2HPO_4, 4 mg of $FeCl_3 \cdot 6H_2O$, and 40 mg of $MgSO_4 \cdot 7H_2O$ (pH 6.6), and suspended in it at a final concentration of 50 mg/mL wet weight cells (or 25 mg/mL for *Arthrobacter* sp. N2). Through an 0.2-μm pore size membrane, stock solutions of test compounds or 3,4-dichloroaniline in DMSO (40 mg/mL) were filter sterilized. The final xenobiotic concentration was 40 μg/mL. Incubation of cells in the same conditions in the presence of DMSO alone served as a negative control, while solutions of the xenobiotic in Knapp buffer at a final concentration of 40 μg/mL were used for abiotic references. The inhibition (EC_{50}) of bioluminescence of *Vibrio fischeri* by test compounds was measured after 5, 15, and 30 min exposure. For each compound, assays were performed four times with a Microbics M 500 analyzer coupled to a PC computer using 500 DOS software for Microtox.

16.18 DNA PIEZOELECTRIC BIOSENSOR ASSAY[24]

At room temperature and in the dark, the freshly cleaned crystal was immersed in an unstirred 1 mM ethanolic solution of 11-mercaptoundecanol for 48 h, washed with ethanol and Milli-Q water, and sonicated for 10 min in ethanol to remove the excess of thiol. At room temperature, the hydroxylic surface was treated for 4 h with a 600 mM solution of epichlorohydrin in a 1:1 mixture of 400 mM NaOH and bis-2-methoxyethyl ether, washed with water and ethanol, immersed in a basic dextran solution (3 g dextran in 10 mL of 100 mM NaOH) for 20 h, further functionalized with a carboxymethyl group using bromoacetic acid (1 M solution in 2 M NaOH) for 16 h, washed with Milli-Q water, and placed in the cell. Prior to covalent coupling, the surface of the crystal was activated with 200 mL solution of 50 mM *N*-hydroxysuccinimide (NHS, Fluka, Buchs, Switzerland) and 200 mM 1-ethyl-3-(3-dimethylaminopropryl) carbodiimide (EDAC, Sigma-Aldrich, Milan, Italy) in water for 5 min. The activating solution was replaced by a 200 mg/mL solution of streptavidin in 10 mM acetate buffer (pH 5). After 20 min, the residual reacting sites were blocked with 200 mL aqueous ethanolamine hydrochloride (pH 8.6, 1 M). After washing with immobilization buffer, 1 mM biotinylated probe was added (200 mL, in immobilization buffer), proceeding to a 200-min immobilization and for the first cycle of hybridization.

By adding 100 mL oligonucleotide solution in the hybridization buffer to the cell well, the hybridizations with the target oligonucleotides (23-mer) were performed. The reaction was monitored for 20 min, and to remove the unbound oligonucleotide, the crystal was washed with the hybridization buffer. The frequency was recorded until reaching a stable value. The frequency shift was defined as the difference between this stable value and the value before hybridization. Both frequency values were taken when the crystal was in contact with the same hybridization buffer, then the shift was related only to compounds fixed on the crystal during the reaction. In all the experiments, the single-stranded probe was regenerated by a 1-min treatment of 1 mM HCl leading a successive hybridization reaction to be monitored, which could be performed up to 10 times. All the experiments were performed at room temperature.

The bacterial DNA extracted from different *Aeromonas* strains was amplified by PCR. The synthetic oligonucleotide primers for the amplification of a 205-bp fragment of the aer gene of *A. hydrophila* were AERO1 and AERO2 (5′-CCAAGGGGTCTGTGGCGACA-3′ and 5′-TTTCACCGGTAACAGGATTG-3′, Pharmacia Biotech, Uppsala, Sweden). PCR products were examined by gel electrophoresis. For investigating the hybridization reaction with the real samples, the optimized procedure was that 20 μL solution of the DNA fragments obtained from the amplification by PCR were diluted with 80 μL hybridization buffer (final total volume, 100 μL, Sigma-Aldrich, Milan, Italy), the sample was denatured by heating at 95°C for 1 or 5 min, and freezing the sample in ice for 30 s. To the cell well, 100 μL sample was added, and the hybridization reaction was allowed to proceed for 20 min, and the crystal was washed with the hybridization buffer. The frequency was recorded until reaching a stable value. The frequency shift was defined as the difference between this stable value and the value before hybridization. After the hybridization reaction, the probe could be regenerated by 1 mM HCl treatment for 1 min, as for the reaction with the standard solution of the oligonucleotides, and another cycle of hybridization-regeneration could be performed.

16.19 PROTEIN PHOSPHATASE INHIBITION ASSAY AND ELISA OF MICROCYSTINS[25]

1. *Protein Phosphatase Inhibition Assay*: A 200 μL assay system consisted of 50 μL Tris buffer (50 mM, pH 7.0) and $CaCl_2$ (0.1 mM), 120 μL buffer with 1.7 mM fluorescent substrate, 4-methylumbelliferyl phosphate (MUP), 5 μL $NiCl_2$ (40 mM), 5 μL BSA (1 mg/mL of distilled water), and 10 μL enzyme (0.02 U per assay, final concentration, 1.5 nM) with 5 or 10 μL water sample or cyanobacterial extract. By automatic injection of substrate prewarmed together with the reaction mixture at 37°C, the reaction was started and the measurements (at 360 nm and 460 nm) were by real-time kinetics in a microplate reader (BMG Lab Technologies) over a period of 1 h. By linear regression of the reaction curves to give slope/min, activity was determined. From a standard fitted curve generated using microcystin MC-LR at 0.0005, 0.005, 0.025, 0.05, 0.25, 0.5, 2.5, 5.0, 25, and 50 nM, sample concentrations were determined. Working range of this assay was defined as that giving readings 20% to 80% that of the reference standard (no inhibitor present). From dose-response curves of activity versus the log of concentration inhibitor, concentration giving 50% inhibition against the reference value (no toxin standard present) was

determined. Using the Microsoft Excel look-up function, microcystin concentration was determined from a fitted curve.

2. *ELISA*: At 28°C, ELISA plates were coated overnight with OVA-ADDA-hemiglutarylor BSA-ADDA conjugate in sodium carbonate buffer (100 μL/well, 2.5 μg/mL, pH 9.6), washed with PBS, and their additional binding sites were blocked by incubation with ovalbumin (98%). The sheep anti-microcystin antiserum was 1/225,000 dilution, and anti-sheep secondary antibody donkey-anti-sheep-HRP was 1/4000 dilution; both had 1.5 h incubation. After adding 100 μL TMB substrate solution (BioFX TMBW-1000-01, BioFX Laboratories Inc., Maryland, USA), plates were incubated for 15 min. The reaction was stopped by adding 50 μL H_2SO_4 (2 M), and absorbance was determined at 450 nm on Versamax microplate reader. By a four-parameter curve fit using SOFT maxw PRO data analysis software (Molecular Devices Corporation), curve fits were performed. Using log-log transformation of data, working curve of this assay was calculated. The linear region at 20% to 80% of maximum absorbance was defined as the assay working range. Sample concentrations were determined by reference to the working curve for microcystin LR, in which the concentration ranged from 4×10^{-6} to 1 nM.

16.20 *Lepidium sativum* ASSAY FOR MICROCYSTIN TOXICITY[26–30]

Toxic *Microcystis aeruginosa* PCC 7806 was cultured on a large scale in 40-L glass tanks, with sufficient aeration for culture mixing and air supply. Toxin was extracted from the freeze-dried cells using 70% aqueous methanol and determined using the protein phosphatase inhibition assay from the difference in the change in absorbance at 410 nm. Seeds of *Lepidium sativum* were surface sterilized with 5% H_2O_2 for 5 min and washed three times with sterile water for 10 min. The seeds were left in fresh water overnight to germinate and then transferred to nutrient medium containing 6.5% N, 2.7% P, 13% K, 7% Ca, 2.2% Mg, 7.5% S, 0.15% Fe, 0.024% Mn, 0.0024% B, 0.005% Zn, 0.002% Cu, and 0.001% Mo in 0.8% agar. Twelve seeds were arranged in a single container and grown at 27°C under continuous light (20 photons/m^2/s). Prior to pouring, either 1 or 10 μg/mL MCLR toxin extract (Sigma, St. Louis, MO) was added to the medium. The fresh weights, root and leaf lengths were measured daily in 2 days. GST and GPX activities were determined in 3 days.

The stems and roots of 10 plants from each of three containers from each group were prepared for 30% homogenate of plant material in buffer (0.01 M Tris-HCl pH 7.8, 7.5 mM PMSF, 2.5 mM EDTA, 325 μM bestatin, 3.5 μM E-64, 2.5 μM leupeptin, and 0.75 μM aprotinin). The homogenate was centrifuged at 4°C and $12{,}000 \times g$ in a benchtop centrifuge for 2 h, and the supernatant was stored at −80°C for the GST and GPX assay. Each pooled sample of 10 plants was tested in duplicate with three pooled samples at each time point.

In the GST assay, to the supernatant fluid, a cocktail containing 0.1 M potassium phosphate (pH 6.5), 30 mM 1-chloro-2,4-dinitrobenzene, and 20 mM glutathione was added and measured immediately at 340 nm for 4 min. In the GPX assay, to the supernatant fluid, a cocktail of 0.1 mM potassium phosphate buffer (pH 7), glutathione reductase (4 ng/μL final), 10 mM glutathione, 30 mM EDTA, and 1 mM NADPH were added. The reaction was started with the addition of cumene hydroperoxide and measured immediately at 340 nm for 3 min.

16.21 ANTIPROLIFERATIVE ASSAY FOR INTERLEUKIN-4[31]

By adding 40 μL sample/standard to 160 μL HAM's F12K medium (Gibco) containing 10% FCS and antibiotics into the first two wells of alternate rows of a microtiter plate, dilutions for the construction of IL-4 dose-response curves and assay of samples were prepared. All additional wells of the plate contained 100 μL HAM's/FCS. By removing 100 μL from the first two wells and transferring the material to the third and fourth wells of the same row, a twofold dilution series was produced. The sample/standard was mixed, and 100 μL medium was removed and added to the next two wells. This procedure was continued along two rows of the microtiter plate and the last 100 μL was discarded to produce the sample/standard dilution series in 100 μL aliquots/well. Additional negative controls contained 100 μL medium alone.

Human lung carcinoma cells (CCL185, American Type Culture Collection) were grown at 37°C in HAM's Fl2K medium supplemented with 10% FCS and penicillin/streptomycin in a humidified atmosphere containing 5% CO_2. When the density reached 2×10^5 cells/mL, the cells were washed twice with HAM's Fl2K medium and resuspended in HAM's Fl2K medium containing 10% FCS, 100 μg/mL streptomycin, and 100 IU/mL penicillin at a density of 2×10^5 cells/mL. To the previously prepared microtiter plate containing the dose-response curve and sample dilution series, 100 μL aliquots of the cell suspension were added. At 37°C, the microtiter plate was incubated in a 5% CO_2 humidified atmosphere for 16 h. In cell proliferation assay, 0.5 μCi of [^{3}H]thymidine was added/well, and the microtiter plate returned to the incubator for another 4 h incubation. The cells were harvested onto filter mats, and using a scintillation counter, the radioactivity incorporated into DNA was measured.

REFERENCES AND NOTES

1. K.H. Engler, R. Coker, I.H. Evans. A novel colorimetric yeast bioassay for detecting trichothecene mycotoxins. J Microbiol Methods 35 (**1999**) 207–218.
2. T. Tsukatani, H. Suenaga, T. Higuchi, T. Akao, M. Ishiyama, K. Ezoe, K. Matsumoto. Colorimetric cell proliferation assay for microorganisms in microtiter plate using water-soluble tetrazolium salts. J Microbiol Methods 75 (**2008**) 109–116. *Note*: With a membrane filter (0.2 μm), 11.1 mM tetrazolium salts dissolved in distilled water was filter sterilized. Electron mediators were dissolved either in DMSO or in distilled water to prepare 8.0 mM solution and mixed with the tetrazolium salt solution at 1/9 ratio, as the detection reagent consisted of 10 mM tetrazolium salts and 0.8 mM electron mediators. *Candida utilis* (NBRC0626), *Saccharomyces cerevisiae* (NBRC2347), *Zygosaccharomyces rouxii* (NBRC0505), *Bacillus cereus* (NBRC13494), *Corynebacterium glutamicum*

(NBRC12168), *Lactobacillus casei* (NBRC15883), *Micrococcus luteus* (NBRC13867), *Staphylococcus aureus* (NBRC12732), *Staphylococcus epidermidis* (NBRC12993), *Acetobacter sp.* (NBRC3283), *Escherichia coli* (NBRC3972), *Klebsiella pneumoniae* (NBRC3512), *Proteus mirabilis* (NBRC13300), *Pseudomonas aeruginosa* (NBRC13275), *Salmonella enteritidis* (NBRC3313), *Salmonella typhimurium* (NBRC12529), *Serratia marcescens* (NBRC102204), *Vibrio parahaemolyticus* (NBRC12711), *Bacillus subtilis* (JCM1465), *Enterococcus faecalis* (JCM5803), *Yersinia enterocolitica* (JCM7577) and *Listeria monocytogenes* (ATCC15313) were used in this assay.

3. N. Hontzeas, K. Hafer, R.H. Schiestl. Development of a microtiter plate version of the yeast DEL assay amenable to high-throughput toxicity screening of chemical libraries. Mutat Res 634 (**2007**) 228–234.
4. M. Liu, V. Seidel, D.R. Katerere, A.I. Gray. Colorimetric broth microdilution method for the antifungal screening of plant extracts against yeasts. Methods 42 (**2007**) 325–329.
5. J.P. Zanon, M.F.S. Peres, E.A.L. Gattás. Colorimetric assay of ethanol using alcohol dehydrogenase from dry baker's yeast. Enzyme Microbial Technol 40 (**2007**) 466–470.
6. J. Tamarit, V. Irazusta, A. Moreno-Cermeño, J. Ros. Colorimetric assay for the quantitation of iron in yeast. Anal Biochem 351 (**2006**) 149–151.
7. T. Kuda, K. Shimizu, T. Yano. Comparison of rapid and simple colorimetric microplate assays as an index of bacterial count. Food Control 15 (**2004**) 421–425.
8. A.M. Gonzaulez, L. Beniutez, T. Soto, J.R. Delecea, J.C. Gutieurrez. A rapid bioassay to detect mycotoxins using a melanin precursor overproducer mutant of the ciliate tetrahymena thermophila. Cell Biol Int 21(**1997**) 213–216.
9. A.R. Kaney, V.J. Speare. An amicronuleate mutant of Tetrahymena thermophila. Exp Cell Res 143 (**1983**) 461–467.
10. V.R. Dayeh, S.L. Chow, K. Schirmer, D.H. Lynn, N.C. Bols. Evaluating the toxicity of Triton X-100 to protozoan, fish, and mammalian cells using fluorescent dyes as indicators of cell viability. Ecotoxicol Environ Safety 57 (**2004**) 375–382.
11. Y. Cetin, L.B. Bullerman. Cytotoxicity of Fusarium mycotoxins to mammalian cell cultures as determined by the MTT bioassay. Food Chem Toxicol 43 (**2005**) 755–764.
12. M.B. Hansen, S.E. Nielsen, K. Berg. Re-examination and further development of precise and rapid dye method for measuring cell growth/cell kill. J Immunol Methods 119 (**1989**) 203–210.
13. F.E. Saadé, G.B. Dunphy, R.L. Bernier. Response of the carrot weevil, Listronotus oregonensis (Coleoptera: Curculionidae), to strains of Bacillus thuringiensis. Biol Control 7 (**1996**) 293–298.
14. F.E. Saadé. Evaluation of strains of bacillus thuringiensis as biological control agents of the adult stages of the carrot weevil, listronotus oregonensis (coleoptera; curculionidae). M.Sc. Thesis, National Library, Ottawa (**1993**).
15. J.L. Gringorten, D.P. Witt, R.E. Milne, P.G. Past, S.S. Sohi, K. Van Frankenhyyzen. An in vitro system for testing bacillus thuringiensis toxins: the lawn assay. J Invertebr Pathol 56 (**1990**) 237–242.
16. M. Bradford. A rapid and sensitive method for the quantification of microgram quantities of protein utilizing the principle of protein-dye binding. Anal Biochem 72 (**1976**) 248–254.

17. K. van Frankenhuyzen, J.L. Gringorten. Frass failure and pupation failure as quantal measurements of bacillus thuringiensis toxicity to lepidoptera. J Invertebr Pathol 58 (**1991**) 465–467.
18. M.A. Andersson, P. Hakulinen, U. Honkalampi-Hämäläinen, D. Hoornstra, J.C. Lhuguenot, J. Mäki-Paakkanen, M. Savolainen, I. Severin, A.L. Stammati, L. Turco, A. Weber, A. von Wright, F. Zucco, M. Salkinoja-Salonen. Toxicological profile of cereulide, the bacillus cereus emetic toxin, in functional assays with human, animal and bacterial cells. Toxicon 49 (**2007**) 351–367.
19. F. Abbondanzi, A. Cachada, T. Campisi, R. Guerra, M. Raccagni, A. Iacondini. Optimization of a microbial bioassay for contaminated soil monitoring: bacterial inoculum standardization and comparison with Microtox® assay. Chemosphere 53 (**2003**) 889–897.
20. T. Petänen, M. Lyytikäinen, J. Lappalainen, M. Romantschuk, J.V.K. Kukkonen. Assessing sediment toxicity and arsenite concentration with bacterial and traditional methods. Environ Pollut 122 (**2003**) 407–415.
21. T. Petänen, M. Romantschuk. Toxicity and bioavailability to bacteria of particle-associated arsenite and mercury. Chemosphere 50 (**2003**) 409–413.
22. B. Delaney, K. Phillips, C. Vasquez, A. Wilson, D. Cox, H.B. Wang, J. Manthey. Genetic toxicity of a standardized mixture of citrus polymethoxylated flavones. Food Chem Toxicol 40 (**2002**) 617–624.
23. C. Tixier, M. Sancelme, S. Aït-Aïssa, P. Widehem, F. Bonnemoy, A. Cuer, N. Truffaut, H. Veschambre. Biotransformation of phenylurea herbicides by a soil bacterial strain, Arthrobacter sp. N2: structure, ecotoxicity and fate of diuron metabolite with soil fungi. Chemosphere 46 (**2002**) 519–526.
24. S. Tombelli, M. Mascini, C. Sacco, A.P.F. Turner. A DNA piezoelectric biosensor assay coupled with a polymerase chain reaction for bacterial toxicity determination in environmental samples. Anal Chim Acta 418 (**2000**) 1–9.
25. D.O. Mountfort, P. Holland, J. Sprosen. Method for detecting classes of microcystins by combination of protein phosphatase inhibition assay and ELISA: comparison with LC-MS. Toxicon 45 (**2005**) 199–206.
26. M.M. Gehringer, K. Vijayne, C. Nadya, T.G. Downing. The use of Lepidium sativum in a plant bioassay system for the detection of microcystin-LR. Toxicon 41 (**2003**) 871–876.
27. M.M. Gehringer, K.S. Downs, T.G. Downing, R.J. Naudé, E.G. Shephard. An investigation into the effect of selenium supplementation on microcystin hepatotoxicity. Toxicon 41 (**2003**) 451–458.
28. C.J. Ward, K.A. Beattie, E.Y.C. Lee, G.A. Codd. Colorimetric protein phosphatase inhibition assay of laboratory strains and natural blooms of cyanobacteria: comparisons with high performance liquid chromatographic analysis for microcystins. FEMS Microbiol Lett 153 (**1997**) 465–473.
29. W.H. Habig, M.J. Pabst, W.B. Jakoby. Glutathione *S*-transferases: the first step in mercapturic acid formation. J Biol Chem 249 (**1974**) 7130–7139.
30. A. Wendel. Glutathione peroxidase. Methods Enzymol 77 (**1981**) 325–332.
31. L.A. Page, M.S. Topp, R. Thorpe, A.R. Mire-Sluis. An antiproliferative bioassay for interleukin-4. J Immunol Methods 189 (**1996**) 129–135.

17

METHODS AND APPLICATIONS OF TOXICITY ASSAYS FOR CHEMICALS

Shiqi Peng

Environmental chemical contaminants such as toxic agents may be suspended in air, absorbed in particulate matter (SPM) and foods, dissolved in water, and thus may result in a large number of toxic actions such as DNA damage, decline in lung function, imbalance of immune function, activation of AhR, abnormal reproduction, and urinary bladder injuries. Specific lesions for cells induced by DNA damaging chemicals may lead to cell death or induce error-prone repair pathways leading to mutagenesis and cancer induction; using the model of SPM and ALIC of human alveolar type II cell line (A549), SPM toxicity can be evaluated; exposure to extremely low concentrations of 2,3,7,8-TCDD results in activation of AhR and the induction of expression of a battery of genes; differential follicle counts may provide a sensitive means of estimating the extent of ovarian toxicity in females exposed to xenobiotics; urinary bladder injuries induced by repeated oral administration of pharmaceuticals may be epithelial ulceration with edema and hemorrhage in the lamina propria and muscle layer of the urinary bladder; environmental chemical contaminants disrupt mammalian peptide signal transduction pathways via the interaction with the endocrine system in humans and various wildlife species; and the cellular production of green fluorescent protein (GFP) is a function of nitrate concentration. Because of these facts, a series of assays has been established, and these are widely used for detecting environmental chemical contaminants. In this chapter, 31 assays are presented: lux-fluoro assay for combined genotoxicity and cytotoxicity of chemicals,[1] flow

Pharmaceutical Bioassays: Methods and Applications. By Shiqi Peng and Ming Zhao

cytometry and microscopy based assay,[2] cell transformation assay,[3,4] air-liquid interface culture (ALIC)-based assay for toxicity of chemicals in SPM,[5–7] immunization in the murine assay,[8] immunotoxicological functional assay,[9] PWM-induced immunoglobulin M (IgM) assay for toxicity of chemicals,[10–12] TCDD-induced toxicity assay,[13] H4IIE EROD assay,[14] rat mutation assay,[15] aryl hydrocarbon receptor (AhR) and related assay,[16] ectonucleotidase expression assay,[17] toxicity equivalent quantity (TEQ) assay,[18] enzyme lacking AA epoxygenase activity assay,[19] Japanese medaka embryo-larval assay,[20] GFP based cell assay,[21–26] reproductive assessment by continuous breeding (RACB) assay for ovarian toxicity of xenobiotics,[27,28] uroepithelial cell assay,[29,30] male rat systemic toxicity assay,[31] 28-day oral toxicity assay,[32,33] endocrine disruptors assay,[34] mating efficiency assay,[35–37] soils assay for contaminants and toxicity,[38] rice nitrate reductase activity assay,[39] *Tradescantia*-micronucleus assay,[40] nitric oxide neurotoxicity assay,[41] EBV-transformed human Burkitt's lymphoma cell assay,[42] murine bone marrow assay for hemopoietic and osteogenic toxicity,[43] pnar-gfp assay for carcinogenic toxicity of nitrate,[44–47] brine shrimp (*Artemia salina*) assay,[48–54] and DPPH radical scavenging property assay.[49,55]

17.1 LUX-FLUORO ASSAY FOR COMBINED GENOTOXICITY AND CYTOTOXICITY OF CHEMICALS[1]

The plasmid pPLS-1 (DSM10333), the genotoxicity sensing reporter component of the test panel, which carried the luxCDABFE genes downstream of a strong SOS-dependent promoter, was constructed according to the literature. The plasmid pGFPuv, which carried the optimized "cycle 3" variant of GFP in frame with the lacZ initiation codon (Clontech Laboratories Inc., CA, USA), was the cytotoxicity sensing reporter component of the test panel. The plasmids pPLS-1 and pGFPuv were used to transform the strain *Salmonella typhimurium* TA1535 (ATCC: *Salmonella choleraesuis* subsp. *choleraesuis* strain TA1535) according to the modified "Hanahan" procedure. The bacteria were grown at 37°C in NB-medium supplemented with 50 μg/mL ampicillin for selection of plasmid carrying cells. The sample containing DNA-damaging agents was dissolved in distilled water at high concentrations and diluted for the test. Each well of White LB96P-CMP Mikro Lumat Plates with a transparent bottom contained 10 μL of solvent or different concentrations of the test sample. The agar plate containing a single colony of bacteria, 10 mL of NB-medium with 50 μg/mL ampicillin was shaken on a rotary shaker at 37°C for 16 h. The bacteria were diluted (1 : 50) in fresh warm NB-medium containing 50 μg/mL ampicillin and incubated at 37°C until the absorbance at 600 nm (A_{600}) reached 0.2–0.3. From this culture, 90 μL aliquots were added to each well of White LB96P-CMP Mikro Lumat Plates. The microplate was covered with a gas-permeable self-adhesive seal and placed into the temperature-controlled microplate reader (EG&G Wallac, Perkin Elmer, USA). The microplate reader was programmed to repeat the measurement cycle at 30°C for up to 8 h of continuous incubation with a duration of about 10 min per measurement cycle (total of 50 cycles) to successively read the absorbance, luminescence, and fluorescence. The measurement cycle of each well was started with orbital shaking for 120 s and followed by luminescence measurement without filter for 0.2 s, absorbance measurement at 490 nm (20 nm band width) for 0.1 s, and fluorescence measurement (excitation at 405 nm and emission at 510 nm) for 0.1 s.

17.2 FLOW CYTOMETRY AND MICROSCOPY BASED ASSAY[2]

Using traditional protocols for assessing micronuclei (MN) frequency, from each mouse and rat, peripheral blood smears were prepared either by snipping the tail tip and squeezing a drop of blood onto each of two labeled microscope slides or by collection of blood drops from the vena cava after euthanasia. Mice or rats were anesthetized using CO_2 asphyxiation, the femurs were removed and cleaned of tissue, the bone ends were snipped off, and the contents of the femurs were immediately flushed onto clean and labeled microscope slides using phosphate-buffered saline to obtain bone marrow. Squash preparations were made to create bone marrow cell monolayer. Blood and bone marrow slides were air-dried, fixed in absolute methanol for 5–10 min, air-dried again, stained with acridine orange, a fluorescent DNA and RNA specific stain, and coded before scoring. For each mouse or rat, using epi-illuminated fluorescence microscopy at 450- to 490-nm excitation and 520-nm emission, 2000 uniformly stained reticulocytes (RETs) per tissue were scored at 1000-fold. In addition, 1000 total RETs and mature were scored on peripheral blood slides to determine %RET; as a measure of chemical-induced bone marrow toxicity, this value was typically less than 5%. In bone marrow, 200 total RETs were scored per animal to determine %RET (typically 40% to 60%).

After euthanasia, 60–120 μL of mouse or rat blood sample was collected from the vena cava, diluted in sodium heparin solution, fixed in ultracold methanol, and then immediately stored at −80°C until flow cytometric (FCM) assayed on a FACSCalibur flow cytometer (Becton Dickinson, San Jose, CA). To consistently define the MN analysis windows and to establish proper daily photomultiplier tube voltages and compensation, malaria-infected mouse cells were used as a reference standard. RETs were identified as CD71-positive (i.e., presence of an active transferrin receptor on the cell surface), whereas mature erythrocytes were identified as CD71-negative. For rat samples, FCM analysis was restricted to the youngest RET (i.e., the subpopulation of erythrocytes with the highest CD71 expression). Using DNA staining dye propidium iodide (PI) in conjunction with RNase treatment, MN was detected. Thus, MN-RET expressed high levels of CD71 (CD71-positive) and PI-associated fluorescence, whereas MN-erythrocytes were negative for CD71 (CD71-negative) and show PI-associated fluorescence. For presence of MN, 10,000 to 20,000 CD71+ RETs were scored per mouse or rat, and approximately 1×10^6 total erythrocytes were counted to determine %RET as a measure of chemical-induced bone marrow toxicity. To potentially evaluate the mechanism of MN formation (aneugenic vs. clastogenic activity), the intensity of the PI-associated fluorescence was determined for 100 MN-RETs from each mouse or rat within a treatment group for each of the four known inducers of MN. The PI-associated fluorescence was proportional to DNA content of the MN. The fluorescence intensity histogram was recorded for each mouse or rat sample, and the median of these values (median channel, or the channel number that divides the single parameter histogram into two parts, each containing an equal number of data points) was selected to provide a quantitative description of average DNA content of a sample's MN. MN with higher DNA content were presumed to contain whole chromosomes, whereas MN with lower DNA content were presumed to contain chromosome fragments. Finally, the mean $\pm$ SD of the median channel values for each dose group was calculated and plotted.

17.3 CELL TRANSFORMATION ASSAY[3,4]

To achieve the desired 1000 total colonies per treatment group, a transformation assay was typically run by conducting two individual trials of 20 cultures per treatment group with 25 to 45 colonies per plate (approximately 500–900 colonies). The rationality of performing the assay in two trials came down to practical considerations, i.e., the size of the manageable experiment able to generate the recommended 1000 or more colonies per treatment group. Upon the treatment with test chemical, the cultures were either re-fed with control medium after a 24-h exposure and then left undisturbed for the 7-day clonal expression period or exposed to chemical for the entire 7-day clonal expression period, and finally rinsed, fixed, stained, and scored for total colony number and identification of morphologically transformed (MT) colonies with a $\times 5$ to $\times 45$ magnification stereo microscope. Normal colonies contained cells in a monolayer with an organized and often flowing pattern of growth with minimal cell stacking, particularly where the cells were at a confluent density. On the other hand, some normal colonies could exhibit a random growth pattern at the outermost perimeter of the colony, where the cells were not confluent. Morphologically, MT colonies contained cells in an extensive random-oriented, 3D, and stacked growth pattern, with crisscrossing cells at the perimeter and in the center of the colony. Cells in MT colonies frequently were more basophilic (appear as darker blue–black compared with the lighter red–blue of normal colonies) and had higher nuclear/cytoplasm ratios than their normal counterparts. For each treatment group within a trial, mean number of colonies/culture dish, total number of colonies/test group, mean plating efficiency (PE $\pm$ SEM), relative plating efficiency (RPE), number of morphologically transformed (MT) colonies, and MT frequency were calculated, where PE $=$ (number colonies per dish/number of cells seeded) $\times$ 100, RPE $=$ (test group PE/solvent control PE) $\times$ 100, and MT frequency $=$ (number of MT colonies/total number of colonies) $\times$ 100. These values were calculated for the individual trials and on pooled data sets for each treatment group from both trials.

17.4 ALIC-BASED ASSAY FOR TOXICITY OF CHEMICALS IN SPM[5–7]

A549 cells (Riken Gene Bank, Tukuba, Japan) and HepG2 cells (Japanese Cancer Research Bank, Tokyo, Japan) were cultured in DMEM (Sigma-Aldrich Chem., St. Louis, MO) supplemented with 10% FBS (Filtron, Altona, Australia), 20 mM hydroxyethyl-piperazine-*N*-2-ethanesulfonic acid (Dojindo, Kumamoto, Japan), penicillin (100 IU/mL), streptomycin (100 μg/mL), amphotericin B (0.25 μg/mL), and 1.0% nonessential amino acid solution (only for the HepG2 culture medium). Both types of cells were subcultivated using 0.25% trypsin and 0.02% EDTA in PBS. A549 cells collected by trypsinization were seeded at 1.0×10^5 cells/cm^2 onto the membrane culture insert precoated with a 0.03% type collagen solution. Initially, to both the apical (Ap) and basolateral (BL) sides, the medium was added. Using a phase-contrast microscope, the formation of the cell sheet was assessed. Based on the measurement of TER, the development of tight junctions in the cell layers with a Millicell-ERS was evaluated. A549 cells were cultured in a liquid phase until the TER showed constant and saturated values (45–50 Ωcm^2 without the blank value of the membrane itself), the culture medium in the Ap side was removed, and the ALIC was started. The chemical-free suspended particulate matter (SPM) samples were sterilized by ethylene oxide gas (EOG) with a commercially available sterilization system (SEMMEL-380B, JMC-Ikiken, Tokyo, Japan). After being measured carefully with a digital weighting machine, the EOG-sterilized SPM samples were loaded directly to the Ap side of an A549 cell layer in ALIC. A549 and HepG2 cells were cultured in 12-well plates for measuring the survivability and ethoxyresorufin O-deethylase (EROD) capacity using extracts of the SPM samples. The cells were seeded at 1.0×10^5 cells/cm^2 and used after they reached confluence. In the preparation of the extracts of the SPM samples, in a 1.5-mL microtube, the suspension of 24 mg of each sample and 200 μL of DMSO was sonicated for 15 min, centrifuged for 10 min at 15,000 rpm, and the supernatant was diluted with culture medium so that the final concentration of DMSO in the culture medium was 0.5% and the highest concentration of SPM in the culture medium was 600 μg SPM/mL culture medium, which corresponded with the cell-surface-area-based load of 158 μg SPM/cm^2 cell surface.

The SPM sample was suspended in 100 mL of dichloromethane and sonicated for 15 min. The SPM particles were filtered with a 0.1-μm pore size membrane. The filtrate was evaporated and the residue was resolved in acetonitrile. The concentrations of the chemicals in the SPM extracts were determined by use of multicolumn HPLC/spectrofluorometer/computer system. To standardize the biological effect of the chemical on the EROD capacity of the HepG2 cells, the chemically derived induction equivalent (EQ) was employed and calculated according to $EQ = S([PAH]_1 EF_1 + \cdots + [PAH]_n EF_n)$, where the equivalency factor (EF) values of the individual chemical was referred to using the same reference.

After 48 h of SPM exposure, each culture was rinsed twice with PBS. To the Ap and BL side, 500 μL and 1.5 mL of substrate solution of acid phosphatase was added, respectively, in the case of the cell culture inserts. For the monolayer culture in the 12-well plates, 1 mL of AP solution was added. After 2 h of incubation at 37°C, the absorbance was measured at 405 nm using a UV spectrophotometer. The net absorbance was determined by subtracting a control value (cell-free well) from the measured values, and the cell survivability was calculated by referencing a predetermined calibration curve.

The inducibility of cytochrome P4501A1/2 was determined in terms of the EROD capacity. A549 and HepG2 cells cultured in 12-well plates were washed twice with

PBS and loaded with 10 μM 7-ethoxyresorufin in the presence of 10 μM dicumarol in culture medium. After 1 h of incubation, the fluorescence intensity was measured with a fluorescence spectrophotometer (Shimadzu RF-1500, Shimadzu) at an excitation wavelength of 530 nm and an emission wavelength of 585 nm to detect the resorufin formed. The intensity was calibrated to the resorufin concentration using a standard curve determined using an authentic resorufin.

17.5 IMMUNIZATION IN THE MURINE ASSAY[8]

Toxicity represented by lethality, local pain as assessed by vocal response to injection, local swelling, loss of hair, and skin lesion was assayed in Balb/c female mice. Test compounds dissolved in sterile saline were injected subcutaneously on the back of the mice as three doses at 1-week intervals. After the last dose, the mice were monitored for 7 days. Pyrogen-free sterile saline was used as control. Mice were immunized with three doses of fucose-mannose ligand (FML) antigen of *Leishmania donovani* (150 μg) and test compounds in sterile saline solution through the subcutaneous (sc) route in the back of 2-month-old Balb/c mice at 1-week intervals. Saline and test compound adjuvant-treated mice were included as controls. Two identical experiments were performed. In the isolation and chemical characterization of FML obtained from stationary-growth phase promastigotes of *Leishmania donovani* Sudan (LD 1S/MHOM/SD/00-strain 1S), promastigotes were submitted to an aqueous extraction, heated to inactivate, and centrifuged. The aqueous supernatant was lyophilized and fractionated by gel filtration on a Bio-Gel P-10 column yielding the FML glycoprotein complex in void volume. After complete immunization, mice's sera were collected for 7 days and analyzed by FML-ELISA.

With FML-ELISA, the anti-FML antibody levels in pools of all the vaccinated and control groups were assayed using 2 μg of antigen per well and goat anti-mouse IgG or goat anti-mouse IgG1, IgG2a horseradish peroxidase-conjugated antibodies (Southern, Biotechnology Associates, Birmingham, AL, USA) in a 1:4000 dilution in blocking buffer. The reaction was developed with *o*-phenyldiamine, interrupted with 1 N sulfuric acid, and monitored at 492 nm. Sera were analyzed by double-blind tests and in triplicate, in each test positive and negative control sera were included, and results were expressed as absorbance values at 492 nm of the 1/100 diluted pool of sera obtained for each treatment, 7 days after the complete immunization.

17.6 IMMUNOTOXICOLOGICAL FUNCTIONAL ASSAY[9]

Adult male CD rats (4 weeks of age, body weights 65–90 g, quarantined for 1 week prior to use) were administered CCl_4 (12.5 or 25 mg/kg) or vehicle (VH, corn oil) daily excluding weekends by oral gavage for 30 or 90 days. Rats were routinely administered approximately 3.0 mL of either CCl_4 or vehicle and fasted for approximately 16 h prior to blood sample collection. One day prior to necropsy, blood was collected, for hematologic and clinical chemical measurements, from the orbital sinus of each rat while the rat was under light carbon dioxide anesthesia. Number of erythrocytes

(RBC), leukocytes (WBC), platelets (PLAT), hemoglobin (Hb) concentration, hematocrit, mean corpuscular volume, mean corpuscular hemoglobin, and mean corpuscular hemoglobin concentration were measured on a Serono Baker 9000 hematology analyzer (Allentown, PA) or calculated. Relative numbers of neutrophils (Neut), band neutrophils (Band), lymphocytes (Lymph), atypical lymphocytes (Alym), monocytes (Mono), eosinophils (Eosin), basophils (Baso), and reticulocyte counts were determined on a Hematrak Automated Differential System cell counter (Geometric Data, King of Prussia, PA). From the leukocytic data, absolute values for the various types of leukocytes were calculated. With a Hitachi 717 clinical chemistry analyzer (Boehringer Mannheim, Indianapolis, IN) using Boehringer Mannheim reagents, serum was evaluated for the activities of alkaline phosphatase, alanine aminotransferase (ALT), aspartate aminotransferase, sorbitol dehydrogenase (SDH), and creatine kinase, and for concentrations of urea nitrogen, total protein, albumin, triglycerides, creatinine, bilirubin, cholesterol, glucose, calcium, sodium, potassium, phosphate, and chloride. From the total protein and albumin concentrations, serum globulin concentration was calculated.

After exposure to CCl_4 (on days 31 and 91) and after a 16-h fast, rats were sacrificed by carbon dioxide anesthesia. After exsanguination, rats were necropsied. From the rats, representative samples of skin, aorta (thoracic), heart, trachea, lungs, salivary glands, esophagus, stomach, liver, pancreas, small intestine (duodenum, jejunum, and ileum), large intestine (cecum, colon, and rectum), kidneys, bladder, pituitary, thyroid-parathyroid, adrenals, prostate, testes, epididymides, seminal vesicles, brain, spinal cord, peripheral nerve (sciatic), bone (sternum and femur), skeletal muscle (thigh), eyes, exorbital lacrimal glands, harderian glands, bone marrow (sternum and femur), lymph nodes (mediastinal, mesenteric, and mandibular), spleen, thymus, and all gross lesions were collected. Due to its increased susceptibility to autolysis, the retina of the eye was evaluated microscopically as a separate organ. Selected gross lesions were saved but not processed for microscopic evaluation, and a microscopic diagnosis would not be additive for them (e.g., osteoarthritis, pododermatitis, chronic dermatitis of the tail, urinary calculi, and deformity of the teeth, toe, tail, or pinna). The brain, heart, liver, spleen, thymus, kidneys, and testes were weighed. The ratios of organ weight/final body weight and organ weight/brain weight were calculated. Testes, epididymides, and eyes were fixed in Bouin's solution, whereas all other tissues were fixed in 10% neutral buffered formalin. Tissues collected from rats in all groups were embedded in paraffin, cut at a nominal thickness of 5 mm, stained with hematoxylin and eosin (H&E), and examined microscopically.

To assess spleen cell numbers, each spleen was bisected and one-half placed on ice in 6 mL of HBSS (Gibco, Grand Island, NY). From each one-half spleen in HBSS containing 1.0 M HEPES (Gibco), a single cell suspension was prepared by first cutting the spleens into several pieces and then gently mashing the pieces between the frosted ends of two glass slides. Spleen cell numbers were determined using a Serono Baker 9000 hematology analyzer and multiplied by the total spleen weight/weight of the spleen section to obtain cell number/total spleen.

Using mouse anti-rat monoclonal antibody (mAb) clones and a direct immunofluorescent staining procedure, lymphocytes were labeled. On a Becton–Dickinson

FACScan flow cytometer, immune cell staining data was acquired and analyzed. Using mAb clone OX19 (anti-CD5, pan T-cell marker) conjugated with phycoerythrin (PE), total T cells were quantitated, using mAb clone OX12 (anti-rat kappa light chain) conjugated with FITC, total B cells were quantitated, using dual immunofluorescent staining with FITC conjugated mAb clone W3/25 (anti-CD4) and PE conjugated mAb clone OX19, T_h cells were quantitated, and using dual staining with FITC conjugated mAb clone OX8 (anti-CD8) and PE conjugated mAb clone OX19 (Serotec Ltd., Bicester, Oxon, UK), $T_{cyt/sup}$ cells were quantitated. Mouse IgG1 (clone F8-11-13) conjugated with FITC or PE and mouse IgG2a (clone BZ-2) conjugated with FITC served as negative controls.

Variance was separately analyzed on the 30- and 90-day data. For parametric data, an overall F-test was performed to test the differences between various groups. In the cases that significant differences were observed, multiple Student's *t*-tests were carried out. Some data may also transform to log, square-root, and log-log to obtain normality or homogenous variances. Kruskal-Wallis test followed by Dunn's multiple comparison procedure was used for nonparametric data. For eosinophil counts and serum creatinine concentrations having numerous ties, permutation methods were used, and the software package StatXact used for Mann-Whitney or Wilcoxon tests. With a one-way analysis of variance, mean final body weights, mean organ weights (absolute and relative to body and brain weights), and spleen cell numbers were analyzed. With Dunnett's test, pairwise comparisons between test and control groups were made. To compare SRBC-specific IgM levels of various treatment groups receiving SRBC, the untreated and vehicle controls were compared using Mann-Whitney test. For both the 30- and 90-day data, the UT and VH groups were not significantly different and were combined for subsequent analyses. For the 30-day data, to determine which dose groups were significantly different from the combined control groups, Jonckheere's trend test was used in a step-down manner. Because the 90-day data departed from a monotone dose-response, following a natural log transformation of the data, Dunnett's test was used. Groups differing at the level of $p \leq 0.05$ were considered significant.

17.7 PWM-INDUCED IgM ASSAY FOR TOXICITY OF CHEMICALS[10–12]

The toxicants-containing sample was dissolved in Ca^{2+} and Mg^{2+}-free PBS or DMSO or in the vehicle (10 μL each). Spleen cells aseptically removed from BALB/cAnN mice (8–9 weeks old) were washed in HBSS supplemented with 10 mM HEPES buffer (pH 7.4) and diluted with RPMI 1640 medium containing 10% heat-inactivated FBS, 2 mM L-Gln, 0.4 M sodium pyruvate, 20 mM β-mercaptoethanol, nonessential amino acids, and antibiotics to make cell suspensions of 10^6 cells/mL. Cells were cultivated in 24-well culture plate under 5% CO_2-air and the presence of peripheral blood mononuclear cells (PWM) at 37°C for 2–6 days, supplemented with sample containing chemical toxicants at the beginning of culture. For total IgM assay, 700 μL of media were collected, and the residual cells in 300 μL of media were used for cell proliferation assay.

At 4°C with 100 μL of rabbit anti-mouse IgM antibody (IgG, Caltag Labs., USA) dissolved in PBS (pH 7.4), 96-well microtiter immunoplates were precoated overnight, washed with PBS containing 0.05% Tween-20, and at 4°C with 150 μL of PBS containing 1% BSA blocked overnight. After washing with PBS, at room temperature into each well 100 μL aliquots of culture media (usually 11-fold diluted media) or reference mouse serum or standard mouse IgM diluted (Seikagaku Corp., Tokyo, Japan) with 1% BSA-PBS were added, left for 60 min, washed, 100 μL aliquots of horseradish peroxidase-conjugated goat anti-mouse IgG in PBS were added, incubated for another 60 min, 150 μL aliquots of substrate solutions containing 0.4% *o*-phenylenediamine and 0.01% hydrogen peroxide in 0.1 M citrate–0.2 M phosphate buffer were added, incubated for 15 min, and 50 μL of H_2SO_4 (0.5 M) was added to terminate the enzyme reaction. The optical density at 490 nm of each well was read by use of a microplate reader to determine the total IgM.

After adding 30 μL of TetraColar One reagent, 300 μL of medium was at 37°C under 5% CO_2-air incubated for 2 h. The OD (NJ-2000, Intermed Co., Tokyo, Japan) at 450 nm of the medium containing the water-soluble tetrazolium salt produced by the enzyme reaction was read, and the cell number in cultures was measured by MTT assay for mitochondrial dehydrogenase activity in viable cells.

17.8 TETRACHLORODIBENZO-*p*-DIOXIN (TCDD)-INDUCED TOXICITY ASSAY[13]

At 37°C in a humidified atmosphere containing 5% CO_2 in DMEM supplemented with 5% FBS, 1% antibiotics, and 0.6 μg/mL insulin, the human breast carcinoma MCF-7 cells were grown, while human breast carcinoma MDA-MB-231 cells were cultured at 37°C in Leibovitz's L-15 medium supplemented with 10% FBS, 1% antibiotic-antimycotic, and 0.06 mg/mL insulin.

1. *Sulforhodamine B (SRB) Assay*: This assay was used to determine survival of cells after the exposure to the solution of 2,3,7,8-TCDD in 0.1% DMSO. MCF-7 and MDA-MB-231 cells (Culture Collection and Research Center, Hsinchu, Taiwan) were seeded in 96-well plates (8×10^3 cells/well). By estimating the cellular proteins the total cell number was measured. The samples were determined by a spectrophotometer and compared to the values of controls. At 492 nm, visible absorbance was recorded in a 96-well plate reader.

2. *2′,7′-Dichlorofluorescin Diacetate (DCFH-DA) Assay*: This assay was used to determine the formation of ROS in cells after the exposure to TCDD (Chem. Service Inc. Company, PA, USA). MCF-7 and MDA-MB-231 cells were seeded in black 96-well plates (8×10^3 cells/well), the media was discarded and replaced with Earle's Balanced Salt Solution containing 10 μM DCFH-DA, TCDD was added to the wells, and ROS generation relative to controls was determined at 30-min intervals by a fluorescent plate reader (485 nm λex, 535 nm λem). If necessary, 2 h prior to the treatment, specific AhR inhibitor α-naphthoflavone (α-NF; 0.15 μM) and ERα inhibitor 4-hydroxyltamoxifen (4-OHT; 100 nM) were applied and kept in the

medium during TCDD exposure until the cells were analyzed, while H_2O_2 (20 mM) was used as a positive control. After 2 h exposure, 15- and 18-fold increases in ROS formation in MCF-7 and MDA-MB-231 cells over control were induced.

3. *o-Phthalaldehyde Assay*: Cells seeded in the 96-well black plates were treated with TCDD for 0.5 and 2 h, to the medium 10 μL of trichloroacetic acid (62.5%) was added, incubated for 10 min, to each well 40 μL of sodium phosphate (1 M, pH 7) was added, the reaction was carried out for 20 min, and to each well 100 μL of sodium phosphate solution (0.16 M) containing 37.5 mM *o*-phthalaldehyde was added. After 30-min room temperature reaction in the dark, the fluorescence was recorded (355 nm λex, 460 nm λem) by fluorometer and was compared with the values of control. NEM (100 μM) was used as a positive control, which induced ca. 50% decreases in GSH depletion in MCF-7 and MDA-MB-231 cells over control after 30 min exposure.

4. *Intracellular NADPH Assay*: Through its reduction to a yellow-colored formazan dye and determined periodically by a spectrophotometer, a water-soluble tetrazolium salt was used to monitor the amount of NADPH. Cells seeded in 96-well plates (8×10^3 cells/well) were treated with TCDD for 2 and 5 h, the reaction medium was discarded, the wells were washed with PBS, 100 μL of fresh medium and 1/10 volume of CCK-8 solution were added, cells were cultured for another 4 h and examined by a spectrophotometer at 30-min intervals. With 650 nm as a reference filter, visible absorbance was recorded at 450 nm in a 96-well plate reader. By comparing the absorbance of a well containing cells treated with naphthalene quinonoids against the control, which was treated with DMSO only, the decrease in the intracellular NADPH was assessed. If necessary, 2 h prior to the treatment, specific poly(ADP-ribose) polymerase-1 (PARP-1) inhibitors including 3-AB (15 mM), BA (5 mM), and coumarin (1.5 mM) were applied and kept in the medium during TCDD exposure until the cells were analyzed.

5. *Enzyme Cycling Assay*: Cells (7.5×10^5 cells/dish) treated with TCDD for 5 h in 3.5-cm dishes were washed with PBS, trypsinized, harvested, transferred to microcentrifuge tubes, centrifuged, the supernatants were removed, the pellets were resuspended in 100 mM potassium phosphate buffer containing 3% trichloroacetic acid and put on ice for 30 min, and the samples were centrifuged at 12,000 × *g*. The acid-soluble fractions were neutralized with 800 mM KOH containing 200 mM Tris, the supernatants were mixed with a reaction medium containing 0.09 mM WST-8, 3.75 μM methoxy PMS, 15 U/mL alcohol dehydrogenase, and 120 mM ethanol in 100 mM potassium phosphate buffer in 96-well plates, and the plates were at 37°C in the dark incubated for 30 min. By comparing the NAD^+ content of a well containing cells treated with TCDD against that of a well with cells treated with DMSO (0.1%) alone, the decrease in the intracellular NAD^+ was assessed.

6. *Western Blot Analysis*: The treated cells were washed three times with PBS, centrifuged at 8000 × *g* for 5 min, soaked in liquid N_2, cell pellets were thawed and lysed in 80 μL of protein extraction buffer (1 M Tris-HCl, pH 7.9, 3 M NaCl, 1% aprotinin, 2 mM phenylmethylsulfonyl fluoride, 5 mM dithiothreitol) for 30 min on ice, and the extracts were centrifuged at 10,000 × *g* for 30 min. With the Bio-Rad protein

assay kit, protein concentration was measured, proteins were loaded at 50 μg/lane and separated by SDS-PAGE and blotted. Anti-PARP and anti-pADPr monoclonal antibodies (Abcam, Cambridge, UK) were used as 1:1000 dilutions. The secondary antibodies, AP-coupled anti-mouse antibody, were incubated at room temperature for 2 h as 1:5000 dilutions. By incubating with chemiluminescence detection system, immunoreactive bands were visualized. To correct for differences in protein loading, the membranes were also probed with anti-GAPDH antibody.

7. *Comet Assay for Quantifying DNA Strand Breaks*: After exposure, cells were mixed with 1.2% low melting agarose, added to slides coated with 1% agarose, denatured with lysis buffer (pH 13) overnight, subjected to electrophoresis at 25 V/300 mA for 25 min, and stained with YoYo-1. Using Comet Scoring software, the tail moment of 50 randomly selected cells was analyzed from each slide.

17.9 H4IIE EROD ASSAY[14]

The wild-type (H4IIE) rat hepatoma cells grown in 75 cm^2 Nunc TC flasks (Nalgene Nunc International, NY, USA) at 37 ± 2°C in a humidified atmosphere (95% air, 5% CO_2) and not passaged more than 10 times were seeded with 4×10^4 cells on the 96-well culture plates and incubated for 24 h. The plated H4IIE cells were dosed in duplicate either with 1 μL of DMSO, each of a nine-dose dilution series of TCDD (0.02–5 nM, ChemService, West Chester, USA) or each with 1 μL of nine-dose dilution series of air extracts (DCM and MeOH) and incubated for 24 h. After incubation, the H4IIE cells were washed three times with phosphate buffer (K_2HPO_4, KH_2PO_4, pH 8.0). To each well, a 100 μL aliquot of 5 AM 7-ethoxyresorufin in 0.1 M HEPES (pH 7.5) was added. On a fluorometer (BMG LabTechnologies, Offenburg, Germany) at excitation/emission wavelengths of 544/590 nm (2-nm bandwidths), between 2 and 12 min, produced resorufin was measured. With a standard curve of resorufin (5 nM to 5 μM), each assay was preceded to quantify the amount of fluorescence produced. To derive three ED_{50} values, each EROD assay was run in triplicate. Based on a logit function for continuous response data using SigmaPlot (v7, Jandel Scientific Software), ED_{50} values were calculated. Prior to protein determination, all plates were frozen at −80°C. Specific 7-ethoxyresorufin-O-deethylase (EROD) activity was expressed as pmol resorufin mg $protein^{-1}$ min^{-1}. Biological TCDD equivalent concentration (Bio-TEC) was calculated by dividing the ED_{50} value of TCDD (pg) by the ED_{50} of DCM or MeOH extracts, namely Bio-TEC (pg/m^3) = $ED_{50\text{-TCDD}}$ (pg)/$ED_{50\text{-extract}}$ (m^3). The detection limit of this assay was 0.1 pg TCDD/m^3.

17.10 RAT MUTATION ASSAY[15]

Twenty male and twenty female lacI transgenic Big Blue rats (7 weeks old) were divided into control, TCDD, AFB1 (Sigma-Aldrich Chem., St. Louis, MO), and combination treatment of both TCDD and aflatoxin-B1 (AFB1) groups, which were housed individually in shoebox cages and maintained on a diet of PMI Lab Diet rodent formula *ad libitum*. Chemicals were administered orally.

TCDD (1 mg) was dissolved in 1 mL of acetone and 0.1 mL of this solution mixed with 49.9 mL of corn oil by sonication to prepare a 2 μg/mL TCDD solution. An 0.5 mg/mL AFB1 solution was prepared by sonicating 5 mg of AFB1 in 10 mL of corn oil. All rats received either 2 μg/mL TCDD or corn oil twice each week for 6 weeks, which was followed in the last week with either 0.5 mg/kg AFB1 or corn oil. A 2-week fixation period allowed any DNA adducts to become mutations. By CO_2 asphyxiation and cervical dislocation, rats were sacrificed, livers were removed, rinsed with lactated Ringer's solution, flash frozen under liquid nitrogen, and stored at −80°C. Genomic DNA was packaged into bacteriophage lambda particles and infected into SCS-8 bacterial host cells in the presence of X-gal on 25 cm × 25 cm plates. By direct assay counts, total plaque forming units (pfu) were determined, and using a white fluorescent light box with a Big Blue plaque reading enhancer, the plates were screened for mutants. Blue lacI mutants were cored and re-plated at low density for purification and verification. By PCR, lacI gene was amplified from the bacteriophage DNA, and the resulting 1400 base pair lacI gene fragment, which included lacI promoter and the lacZ operator region, was purified using Promega Wizard PCR Prep DNA purification cartridges. On a Licor Long ReadIR 4200 automated sequencer, DNA sequencing was performed. The forward sequencing primer was located at positions 103 to 82, and the reverse sequencing primer was located at positions 1187 to 1208. Sequencing was simultaneously done from both directions. Mutation information was managed with custom-designed software, and to determine statistical significances, the sequence spectra were compared using the Adams Skopek method.

17.11 AhR AND RELATED ASSAY[16]

Using QuikChange site-directed mutagenesis kit (Stratagene, Amsterdam, The Netherlands), CYP1A1 mutants F127A, Y236A, F240G, F240A, F240S, F240V, F240I, F240L, I243A, Y246A, I326A, L337A, and I390A were generated by PCR mutagenesis. With DNA sequencing, all the nucleotide sequences of the mutated rat CYP1A1 cDNAs were confirmed. The coexpression plasmids were constructed for each mutant and yeast NADPH-P450 reductase. GAPDH promoter derived from *Zygosaccharomyces rouxii* was used for the expression of native and mutants CYP1A1.

In a synthetic minimal medium containing 8% glucose, 5.4% yeast nitrogen base without amino acids, and 160 μg/mL histidine, the recombinant *Saccharomyces cerevisiae* cells expressing rat CYP1A1 or its mutant were cultivated. From recombinant *Saccharomyces cerevisiae* cells, microsomal fraction was prepared.

With a Shimadzu UV-2200 spectrophotometer (Kyoto, Japan) and according to the procedure, the reduced CO-difference spectra were measured. From the reduced CO-difference spectrum, using a difference of the extinction coefficients at 446 nm and 490 nm of 91 mM^{-1} cm^{-1}, the concentration of native CYP1A1 or its mutants was determined.

The solutions of substrates DD, 1-monochloroDD (1-MCDD), and 2-monochloroDD (2-MCDD) in ethanol, 2,3-DCDD, 2,7-DCDD, and 2,3,7-TriCDD in acetone, and 2,3,7,8-TCDD in DMSO were prepared. The reaction mixture contained 1.56 μM 2,3,7,8-TCDD or each of 10 μM polychlorinated dibenzo-p-dioxins (PCDDs), 1% organic solvent, the microsomes containing 30–100 nM of rat CYP1A1 or its mutants, and 50 mM potassium phosphate buffer (pH 7.4). By adding NADPH at a final concentration of 0.5 mM, the reaction was initiated. After varying time intervals, aliquots of the reaction mixture were collected and extracted with four volumes of chloroform/methanol (3/1, v/v), and the organic phase was recovered and dried up. The resulting residue was dissolved in acetonitrile for HPLC analysis with YMC-Pack ODS-AM (4.6 mm × 300 mm, YMC, Tokyo, Japan). A mobile phase consisting of a linear gradient of 20% to 100% acetonitrile aqueous solution containing 0.01% TFA for 25 min followed by 100% acetonitrile for 10 min was used. A flow rate of 1.0 mL/min was used, and the elution was detected by measuring absorption at 227 nm, and the column temperature was maintained at 40°C.

Using a Finnegan mat Thermo Quest GC with EI mode, the isolated metabolite from HPLC effluents was analyzed with GC mass spectrometry. The GC column was Chrompack Cp-Sil 24CB-MS (0.32 mm × 30 m). The vector pGV-P's SV40 promoter region was replaced with five tandem repeats of XRE followed by the nucleotides −164 to +53 of rat glutathione *S*-transferase Ya subunit gene, and the resultant plasmid was introduced into mouse hepatoma Hepa-1clc7 cells. Using the stable transformant, AhR assay was performed. After adding each of polychlorinated dibenzo-p-dioxins (PCDDs) and their metabolites to the cell culture, luciferase activity was measured with luciferase assay system.

17.12 ECTONUCLEOTIDASE EXPRESSION ASSAY[17]

Based on the relationships of ectonucleotidases, ectonucleoside triphosphate diphosphohydrolases (E-NTPDase) and ectopyrophosphatase/phosphodiesterases (E-NPP enzymes) summarized in Fig. 17.1, the following assays were established.

1. *Cell Culture*: Murine Hepa1c1c7 cells and chicken hepatoma LMH cells (ATCC, Manassas, VA) were grown in α-MEM and Waymouth's MB 752/1

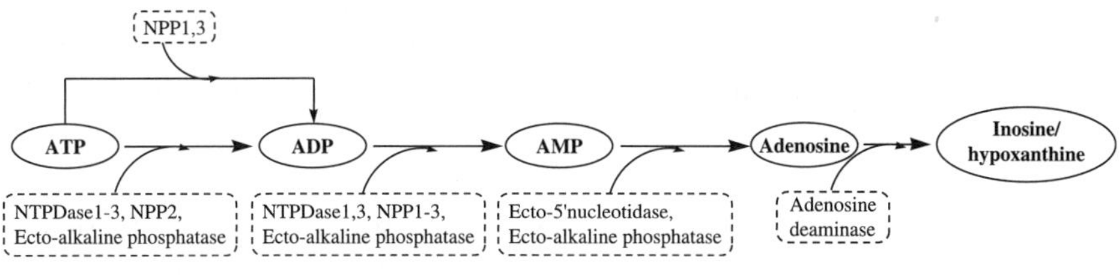

Figure 17.1 Metabolism of ATP and ADP as well as catalyzing of ectonucleotidases.

(Gibco BRL Life Technologies, Gaithersburg, MD), respectively, both were supplemented with 10% FBS, penicillin (100 IU/mL), and streptomycin (100 μg/mL), and were maintained at 37°C in 5% CO_2/95% air. Human hepatoma cells HuH-7, murine Hepa1-6, and human hepatoma cells HepG2 cells were grown in DMEM and MEM, respectively, both were supplemented with 10% FBS, penicillin (100 IU/mL), streptomycin (100 μg/mL), and amphotericin B (0.25 μg/mL), and were maintained at 37°C in 5% CO_2/95% air. Passage 1 or 2 HUVECs were grown to 70% confluence in medium M199 supplemented with L-Gln, penicillin, streptomycin, and 20% FBS. As representatives of non-transformed cells and highly sensitive to TCDD passage 1, chick embryo hepatocytes obtained from 14- to 15-day-old chick embryos were maintained in Ham's medium supplemented with penicillin (100 IU/mL), streptomycin (100 μg/mL), and 2% FBS, whereas passage 1 cardiac myocytes prepared from 8- to 10-day-old chick embryos were maintained in MEM supplemented with penicillin (100 IU/mL), streptomycin (100 μg/mL), and 10% FBS. Cell lines were generally grown to about 70% confluence, and primary cultures were maintained for 2 to 3 days before treatment. Hepatocytes (1×10^6 cells/well), cardiac myocytes (0.2×10^6 cells/well), and HUVECs (1.24×10^4 cells/well) were plated in 24-well plates. Prior to treatment, medium was removed, and cells were treated for 24–48 h with medium containing TCDD (NCI Chemical Carcinogen Repository, Kansas City, MO) or with an equivalent amount of the vehicle dioxane (controls, 0.002% or less). TCDD was used at the concentration giving the maximum CYP1A induction to ensure that the condition for measuring ATP and ADP metabolism was capable of activating the Ah receptor. Prior to biochemical assays, medium was removed, and cells were washed twice with medium without any additions, while in the case of the ENPP and ectoalkaline phosphatase assays, medium was removed and cells were washed twice with phosphate-free buffer. After the assays, cells were washed twice with PBS and lysed with 0.1 M NaOH.

2. *Measuring Extracellular ATP and ADP Metabolism*: To washed cells at 37°C, 0.1 mL of assay buffer [15 mM bistrispropane, 134 mM NaCl, 5 mM glucose, 10 μM Ap5A (P^1,P^5-di[adenosine-5]pentaphosphate), 3 mM $CaCl_2$, and 3 mM $MgCl_2$, pH 8.0] containing 0.05–2 mM [^{14}C]-ATP (specific activities, 56 mCi/mmol) or 0.05–2 mM [^{14}C]-ADP (specific activities, 56.7 mCi/mmol) was added and incubated for 5 min to perform the measurement of extracellular ATP and ADP metabolism. In this measurement, Ap5A served to inhibit adenylate kinase, a cytosolic enzyme catalyzing the reversible transphosphorylation reaction of 2 ADP $\leftrightarrow$ AMP + ATP. Via placing the cells on ice and adding 0.02 mL of "stop solution" containing 160 mM disodium EDTA (pH 7.0) and 17 mM ADP, further metabolism of substrate was halted. ^{14}C-labeled metabolites separated on thin layer chromatography (TLC) were measured through radio TLC scanning with InstantImager. From the results, the values of the reaction mixtures incubated without cells were subtracted as blanks. ATP and ADP metabolism were assayed for each cell type in two to three independent experiments, including two to four replicate wells at each substrate concentration.

3. *E-NPP Activity Assay*: Activity was assayed based on the conversion of thymidine 5′-monophosphate *p*-nitrophenyl ester (TMPNP, Sigma-Aldrich Chem.,

St. Louis, MO) to *p*-nitrophenol E-NPP. After washing the cells with phosphate-free buffer (15 mM Tris, 134 mM NaCl, 3 mM $CaCl_2$, 3 mM $MgCl_2$, pH 8.0), the medium was replaced with 0.5 mL of TMPNP (0.05, 0.1, 0.2, 0.5, and 1 mM final concentrations) in the same buffer, and the cells were incubated at 37°C for 30 min. The preliminary experiments had shown that under these conditions, the reaction rates were linear. From each of two wells for each substrate concentration, duplicate aliquots (0.2 mL) were collected. By adding 0.25 mL of NaOH (0.2 M), each reaction was stopped. Aliquots were centrifuged at 4°C and 1000 × *g* for 5 min, and on a Cary 400 Bio UV-Vis spectrophotometer, the absorbance of *p*-nitrophenol in supernatants was measured at 400 nm, with reference to a standard curve for *p*-nitrophenol. As blanks, the values of the reaction mixtures incubated without cells were subtracted. For each cell type, at least three independent experiments were needed.

4. *Ectoalkaline Phosphatase Activity Assay*: Based on the conversion of *p*-nitrophenyl phosphate to *p*-nitrophenol, ectoalkaline phosphatase activity was assayed. After washing the cells with phosphate-free buffer (15 mM Tris, 134 mM NaCl, 3 mM $CaCl_2$, 3 mM $MgCl_2$, pH 8.0), the medium was replaced with 0.5 mL of *p*-nitrophenyl phosphate (0.05 and 0.5 mM final concentrations) in the same buffer, and the cells were incubated at 37°C for 30 min. From each of three wells, for each substrate concentration, duplicate 0.2 mL aliquots were collected, and to each of them 0.25 mL of 0.2 M NaOH was added. After centrifugation (1000 × *g*, 4°C, 5 min), *p*-nitrophenol in the supernatant was measured as above. As blanks, the values of the reaction mixtures incubated without cells were subtracted.

5. *EROD Activity*: Cells in 24-well plates were washed, the medium was replaced with 0.5 mL of Ham's medium containing 2% FBS, 4 μM 7-ethoxyresorufin, and 10 μM dicumarol (to inhibit resorufin reductase), and incubated at 37°C for 30 min. The preliminary experiments had shown that under these conditions, resorufin production was linear. After adding 0.25 mL of acetone, two 0.2 mL aliquots from each well were at 4°C and 1000 × *g* centrifuged for 5 min. At excitation wavelengths of 558 nm and emission wavelengths of 590 nm, resorufin fluorescence in the supernatant was measured. For selected samples, fluorescence emission spectra were recorded to check the characteristic resorufin spectrum at 590 nm. From the results, the mean values for duplicate blanks with all of the components of the reaction mixture and without cells were subtracted. To determine the concentration of TCDD needed for maximal induction of EROD in each cell type, preliminary experiments were performed.

6. *SDS-PAGE and Western Immunoblotting*: Cells were cultured in 6- or 12-well plates, medium was removed, cells were treated with TCDD or dioxane (solvent), washed with PBS, removed by scraping into sample buffer, and heated to 100°C for 10 min. For studies involving chick monoclonal antibodies, sample buffer contained no β-mercaptoethanol (0.125 M Tris-HCl, pH 6.8, 4% SDS, 20% glycerol, 0.002% bromophenol blue). Before transfer to nitrocellulose and blotting with monoclonal antibody, MAB10 (against chick gizzard ectoATPase) or MC 18 (against chick oviduct/liver/stomach ectoATPDase) proteins were separated on a 4% to 20% nonreducing gel. In parallel with each immunoblotting experiment, a group of

cells was plated in 24-well dishes, treated with TCDD or dioxane, and used for measuring EROD activity to check the TCDD responsiveness of the cells used for immunoblots. By chemiluminescence and band intensity measured with an AlphaImager 2000 system (Alpha Innotech, San Leandro, CA), immunoreactivity was detected.

7. *RT-PCR of Mouse and Human NTPDase2*: Mouse Hepa1c1c7 and Hepa1-6 cells and human HuH-7 and HepG2 cells were grown in 10-cm dishes. Using RNA STAT-60 (Friendswood, TX), total RNA was extracted from the cells. From the total RNA, according to the manufacturer's instructions, by use of reverse transcriptase M-MLV first-strand cDNA synthesis was performed. Incubation mixtures consisted of 4 μL of 5× incubation buffer, 1 μg of total RNA, 2 μL of 10× concentrated hexanucleotides, 1 μL of dNTPs (10 mM), 1 μL of RNase inhibitor (40 U/μL), 2 μL M-MLV RT (20 U/μL), and DEPC-water up to 20 μL and were incubated at 37°C for 1 h. For RT-PCR amplification, specific primers selected from the mouse and human NTPDase2 and mouse and human GAPDH genes using the Lasergene Navigator (DNA STAR) program were 5′-GCTCCTGGGTTGCTCTCCTG-3′ and 5′-AGTCTCTGGTGCTTGCCTTTCTAA-3′ for mouse NTPDase2, 5′-TGGAGGCAGCCGCAGTGAATGT-3′ and 5′-GGAGGCGAAGAGCAGCAGGAGGAC-3′ for human NTPDase2, 5′-CTGACGTGCCGCCTGGAGAAA-3′ and 5′-TGTTGGGGGCCGAGTTGGGATAG-3′ for mouse GAPDH, and 5′-CTCCGGGAAACTGTGGCGTGATG-3′ and 5′-TGTGCTCTTGCTGGGGCTGGTG-3′ for human GAPDH. PCR amplification was performed in reaction mixtures containing 1 μL of single-strand cDNA, 2 μL of relevant sense and antisense primers (200 nM final concentration), and 45 μL of Platinum PCR Supermix containing all of the other components needed for the PCR reactions. The conditions 94°C × 1 min and 62°C × 30 s and 72°C × 1 min for up to 35 cycles for mouse NTPDase2, 94°C × 1 min and 66°C × 30 s and 72°C × 1 min for up to 40 cycles for human NTPDase2, 94°C × 1 min and 62°C × 30 s and 72°C × 1 min for up to 25 cycles for mouse GAPDH, 94°C × 1 min and 64°C × 30 s and 72°C × 1 min for up to 25 cycles for human GAPDH were used. By electrophoresis on a 1.5% agarose gel and visualized with ethidium bromide using a UV transluminator, PCR products were separated. The expected sizes of the PCR products were 382, 314, 347, and 469 bp for mouse NTPDase2, human NTPDase2, mouse GAPDH, and human GAPDH, respectively. Using NIH Image 1.60.1 software, the densitometry values of the PCR signals were measured, of which the values for NTPDase2 were normalized against those for GAPDH. To ensure the comparison of the PCR results for control and TCDD-treated samples were made prior to reaching PCR plateau levels, PCR products were obtained and analyzed from multiple cycles during the course of amplification for each cell line.

17.13 TOXICITY EQUIVALENT QUANTITY (TEQ) ASSAY[18]

1. *ELISA*: In 0.1 M carbonate bicarbonate buffer (pH 9.6), microtiter plates were coated overnight at 4°C with 100 μL/well of the appropriate coating antigen

concentration. Standards consisting of a 1/1/2 (v/v/v) solution of DMSO/MeOH/100 ppm Triton X-100/PBS (pH 7.5) containing 2 mg/mL BSA were prepared. After an initial blocking step with BSA-PBS and a wash step, 50 μL of standards was added. To each sample, wells contained 25 μL of PBSB, 25 μL of human milk sample in DMSO-MeOH, and then 50 μL of antiserum diluted in PBSB was added to have a final ratio of DMSO-MeOH/PBSB of 1/3. The plates were incubated for 90 min, washed, 100 μL of goat anti-rabbit antibody coupled to horseradish peroxidase (diluted in PBS + 0.05% Tween-20) was added, incubated for 60 min, washed with wash buffer, and 100 μL of enzyme substrate containing TMB (Sigma, St. Louis, MO) was added, incubated for 20 min, and the color reaction was stopped by adding 50 μL of sulfuric acid (2 M). The resultant color was measured at 450 nm on a model 550 Microplate Reader in single-wavelength mode, and based on a standard curve derived from a fit of absorbance versus the logarithm of concentration, the dioxin levels in the human milk samples were calculated.

2. *Gas Chromatography/Mass Spectrometry (GC/MS)*: Using a JEOL JMS-700 MStation mass spectrometer (J&W Scientific) coupled to a HP-6890 high-resolution gas chromatography (HRGC) with a capillary column of DB-17HT (30 m × 0.25 mm i.d., film thickness 0.15 μm), GC/MS was performed for the isomer-specific separation. The GC program was 150°C for 1 min, raised 20°C/min to 220°C and subsequently 4°C/min to 280°C, maintained for 11.5 min, with helium as carrier gas, 280°C injector temperature, and 280°C GC/MS interface temperature. In the selected ion monitoring mode with mass resolution of 10,000, electron impact ionization energy of 38 eV, and ion source temperature of 260°C the MS was operated. Quantification was done by the isotopic dilution method, that is, PCDD/Fs congeners were quantitated by comparison with their respective reference $^{13}C_{12}$-labeled standards by two methods. In one method, one kind of stable isotope in each congener of PCDD/Fs was used as the internal standard, and in the other method, $^{13}C_{12}$-1,2,3,4-TCDF was for tetra- to hexa-CDD/Fs, and $^{13}C_{12}$-OCDD for hepta- to octa-CDD/Fs, and then the PCDD/Fs concentrations were calculated by fat basis. Using WHOTEF, the toxicity equivalent quantity (TEQ) was calculated.

17.14 ENZYME LACKING AA EPOXYGENASE ACTIVITY ASSAY[19]

1. *Embryos and Tissue Processing*: Fertilized White Leghorn chicken eggs were incubated at 37°C in high humidity. Fifteen-day-old embryos (hatching occurs at 21 days) were treated with a maximum inducing dose for hepatic cytochrome P-450 1A (CYP1A) of 2,3,7,8-TCDD in 0.005 mL of dioxane (1 nmol/egg) or with dioxane alone. Through a hole in the shell, drugs were injected into the fluids surrounding the embryo. After 16 h TCDD exposure, organs were removed and promptly fixed in 4% paraformaldehyde (Prill, Electron Microscopy Sciences, Fort Washington, PA) in PBS (150 mM NaCl in 10 mM Na_2HPO_4, pH 7.4) at room temperature for 1 h and then overnight at 4°C. Organs were dehydrated successively in 35%, 50%, 70%, and 90% ethanol for 1-h periods, in 100% ethanol for two 45-min periods, in ethanol/xylene (1/1, v/v) for 30 min, and in xylene for 30 min. They were placed

in a 1/1 (v/v) mixture of xylene and paraplast for 15 min, and in aliquots of fresh paraplast successively for 1 h and two 30-min periods, all at 56°C. Serial 5-mm sections were cut and mounted on Superfrost/Plus slides (Fisher Scientific, Pittsburgh, PA). On removal of paraffin, the sections were placed successively in xylene for two 5-min periods, in 100% ethanol for two 3-min periods, and in 95%, 70%, and 50% ethanol for successive 1-min periods for rehydration, stained with Harris' hematoxylin and eosin, or analyzed by *in situ* hybridization or immunohistochemistry.

2. *Riboprobe Synthesis*: By passage through 0.2-mm pore nylon filters (Nalgene, Rochester, NY), all solutions were made RNase free. As CYP1A4 and 1A5 cDNAs shared 80% nucleic acid identity within their open reading frames, 3′-untranslated regions (3′UTRs) were used to derive gene-specific riboprobes for mRNA expression analysis. For CYP1A4 probe, by digesting a partial cDNA clone containing the 3′ 1.2 kb of CYP1A4′ with AvaI and HindIII restriction enzymes (Boehringer Mannheim, Indianapolis, IN), plasmid $pTCDD_{AHH}3'$,2 containing 3′-most 540 bp of CYP1A4 cDNA was generated. Using T4 DNA ligase, the fragment was ligated into pGEM4Z. Plasmid pTCDDAA7 containing 3′-most 415 bp of a partial CYP1A5 clone in pExLox was used for CYP1A5 probe. From linearized plasmids $pTCDD_{AHH}3'$ or $pTCDD_{AA}7$, using a SP6/T7 Transcription Kit, sense and antisense probes for CYP1A4 and CYP1A5 were generated. The used restriction enzymes and promoters were sense: CYP1A4, AvaI, T7 promoter; CYP1A5, EcoRI, SP6 promoter; and antisense: CYP1A4, HindIII, SP6 promoter; CYP1A5, HindIII, T7 promoter. The reaction mixtures consisted of 0.01 M DTT, 20 U RNase inhibitor, 10 nmol of each of ATP, CTP, and GTP, 1 mg of linearized plasmid, 100 μCi [^{35}S]UTP, 10 U of RNA polymerase, and 1× transcription buffer in a total volume of 0.02 mL. Reactions using T7 polymerase and SP6 polymerase (Boehringer Mannheim) proceeded at 37°C and 40°C, respectively, for 30 min, and additional RNA polymerase (10 U) was added. After 70 min, riboprobes were at 37°C with 2U DNase treated for 10 min, and 20 μg of yeast tRNA in 1.5 mM DTT was added. Total 0.15 mL of riboprobes was extracted twice with 0.15 mL of phenol/SEVAG (1/1, v/v; SEVAG consisting of chloroform/isoamyl alcohol, 24/1, v/v) and once with SEVAG alone. From an 0.001 mL aliquot of the aqueous phase, total counts per minute (cpm) were calculated. RNA was precipitated at −70°C with 2.5 volumes ethanol and 0.25 volumes ammonium acetate (7.5 M), rinsed with 70% ethanol, resuspended in 0.2 mL of DTT (10 mM), which was repeated three times, and the precipitates were finally resuspended in 0.05 mL of DTT (10 mM). An 0.001 mL aliquot was counted to measure the incorporating radiolabeled nucleotide. The values were typically between 20% and 50% of the added cpm. Probes were stored at −20°C. To determine the specificity of each riboprobe, using the Genius 4 System (Boehringer Mannheim), the linearized $pTCDD_{AHH}3'$ and $pTCDD_{AA}7$ were *in vitro* transcribed. The transcribed riboprobes resolved on a 1.5% agarose gel in 0.45 M Tris-borate, 0.01 M EDTA (pH 8.0) were transferred to Nytran membranes in 10× SSC (20×, 175.3 g NaCl, 88.2 g sodium citrate in 1 L dH_2O, pH 7.0). The membranes were baked at 80°C in a vacuum oven for 1 h. Using the Random Primed DNA labeling kit, cDNA probes consisting of either 540 bp AvaI/HindIII fragment from $pTCDD_{AHH}3'$ or 415 bp EcoRI/HindIII fragment from $pTCDD_{AA}7$ were radiolabeled

with [^{32}P]dCTP and hybridized to the membrane-bound riboprobes at 42°C for 15 h in 50% formamide, 5× Denhardt's (50×, 5 g Ficoll, 5 g polyvinylpyrrolidone, 5 g BSA fraction V, in 500 mL dH_2O), 0.25% SDS, 2 mg/mL herring sperm DNA, 5× SSPE (20×, 3 M NaCl, 0.2 M NaH_2PO_4, 20 mM EDTA, pH 7.4). Membranes were at room temperature washed twice in 2× SSPE/0.1% SDS and twice in 0.1× SSPE/0.1% SDS, and at 50°C twice in 0.1× SSPE/0.1% SDS and exposed to film (X-OMAT; Kodak, Rochester, NY). From a p-ventricular myosin heavy chain (pVMHC1) plasmid containing 3′-most 3 kb of the VMHC1 cDNA, VMHC1 subcloned into the EcoRI site of pGEM4Z sense and antisense probes was generated. The antisense probe was made using the T7 promoter following linearization with NdeI. Using the SP6 promoter and linearization with BamHI, the sense probes were made.

3. *In Situ Hybridization*: Deparaffinized and rehydrated tissue sections were at room temperature in 4% paraformaldehyde/PBS fixed for 20 min. After rehydration, sections were at room temperature in 4% paraformaldehyde in PBS fixed for 20 min, rinsed with 2× SSPE, and placed in 0.1 M Tris (pH 7.5)/0.01 M EDTA at 37°C for 30 min. Proteinase K (20 mg/mL final concentration) was added for the permeabilization of the tissues. Slides were with 2× SSPE rinsed, in 0.2 N HCl incubated for 15 min, once with 2× SSPE rinsed, twice with 0.1 M triethanolamine (pH 8.0) rinsed, and with triethanolamine plus 0.25% acetic anhydride blocked for two 5-min periods. Before hybridization, sections were stored in 2× SSPE for up to 3 h, and prior to adding the probe, in distilled H_2O rinsed. To each slide and a coverslip placed over the sections, 50 μL of probe at a specific activity of 50,000 cpm/mL hybridization solution (50% formamide, 0.3 M NaCl, 10% dextran sulfate, 20 mM Tris of pH 8.0, 5 mM EDTA, 10 mM $NaPO_4$ of pH 8.0, 13 Denhardt's, 100 mM DTT, 500 mg/mL yeast RNA, heated to 60°C) was added. Probes were at 60°C in a moist chamber containing 50% formamide, 0.3 M NaCl, 10 mM Tris (pH 8.0), 5 mM EDTA, and 10 mM $NaPO_4$ (pH 8.0) hybridized for 15 to 20 h. Slides were in washing buffer (50% formamide, 2× SSC, 10 mM β-mercaptoethanol) rinsed twice, placed in fresh buffer at 65°C for 30 min, washed for two 10-min periods at 37°C in 0.5 M NaCl, 10 mM Tris (pH 7.5), 5 mM EDTA, and once for 10 min at room temperature. Tissue sections were treated at 37°C with 20 mg/mL RNase A in 4× SSPE for 30 min, with 4× SSPE rinsed, at 65°C in washing buffer incubated for 30 min, at 37°C in 2× SSC rinsed, and in 0.1× SSC for 15 min dehydrated in 100% ethanol, air dried, and exposed to DuPont MRF 34 clear film for 3 days to determine the intensity of hybridization and plan for the subsequent length of exposure to emulsion. Prior to dipping slides in emulsion, sections were in 100% ethanol rinsed for 15 min, rehydrated successively for 15-min periods in 95%, 70%, and 50% ethanol in 0.3 M ammonium acetate/1% glycerol, and rinsed twice with 0.3 M ammonium acetate/1% glycerol. Slides were dipped in a solution of 0.6 M ammonium acetate/2% 1:1(v/v) mixture of glycerol and emulsion (prewarmed to 42°C, Rochester, NY), air dried overnight, and at 4°C exposed for 5–10 days. In Kodak D19 developer, slides were developed, fixed, with hematoxylin and eosin counterstained, and mounted in Permount for examination by light microscopy.

4. *Immunohistochemistry*: Endogenous peroxidase activity in deparaffinized and rehydrated tissue sections was with 0.6% hydrogen peroxide in methanol at

room temperature quenched for 1 h. Using the Vectastain Elite ABC Kit with goat anti-rabbit IgG as the secondary antibody and DAB tetrahydrochloride as the peroxidase substrate (Vector Laboratories, Burlingame, CA), immunostaining was performed. Immunopurified polyclonal antisera to CYP1A4 and CYP1A5 that were specific for their respective antigens were used as the primary antibodies. Three to five serial dilutions (undiluted at 1 : 10, 1 : 50, 1 : 100, and 1 : 1000) of each antiserum were used for every organ. The reactions performed without primary antiserum were completely negative. Sections were dehydrated in xylene, mounted using Permount (Fisher Scientific), and photographed under light microscopy using Kodak 64T film. Alkaline phosphatase served as a marker for proximal kidney tubules. Deparaffinized and rehydrated kidney sections were with CT.3 (100 mM Tris-HCl of pH 9.5, 150 mM NaCl, 50 mM $MgCl_2$, 0.3% Triton X-100) rinsed for two 5-min periods, then once with CT (100 mM Tris-HCl of pH 9.5, 150 mM NaCl, 50 mM $MgCl_2$) alone. Sections were reacted with 5-bromo-3-indolylphosphate/nitroblue tetrazolium (Calbiochem, San Diego, CA), the alkaline phosphatase substrate, which produced a blue precipitate when exposed to active enzyme. Two entirely independent experiments, with essentially the same results, were carried out for both the *in situ* hybridizations and immunohistochemistry, in each of which multiple sections from two control and two treated embryos were examined. As indices of CYP1A4 and CYP1A5, cytochrome P450-dependent arachidonic acid metabolism and EROD activities were assayed, respectively. The enzyme activities in homogenates of organs from 18-day-old embryos that had been treated with TCDD (1 nmol in 5 μL of dioxane) or with the vehicle alone for 24 h were assayed. Before excision, by perfusing with 4.0 mL of KPO_4 (0.1 M, pH 7.4) injected through the right ventricle, blood was removed from organs. Like tissues were pooled, homogenized in 3 volumes of the 0.1 M KPO_4, and stored at −70°C until assayed. Reaction mixtures for arachidonic acid metabolism consisted of 1 mM NADPH, 0.05 U isocitric dehydrogenase, 10 mM isocitric acid, 10 mM magnesium chloride, 3 mg of wet weight of organ homogenate, 30 mM [^{14}C] AA (54.6 mCi/mmol, New England Nuclear, Boston, MA), and 0.1 M KPO_4 (pH 7.4) having a total volume of 0.25 mL. Reactions were also carried out without NADPH. Organs not inducting arachidonic acid metabolism were also tested at concentrations of 25 mg of wet weight equivalent per reaction mixture to be sure that low levels of induction were not being missed. Following at 37°C in a shaking water bath in air incubation for 10 min, reactions were stopped with 0.1 mL of glacial acetic acid. Arachidonic acid metabolites were extracted and resolved by reverse-phase HPLC using a C18 m-Bondapak column and a linear gradient of 50% acetonitrile in H_2O plus 0.1% acetic acid to 100% acetonitrile plus 0.1% acetic acid, at a flow rate of 1 mL/min. Only NADPH-dependent peaks were considered as indicative of CYP-dependent arachidonic acid metabolism. Based on their retention times, co-migration of EETs with standards and verification by GC/MS arachidonic acid epoxygenase products (EETs and EET-diols) were identified. Organ homogenates were at 37°C with 2 mM 7-ethoxyresorufin, 1 mM NADPH, 10 mM dicumerol, and 0.1 M Tris-HCl (pH 8.3) in a total volume of 0.5 mL incubated for 5 min, after which an equal volume of acetone was added. The tubes were centrifuged, and resorufin in the supernatants was measured by fluorescence (excitation at

558 nm, emission at 586 nm). To ascertain whether the fluorescence at 586 nm in fact represented the characteristic resorufin emission peak, the emission spectra of representative samples were recorded. Fluorescence of a sample of quinine sulfate previously standardized against a known concentration of resorufin was used to determine the amount of formed resorufin.

17.15 JAPANESE MEDAKA EMBRYO-LARVAL ASSAY[20]

Approximately 50 to 70 adult Japanese medaka (*Oryzias latipes*), golden strain (housed in 38- or 57-L glass aquaria containing water that was filtered in-line by a sand filter, two 25-μm particle filters, and an activated carbon filter, flowed through an 8 W ultraviolet sterilizer, and was distributed to individual aquarium at a flow rate up to 7600 mL/h and maintained at 25 ± 2°C using a central heater and an individual heater) were fed Tetramin tropical fish food (Tetra Werke, Melke, Germany) and newly hatched brine shrimp, *Artemia salina*, three to four times a day and maintained on an artificial photoperiod of 16-h light and 8-h dark.

Within 1 to 2 h after fertilization, eggs were collected from female medaka. For embryo larval toxicity tests, a static nonrenewal exposure assay system was used. Thirteen to 18 blastula-stage eggs were individually placed into 1 mL of five to six different concentrations of [^{3}H]-TCDD (0.9–22.5 pg/mL), 3,3′,4,4′-tetrachlorobiphenyl (50–2000 ng/mL), 2,3,3′,4,4′-pentachlorobiphenyl (20–2000 ng/mL), 3,3′,4,4′,5-pentachlorobiphenyl (25–1000 pg/mL), 2,2′,3,4,4′,5′-hexachlorobiphenyl (20–2000 ng/mL), and 2,2′,4,4′,5,5′-hexachlorobiphenyl (20–2000 ng/mL) (Cambridge Isotope Laboratories, Andover, MA) dissolved with 0.02% to 0.08% acetone in rearing solution within a 2-mL Teflon-capped vial. The rearing solution consisted of 0.10 g of NaC1, 0.03 g of KC1, 0.04 g of $CaC1_2$, 0.16 g of $MgSO_4$, and 100 mL of deionized water. The exposed eggs were at 25 ± 0.5°C in an incubator kept and examined for toxic effects under a dissecting microscope daily until death or 3 days posthatch. All assays were repeated three times, and rearing solution and vehicle solvent controls (0.02% to 0.08% acetone) were used.

One milliliter of three different concentrations of [^{3}H]-TCDD (6.13, 10.02, and 12.15 pg/mL) and 0.02% acetone in rearing solution was added to 60–90 individual 2-mL Teflon-capped vials containing newly fertilized eggs, which were at 25 ± 0.5°C in an incubator kept. The amounts of [^{3}H]-TCDD distributed into chorionated embryo/mg chorionated embryo were measured on different exposure days, and the amounts of [^{3}H]-TCDD distributed into newly hatched larva/rag newly hatched larva were measured on exposure day 10. By a Tracor Mark III liquid scintillation counter (Tracor Analytic, ELK Grove Village, IL), [^{3}H]-TCDD amount was measured. After subtraction of background level, disintegrations per minute of samples were converted to concentration equivalents. Composites of embryos and newly hatched larvae were weighed, digested with 1 mL of sodium hydroxide (1 N) for 24 h, and neutralized with 50 μL of glacial acetic acid. Prior to liquid scintillation counting, 10 mL of ScintiVerse BD was added to the water or the digested tissue samples.

17.16 GREEN FLUORESCENT PROTEIN-BASED CELL ASSAY[21–26]

As seen in Fig. 17.2, pGreen1.1 was created by excising the 1846 base pair (bp) Hind III fragment from the plasmid pGudLuc1.1. This fragment contained the 480 bp dioxin-responsive domain from the mouse CYP1A1 gene inserted upstream of the mouse mammary tumor virus (MMTV) promoter, and it conferred dioxin responsiveness upon the MMTV promoter and adjacent reporter gene. This fragment was inserted into the Hind III site immediately upstream of the enhanced green fluorescent protein (EGFP) reporter gene in the plasmid pEGFP-1 (Clontech, Palo Alto, CA).

The mouse hepatoma (Hepa1c1c7) cell line was maintained in nonselective media (αMEM containing 10% fetal bovine serum). Plates of cells (approximately 80% confluent) were transfected with the construct pGreen1.1 (20 μg) using polybrene. After growth for 24 h, cells were split 1 to 10 and replated into selective media. After growth for 4 weeks, resistant clones were isolated and screened for EGFP assay.

Male Hartley guinea pigs (250–300 g, Charles River Breeding Laboratories, Wilmington, DE) were exposed to 12 h of light and 12 h of dark daily and were allowed free access to food and water. Hepatic cytosol was prepared in HEDG buffer [25 mM HEPES, pH 7.5, 1 mM EDTA, 1 mM DTT, and 10% (v/v) glycerol] and stored at −80°C for use. Protein concentrations were determined using bovine serum albumin as the standard.

H1G1.1c3 cells were initially plated into 6-well culture plates and treated with DMSO (1% maximum final concentration) or TCDD (1 nM in DMSO) at 37°C for 24 h. Cells were harvested by scraping into lysis buffer (50 mM NaH_2PO_4, 10 mM Tris-HCl pH 8, 200 mM NaCl) and lysed by repeated passage through a 27-gauge needle. Samples were centrifuged, and the fluorescence of an aliquot of the supernatant was determined using excitation wavelengths of 460 nm and emission wavelengths of 510 nm. In the microtiter plate analysis of EGFP, the intact cells were plated into black and clear-bottomed 96-well tissue culture dishes at 7.5×10^4 cells/well and allowed to attach for 24 h. The selective media were changed to

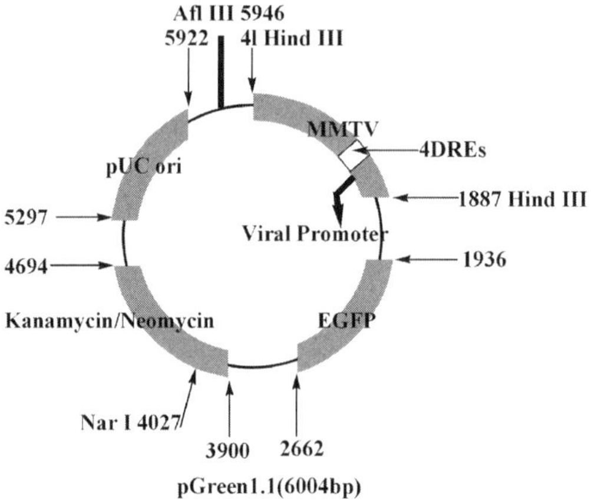

Figure 17.2 Expression vector pGreen1.1.

100 μL of nonselective media containing the test chemical or DMSO (1% final solvent concentration). EGFP levels of the intact cells in the nonselective media were measured at the indicated time points, using a Fluostar microtiter plate fluorometer with an excitation wavelength of 485 nm (25-nm bandwidth) and an emission wavelength of 515 nm (10-nm bandwidth). To normalize results between experiments, the instrument fluorescence gain setting was adjusted so that the level of EGFP induction by 1 nM TCDD produced a relative fluorescence of 9000 relative fluorescence units (RFUs). After growing on 25-mm round cover slips for 24 h and treating with DMSO or 1 nM TCDD for 48 h, the media of H1G1.1c3 cells was replaced with PBS and cell fluorescence was visualized using a fluorescence microscope with a 490-nm excitation filter and a 535-nm emission filter to photograph the cells.

Recombinant mouse hepatoma (H1L1.1c2) cells grown in 24-well microplates were incubated with DMSO (10 μL/mL), the indicated concentration of TCDD-containing sample in DMSO, or its related chemical-containing sample in DMSO at 37°C for 4 h. After incubation, luciferase activity of cells in each well was determined and normalized to sample protein concentration using fluorescamine with bovine serum albumin as the standard.

The complementary pair of 5′-GATCTGGCTCTTCTCACGCAACTCCG-3′ and 5′-GATCCGGAGTTGCGTGAGAAGAGCCA-3′ (corresponding with the AhR binding site of DRE3 and designated as the DRE oligonucleotide) was radiolabeled with [γ^{12}P]ATP (6000 Ci/mmol). Gel retardation analysis of cytosolic AhR complexes transformed *in vitro* with TCDD containing sample (20 nM) or the related chemical containing sample was carried out and protein-DNA complexes were visualized by autoradiography. The amount of [^{32}P]-labeled DRE presented in the induced protein-DNA complex was determined using a Molecular Dynamics phosphorimager.

17.17 RACB ASSAY FOR OVARIAN TOXICITY OF XENOBIOTICS[27,28]

RACB protocol consisted of Task 1 (initial dose-setting study), Task 2 (continuous breeding phase), Task 3 (crossover mating trial), and Task 4 (F_1 offspring dose). Task 2 included control (n = 40 animals/sex) and up to three treatment groups (n = 20 animals/sex). In Task 2, F_0 (parental) rodents were exposed to a chemical-containing sample during a 7-day premating period, randomly assigned to a mating pair, and treated with the same chemical-containing sample and dose throughout a 98-day period of continual cohabitation (during which multiple litters are born). To encourage immediate remating, in early pregnancies neonates were removed from the dam within 12 h. The principal measure of toxicity in Task 2 was aberrant reproductive performance in F_0 rodents as indicated by alterations in the number of litters per breeding pair or neonatal body weight and sex ratio. If a positive toxicity was observed in Task 2, Task 3 was performed to determine whether males or females were more sensitive. For this phase, high-dose F_0 mice of each sex were paired to control F_0 mice of the opposite sex, and reproductive performance was compared with matings between control F_0 males and females to determine the affected sex(es). If a negative toxicity was observed in Task 2, Task 3 was omitted. After 98 days, breeding pairs were separated and continuously treated until the F_1 generation was delivered and weaned. In Task 4, the F_1 offspring of Task 2 parents were dosed until 74 $\pm$ 10 days of age, at which time male and female animals from the same treatment group but different litters were mated (n = 20 per sex per group) to generate F_2 litters. The F_1 offspring had been exposed to the chemical as gametes and as young adults during prenatal and postnatal development.

Ovaries from Task 1 (approximately 50 days old), F_0 parents from negative Task 2 (approximately 215 days old), F_0 parents from Task 3 (approximately 240 days old), or F_1 offspring from Task 4 (approximately 120 days old) were removed at necropsy, trimmed of fat, fixed by immersion in Bouin's solution for 12–24 h, and transferred to 70% ethanol for storage and transport. Intact ovary from each animal was dehydrated in graded alcohols and xylene, embedded in a longitudinal orientation in separate paraffin blocks, and sectioned serially at 6 μm (approximately 400 sections per ovary). Two rows of five sections retained in sequence were applied to each slide (approximately 40 slides per ovary) and stained with hematoxylin and eosin.

With few exceptions, counts were gathered for ovaries from 10 mice per group. Ovaries were available from only five to six treated animals in F_1 offspring from two Task 4 EGME assays and from nine treated animals from Task 2 oxalic acid assay. To produce an equal number of control tissues for analysis from untreated animals of the same assays, the ovaries were also chosen randomly. Sections from each ovary were examined in order beginning with the first section of the first slide. Typically, follicles were encountered from at least the third or fourth section (a distance of 18 to 24 μm into the ovary), categorized, and enumerated. Differential counts were made from every 10th section (e.g., the third or fourth sections of all other slides) or approximately 40 sections per ovary. This procedure provided a nonrandom 10% sample with each counted section separated by a distance of approximately 60 μm.

The ovarian follicles were categorized to small, growing, and antral ones by (1) the relative cross-sectional diameter of the follicle as measured from the outer margins of the granulosa cell layers, (2) the number of granulosa cell layers, and (3) the nature of the antral space. Small follicles approximately 20 μm in mean diameters consisted of an isolated oocyte single layer of granulosa cells or an oocyte surrounded by a partial or unbroken single layer of granulosa cells. Growing follicles, 20 to 70 μm, had an oocyte surrounded by a multilayered, solid mantle of granulosa cells. Antral follicles, more than 70 μm, were characterized by a central oocyte and fluid-filled space bordered by hundreds of layered granulosa cells. Each counting session was limited to no more than 3 h to limit fatigue.

17.18 UROEPITHELIAL CELL ASSAY[29,30]

Male beagle dogs (31 months old) were individually housed at 23 $\pm$ 2°C and a relative humidity of 60 $\pm$ 20% with a 12-h light/dark cycle. The dogs from a control group were killed by exsanguination under sodium pentobarbital anesthesia (25 mg/kg, iv). The bladder was aseptically excised, lengthwise incised, and at 4°C washed three times with Krebs solution containing 110 mM NaCl, 5.8 mM KCl, 25 mM $NaHCO_3$, 1.2 mM KH_2PO_4, 2.0 mM $CaCl_2$, 1.2 mM $MgSO_4$, and 11.1 mM glucose (pH 7.4). The bladder was trimmed to remove excess fatty tissues and transferred, mucosal side down, to a metal rack with 10 sharp metal pins along each edge. The bladder was placed in the Krebs solution at 4°C, and the smooth muscle layer was carefully removed. The tissue was stretched, mucosal side up, across the metal pins on a $10 \times 10/cm^2$ plate and incubated in MEM (Life Technologies Inc., Grand Island, NY) containing 1% (v/v) penicillin/streptomycin/Fungizone (PSF), 2.5 mg/mL dispase, and 20 mM MEM/PSF/dispase solution (pH 7.4, Life Technologies Inc.) at 4°C for 24 h. After incubation, the MEM/PSF/dispase solution was aspirated, the stripped mucosa was transferred to a sterile 150-mm culture dish, and with cell scrapers uroepithelial cells were scraped from the connective tissues. The scraped cells were suspended in 20 mL of trypsin-EDTA (0.25% trypsin and 1 mM EDTA · 4Na), and incubated at 37°C for 30 min. Later, the single cell suspension was diluted with MEM containing 1% PSF, 5% FBS, and 20 mM of MEM/PSF/FBS solution (pH 7.4) to 50 mL in a sterile tube and spun down at 4°C and 1000 rpm with a centrifuge (KUBOTA 8800, KUBOTA Corporation, Tokyo, Japan) for 5 min. The resulting supernatant was aspirated carefully, and the cells were suspended in 50 μL of MEM/PSF/FBS solution. The cells were rewashed in 50 mL of keratinocyte-SFM medium (Life Technologies Inc.) and resuspended (6.0×10^5 to 7.0×10^5 cells/mL final concentration).

The collagen solution was prepared by mixing 5 mg of type IV collagen (Sigma-Aldrich Chem., St. Louis, MO), 100 μL of glacial acetic acid, and 50 mL of distilled water and at 4°C kept overnight without stirring. The collagen solution was sterilized with an 0.22-μm bottle top filter and stored at 4°C. The keratinocyte-SFM medium was added to both chambers with a 12-mm Transwell filter and at 37°C incubated for 2 h. Prior to use, the collagen solution was diluted 1:9 with 10 mM Na_2CO_3-HCl (pH 9.0), and 500 μL of resultant solution was added to each apical chamber after aspirating the keratinocyte medium and then incubated for 1 h at 37°C. Before plating, the collagen solution was aspirated, 0.5 mL of cell suspension was added to the apical chamber, and 1.5 mL of keratinocyte medium was added to the basal chamber. After 37°C and 3-day incubation, when the TER reached levels of approximately 1000 Ωcm^2 or higher, the apical and basal media were aspirated and replaced with 0.5 and 1.5 mL of the keratinocyte medium containing 1 mM $CaCl_2$ (KM/Ca solution), respectively. TER was measured with an epithelial voltohmmeter. Prior to use, the electrodes were sterilized by immersing them in 70% ethanol and washed with sterile PBS. Calculations of Ωcm^2 were made by subtracting blank values from that of all samples and multiplying by the area seeded with cells.

Cells cultured on a 12-mm Transwell filter (Corning Coaster Corporation, Cambridge, MA) were fixed in 2% glutaraldehyde in 0.1 M phosphate buffer (pH 7.4), rinsed with 8.2% sucrose, postfixed in 1% OsO_4 in the buffer, dehydrated through an ascending alcohol series, and embedded in epoxy resin. For light microscopic examination, semithin sections were made and stained with 1% toluidine blue. To observe cells with a transmission electron microscope, ultrathin sections were stained with uranyl acetate and lead citrate. For immunofluorescence staining, cells cultured on a 12-mm Transwell filter were fixed with a HISTOCHOICE at room temperature for 1 min, followed by acidic methanol (95% methanol and 5% glacial acetic acid) at -20°C for 15 min. The fixed cells were incubated with anti-ZO-1 rabbit polyclonal antibody diluted 1:50 with PBS for 1 h and incubated with FITC-conjugated donkey anti-rabbit IgG diluted 1:40 with PBS for 30 min, or incubated with anti-E-cadherin mouse monoclonal antibody (Transduction Laboratories, San Diego, CA) 1:100 with PBS and then incubated with FITC-conjugated goat anti-mouse IgG (Amersham Pharmacia Biotech UK Ltd., Buckinghamshire, England) diluted 1:10 with PBS. To observe cells with a confocal laser scanning microscope, the cells were washed three times for 5 min with PBS, and the cell-grown Transwell filters were cut, transferred to a slide glass, and mounted with the mounting medium.

After plating on a 12-mm Transwell filter, the changes in TER were serially measured over 20 days. The medium was replaced every 3 days. When TER reached 10,000 Ω cm^2 or more, the cultured cells were observed by light and electron microscopy, and their ZO-1 and E-cadherin were checked with a confocal microscope. Furthermore, effects of cytochalasin-B, which depolymerizes actin microfilaments and consequently opens the tight junction, on TER and ZO-1 were examined to validate this culture system. Cytochalasin-B was dissolved in DMSO, diluted with the KM/Ca solution to make concentrations of 1.6, 4, and 10 μM, and sterilized with an 0.22-μm membrane filter (Millex-GV, Millipore Co., Bedford, MA). The final DMSO concentration was 0.08% in the 1.6 μM solution, 0.2% in the 4 μM solution, or 0.5% in the 10 μM solution. DMSO diluted 1:200 with KM/Ca solution served as a negative control. TER was measured at 0, 1, 2, 4, 8, 24, and 48 h after exposure, and immunofluorescence for ZO-1 was observed 48 h after exposure.

Cells with 10,000 Ωcm^2 of TER or more were used for this assay. Compounds were dissolved in DMSO, diluted with KM/Ca solution to make concentrations of 0.8 and 2 mM, and sterilized with an 0.22-μm membrane filter. The final DMSO concentration ranged between 0.08% and 0.2%, and diluted DMSO acted as a negative control. TER was measured at 0, 1, 2, 4, 8, 24, 48, 72, and 120 h after exposure, and immunofluorescence for ZO-1 was observed 48 and 120 h after exposure.

17.19 MALE RAT SYSTEMIC TOXICITY ASSAY[31]

1. *Animals*: Male albino rats (age 32 days) were housed in a well-ventilated animal house with 12-h light/12-h dark schedule, fed with a balanced animal feed (Ashirwad Animal Feed Industries, Punjab, India), and had access to normal drinking water at all times and were acclimatized to the animal house condition for 10 days prior to the experiments. Histopathologic sections of liver and kidney from wastewater treatment plants (WWTP) effluent treated intact rats were fixed in Bouin's solution, dehydrated by upgrading from 30% to 100% series of alcohol and to xylene each for 1 h followed by cutting the section in paraffin blocks to 5 μm thickness and staining the section in hematoxylin and eosin.

2. *Hormone Analysis*: For determining serum levels of testosterone, luteinizing hormone (LH), and follicle-stimulating hormone (FSH) of completely treated rats,

by cardiac puncture from all intact vehicle-treated and 200 mL equivalent of water sample-treated rats, blood was collected and allowed to clot at room temperature for 30 min, the serum was collected and centrifuged at 2000 × *g* for 10 min. Using the commercial enzyme immunoassay kits (Omega Diagnostics, UK) according to the manufacturers' instructions, the clear supernatant was assayed for hormone levels.

3. *Steroidogenic Enzyme Activity In Vitro*: The testis removed from the different groups of intact rats were homogenized in 20% spectroscopic-grade glycerol containing 5 mM potassium phosphate and 1 mM EDTA and centrifuged at 10,000 × *g* and 4°C for 10 min. In 3β-hydroxysteroid dehydrogenase (3β-HSD) activity assay, 3 mL of incubation mixture consisted of 1 mL aliquot of the supernatant, 100 μM sodium pyrophosphate buffer (pH 8.9), 0.9 mL of double distilled water, and 30 μg of DHEA. At 25°C after the adding 0.5 μM NAD^+ to the incubation mixture, against a blank without NAD^+ enzyme activity was measured. In 17β-hydroxysteroid dehydrogenase type III (17β-HSD3) activity assay, 3 mL of incubation mixture consisted of 1 mL aliquot of the supernatant, 400 μM sodium pyrophosphate buffer (pH 10.2), 25 mg of bovine serum albumin, and 0.3 μM testosterone. After adding 1.1 μM NADP to the incubation mixture, against a blank without NADP enzyme activity was measured. With this experiment, the activity of 17β-HSD-catalyzed reverse reaction (conversion of testosterone to androstenedione) was determined, while 17β-HSD-catalyzed forward reaction (conversion of androstenedione to testosterone) was performed under almost similar condition using 50 mM phosphate buffer, 30 nM androstenedione, and 7 mM NADPH. One unit of enzyme activity equaled a change in the absorbance of 0.001 U/min at 340 nm.

4. *Semiquantitative RT-PCR*: Total RNA extracted from the testes and adrenal of the intact rats treated with 200 mL equivalent of water samples were quantified. Equal amount of RNA was transcribed with the help of the RT-PCR kit (Bangalore Genei, Bangalore, India) according to the manufacturer's instructions. Denaturing at 94°C for 60 s, annealing at 55°C for 30 s and by extension at 72°C for 60 s and running 30 cycles for amplification, PCR was performed. The primer sequences were $P450_{SCC}$-F CGCTCAGTGCTGGTCAAAA, $P450_{SCC}$-R TCTGGTAGACGGCGTCGAT, $P450_{C17}$-F GACCAAGGGAAAGGCGT, $P450_{C17}$-R GCATCCACGATACCCTC, 3β-HSD-F CCGCAAGTATCATGACAGA, 3β-HSD-R CCGCAAGTATCATGACAGA, 17β-HSD-F TTCTGCAAGGCTTTACCAGG, 17β-HSD-R ACAAACTCATCGGCGGTCTT, AR-F TTACGAAGTGGGCATGATGA, AR-R ATCTTGTCCAGGACTCGGTG, GAPDH-F AGACAGCCGCATCTTCTTGT, and GAPDH-R CTTGCCGTGGGTAGAGTCAT. On 2% agarose gel, the PCR products were separated and visualized in a gel documentation system. With a gel analyzer, the intensity of the bands on gels was converted into a digital image, the amounts of RT-PCR products were quantified with Scion Images software (Scion Corporation, Fredrick, MD, USA), and PCR products of GAPDH gene were used as internal standards.

5. *Serum AP, Acid Phosphatase, SGPT, and SGOT Level Tests*: One day before euthanizing, from tail vein of all the intact treated and control rats serum was collected and subjected to serum alkaline phosphatase, acid phosphatase, SGOT, and SGPT assays.

6. *Reversed Phase HPLC Fractionation of the Effluents*: On a Synergi Max-RP C12 column with 4-μm particle size, high performance liquid chromatography (HPLC) fractionation of WWTP influent and effluent extracts was performed. The HPLC system had Agilent G1312A binary gradient pump, with a gradient consisting of 0.1% formic acid (v/v) in water (A) and 100% methanol (B) at a flow rate of 700 μL/min and following 5% B held for 3.5 min, increased linearly to 80% by 10 min and held for 3 min, and stepped to 100% and held for 8 min. To bring the total run time per sample to 30 min, at the beginning of each run a 9-min equilibration step at 5% B was used. After injection of 10 μL of influent and effluent extracts, the compounds were monitored by fluorescence with 229 nm excitation/310 nm emission or 210 nm wavelength.

17.20 28-DAY ORAL TOXICITY ASSAY[32,33]

1. *Chemical and Animals*: Androgen agonist 17β-methyltestosterone was (Sigma-Aldrich, St. Louis, MO, USA) suspended in aqueous solution with 0.5% methylcellulose to produce the required dosing concentrations (w/v) and stored in air-tight light-resistant containers at 4°C when not in use, which was demonstrated to be chemically stable for at least 21 days. Male and female 5-week-old specific pathogen free Wistar rats were housed individually in suspended stainless steel wire mesh cages and provided *ad libitum* with certified rodent pellet diet (Pietrement, France) and filtered tap water. At the start of treatment, rats were 7 weeks old, their room was maintained at 22 ± 2°C, a relative humidity of approximately 55%, a 12-h light/dark regime, and an air exchange rate of $15\,h^{-1}$. Clinical signs were recorded daily for all rats.

Detailed physical examinations were given weekly during the treatment period and the nature, onset, severity, reversibility, and duration of clinical signs were recorded. Individual body weight and food consumption of the rats were measured throughout the treatment period. During the acclimatization phase and the 4th week of treatment, a functional observation battery (FOB) consisting of assessment of grip strength, landing foot splay distance, and exploratory motor activity, together with the testing of grasping, righting, corneal, pupillary, auditory startle, and head shaking reflexes was performed. In the 4th week of the experiment, in order to have females in the same stage of estrus (diestrus) at the time of blood sampling and sacrifice, the estrous cycle of all females was determined daily by vaginal smears for at least five consecutive days. Females were necropsied between assay days 29 and 33 and were acyclic. At the histopathologic, examination the stage of estrus was confirmed.

2. *Assessed Parameters*: Red cell count, hemoglobin, hematocrit, mean corpuscular volume, mean corpuscular hemoglobin, mean corpuscular hemoglobin concentration, white cell count, differential count and platelet count, total bilirubin, glucose, urea, creatinine, total protein, albumin, total cholesterol, triglycerides, chloride, sodium, potassium, calcium, inorganic phosphorus concentrations, aspartate aminotransferase, and alkaline phosphatase were assessed. For both sexes TSH, T3, T4, and LH levels were measured, in addition, testosterone levels and estradiol

levels were measured for males and females, respectively. In hormonal evaluation, specific radioimmunoassay kits were used.

3. Between experimental day 29 and 33, all rats were necropsied, while females were sacrificed when they were in diestrus. The necropsy included examining fresh adrenal gland, brain, epididymis (right), heart, kidney, liver, ovary, pituitary gland, prostate (dorsolateral and ventral part separate), spleen, testis, thymus, thyroid (with parathyroid), uterus (including cervix), and seminal vesicle (with coagulating glands). Besides, paired organs were weighed together. By immersion in neutral buffered 10% formalin, adrenal gland, bone (sternum), bone marrow (sternum), brain, cecum, colon, duodenum, heart, ileum, jejunum, kidney, liver, lung, mammary gland, mesenteric lymph node, ovary, pancreas, parathyroid gland, pituitary gland, prostate (ventral), prostate (dorsolateral), rectum, sciatic nerve, seminal vesicles, skin, spinal cord, spleen, stomach, submaxillary gland, submaxillary lymph node, thymus, thyroid, trachea, urinary bladder, uterus, and vagina were fixed. In Davidson's fixative, the testis and right epididymis were fixed. All samples were embedded in paraffin wax, and histologic sections were stained with hematoxylin and eosin. For rats from all groups, adrenal gland, kidney, liver, thymus and thyroid, right epididymis, mammary gland, ovary, prostate (dorsolateral part and ventral part), seminal vesicle (including coagulating glands), testis, vagina, uterus (including cervix), and macroscopic abnormalities were examined. For rats from high dose of each sex and the control groups, brain, heart, lung, pancreas, parathyroid gland, pituitary gland, skin, and spleen were examined.

4. *Spermatology*: At necropsy, the left epididymis of each male was removed, while the caudal epididymis part was dissected, weighed, placed into 10 mL of prewarmed (37°C) medium 199 for sperm collection, and total number of homogenization-resistant sperm was counted (million/g of caudal epididymis). The morphology of 200 formalin-fixed, eosin-Y-stained spermatozoa was assessed for head, flagellum, and cytoplasmic droplet abnormalities.

5. *In Situ Detection and Enumeration of Apoptotic Germ Cells in the Testis*: In the *in situ* end labeling of fragmented DNA in testes, the TUNEL method was followed. With 0.01 M citrate buffer (pH 6.0), deparaffinized sections were incubated at 97°C for 40 min to digest nuclear proteins, treated with 2% hydrogen peroxide for 5 min to inactivate endogenous peroxidase activity, at 37°C incubated with 0.3 U/μL terminal deoxynucleotidyltransferase (TdT, Amersham Pharmacia Biotech, Piscataway, NJ) and biotin-16-deoxy (d)-UTP (Roche Molecular Biochemicals, Mannheim, Germany) for 60 min, giving 10U/0.2 μL TdT, and at room temperature treated with 300 mM NaCl and 30 mM sodium citrate for 15 min. By treatment with 2% bovine serum albumin at room temperature for 10 min, nonspecific binding sites were blocked. Finally, sections were incubated with avidin peroxidase complex (extravidin-peroxidase, Sigma-Aldrich, St. Louis, MO, USA) in a 1 : 10 ratio at 37°C for 30 min, and with diaminobenzidine and hydrogen peroxide, the peroxidase reaction was visualized. Having a light counterstain with hematoxylin, sections were dehydrated and mounted. In negative control, no TdT enzyme was used. For apoptotic testicular germ cells, *in situ* quantitative evaluation was done at

400-fold magnification, on a total number of 80 perpendicularly sectioned tubules per animal. In each testis, 10 randomly selected tubules at stages I–VI, VII–VIII, IX–XIII, and XIV of the rat spermatogenic cycle were examined. The incidence of germ cell apoptosis (apoptotic index) was expressed as the number of apoptotic germ cells per 100 Sertoli cells.

17.21 ENDOCRINE DISRUPTORS ASSAY[34]

Reproductive effects such as vitellogenin (VG) induction and multigeneration reproduction of E2 exposure were evaluated through the diet. In the evaluation of VG induction, male and female adult Japanese medaka were exposed to E2 diet at a dose of 5 mg/kg for 10 days. At the end of the exposure, male and female adult Japanese medaka were anesthetized to prepare hepatic cytosolic VG samples. As an exclusively female precursor to the egg-yolk proteins in fish, hepatic VG was assayed with Western blot using a monoclonal anti-striped bass (*Morone saxatilis*) antibody. From adult Japanese medaka, microsome was prepared and the isolated total hepatic cytosolic protein samples were separated on SDS-PAGE (6.25%). In the evaluation of multigeneration reproduction, juvenile medaka were exposed to E2 diet at a dose of 0.05, 0.5, and 5.0 mg/kg of 5% body weight per day for 170 days [184 days post-hatch (DPH)]. Growth end points and biochemical end points (VG production and hepatic microsomal testosterone metabolism) were examined.

17.22 MATING EFFICIENCY ASSAY[35–37]

Yeast *Saccharomyces cerevisiae* strains W303A (MATa; ade2, his3, trp1, ura3) and 144-3A (MATa; ura3, leu2, his4) were grown overnight at 30°C in YPD medium (10 g of Bacto-yeast extract, 20 g of Bacto-peptone per liter, and 2% glucose). For the pheromone signaling pathway experiment, the W303A strain with pSL307 plasmid fused to the FUS1-lacZ was grown overnight at 30°C in SD medium (0.67% yeast nitrogen base and 2% glucose/L) supplemented with adenine, L-histidine, and L-tryptophan. The overnight culture (1/10) was transferred to fresh YPD or SD medium and incubated at 30°C for 3 h. Cultures that were adjusted at the logarithmic phase (OD = 0.8 to 1.0) of growth were harvested and subsequently used at 30°C.

As depicted in Fig. 17.3, α-cells (W303A and W303A with fused plasmid pSL307) were incubated in the absence or presence of chemicals in YPD (10 g bacto-yeast extract, Wako Chemicals) or SD medium (0.67% yeast nitrogen base, Difco Laboratories) for 60 min. Harvested cells were washed three times with distilled water and incubated in fresh YPD or SD medium with 5 mM α factor (WHWLQLKPGQPMY, Sigma Chemical, Inc.) for 120 min. Using a microscope, shmoo formation was observed. Using a β-galactosidase activity assay, the pheromone response pathway in the W303A strain with pSL307 fused to the FUS1-lacZ reporter gene was quantified. Yeast cells were at 3000 × *g* centrifuged and harvested. Pellets were washed twice in distilled water and resuspended in 1.0 mL of Z buffer

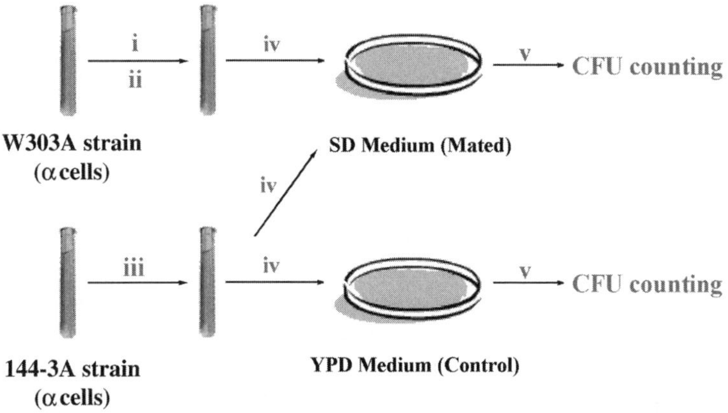

Figure 17.3 Diagram of the yeast quantitative mating efficiency assay. (i) Treated with chemicals in YPD medium for 60 min at 30°C. (ii) Diluted to 10^{-4}. (iii) Incubated in YPD medium for 60 min at 30°C. (iv) 0.1 mL. (v) Incubation for 2–3 days.

(21.5 g $Na_2HPO_4 \cdot 12H_2O$, 6.2 g $NaH_2PO_4 \cdot 2H_2O$, 0.75 g KCl, 0.246 g $MgSO_4 \cdot 7H_2O$, 2.7 mL β-mercaptoethanol per liter, pH 7.0) with glass beads. Three drops of chloroform and two drops of 0.1% SDS were added, followed by vortexing for 10 s and incubation at 28°C for 5 min. The reaction was started by adding 0.2 mL of 4 mg/mL *p*-nitrophenyl-β-D-galactopyranoside (ONPG) at 28°C and stopped by adding 0.5 mL of 1 M Na_2CO_3. Cells were removed by centrifugation and measured at 420 nm. In the mating efficiency assay, diluted a cells (W303A strain) were treated with chemicals containing sample in YPD medium for 60 min and spread on both SD and YPD media. The colony-forming units (CFUs) of *a* cell incubation on YPD medium for 2–3 days gave the cardinal number of mating efficiency. Simultaneously, α cells (144-3A strain) were incubated on YPD medium for 60 min and streaked with diluted *a* cells on SD medium. Conjugated diploid *a*/α cells on SD medium complemented the auxotrophy of each haploid cell.

17.23 SOILS ASSAY FOR CONTAMINANTS AND TOXICITY[38]

Using a stainless steel hand trowel, 2 kg of soil for toxicity testing was collected, bagged, sealed, and returned to the laboratory on ice. With disposable scoops modified from plastic syringes, subsamples for metals and volatile organic compounds (VOCs) were taken in the field. For VOC analysis, 10 g subsamples were placed in glass VOC vials, while for metals analyses approximately 50 g of soil was placed in a separate containers. Analytical samples were shipped on ice to a laboratory. Within 10 days, all samples were analyzed; this was well within acceptable holding times (6 months for metals and 14 days for VOCs). Within 14 days, toxicity test was started. Purge and trap GC/MS was used for 31 organic contaminants in soils. After acid digestion, metals (As, Ba, Cd, Cr, Pb, Se, Ag) in soil were quantified using inductively coupled plasma–atomic emission spectrometry (ICP-AES). Via cold vapor atomic absorption, mercury in soil was analyzed.

All earthworms (*Eisenia fetida*) used in toxicity assays and maintained at 20°C to 22°C were clitellid adults with a minimum weight of 300 mg. Tests consisted of three replicates of 10 organisms per treatment. Treatments included seven flooded soils, soil from the background location, positive control, and negative control. Exposure period was 30 days with mortality as the end point, additional sublethal end points included weight loss, burrowing time, and metal bioaccumulation. The remaining soil was divided into 220 g of subsamples and wetted to 35% moisture. An artificial soil was mixed as positive and negative controls. Artificial soil was a hydrated dry weight mixture of 70% sand, 20% kaolin clay, and 10% peat moss. Negative control was set for providing evidence of the health and relative quality of the test organisms, determining the suitability of test conditions and handling procedures and providing a basis for data interpretation. Positive control was involved to ensure that in the presence of a known toxicant, test organisms would respond predictably. As positive control, 65 mg of 2-chloroacetamide was included in 1000 g of dry artificial soil. 2-Chloroacetamide was commonly used as a reference toxicant for *Eisenia fetida*, and 6.5% concentration was slightly greater than the historical LC_{50} for in-house cultures. 2-Chloroacetamide was added to the laboratory water for 35% hydration of the artificial soil. Hydrated soil was divided into three subsamples of 220 g, placed in glass test chambers, and allowed to equilibrate at 21 ± 2°C for 24 h. From the culture, 10 adult clitellid earthworms were selected, carefully freed of debris and excess moisture and weighed as a group before being placed on the surface of soil in each test chamber. Burrowing time was recorded, test chambers were loosely covered with ventilated jar lids and placed in an environmental chamber with continuous lighting and a controlled 21 ± 2°C. On days 7 and 14, test organisms were examined for mortality. On day 14, test organisms were weighed and returned to test containers. After 30 days, test organisms were examined for mortality, transferred to moist filter paper in Petri dishes for 24 h to purge gut contents, frozen, and the metal content of the earthworm tissue was measured. Worm tissue was pooled across replicates, digested with nitric acid and hydrogen peroxide and analyzed via ICP-AES.

To simulate the exposure of nematodes to pore water after heavy rainfall or another flood event, *Caenorhabditis elegans* testing was designed. A 50 g subsample of each soil collected for toxicity testing was added to 100 mL of Nanopure water. On a magnetic stir plate, these slurries were mixed for 1 h and filtered through course filter paper (P8) into glass bottles. The tests used 12-well cell culture plates with 1 mL of soil extract per well. Four-day-old, age-synchronized adult nematodes were transferred into the solutions, and test vessels were stored in a dark environmental chamber at 20 ± 1°C. Experimental design included three replicates per sample, 9–11 test organisms per replicate, and 48 h exposure period with mortality as the end point. Via a Metrohm Peak 761 Compact Ion Chromatograph using a Metrosep Supp 5-250 anion column, subsamples of simulated pore water were analyzed for anion concentrations. Mean *Eisenia fetida* weight changes and burrowing times for soil samples from each location were compared using Kruskal-Wallis and Mann-Whitney U statistical tests run on SPSS software. Mortality of *Eisenia fetida* and *Caenorhabditis elegans* in each test soil was compared with that of negative control using Fisher's exact test run on Toxstat software (Version 3.4).

17.24 RICE NITRATE REDUCTASE ACTIVITY ASSAY[39]

Using chilled mortar and pestle in 3 mL of extraction buffer containing 50 mM HEPES-KOH (pH 7.5), 10 mM $MgCl_2$, 10 mM flavin adenine dinucleotide (FAD), 1 mM DTT, insoluble polyvinyl pyrrolidone (PVP 1%, w/v), casein (0.05%, w/v), and 1% Triton X-100, about 1 g of root/shoot samples were homogenized, centrifuged at 4°C and 22,000 × *g* for 10 min. The clear supernatant was dialyzed in cold cellophane membranes for 3–4 h against the extraction buffer, the enzyme extract was desalted and added to an incubation mixture containing 50 mM HEPES-KOH (pH 7.5), 10 mM $MgCl_2$, 10 mM FAD, 1 mM DTT (for determination of NR act), or at 20°C with 50 mM HEPES-KOH (pH 7.5), 10 mM $MgCl_2$, 10 mM FAD, 1 mM DTT, and 15 mM EDTA (for determination of NR max) preincubated for 15 min. By adding 10 mM KNO_3 and 0.2 mM NADH and at 30°C incubated for 30 min, the reaction was started, and by adding 0.1 M Zn-acetate, the reaction was stopped. After 15 min, by adding 2 mL of color developing reagent [1% sulfanilamide and 0.2% *N*-(1-naphthyl)ethylenediamine dihydrochloride (NEDH), each prepared in 1.5 M HCl and mixed in equal volumes], produced nitrite was at 540 nm measured colorimetrically. Using BSA (Sigma-Aldrich, St. Louis, MO, USA) as standard, protein was estimated.

In addition to the normal ingredients of the assay medium, reaction mixture contained increasing concentrations of H_2O_2 (0–50 mM), NaCl (0–600 mM), PEG-6000 (0–80%), and metal salts (0–4 mM). To observe the possible protective roles of osmolytes proline, glycine betaine, and sucrose on nitrate reductase (NR) activity, 1 M each osmolyte was added to the reaction mixture, in which either 1 mM H_2O_2, 300 mM NaCl, 40% PEG, or 2 mM metal was maintained. Contents were at 40°C kept for 45 min to assay NR activities. By constructing Lineweaver-Burk (L-B) plots, value of Michaelis constant (Km) and Vmax for NR under all conditions were determined with respect to nitrate.

Proline in roots and shoots of the seedlings was estimated. In 10 mL of 3% aqueous sulfosalicylic acid, 500 mg of fresh samples were homogenized and centrifuged at 22,000 × *g* for 5 min. To 2 mL of supernatant, 2 mL of acid ninhydrin and 2 mL of glacial acetic acid were successively added, and the contents were at 100°C in a water bath refluxed for 1 h. In a test tube, the mixture was extracted with 10 mL of toluene by vigorous stirring. On an ELICO CL-24 spectrophotometer (Hyderabad, India) at 515 nm, the chromophore absorption was read. L-Proline was used to prepare standard curve. The amount of proline in the sample was calculated in mg (proline)/g fresh weight. To estimate glycine betaine, roots and shoots samples were oven dried at 80°C for 4 days. About 500 mg of dried samples were ground to fine powder and at 25°C with 20 mL of water shaken for 24 h. Contents were filtered and stored in the freezer until analyzed. Aliquots containing 0.5 mL of thawed extract were with

0.2 N H_2SO_4 diluted to 1 : 1 and cooled in ice water for 1 h. After adding 0.2 mL of cold KI-I_2 reagent (dissolving 15.7 g iodine and 20.0 g KI in 100 mL water), the mixture was stirred gently, stored at 0°C to 4°C for 16 h, centrifuged at 16,000 rpm for 15 min, and supernatant was aspirated with a fine-tipped glass tube. Because the solubility of complex in the acid reaction mixture increased markedly with cold, conditions were maintained throughout the process. Tubes to be kept cold until the periodide complex was separated from the media was important. From the acid media, periodide crystals were separated and dissolved in 9 mL of 1,2-dichloroethane. For complete solubilization of crystals in the developing solvent, vigorous shaking was needed. After 2 h, absorbance was recorded at 365 nm. To prepare standard curve, glycine betaine was used in the range 50–200 mg/mL, and 1 N H_2SO_4 was used as solvent. Amount of glycine betaine (Sigma-Aldrich, St. Louis, MO, USA) was expressed as mg/g dry weight of samples. Using anthrone reagent, sucrose in roots and shoots was estimated. Root/shoot samples were oven dried at 70°C for 24 h.

17.25 *Tradescantia*-MICRONUCLEUS ASSAY[40]

Before assay, the water samples were adjusted near pH 7 with HCl or NaOH. In the *Tradescantia*-micronucleus Trad-MCN tests, *Tradescantia* clone 03 was used. In a 500-mL beaker, plant cuttings bearing young inflorescences (15 cuttings per experimental group) were treated with water samples for 6 h. The plant cuttings of treating and control groups were placed under artificial light for a 24-h recovery period. In this assay, eight treating (two groups for each of the four water samples), one negative control (clean tap water), and three positive control (HgCl of 30, 150, and 300 mg/L) groups were involved. A standard protocol including treatment, fixation, slide preparation, and scoring micronuclei frequencies was used for Trad-MCN assay. From the MCN frequencies scored (1500 tetrads scored per group), means and standard deviations of each group were obtained, and Dunnett's *t*-test was used.

17.26 NITRIC OXIDE NEUROTOXICITY ASSAY[41]

Isolated monoclonal antibody (IgG) to plasminogen activation inhibitor-1 was dissolved in PBS at 1 mg/mL and stored at −80°C before use. By mixing 10 mL of stock solution of antibody with 10 mL of ammonium acetate buffer (50 mM, pH 5) and adding various concentrations of peroxynitrite, the reaction was carried out. The samples were analyzed on an Applied Biosystems analytical capillary electrophoresis system model 270A with 70-cm fused silica capillary column. The buffer consisted of 50 mM sodium tetraborate and 25 mM lithium chloride. At 25 kV, capillary electrophoretic separation was performed. Under vacuum, sample was injected for a period of 3 s at about 6 nL/s. The analytes were monitored at 214 nm.

Using Control's GP 422, amino acid in the reaction mixtures was analyzed. To each 0.5 mL of acid hydrolysis mixture in a 2-mL glass ampoule, 100 μL aliquot of the nitrated antibody was added. Acid used for hydrolysis consisted of 6 N HCl and trifluoroacetic acid containing 1% phenol in a 2:1 ratio. Prior to use, this mixture was sparged with helium. After a freeze/thaw cycle, the ampoules sealed under vacuum samples were heated at 105°C to 110°C for 24 h, cooled, unsealed, on a Speed-Vac taken to dryness, reconstituted in 250 μL of Beckman Na-S buffer, and analyzed on the Beckman 7300 Amino Acid Analyzer. Corrected amino acid compositions were obtained versus standards hydrolyzed in the same manner.

In 0.5 mL of HCl (6 N), 1 mg of plasminogen activator inhibitor-1 monoclonal antibody was hydrolyzed for 18 h, and using standard technique, amino acid content was analyzed. Nitrotyrosine in hydrolysate was analyzed on HPLC. The mobile phase consisted of monochloroacetic acid (14 mg/mL), acetonitrile (7%), and tetrahydrofuran (3%). After clearing the column, the effluent was passed through a guard cell set for reduction (−2000 mV) and analyzed with an oxidative electrochemical cell (450 mV).

A specialized technique was used determining cysteine content. In the presence of dithiodiglycolic acid, samples were hydrolyzed *in vacuo* at 120°C for 16 h, derivatized with phenylisothiocyanate, and the resulting phenylthiocarbamyl amino acid was separated on RP-HPLC. Amino acid residue number was based on a typical molecular weight for IgG of 160,000.

17.27 EBV-TRANSFORMED HUMAN BURKITT'S LYMPHOMA CELL ASSAY[42]

EBV-transformed human Burkitt's lymphoma cells (Brown University) cultured in RPMI 1640 media supplemented with 10% FCS and supplied with HEPES buffer (pH 7.2, Gibco Inc., Grand Island, NY) were treated with potassium chromate for 18 h, *cis*-platinum for 16 h, all other metal salts for 6 h, lysed in 0.5% SDS solution, frozen at −70°C, thawed at 37°C, the DNA was sheared by passing the cell lysates four times through a 21-gauge needle, the lysates were expelled into a tube by applying medium pressure, extensive foaming was avoided, and 0.5 mL of 100 mM KCI and 20 mM Tris (pH 7.5) were added. By vortexing the tube for 5 s at maximal speed, the content was mixed and the tubes were at 65°C heated for 10 min. From the water bath, samples were removed, inverted three times, and placed on ice for 5 min to form potassium dodecyl sulfate precipitates. At 4°C by centrifugation at 6000 × *g* for 5 min, the precipitates were collected. By brief vortexing at the highest setting, the supernatant was removed and the pellets were resuspended in 100 mM KC1 and 20 mM Tris-HCl (pH 7.5). The samples were at 65°C heated for another 10 min, washed, and heated, which steps were repeated two times. With treating 0.2 mg/mL proteinase K in 0.5 mL of solution containing 100 mM KCl, 20 mM Tris-HCl (pH 7.5), and 10 mM EDTA, protein-linked DNA was released from the final K-SDS precipitates. The samples were at 50°C incubated for 3 h. To each tube, 50 μL of BSA (4 mg/mL) was added, the tubes were placed on ice, at 4°C and 12,000 × *g* centrifuged for 10 min and using Hoechst 33258 the supernatant was quantitatively analyzed for DNA. A set of standard solutions of 100, 200, 500, 1000, 2000, and 5000 ng/mL DNA were prepared. Standard DNA (0.5 mL) or the entire supernatant from the potassium dodecyl sulfate pellet was mixed with 0.5 mL of freshly prepared Hoechst dye reagent (250 ng/mL), the samples were placed in the dark for 10 min, fluorescence was assessed by excitation at 365 nm, and the emitted light was measured at 450–460 nm. A BSA tube was used as the standard blank.

Using Trypan blue, initial screening of short-term viability in the lymphoma cells exposed to the agents was assessed. With a hemocytometer, the number of cells excluding Trypan blue were determined. The cells exposed to the chemicals for DNA-protein cross-link assay were also used for determining long-term viability. After metal exposure, cells were allowed to proliferate, and using a hemocytometer, the number of cells excluding Trypan blue were determined at intervals of selected days, which depended upon cell proliferation as well as Trypan blue exclusion.

17.28 MURINE BONE MARROW ASSAY FOR HEMOPOIETIC AND OSTEOGENIC TOXICITY[43]

For bone-forming cultures, the 8- to 10-week-old BALB/c mice received a tail vein injection with 150 mg of 5-fluorouracil (Roche Diagnostics Ltd., UK)/kg body weight. Five days later, marrow cells from at least six mice were pooled and dispersed into a single cell suspension.

By a colony-forming unit/granulocyte macrophage (CFU-GM) assay, the proliferation of GM precursor cells was scored. Cells (5×10^4) were grown in a semisolid agar culture supplemented with 2 mM β-mercaptoethanol and a colony-stimulating factor derived from serum of mice receiving endotoxin. After incubation at 37°C and 5% CO_2 for 7 days, the colony numbers of granulocytes and/or macrophages were counted. After incubating marrow cells on confluent stromal cell layers at 33°C and 7% CO_2 in medium supplemented with 0.1 μM hydrocortisone sodium hemisuccinate and 0.1 mM β-mercaptoethanol, the proliferation of primitive pluripotent stem cells with *in vivo* hemopoietic repopulating capacity was measured. Half of the medium was replaced at 1-week intervals. The CFU-GM presented in these co-cultures after 4 week incubation were derived from primitive hemopoietic stem cells, which was a measure of proliferation capacity for early hemopoietic stem cells.

By a CFU/stromal stem cell (CFU-F) assay, the proliferation of stromal stem cells was scored. Cells (2×10^6) were incubated in liquid medium at 37°C and 5% CO_2, and the medium was replaced at 4-day intervals. After 10 days, colonies of adherent fibroblastoid cells were counted. A modified Eagle's medium with 10% horse serum, 10% FCS, 1% penicillin-streptomycin, and 1% L-Gln was used for these stem cell assays. After 4 week incubation of marrow on confluent stromal layers, hemopoietic differentiation of primitive stem cells was measured. Adherent cells were removed, and the cells were incubated with monoclonal rat anti-mouse antibodies for 30 min on ice for fluorescence labeling. The cells were washed, incubated with fluorescein isothiocyanate-conjugated rabbit anti-rat immunoglobulin serum, and analyzed on a FACSStar Plus flow cytometry system. After 4-week growth of marrow cells in Iscove's medium supplemented with 10% fetal calf serum, L-Gln, 1% penicillin-streptomycin, ascorbic acid (100 pg/mL), and β-glycerophosphate

(0.6 g/100 mL), osteogenic differentiation was tested. Cells (1.25×10^5 cells/200 μL) were plated in flat-bottomed 96-well plates, and 50% of the medium was replaced with fresh medium at 3-day intervals. After 4 weeks, collagen synthesis was measured according to the incorporation of [^{3}H]proline into collagen digestible protein, calcium content was measured based on a calorimetric reaction with *o*-cresolphthalein-complexon, and DNA content was measured with fluorescence after binding to propidium iodide.

To hemopoietic cultures, osteogenic cultures and undifferentiated 3T3 cells grown in 96-well plates $Pb(NO_3)_2$, catechol, hydroquinone, benzene, and phenol (Sigma-Aldrich, St. Louis, MO, USA) were added for 3 days exposure, and the number of 3T3 cells was measured by the uptake of neutral red. This assay was included as a reference for the concentration inducing cytotoxicity after 3-day exposure by preventing cell proliferation in a nonspecialized cell. In the stem cell proliferation assays, the marrow cells were exposed for 3 days only. In the bone-forming cultures, the colony forming was exposed at the onset of osteogenic differentiation (day 14–17).

Besides, each assay was repeated at least three independent times, and triplicate, quadruplicate, and at least six replicate cultures were used for each assay with 3T3 cells, colony-forming assays, and bone-forming cultures, respectively. Seven $\log_2$ dilutions or $\log_{1.6}$ dilutions were tested for each compound. By logistic regression (Graphpad Inplot), the concentration that caused 50% inhibition (IC_{50}) was calculated.

17.29 PNAR-GFP ASSAY FOR CARCINOGENIC TOXICITY OF NITRATE[44–47]

Escherichia coli DH5α (Φ80dlacZΔM15 recA endA gyrA thi hsdR supE relA deoR Δ[lacZYA-argF]), pUC18 (for routine cloning procedures, Roche Diagnostics Ltd., UK), pCRII (for TA cloning of PCR products, Invitrogen), pGreen-TIR containing the gfp gene, and pMV4 containing the narG promoter were used. Minimal medium (pH 7.0) contained 3.9 mM KH_2PO_4, 6.1 mM K_2HPO_4, 1.5 mM $(NH_4)_2SO_4$, 0.2 mM $MgSO_4 \cdot 7H_2O$, 22.4 μM $MnSO_4 \cdot 4H_2O$, 0.9 μM $FeSO_4 \cdot 4H_2O$, 4.4 μM $CaCl_2 \cdot 2H_2O$, 400 μg/mL casamino acids, 14.8 μM thiamine hydrochloride, and 22.2 mM glucose. LB medium (pH 7.5) contained 10 mg/mL bactotryptone, 5 mg/mL yeast extract, and 10 mg/mL NaCl. Cultures of *Escherichia coli* DH5α (pPNARGFP) were supplemented with 100 μg/mL ampicillin.

Primers J1 (5′-CATCGAATTCTCCTGTGGGAGCCT-3′) and J2 (5′-CTGGCATGCATTCACTTGCCGCCTT-3′) were designated for the amplification of the nar promoter. J1 contained an artificial EcoRI site (GAATTC) and anneals to nucleotides +3 to −23 in the nar region, where +1 was the first nucleotide of the start codon. J2 contained an artificial SphI site (GCATGC) and anneals to nucleotides −438 to −414. Plasmid DNAs were purified using Qiagen Plasmid Midi Kits or Wizard Plus SV miniprep kits (Promega, USA). DNA sequencing was done by MWG Biotech.

The whole bacterial cells were measured with an excitation wavelength of 480 nm (10 nm bandwidth) and an emission wavelength of 510 nm (5 nm bandwidth) on a fluorescence spectrometer. On a fluorescence microscope equipped with epifluorescence optics, the nitrate-induced fluorescence image of *Escherichia coli* DH5α (pPNARGFP) was obtained by use of BP465-495 excitation filter and a Ba 520 emission filter. The images were captured using a Cool-snap digital camera and Metamorph software system. The measurements of optical density were performed on a spectrophotometer at 600 nm.

17.30 BRINE SHRIMP (*Artemia salina*) ASSAY[48–54]

1. *Artemia salina Hatching Assay*: *Artemia salina* cysts (20–30 individuals/mL) were incubated in sterile polystyrene 24-well microplates containing 1 mL of oxygen-saturated incubation medium, either FSW, solvent blank, test compounds, or standards, and media was renewed daily. According to a random stratified design, the plates were stocked with four replicate incubations per treatment. The plates were sealed, incubated at 15°C, aerated twice daily for a period of 72 h until hatching had been completed, and to each well 0.5 mL of formaldehyde (16%) was added to kill and fix any hatched nauplii. By the contents of individual wells being transferred to a solid watch glass and observed under a binocular dissection microscope, hatching success was assessed. The number of hatched nauplii was compared with the totally stocked cysts number.

2. *Artemia salina 24 and 72 h Survival Trials*: By vigorously aerating 1000 mL of fresh water, *Artemia salina* cysts were at 18°C in a conical flask induced for 24 h to hatch. Using a fine mesh sieve, the newly hatched nauplii were concentrated into 100 mL, from which aliquots of 50 μL (approximately 30–50 nauplii) were pipetted directly into 24-well plates containing 2 mL of oxygen-saturated filtered seawater (FSW), solvent blank, test compounds, or standards. The plates were sealed, incubated at 15°C for 24 or 72 h with additional aeration occurring after 12 h, and media was renewed daily. By scoring the number of dead nauplii lying on the bottom of each plate using a Zeiss inverted microscope, survival was assessed. Once dead nauplii counts had been taken, 0.5 mL of formaldehyde (16%) was added for killing all remaining nauplii, and the contents of each well were transferred to a solid watch glass for enumeration.

17.31 DPPH RADICAL SCAVENGING PROPERTY ASSAY[49,55]

This assay was based on the reduction of DPPH [2,2-diphenyl-1-picrylhydrazyl or 2,2-diphenyl-1-(2,4,6-trinitrophenyl)-hydrazyl, Fluka Chemie AG, Buckinghamshire, England] radical with TLC autographic assay. After developing and drying, TLC plates (with amounts of sample ranging from 0.1 to 100 μg) were sprayed with 0.2% (2 mg/mL) solution of DPPH in methanol and examined half an hour after spraying. On TLC plates, active compounds appeared as yellow spots against a purple background. In each well, 1 μL of solution of DPPH in methanol (500 μM, 0.2 mg/mL) was mixed with 1 μL of various concentrations of solution of test compounds in methanol, kept in the dark for 30 min, and absorbance monitored at 517 nm.

To determine the absorbance of DPPH before interacting with test compounds, a blank experiment was performed. Sample amount in μg/mL at which the absorbance at 517 nm decreased to half its initial value was defined as the IC_{50} value of the compound.

REFERENCES AND NOTES

1. C. Baumstark-Khan, A. Rode, P. Rettberg, G. Horneck. Application of the Lux-Fluoro test as bioassay for combined genotoxicity and cytotoxicity measurements by means of recombinant Salmonella typhimurium TA1535 cells. Anal Chim Acta 437 (**2001**) 23–30.
2. K.L. Witt, E. Livanos, G.E. Kissling, D.K. Torous, W. Caspary, R.R. Tice, L. Recio. Comparison of flow cytometry- and microscopy-based methods for measuring micronucleated reticulocyte frequencies in rodents treated with nongenotoxic and genotoxic chemicals. Mutat Res 649 (**2008**) 101–113.
3. R.A. LeBoeuf, G.A. Kerckaert, M.J. Aardema, D.P. Gibson, R. Brauninger, R.J. Isfort. The pH 6.7 syrian hamster embryo cell transformation assay for assessing the carcinogenic potential of chemicals. Mutat Res 356 (**1996**) 85–127.
4. R.J. Isfort, R.A. LeBoeuf. Application of in vitro cell transformation assays to predict the carcinogenic potential of chemicals. Mutat Res 365 (**1996**) 161–173.
5. K. Shimizu, O. Endo, S. Goto, A. Sakoda, Y. Ono, Y. Sakai. Bioassay-based evaluation of toxicity of suspended particulate matter in humans: integrated uses of alveolar cells (A549) in air-liquid interface culture and hepatocarcinoma cells (Hep G2). Biochem Eng J 22 (**2004**) 1–9.
6. M. Koyano, O. Endo, S. Goto, K. Tanabe, S. Koottatep, H. Matsushita. Carcinogenic polynuclear aromatic hydrocarbons in the atmosphere in Chiang Mai, Thailand, Jpn J Toxicol Environ Health 44 (**1998**) 214–225.
7. K.L. Willett, R.P. Gradinali, J.L. Sericano, T.L. Wade, S.H. Safe. Characterization of the H4IIE rat hepatoma cell bioassay for evaluation of environmental samples containing polynuclear aromatic hydrocarbons (PAHs). Arch Environ Contam Toxicol 32 (**1996**) 442–448.
8. B.P. da Silva, J.B.R.C. Soares, E.P. de Souza, M. Palatnik, C.B.P. de Sousa, J. Parente. Pulcherrimasaponin, from the leaves of Calliandra pulcherrima, as adjuvant for immunization in the murine model of visceral leishmaniasis. Vaccine 23 (**2005**) 1061–1071.
9. G.S. Ladics, C. Smith, G.S. Elliott, T.W. Slone, S.E. Loveless. Further evaluation of the incorporation of an immunotoxicological functional assay for assessing humoral immunity for hazard identification purposes in rats in a standard toxicology study. Toxicology 126 (**1998**) 137–152.
10. K. Takahashi, M. Ohsawa, H.A. Utsumib. Simple bioassay for evaluating immunotoxic properties of chemicals by use of in vitro antibody production system. J Health Sci 48 (**2002**) 161–167.
11. M. Ohsawa, K. Masuko-Sato, K. Takahashi, F. Otsuka. Strain differences in cadmium-mediated suppression of lymphocyte proliferation in mice. Toxicol Appl Pharmacol 84 (**1986**) 379–388.
12. T. Mosmann. Rapid colorimetric assay for cellular growth and survival: application to proliferation and cytotoxicity assays. J Immunol Methods 65 (**1983**) 55–63.

13. P.H. Lin, C.H. Lin, C.C. Huang, M.C. Chuang, P. Lin. 2,3,7,8-Tetra-chlorodibenzo-p-dioxin (TCDD) induces oxidative stress, DNA strand breaks, and poly(ADP-ribose) polymerase-1 activation in human breast carcinoma cell lines. Toxicol Lett 172 (**2007**) 146–158.
14. L.E. Brown, K.R. Trought, C.I. Bailey, J.H. Clemons. 2,3,7,8-TCDD equivalence and mutagenic activity associated with PM10 from three urban locations in New Zealand. Sci Total Environ 349 (**2005**) 161–174.
15. A.S. Thornton, Y. Oda, G.R. Stuart, J. Holcroft, J.G. de Boer. The dioxin TCDD protects against aflatoxin-induced mutation in female rats, but not in male rats. Mutat Res 561 (**2004**) 147–152.
16. R. Shinkyo, T. Sakaki, T. Takita, M. Ohta, K. Inouye. Generation of 2,3,7,8-TCDD-metabolizing enzyme by modifying rat CYP1A1 through site-directed mutagenesis. Biochem Biophys Res Commun 308 (**2003**) 511–517.
17. E. Wood, M.J. Broekman, T.L. Kirley, S. Diani-Moore, M. Tickner, J.H.F. Drosopoulos, N. Islam, J.I. Park, A.J. Marcus, A.B. Rifkind. Cell-type specificity of ectonucleotidase expression and upregulation by 2,3,7,8-tetrachlorodibenzo-p-dioxin. Arch Biochem Biophys 407 (**2002**) 49–62.
18. Y. Sugawara, K. Saito, M. Ogawa, S. Kobayashi, G. Shan, J.R. Sanborn, B.D. Hammock, H. Nakazawa, Y. Matsuki. Development of dioxin toxicity evaluation method in human milk by enzyme-linked immunosorbent assay-assay validation for human milk. Chemosphere 46 (**2002**) 1471–1476.
19. M. Gannon, D. Gilday, A.B. Rifkind. TCDD induces CYP1A4 and CYP1A5 in chick liver and kidney and only CYP1A4, an enzyme lacking arachidonic acid epoxygenase-activity, in myocardium and vascular endothelium. Toxicol Appl Pharmacol 164 (**2000**) 24–37.
20. Y. Kim, K.R. Cooper. Toxicity of 2,3,7,8-tetrachlorodibenzo-p-dioxin (TCDD) and polychlorinated biphenyls (PCBs) in the embryos and newly hatched larvae of the Japanese medaka (Oryzias latipes). Chemosphere 39 (**1999**) 527–538.
21. S.R. Nagy, J.R. Sanborn, B.D. Hammock, M.S. Denison. Development of a green fluorescent protein-based cell bioassay for the rapid and inexpensive detection and characterization of Ah receptor agonists. Toxicol Sci 65 (**2002**) 200–210.
22. P.M. Garrison, K. Tullis, J.M. Aarts, A. Brouwer, J.P. Giesy, M.S. Denison. Species-specific recombinant cell lines as bioassay systems for the detection of 2,3,7,8-tetrachlorodibenzo-p-dioxin-like chemicals. Fundam Appl Toxicol 30 (**1996**) 194–203.
23. M.S. Denison, L.M. Vella, A.B. Okey. Structure and function of the Ah receptor for 2,3,7,8-tetrachloro-dibenzo-p-dioxin: species difference in molecular properties of the receptors from mouse and rat hepatic cytosols. J Biol Chem 261 (**1986**) 3987–3995.
24. M.M. Bradford. A rapid and sensitive method for the quantitation of microgram quantities of protein utilizing the principle of protein-dye binding. Anal Biochem 72 (**1976**) 248–254.
25. M.S. Denison, D. Phelan, G.M. Winter, M.H. Ziccardi. Carbaryl, a carbamate insecticide, is a ligand for the hepatic Ah (dioxin) receptor. Toxicol Appl Pharmacol 152 (**1998**) 406–414.
26. M.H. Ziccardi, I.A. Gardner, M.S. Denison. Development and modification of a recombinant cell bioassay to directly detect halogenated and polycyclic aromatic hydrocarbons in serum. Toxicol Sci 54 (**2000**) 183–193.

27. B. Bolon, T.J. Bucci, A.R. Warbritton, J.J. Chen. Differential follicle counts as a screen for chemically induced ovarian toxicity in mice: results from continuous breeding bioassays. Fundam Appl Toxicol 39 (**1997**) 1–10.

28. T. Pedersen, H. Peters. Proposal for a classification of oocytes and follicles in the mouse ovary. J Reprod Fertil 17 (**1968**) 555–557.

29. I. Yoshikazu, J. Toshimasa, F. Kazuhisa. Effect of nefiracetam, a neurotrans- mission enhancer, on primary uroepithelial cells of the canine urinary bladder. Toxicol Sci 72 (**2003**) 164–170.

30. S.A. Lewis, J.W. Hanrahan. Physiological approaches for studying mammalian urinary bladder epithelium. In: Methods in Enzymology (S. Fleischer and B. Fleischer, Eds.) 192 (**1990**) 632–650. Academic Press.

31. V. Kumar, A. Chakraborty, G. Viswanath, P. Roy. Androgenic endocrine disruptors in wastewater treatment plant effluents in India: their influence on reproductive processes and systemic toxicity in male rats. Toxicol Appl Pharmacol 226 (**2008**) 60–73.

32. S. Wason, G. Pohlmeyer-Esch, C. Pallen, X. Palazzi, G. Espuña, R. Bars. 17α-Methyltestosterone: 28-day oral toxicity study in the rat based on the "Enhanced OECD Test Guideline 407" to detect endocrine effects. Toxicology 192 (**2003**) 119–137.

33. H.P. Gelbke, M. Kayser, A. Poole. OECD test strategies and methods for endocrine disruptors. Toxicology 205 (**2004**) 17–25. *Notes*: 1. Uterotropic assay: In uterotropic assay for *in vivo* estrogenicity or antiestrogenicity, either immature intact or adult ovariectomized female rats or mice were used. Besides effects on other estrogen-dependent parameters, administration of chemicals with estrogenic activity increased the uterine weight, while antiestrogenic chemicals diminished this estrogen-dependent weight increase for intact immature or ovariectomized adult females. Validation was primarily done with intact immature rats, mice as well as adult ovariectomized animals. 2. Hershberger assay: In the Hershberger assay for *in vivo* screening (anti)androgenicity of chemicals, castrated immature male rats were used. Androgenic chemicals increased the weight of androgen-dependent tissues, e.g., the male accessory organs like the prostate, while (anti)androgenic chemicals diminished this weight increase for testosterone-treated immature castrated male rats. 3. Enhanced subacute test (TG 407): The 28-day (subacute) test in rats was a valuable method to identify the toxicologic profile of chemicals after an administration period of 28 days. To detect specific endocrine-related adverse effects and general toxicity, enhanced TG 407 had the potential. In certain circumstances, it was possible to determine if a substance represented a true endocrine disruptor with a primary mode of toxicologic action via an effect on endocrine systems or if the endocrine effects were secondary to other toxicologic end points. This type of insight into toxicologic modes of action can never be derived from *in vitro* assays or short-term *in vivo* endocrine screening assays. To further substantiate the validity of enhanced TG 407, a comparison was carried out with results obtained by tests with longer and/or *in utero* exposure.

34. P.J. Patyna, R.A. Davi, T.F. Parkerton, R.P. Brown, K.R. Cooper. A proposed multigeneration protocol for Japanese medaka (Oryzias latipes) to evaluate effects of endocrine disruptors. Sci Total Environ 233 (**1999**) 211–220.

35. K. Fujita, M. Nagaoka, Y. Komatsu, H. Iwahashia. Yeast pheromone signaling pathway as a bioassay to assess the effect of chemicals on mammalian peptide hormones. Ecotoxicol Environ Safety 56 (**2003**) 358–366.

36. C. De Virgilio, N. Burckert, W. Bell, P. Jeno, T. Boller, A. Wiemken. Disruption of TPS2, the gene encoding the 100-kDa subunit of the trehalose-6-phosphate synthase/phosphatase complex in Saccharomyces cerevisiae, causes accumulation of trehalose-6-phosphate and loss of trehalose-6-phosphate phosphatase activity. Eur J Biochem 212 (**1993**) 315–323.
37. J.H. Miller. Experiments in Molecular Genetics. Cold Spring Harbor Laboratory, Cold Spring Harbor, NY (**1972**) 352–355.
38. S.M. Harmon, D.E. Wyatt. Evaluation of post-Katrina flooded soils for contaminants and toxicity to the soil invertebrates Eisenia fetida and Caenorhabditis elegans. Chemosphere 70 (**2008**) 1857–1864.
39. P. Sharma, R.S. Dubey. Modulation of nitrate reductase activity in rice seedlings under aluminium toxicity and water stress: role of osmolytes as enzyme protectant. J Plant Physiol 162 (**2005**) 854–864.
40. G. Yang. Tradescantia-micronucleus assay on the water quality of lake Hongzhe in Jiangsu Province, China. Mutat Res 426 (**1999**) 155–157.
41. J.S. Althaus, G.J. Fici, S.M. Plaisted, F.J. Kezdy, C.M. Campbell, J.G. Hoogerheide, P.F. Von Voigtlander. Protein nitration by peroxynitrite: a method for monitoring nitric oxide neurotoxicity. Microchem J 56 (**1997**) 155–164.
42. M. Costa, A. Zhitkovich, M. Gargas, D. Paustenbach, B. Finley, J. Kuykendall, R. Billings, T.J. Carlson, K. Wetterhahn, J. Xu, S. Patierno, M. Bogdanffy. Interlaboratory validation of a new assay for DNA-protein crosslinks. Mutat Res 369 (**1996**) 13–21.
43. G.E.R. Schoeters, F.V. Plaetse, H. Leppens, P.V. Vlasselaer, R.V. Den Heuvel. Haemopoietic and osteogenic toxicity testing in vitro using murine bone marrow cultures. Toxicol In Vitro 9 (**1995**) 421–428.
44. C.J. Taylor, L.A. Bain, D.J. Richardson, S. Spiro, D.A. Russellb. Construction of a whole-cell gene reporter for the fluorescent bioassay of nitrate. Anal Biochem 328 (**2004**) 60–66.
45. W.G. Miller, S.E. Lindow. An improved GFP cloning cassette designed for prokaryotic transcriptional fusions. Gene 191 (**1997**) 149–153.
46. E.J. Sodergren, P.Y. Hsu, J.A. DeMoss. Roles of the narJ and narI gene products in the expression of nitrate reductase in Escherichia coli. J Biol Chem 263 (**1988**) 16156–16162.
47. S.F. Li, J.A. DeMoss. Promoter region of the nar operon of Escherichia coli: nucleotide sequence and transcription initiation signals. J Bacteriol 169 (**1987**) 4614–4620.
48. G.S. Caldwell, M.G. Bentley, P.J.W. Olive. The use of a brine shrimp (Artemia salina) bioassay to assess the toxicity of diatom extracts and short chain aldehydes. Toxicon 42 (**2003**) 301–306.
49. M. Shoeb, S.M. MacManus, Y. Kumarasamy, M. Jaspars, L. Nahar, P.K. Thoo-Lin, H. Nazemiyeh, S.D. Sarker. Americanin, a bioactive dibenzylbutyrolactone lignan, from the seeds of Centaurea Americana. Phytochemistry 67 (**2006**) 2370–2375.
50. Y. Kumarasamy, P.J. Cox, M. Jaspars, L. Nahar, S.D. Sarker. Isolation, structure elucidation and biological activity of hederacine A and B, two unique alkaloids from Glechoma hederaceae. Tetrahedron 59 (**2003**) 6403–6407.
51. A.L. Parra, R.S. Yhebra, I.G. Sardiñas, L.I. Buela. Comparative study of the assay of Artemia salina L. and the estimate of the medium lethal dose (LD50 value) in mice, to determine oral acute toxicity of plant extracts. Phytomedicine 8 (**2001**) 395–400.
52. M. Hartl, H.U. Humpf. Toxicity assessment of fumonisins using the brine shrimp (Artemia salina) bioassay. Food Chem Toxicol 38 (**2000**) 1097–1102.

53. M. Pelka, C. Danzl, W. Distler, A. Petschelt. A new screening test for toxicity testing of dental materials. J Dentistry 28 (**2000**) 341–345.

54. J.J. Hlywka, M.M. Beck, L.B. Bullerman. The use of the chicken embryo screening test and brine shrimp (Artemia salina) bioassays to assess the toxicity of Fumonisin B1 mycotoxin. Food Chem Toxicol 35 (**1997**) 991–999.

55. B.F. Juma, R.R.T. Majinda. Erythrinaline alkaloids from the flowers and pods of Erythrina lysistemon and their DPPH radical scavenging properties. Phytochemistry 65 (**2004**) 1397–1404.

18

METHODS AND APPLICATIONS OF HEPATOXICITY AND HEPATOPROTECTIVE ASSAYS

Shiqi Peng

In contrast with most other organs, the liver by virtue of proliferation and apoptosis comprises a unique and remarkable property (i.e., to precisely control growth and mass). Chemical-induced hepatoxicity is implicated in the onset and progression of hepatic insufficiency, hepatitis, and hepatocarcinoma. For instance, mitochondrial proliferation of chemicals is related to hepatocellular carcinomas; chemical-induced renal adenocarcinoma, hepatic adenoma, and carcinogenicity can promote the formation of preneoplastic foci in rats or mice; the clonal growth of glutathione *S*-transferase (GST-P) enzyme-altered foci and the histopathologic change in the liver of rats or mice may relate to liver carcinogenesis of chemicals; in the liver, injury induced by chemicals, such as CCl_4, BCG, and LPS, the activities of protein, glucose 6-phosphatase, amidopyrine *N*-demethylase, and aniline hydroxylase, the levels of hepatic triglycerides and lipid peroxidation, the concentrations of NO, MDA, and SOD, and the viability of thymocytes can be changed significantly; and in hepatocarcinogenesis induced by chemicals such as chloral hydrate, the body weight of B6C3F1 mice can be decreased significantly. On the other hand, proliferation factors (PFs) and HGF are involved in liver regeneration cascade and

Pharmaceutical Bioassays: Methods and Applications. By Shiqi Peng and Ming Zhao

TGF-β-induced growth inhibition of CCL-64 cells, thus PFs present in the partially hepatectomized rat serum can serve as an index of liver regeneration cascade, and two cytokines (HGF and HGF) may result in an additive effect on proliferation. In the past decade, much research has focused on hepatoxicity and hepatoprotection. As a result, a series of hepatoxicity and hepatoprotective assays has been established. In this chapter, 39 assays are described: GST-P enzyme-altered foci assay,[1–4] GST-P$^+$ and TGF-α$^+$ foci assay,[5] partial hepatectomy assay,[6,7] *in vivo* short-term liver initiation assay in rats,[8] TUNEL related assay,[9] hematopoietic progenitor cells mobilization assay,[10] electrophoretic mobility shift assay,[11,12] telomerase activity assay,[13] von Willebrand factor (vWF) level assay,[14] bromodeoxyuridine incorporation assay,[15] NF-κB activation assay,[16] rat summation of initiation activity assay,[17] hepatoprotective assay,[18,19] cytokine gene expression assay,[20] competitive inhibition assay for immunodominance of O-specific polysaccharides of gram-negative bacilli,[21] marine HP toxin okadaic acid (OA) assay and ELISA,[22] ANTI-GST-CTX MVIIA antibody in omega-conotoxin MVIIA assay,[23] assays for rituximab in patient plasmas,[24] assays of TNF-α, superoxide and thymocyte for liver injury of mice,[25–27] gene transfer of kringle 1-5 and related assays for mice with hepatocellular carcinoma,[28] body weight assay for mice with liver tumor incidence,[29] P450 activity and inducibility in rat hepatocytes for cytotoxicity assay,[30] LDH release in rat hepatocytes for toxicity assay,[31] rainbow trout hepatocytes' EROD activity assay,[32] EROD activity, ROS production and cytotoxic concentration in fish hepatocytes for drug toxicity assays,[33] coho salmon's IGF-I gene sequencing and TaqMan assay,[34] DNA strand break in rat hepatocyte and comet assay,[35] DNA laddering in rat hepatocyte and TUNEL assay,[36] counting increases in viable cell number for hepatocyte proliferation assay,[37,38] assay of HGF and TGF-β in CCL-64 cells,[39] assay for serum IFN-α of patients with chronic hepatitis C,[40] rat medium-term liver, DNA microarray and Cu$^+$-reducing antioxidation assays for DEN-induced hepatocarcinogenesis,[41] rat medium-term liver assay for DEN-induced hepatocarcinogenesis assay,[42] assay for rat carcinogenesis initiated by three carcinogens,[42] assay for rat carcinogenesis initiated by five carcinogens,[43] assay for DNE-initiated rat carcinogenesis,[44,45] assay for normal dose DNE-initiated and low dose DNE-maintained rat carcinogenesis,[44,45] TGF-β-mediated antiproliferation assay,[47] and assay for DNE and HCB-initiated rat carcinogenesis.[48]

18.1 GST-P ENZYME-ALTERED FOCI ASSAY[1–4]

According to the required levels (0%, 0.03%, 0.1%, and 0.3%), the test chemical was incorporated into irradiated (6.0 kGy) powder diet MF (Oriental Yeast Co., Ltd., Tokyo, Japan). The stability of 0.1%, 0.03%, 0.5%, and 5.0% chemical in the prepared diets was previously confirmed for 6 weeks at room temperature. Male F344/DuCrj rats (5 weeks old) were given an approximately 1-week quarantine/acclimation period to monitor health conditions and body weights. The normal rats (6 weeks old) were randomly divided into 8 groups (18 rats each for groups 1–5, 9 rats each for groups 6–8). The rats in groups 1–5 received an injection of the initiator

N-nitrosodiethylamine (200 mg/kg body weight, ip, Tokyo Kasei Kogyo Co., Ltd., Tokyo, Japan). The rats in groups 6–8 received an injection of the vehicle. Two weeks later, the rats received an injection of the test chemical at the desirable dose for a suitable treating period. Three weeks after the beginning of the experiment, all rats were given two-thirds partial hepatectomy. At week 8, all the surviving rats were killed under ether anesthesia. Their organs in the thoracic and abdominal cavities were examined macroscopically. Their livers were immediately excised and weighed to calculate the liver-to-body weight ratio. For the 4- to 5-mm-thick sections from the cranial and caudal parts of the right lateral lobe and the caudal part of the caudate lobe, of all surviving rats were fixed in 10% buffered formalin solution, embedded in paraffin wax, sectioned, and stained immunohistochemically for glutathione *S*-transferase analysis (GST-P, ABC method). All GST-P positive hepatocytic foci larger than 0.2 mm in diameter (the lowest limit for reliable evaluation) were measured using a color image processor, and the numbers and areas of foci/cm^2 of liver section were calculated.

18.2 GST-P$^+$ AND TGF-α^+ FOCI ASSAY[5]

Using a Nikon photomicroscope (Microphot-FXA) connected to a KS-300 apparatus, GST-P$^+$ (positive) foci larger than 0.15 mm in diameter were measured, and the data were expressed as number and area (mm^2) of GST-P$^+$ foci per liver section (cm^2). After partial hepatectomy, sagittal sections of each of the remaining lobes (three sections) of the liver were evaluated per animal. In a special Macro-Stand device (support with Canon TV zoom lens V6 $\times$ 16/16, 100 mm, plus a Canon 58-mm close-up 240 lens connected to a CCD black-and-white video camera module with a Sony DC-777 camera unit) connected to a KS-300, the sections were analyzed. With the aid of the KS-300 system, as indicated above, the number of the clearly discernible TGF-α^+ foci (foci with more than 20 hepatocytes) per cm^2 of liver was also counted. Because their borders frequently were not clearly discernible, their areas were not estimated. According the intensity (weak or strong staining), homogeneity (i.e., diffuse staining of FHA or only in specific regions of the foci) and distribution (i.e., only cytoplasmic or nuclear and cytoplasmic staining) of their immunoreactivity TGF-α positive FAHs were evaluated.

18.3 PARTIAL HEPATECTOMY ASSAY[6,7]

Male F344 rats (30 days old, Harlan-Sprague-Dawley, Indianapolis, IN) acclimated for 4 weeks before starting assay were randomly divided into three groups. At week 0, the rats received a single ip injection of a solution of *N*-nitrosodiethylamine (200 mg/kg, Sigma-Aldrich, St. Louis, MO, USA) in 0.9% saline. Two weeks later, the rats received daily gavage administration of corn oil or 0.1 mmol/kg test compound in a corn oil vehicle through the remainder of the 8-week assay. At week 3, all rats receive a partial hepatectomy. The rats were given food and water, and lighting was set on a 12-h light/dark cycle. On days 23, 26, 28, 47, and 56, at least five rats from each group were sacrificed by aortic exsanguination. Whole livers were removed, and the tissues were fixed in 10% neutral-buffered formalin, embedded in paraffin, serially sectioned at 5 μm, and mounted on microscope slides. Formalin-fixed sections were stained with hematoxylin and eosin for histopathologic examination. BrdU-positive labeling indices (LIs) were quantified by microscopic analysis. The LI were determined by randomly counting the number of positive nuclei per 1000 hepatocytes or number per unit area (mm^2) in sections stained immunohistochemically for BrdU.

18.4 *IN VIVO* SHORT-TERM LIVER INITIATION ASSAY IN RATS[8]

On day 0, male F344 rats (Charles River Japan, Atsugi, Japan) rats were subjected to a two-thirds partial hepatectomy, subdivided into four groups of eight rats each (except for the pipemidic acid low-dose group that consisted of seven rats) in each assay to examine each of the four quinolones, and treated once orally each with vehicle and low, intermediate, and high doses of each quinolone 12 h after the completion of the two-thirds partial hepatectomy. At 10 mL/kg body weight, dosing volumes of the vehicle and suspensions of each quinolone were calculated. After feeding a basal diet for 14 days, the rats were fed with a diet containing 0.015% 2-AAF (Tokyo Kasei Co., Tokyo, Japan) for the following 10 days. On day 19, CCl_4 (Wako Pure Chemicals Industries Ltd., Osaka, Japan) was at 0.8 mL/kg body weight orally administered once to the rats. On day 34, the rats were sacrificed under ether anesthesia, and in 10% neutral-buffered formalin, slices of all liver lobes were fixed. In the immunohistochemical assay, tissues of the fixed liver lobes were embedded in paraffin, sectioned and stained using an antibody GST-P, a marker of preneoplastic lesions in the rat liver, by the avidin-biotin complex method. With a computer-assisted image analyzer (Win Roof; Mitani Corp., Japan), the number and area of GST-P positive foci (>0.1 mm in diameter) and the total area of each liver section examined in each animal were measured, and the mean number and area of GST-P positive foci per unit area of all the liver sections examined per animal were calculated.

18.5 TUNEL RELATED ASSAY[9]

1. *Western Blot Analysis of Liver Tissue*: Liver tissue was in lysis buffer (10 mM Tris pH 7.5, 10 mM NaCl, 0.1 mM EDTA, 0.5% Triton-X 100, 0.02% NaN_3, and 0.2 mM PMSF) homogenized, on ice incubated for 30 min, and centrifuged at 10,000 × *g* for 15 min. Prior to use, all buffers received a protease inhibitor cocktail (1/100 v/v). To assess p53 intracellular distribution, from liver tissue using a commercially available extraction kit (NE-PER, Pierce Biotechnology, Rockford, IL, USA), nuclear and cytoplasmic protein extracts were prepared. The purity of the nuclear and cytoplasmic protein fractions was analyzed by topoisomerase I and β-tubulin Western blot, respectively. Nuclear extracts were further analyzed for p21 protein expression. Using the BCA protein assay (Pierce Biotechnology) and with BSA as standard, protein concentrations were determined. On SDS

polyacrylamide gels (12% SDS-PAGE), 60 μg of p53, 40 μg of p21, and 20 μg of proliferating cell nuclear antigen (PCNA) were separated discontinuously and transferred to a polyvinyldifluoride membrane. After blockade of nonspecific binding sites, membranes were incubated at room temperature with rabbit monoclonal anti-p53 (1:1000), mouse monoclonal anti-p21 (1:100, Santa Cruz Biotechnology), rabbit polyclonal anti-PCNA (1:2000), peroxidase-conjugated goat anti-rabbit IgG antibody (p53: 1:15,000; PCNA: 1:8000), or rabbit anti-mouse IgG antibody (p21: 1:20,000, Santa Cruz Biotechnology) as secondary antibodies for 2 h. By means of luminol-enhanced chemiluminescence (ECL plus, Amersham Pharmacia Biotech, Freiburg, Germany) and exposure of the membrane to a blue light sensitive autoradiography film, protein expression was visualized. Signals were densitometrically assessed and normalized to β-actin signals (mouse monoclonal anti-β-actin antibody, 1:20,000, Sigma-Aldrich, St. Louis, MO, USA), except for p53 and p21 expression in nuclear and cytoplasmic protein extracts.

2. *mRNA Analysis of Liver Tissue*: According to the manufacturer's instructions, using TRIzol Reagent, from 50 mg of liver tissue total RNA was isolated and its concentration was determined spectrophotometrically. Using oligo(dT)18 primer (Biolabs, Frankfurt am Main, Germany) and Superscript II RNaseH-T (Invitrogen, Karlsruhe, Germany), cDNA was prepared by reverse transcription of 2 μg of total RNA. Using Taq polymerase (Perkin-Elmer, Rodgau-Jügesheim, Germany) and primers RT-mHGF 5′-CTGGGGCTACACTGGATTG-3′ and 5′-GATGCTTCAA-ACACACTGGC-3′, by 28 cycles of PCR, mouse hepatocyte growth factor (HGF) was amplified. In the RNA integrity and cDNA synthesis, mouse GAPDH was used as a housekeeping gene, and 5′-AACGACCCCTTCATTGAC-3′ and 5′-TCCACGACATACTCAGCAC-3′ were used as primers. By electrophoresis on 2.0% agarose gels, PCR products were separated. Ethidium bromide–stained bands were visualized by UV illumination and densitometrically quantified. The data reflected the expression of HGF gene product in relation to that of GAPDH.

3. *Histology and Immunohistochemistry of Liver Tissue*: In 4% phosphate-buffered formalin and embedded in paraffin, liver tissue was fixed for 2–3 days. Paraffin-embedded tissue blocks were cut to 4-mm sections and stained with hematoxylin and eosin (H&E). For p53 immunohistochemistry, the primary antibody and secondary goat were a rabbit monoclonal anti-mouse p53-antibody (1:100, Santa Cruz Biotechnology) and anti-rabbit antibody (1:100, Santa Cruz Biotechnology), respectively. 3,3′-Diaminobenzidine was used as chromogen. The sections were counterstained and examined with hemalaun and light microscopy, respectively. For PCNA immunohistochemical demonstration, sections were collected on poly-L-lysine-coated glass slides treated by microwave for antigen unmasking. As primary antibody, rabbit polyclonal anti-PCNA (1:50, Santa Cruz Biotechnology) was incubated at room temperature for 60 min and treated with an alkaline phosphatase-conjugated goat anti-rabbit antibody (1:25). For analysis of BrdU incorporation, sections were incubated with anti-BrdU monoclonal antibody (1:50, Dako Cytomation, Hamburg, Germany) and treated with horseradish peroxidase-conjugated goat anti-mouse immunoglobulin (1:100). Peroxidase binding sites were detected by 3,3′-diaminobenzidine. PCNA- and BrdU-positive cells were counted as cells/mm^2

within 40 consecutive high-power fields (×40/0.65 objective). Using a commercially available *in situ* apoptosis detection kit, TUNEL was analyzed for determining apoptotic hepatocytes. TUNEL-positive nuclei were also as cells/mm^2 counted within 40 consecutive high-power fields (×40/0.65 objective).

18.6 HEMATOPOIETIC PROGENITOR CELLS MOBILIZATION ASSAY[10]

1. *Flow Cytometry*: Cells were incubated with either anti-CD133 MoAb or anti-CD14 MoAb conjugated superparamagnetic beads, washed, and processed through a MACS magnetic separation column. An aliquot of the purified cells was analyzed by flow cytometry. After 5- and 2-min lysis with hemolytic buffer (0.155 M NH_4Cl, 0.012 M $NaHCO_3$, 0.1 mM EDTA, pH 7.2), 1×10^6 cells were incubated with phycoerythrin (PE)-conjugated anti-CD133 monoclonal antibody (Miltenyi Biotec, Bergisch Gladbach, Germany) in combination with either FITC-conjugated anti-CD34 MoAb, anti-CD45 MoAb, anti-CD14 MoAb, or anti-Thy-1 MoAb (anti-HPCA-2, BD Pharmingen, Hamburg, Germany), as well as with FITC-anti-CD34 MoAb in combination with either anti-bcrp-1 MoAb (eBioscience, Vienna, Austria) or PE anti-c-kit (DAKO Cytomation, Hamburg), with isotype-matched mouse immunoglobulins as controls. All incubations were carried out at 4°C in the presence of normal goat serum. Using a FACScalibur flow cytometer (Becton Dickinson, Heidelberg, Germany) and Cell Quest software (Becton Dickinson), two-color flow cytometry was accomplished, each of which included at least 50,000 events. By using isotype controls for PE and FITC, analysis gates were set to allow the lower left panel to have at least 98% of the total cells. After correction for the percentage of cells, the percentage of positive cells was assessed reactive with the respective isotype control.

2. *Clonogenic Assays for Hematopoietic Progenitors*: Immunoselected cells as well as corresponding CD133/CD14-negative cell fractions were cultured in fibronectin-coated chamber slides at a density of 2×10^6 cells/mL in stem cell growth media (methylcellulose; Cell Systems, St. Katharinen, Germany). To media stem cell growth factor (100 ng/mL), hepatocyte growth factor (40 ng/mL), epidermal growth factor (10 ng/mL), and fibroblast growth factor-4 (20 ng/mL) were added. Cells were at 37°C in 5% CO_2 incubated for up to 28 days. On alternate days, additional feeding was performed. By gentle pipetting, the supernatant was removed and replaced with fresh medium. Purified cells (1×10^3 cells/mL) were plated in semisolid growth media supplemented with various hematopoietic growth factors, in duplicate incubated at 37°C in 5% CO_2 and 95% humidity for 14 days, and scored after culture using an inverted microscope.

3. *Immunocytochemical Analysis*: At room temperature, cultured cells were fixed either with 4% paraformaldehyde (Sigma, Deisenhofen, Germany) for 10 min followed by 2-min methanol fixation at −20°C or with ice-cold acetone for 10 min. After blocking with 10% goat serum for 20 min, specimens were incubated overnight at room temperature with the primary antibody. The primary antibodies included anticytokeratin CK 7 (1:50), anti-CK8 (1:50), anti-CK18 (1:50), anti-CK19

(1 : 100), anti-alpha fetoprotein (1 : 100), anti-human albumin (1 : 40, all mouse anti-human and monoclonal), and anti-multi-cytokeratin (rabbit anti-human polyclonal). By using an anti-mouse avidin biotinylated AP complex technique and the Vector Red substrate or an anti-rabbit immuno AP polymer in combination with the New Fuchsin substrate, the reactivity was detected. Specimens were counterstained with hematoxylin. In negative controls, primary antibody was replaced by isotypes or PBS as well as staining of peripheral blood smears. In positive controls, cytospins with freshly thawed primary human hepatocytes were used.

4. *RT-PCR*: Using RNeasy Midi Kit (Qiagen, Hilden, Germany), from postoperatively obtained blood cells, cultured CD133C/CD14C-derived cell populations, and cultured CD133K/CD14K-derived cell populations, total RNA was extracted and subjected to RT-PCR according to the manufacturer's instructions using a cDNA synthesis kit. For subsequent PCR with final concentration of PCR reaction of 10 mM Tris-HCl, 1.5 mM $MgCl_2$, and 50 mM KCl, 1.25 U Taq DNA polymerase in 100 μL of total volume 1 μL of cDNA product was used. Specific primers were CK8-sense 5′-AACAACCTTAGGCGGCAGCT-3′, CK8-antisense 5′-GCCTGAGGAAGTTGATCTCG-3′, CK18-sense 5′-TGGTACTCTCCTCAATCTGCTG-3′, CK18-antisense 5′-CTCTGGATTGACTGTGGAAGT-3′, CK19-sense 5′-ATGGCCGAGCAGAACCGGAA-3′, CK19-antisense 5′-CCATGAGCCGCTGGTACTCC-3′, albumin-sense 5′-TGCTTGAATGTGCTGATGACAGGG-3′ and albumin-antisense 5′-AAGGCAAGTCAGCAGGATCTCATC-3′. The PCR mixture was denatured at 95°C for 2 min, and 34 cycle amplifications (95°C for 1 min, 60°C for 1.5 min, 72°C for 1.5 min) were performed. After amplification, on a 1.5% agarose gel stained with 0.5 mg/mL ethidium bromide, PCR products were electrophoresed.

18.7 ELECTROPHORETIC MOBILITY SHIFT ASSAY[11,12]

In electrophoretic mobility shift assay, frozen liver sections (0.1 g) at −80°C were homogenized with a Dounce homogenizer in 2 mL of buffer A [10 mM HEPES-KOH, pH 7.9, containing 10 mM KCl, 1.5 mM $MgCl_2$, 1 mM dithiothreitol (DTT), 1 mM PMSF, and 500 U/mL Trasylol], allowed to swell for 15 min, centrifuged at 1100 × *g* for 5 min, the pellet was suspended in 1 mL of lysis buffer (buffer A supplemented with 0.1% Triton X-100), allowed to stand for 10 min, centrifuged at 1100 × *g* for another 10 min, the nuclear pellet was suspended in 80 μL of nuclear extraction buffer [20 mM HEPES-KOH, pH 7.9, containing 0.42 M NaCl, 1.5 mM $MgCl_2$, 1 mM DTT, 1 mM PMSF, 500 U/mL Trasylol, 0.2 mM EDTA and 25% (v/v) glycerol], incubated for 30 min and centrifuged at 16,500 × *g* for 20 min to provide nuclear extracts. At room temperature and incubating the nuclear extract in reaction buffer [20 mM HEPES-KOH, pH 7.9, 1 mM EDTA, 60 mM KCl, 10% glycerol, 1 mg of poly (dI-dC)] with the probe (40,000 dpm) for 20 min, binding reactions were performed. On a 4.8% polyacrylamide gel in high ionic strength buffer, products were electrophoresed, and dried gels were analyzed by autoradiography. An NF-κB consensus oligonucleotide (5′-AGTTGAGGGGACTTTCCCAGGC, Promega, Madison, WI, USA) from the mouse immunoglobulin light chain was labeled with [γ-^{32}P]-ATP and T4 polynucleotide kinase. Protein was measured.

18.8 TELOMERASE ACTIVITY ASSAY[13]

In Kontes homogenization tubes and matching pestles (Molecular Dynamics, Sunnyvale, CA), which rotated at 450 rpm, 20 mg of frozen liver specimens were homogenized with cold lysis buffer. Collected cells (approximately 2×10^5) were washed twice with cold PBS, resuspended in 200 μL of lysis reagent, incubated on ice for 30 min, and centrifuged at 48°C and $16{,}000 \times g$ for 20 min to prepare cell extracts. The supernatants were removed, and the cell extracts were transferred to a fresh tube. For each test sample, PCR amplification tubes contained 25 μL of reaction mixture, 1–3 μL of cell extract (1×10^5 to 3×10^5 cells), and sterile water sufficient to provide a final volume of 50 μL. All pipetting steps were performed on ice. Mixtures were at 25°C incubated for 10–30 min to extend and at 94°C incubated for 5 min to inactivate telomerase. Amplification consisted of 30 PCR cycles at 94°C for 30 s, 50°C for 30 s, and 72°C for 90 s (final cycle, 10 min). A PCR product aliquot was denatured, hybridized to a digoxigenin (DIG)-labeled telomeric repeat-specific probe, and bound to a streptavidin-coated 96-well plate. The immobilized PCR product was finally detected with an antibody against digoxigenin that was conjugated to horseradish peroxidase, visualized by a color reaction product using the substrate 3,3′,5,5′-tetramethylbenzidine, and semiquantified photometrically. The ratio for each time point after PHx was compared with that at 0 h. For cell lines' telomerase activity, the activity of whole 293 cells was defined as 1 arbitrary unit (AU).

To construct RNA control, by Spe I a plasmid containing mTERT cDNA fragment was linearized. According to the manufacturer's instructions, RNA transcription from Spe I-digested pCR 2.1 was performed *in vitro* with T7 RNA polymerase. The optical density was measured for transcribed control RNA, and the amount was determined for control RNA. To assess the preparation quality, a control mTERT transcript having 356 base fragments was separated by electrophoresis on 1.5% formaldehyde gel and stained by ethidium bromide. The preparation of the mTERT transcript used in this assay consisted of more than 95% full-length transcripts. Using the High Pure RNA Tissue Kit (Roche Diagnostics GmbH, Mannheim, Germany), total cytoplasmic RNA was isolated from pooled livers of 4–6 mice to measure mTERT, a LightCycler-RNA Amplification Kit SYBR Green I (Roche Diagnostics GmbH, Mannheim, Germany) was used. Forward primer 5′-GGGAGATGGCCAAGA-GCGTCTAAA-3′ and reverse primers 5′-CGGTGGGCTGGTGTTCAAGG-3′ were designed to amplify a 273-bp segment of mTERT coding region. In 20 mL of PCR mixture, LightCycler-RT-PCR Reaction Mix SYBR Green, 5 mmol of $MgCl_2$, 0.25 mmol of forward primer, 0.5 mmol of reverse primer, LightCycler-RT-PCR Enzyme Mix, and template RNA were included. RT-PCR amplification was started at 55°C for 30 min for reverse transcription, at 95°C for 30 s for denaturation, and then at 95°C for 1 s, 57°C for 10 s and 72°C for 10 s to perform 45 cycles of amplification. All reactions were performed in a LightCycler. Using High Pure RNA Isolation Kit, total cytoplasmic RNA was from collected cells isolated. To measure hTERT, a LightCycler-Telo TAGGG hTERT Quantification Kit (Roche Diagnostics GmbH, Mannheim, Germany) was used according to the manufacturer's instructions.

By detecting the length of the telomere repeat array with Southern blot, telomere length was measured. From liver tissues, genomic DNA was extracted and quantified with fluorometry. With 10 U of each restriction enzyme Hinf I and Rsa I, 5 mg of extracted DNA was at 37°C simultaneously digested for 16 h. DNA was extracted once with phenol/chloroform (1 : 1), precipitated with 2.5 volumes of 100% ethanol in the presence of 2.5 mM sodium acetate, air dried, and resuspended in 20 mL of 10 mM Tris-HCl and 1 mM EDTA. Digested DNA (2 mg) was loaded onto a 0.5% agarose gel, electrophoresed in 1× Tris-borate EDTA buffer at 10 V/cm for 4 h alongside molecular size markers, the gel was denatured for 20 min in 0.5 M NaOH and 1.5 M NaCl, neutralized in 0.5 M Tris-HCl (pH 8.0) and 1.5 M NaCl for 10 min, and transferred to a Hybondt membrane by Southern blot. In a Rapid hyb buffer, the filter was hybridized to a ^{32}P-labeled telomeric oligonucleotide probe [γ-^{32}P(CCCTAA)^{3}] (sp act, 3.0×10^8 to 3.4×10^8 cpm/mg DNA), washed in 5× sodium saline citrate (SSC; 750 mM NaCl and 75 mM sodium citrate) with 0.1% sodium dodecyl sulfate twice for 15 min each at room temperature, and in 1 × SSC twice at 42°C for 15 min each according to the manufacturer's instructions. The filter was dried, at 270°C exposed to x-ray film for 3 days, and the x-ray film was scanned with ScanJet 6100C/T and analyzed by NIH image with Power Macintosh G3. In each lane, the maximum radioactivity was measured, and telomere length was calculated by comparing with the appropriate molecular size markers.

18.9 VON WILLEBRAND FACTOR (vWF) LEVEL ASSAY[14]

For light microscopy of stained liver specimens for vWF, positively stained portal vessels and sinusoids appeared as brown linear deposits. Using a computerized image analysis system composed of a 3-chip RGB video camera (Sony, Japan) installed on a light microscope (Zeiss, Germany) and attached to an IBM-compatible personal computer (Pentium II, MMX, 450 MHz, 125 MB RAM) equipped with a frame grabber, the extension of the liver tissue stained by vWF and the staining intensity (density) were analyzed. On a high-resolution 17-inch color monitor, histologic images were captured, digitized, and displayed. In each specimen, an average of five most intensely stained high-power microscopic fields (×400) were separately assessed for the periportal, pericentral, and lobular zones. The histologic images were loaded onto screen buffers that had a resolution of 760 × 570 pixels. Using Image-Pro Plus 4 software (Media Cybernetics, Baltimore, MD, USA), vWF staining parameters were measured automatically. The percentage of stained area per high-power field was defined as vWF staining extension. By measuring the average of gray levels per field, the staining density was calculated, of which the values ranged from 0 (the darkest) to 255 (the brightest).

18.10 BROMODEOXYURIDINE INCORPORATION ASSAY[15]

According to the manufacturer's instructions, using a monoclonal anti-BrdU antibody (clone Bu20A, Dako, Denmark), hepatic sections (5 mm thick) were immunostained for BrdU, which labeling index was obtained by examining 3 to 5 high-power fields in three rats per group, before and 24 h after PH. Each field was divided into four zones, among which zone 1 was closest to the centrolobular vein and zone 4 corresponded with the periportal areas. The ratio between marked cells and total counted cells was defined as labeling index. All slides were analyzed with the same examiner.

18.11 NF-κB ACTIVATION ASSAY[16]

1. *Immunohistochemistry*: From rat liver tissues, cryostat sections 5 mm thick including several sets of mirror consecutive sections were prepared, incubated with a rabbit polyclonal antibody to NF-κB p65 (1:50, Santa Cruz Biotechnology, CA, USA), and subjected to the avidin-biotin complex (ABC) method, double-immunostained for Kupffer cells with a mouse monoclonal ED2 anti-rat macrophage antibody (Serotec, Oxford, UK), and incubated with Alexa 488-conjugated goat anti-mouse IgG (H+L) $F(ab')^2$ (1:500, Molecular Probes, Eugene, OR, USA). Another mirror section was used for identifying sinusoidal endothelial cells with a mouse anti-rat CD31 monoclonal antibody (1:100) and incubated with Alexa 594-conjugated goat anti-mouse IgG (H+L) $F(ab')^2$, (1:500). For HB-EGF assays, the sections were incubated with rabbit polyclonal antiserum (1:200), which recognized the mature and proforms of HB-EGF, and subjected to the ABC method. On another mirror section, sinusoidal endothelial and Kupffer cells were identified. Normal rabbit immunoglobulin and serum served as negative controls.

2. *Southwestern Histochemistry*: Biotinylated synthetic sense DNA was 5′-AGTTGAGGGGACTTTCCCAGGC-3′ containing a consensus sequence for NF-κB. Negative controls included absence of probes and biotinylated mutant NF-κB probes (sense, 5′-AGTTGAGGCTCCTTTCCCAGGC-3′, Tokyo, Japan). The frozen liver sections were at 4°C with 0.2% paraformaldehyde fixed for 20 min and washed three times with HEPES buffer (10 mM HEPES, 40 mM NaCl, 0.1 mM EDTA, pH 7.4), with 0.3% H_2O_2 in methanol endogenous peroxidase activity was blocked for 30 min, at 30°C with 0.1 mg/mL DNase I in HEPES buffer containing 0.25% bovine serum albumin (BSA) and 10 mM $MgCl_2$ incubated for 30 min, washed twice with HEPES buffer containing 10 mM EDTA, at 37°C with biotinylated probes diluted at a concentration of 10 mM in HEPES buffer containing 0.25% BSA and 0.5 mg/mL poly(dI–dC) (Pharmasia LBK, Piscataway, NJ, USA) incubated for 1 h, and then subjected to the ABC method.

3. *Rat Hepatocytes and Nonparenchymal Cells (NPCs) in Primary Culture*: By *in situ* perfusion method and differential centrifugation, adult rat hepatocytes and NPCs were isolated from a normal rat liver. Aliquots of hepatocytes and NPCs (5×10^6 cells/sample) were with WME containing 10% FCS, 0.1 μM insulin, and 0.1 μM dexamethasone incubated overnight and then in the absence of serum for 12 h. Cells were stimulated with 20 ng/mL TNF-α for 0.5–1 h, or preincubated with 50 mM inhibitor of NF-κB activation pyrrolidine dithiocarbamate for 1 h and then were stimulated with 20 ng/mL TNF-α for 0.5–1 h.

4. *NF-κB Binding Activity and HB-EGF Gene Expression*: Nuclear extracts were prepared. Total RNA was extracted from hepatocytes and NPC. NF-κB binding activity and the expression of HB-EGF and β-actin genes were measured by means of EMSA and RT-PCR, respectively.

18.12 RAT SUMMATION OF INITIATION ACTIVITY ASSAY[17]

In experiment I, summation effects of initiation activities of multiple doses of 1,2-dimethylhydradine were studied. After two-thirds partial hepatectomy (PH), single or repeated (four) intragastric applications of total dose of 4 mg/kg were performed. Subsequently, male F344 rats (Charles River Japan, Atsugi, Japan) were fed a 2-week basal diet, and then a 2-week diet containing 0.015% of 2-AAF. As a chemical hepatectomy, 3 weeks after 1,2-dimethylhydradine (Tokyo Kasei Co., Tokyo, Japan) administration, all rats received a dose of CCl_4 (0.8 mL/kg, ig). At the end of week 5, the survivors were killed and their liver slices were fixed in 10% neutral-buffered formalin for immunohistochemical examination of GST-P-positive foci. The single 1,2-dimethylhydradine administration time was timed for 12 h, the most effective point for induction of GST-P-positive foci, after PH. The repeated 1,2-dimethylhydradine administration time (each 1 mg/kg, ig) were timed for 12, 18, 24, and 30 h after PH. In experiment II, summation effects of initiation activities of different chemicals were studied. Twelve hours and 30 h after PH were selected as administration time points for chemicals. Rats were either dosed intragastrically with 1,2-dimethylhydradine 1 mg/kg and then *N*-bis(2-hydroxpropyl)nitrosamine (Tokyo Kasei Co., Tokyo, Japan) 10 mg/kg in turn, or dosed intragastrically with 1,2-dimethylhydradine 1 mg/kg and then diethylnitrosamine 0.3 mg/kg or dosed intragastrically with 1,2-dimethylhydradine 1 mg/kg and then saline, or dosed intragastrically with saline and then *N*-bis(2-hydroxpropyl)nitrosamine 10 mg/kg, or and saline then dosed intragastrically with saline and then diethylnitrosamine 0.3 mg/kg. As a control group for experiments I and II, two-thirds partial hepatectomized rats received no chemicals.

18.13 HEPATOPROTECTIVE ASSAY[18,19]

To induce liver injury, CF rats (150–200 g) and Swiss albino mice (20–30 g) of either sex were administered orally CCl_4 diluted with liquid paraffin. The rats or mice in vehicle control group were orally administered an equal volume of liquid paraffin. In hepatoprotective assays, four suitable doses of test compound and two standard doses of silymarin (25 and 50 mg/kg, po) were fed to a respective group of rats 48 h, 24 h, 2 h before, and 6 h after CCl_4 (0.5 μL/kg, po) intoxication, respectively. Blood was collected from the orbital sinus of all the rats 18 h after CCl_4 administration, then glutamic-pyruvic transaminase (GPT), glutamic oxaloacetic transaminase (GOT), and bilirubin were analyzed. In the posttreatment studies, the same doses of test compound and silymarin as mentioned for hepatoprotective assays were fed to rats 6 h, 24 h, and 48 h after CCl_4 (0.5 mL/kg, po) administration. Blood was collected from all rats 2 h after the last dose of test compound administration and then GOT, GPT, and bilirubin were determined. The test compound and silymarin (50 mg/kg, po) were fed to two different groups of rats at 48 h, 24 h, 2 h before, and 6 h after CCl_4 (100 mL/kg, po) administration, while the remaining two groups served CCl_4 and vehicle as control. Eighteen h after CCl_4 administration, all rats in the four groups were fasted overnight and killed by decapitation. Livers were immediately excised and divided into two parts for preparing homogenate (10%, w/v). One part was homogenized in isotonic sucrose (0.25 M) for determining protein and glucose 6-phosphatase activity, and for preparing microsomes by calcium precipitation to determine amidopyrine *N*-demethylase and aniline hydroxylase activities with spectrophotometric methods, in which NADPH was used instead of the NADPH generating system. Another part was homogenized in isotonic PBS (0.01 M, pH 7.2, 0.15 M, NaCl) for determining hepatic triglycerides and lipid peroxidation. The hepatoprotective activity, expressed as hepatoprotective percentage H, was calculated by $H = [1 - (T - V)/(C - V)] \times 100$, where T is mean value of drug and CCl_4, C is mean value of CCl_4 alone, and V is the mean value of control-treated rats. In the determination of acute toxicity, different groups of mice (each 10) were fed with different doses of test compound, while one group with the same number of mice served as control. The mice were observed continuously for 1 h and then hourly for 4 h for any gross behavioral changes and further up to 72 h for any mortality.

18.14 CYTOKINE GENE EXPRESSION ASSAY[20]

1. *Sample Preparation*: After slaughter, tissue samples from the small intestine were immediately collected, frozen in liquid nitrogen, and kept at −80°C. Using Trizol Reagent (Invitrogen Co., Carlsbad, CA, USA), total RNA was extracted. By spectrophotometry at OD_{260nm}, RNA concentration was tested. By OD_{260}/OD_{280} ratio and integrity from agarose-gel electrophoresis, RNA purity was verified. The ratios ranged from 1.8 to 2.0, and on the gels no indications of RNA degradation were observed.

2. *Primer Design*: Using RT-PCR, gene expression was analyzed. To avoid amplification of genomic DNA and to differentiate cDNA and genomic DNA, primers were designed to be intron-spanning. The Primer3 primer design software and cattle-specific GenBank sequences IL-2 (AF535144), IL-4 (M77120), IL-8 (AF061521), IL-12p35 (AJ271034), IL-13 (NM174089), TNF-α (AF011926), IFN-γ (Z54144), MCP-1 (L32659), MCP-2 (S67956), and MUC-1 (AF399757) were used for primer design. GAPDH and ribosomal protein (RP) L-19 were used as the internal control genes. Using agarose gel electrophoresis and sequence analysis of amplified product (ABI 3100 Automated DNA Sequencer, Applied Biosystems and DYEnamic ET DyeTerminator Sequencing Kit, Amersham Bioscience), specificity of RT-PCR products was validated and documented. In addition, to confirm the specificity of the amplicon, for all runs a LightCycler melting curve was obtained.

3. *RT-PCR*: According to the manufacturer's instructions, from 5.0 μg of total RNA using oligo(dT) primer and Superscript II (Invitrogen Co., Carlsbad, CA, USA), cDNA was reverse-transcribed. Using LightCyclerTM (Roche Diagnostics, Mannheim, Germany) and SYBR Green I dye (Roche Diagnostics, Mannheim, Germany), real-time PCR was performed. Following the specific conditions for each gene, reactions were incubated in a LightCyclerTM. For all genes, fluorescence acquisition temperature was 72°C. To determine amplification efficiency, a cDNA dilution curve was made for all primer pars.

18.15 COMPETITIVE INHIBITION ASSAY FOR IMMUNODOMINANCE OF O-SPECIFIC POLYSACCHARIDES OF GRAM-NEGATIVE BACILLI[21]

Bordetella bronchiseptica ATCC 10580, Rb50, 15374, and 3145, and *Bordetella parapertussis* ATCC 15989 and 12822 (NRC, Ottawa, Canada), were grown on Bordet-Gengou (BG) agar plates, transferred to Stainer-Scholte media, at 37°C incubated for 16–24 h with shaking in flasks, harvested by centrifugation, killed by boiling for 1 h, and stored at −20°C for LPS extraction. *Haemophilus ducreyi* LPS (Teresa Lagergärd, Göteborg, Sweden) was used as a control.

After isolation by hot phenol water extraction and purification by enzyme treatment and ultracentrifugation, two methods were used for LPS degradation. (1) At 100°C, 100 mg of LPS was in 10 mL of 1% acetic acid heated for 60 min, at 4°C and 35,000 rpm ultracentrifuged for 5 h, the carbohydrate containing supernatant was passed through a 1 × 100 cm column of BioGel P-4 in pyridine/glacial acetic acid/water buffer (4/8/988 mL, pH 4.5) and monitored with a Knauer differential refractometer. From the void volume, 28 μg of O-specific polysaccharides (O-SP) were recovered. (2) In 18 mL of solution containing 30% acetic acid/5% sodium nitrite/water (1/1/1), 100 mg of LPS was at room temperature on a magnetic stirrer deaminated for 6 h, ultracentrifuged, and the supernatant was freeze-dried and purified on the BioGel P-4 column. Twenty-three milligrams of the saccharide fraction was designated as O-SP_{deam} and recovered from the void volume. To isolate oligosaccharides for competitive inhibition assays, LPS was dissolved in anhydrous HF (100 mg of LPS, 8 mL of HF, 25°C, 24 h), in a hood and on a plastic Petri dish at room temperature evaporated, the residue was dissolved in pyridine/glacial acetic acid/water buffer (4/8/988 mL), passed through a 2.5 cm × 80 cm column of Sephadex G-50, and further purified by HPLC on 250 × 9.5 mm Phenomenex Aqua column in 0.1% TFA (for the first 10 min) and a gradient of 0–50% acetonitrile in 0.1% TFA. The effluent was monitored at 220 nm. The isolated oligosaccharides contained an average of 15 repeats of diacetamidouronic acid (average molecular mass of 4158 Da, as assayed by MALDI-TOF, Bruker Daltonics, Billerica, MA).

BSA (Sigma, St. Louis, MO) was derivatized to aminooxylated derivatives using a two-step procedure. (1) BSA was treated with succinimidyl 3-(bromoacetamido)propionate (SBAP, Pierce, Pittsburgh, PA) introducing thiol-reactive bromo-acetamido moieties (BSA-Br). (2) BSA-Br was coupled with O-(3-thiopropyl)hydroxylamine to form the aminooxylated protein through stable thioether linkages (BSA-ONH_2). For conjugation, at room temperature, 5 mg of BSA-ONH_2 was reacted with 10 mg of O-SP of *Bordetella bronchiseptica* 10580 (Bb10580), *Bordetella bronchiseptica* Rb50 (BbRb50), or *Bordetella parapertussis* 10595 (Bpp15989) in 1.5 mL of buffer A (PBS, 0.1% glycerol, 5 mM EDTA) at pH 5.7 for 15 h with stirring. The reaction mixture was passed through a 1 × 100 cm Sephadex G-100 column in 0.2 M NaCl, and the void volume fraction was characterized by protein concentration, immunodiffusion, SDS-PAGE, and MALDI-TOF spectroscopy.

BSA-ONH_2 (5 mg) reacted with 10 mg of O-SP_{deam} of the listed strains, and the products were designated as BSA-ONH_2/$Bb10580_{deam}$, BSA-ONH_2/$BbRb50_{deam}$, and BSA-ONH_2/$Bpp15989_{deam}$. Female NIH Swiss Webster mice (5–6 weeks old) were injected sc 3 times at 2-week intervals with 2.5 μg of saccharide as a conjugate in 0.1 mL of PBS. Seven days after the second or third injections, blood was collected from each group of 10 mice. Controls received PBS alone. With heat-killed whole bacteria, hyperimmune sera against *Bordetella bronchiseptica* strains 10580 and Rb50 and against *Bordetella parapertussis* strain 15989 were prepared.

18.16 MARINE HP TOXIN OKADAIC ACID (OA) ASSAY AND ELISA[22]

1. *Mouse Assay*: A 20 g portion of hepatopancreas (HP) was extracted three times with acetone for three times each 50 mL, filtered through a cellulose filter, the combined toxin extract was evaporated, resuspended in 4 mL of Tween-60 (1%) at 5 g of HP/mL, and each one of three mice (Albino Swiss, 18–20 g) was injected intraperitoneally with 1 mL of this solution. The criterion of toxicity promulgated by the EU Decision 2002/225/EC was followed, i.e. the inoculation of each of the mice with an extract equivalent to 5 g of HP resulted in a survival less than 24 h for two out of three mice.

2. *ELISA*: An ELISA test kit was used to analyze samples containing only OA (Institute for Marine Biosciences, Canada) and no dinophysistoxin-1. A portion of mussel hepatopancreas was extracted with a fivefold volume of 90% aqueous methanol, the extract was doubled with pure water and went directly through the ELISA procedure. On semilogarithmic graph paper, the absorbance (%) at 450 nm (Bio-Tek Instruments) was plotted against the concentration of OA (ng/mL). The linearity ranged from 1 to 10 ng/mL, which corresponded with 0.1–1 μg/g HP. Even values as low as 0.05 μg/g HP could be detected with limited reliability.

18.17 ANTI-GST-CTX MVIIA ANTIBODY IN OMEGA-CONOTOXIN MVIIA ASSAY[23]

1. *Expression of pGEX-2T-CTX Plasmid*: Using *Escherichia coli* favorable codons, an artificial DNA sequence encoding omega-conotoxin (CTX) MVIIA was designed. BamHI was introduced to the 5′ terminus, whereas a stop codon and EcoRI site was added to 3′ terminus. With T4 PNK, the synthetic DNA fragments

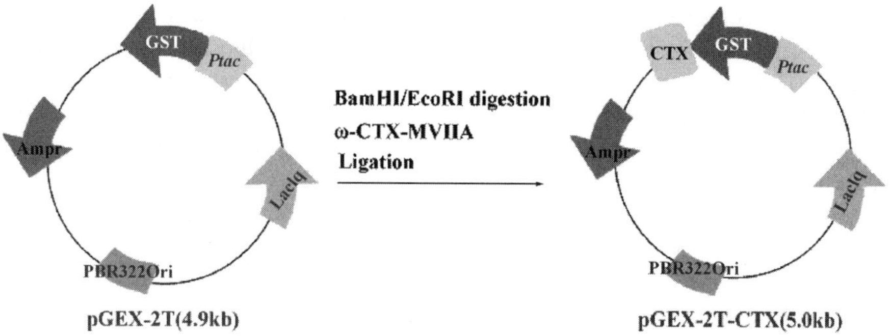

Figure 18.1 Construction of expression plasmid pGEX-2T-CTX MVIIA.

were phosphorylated. The single-strand fragments were annealed to form the paired double strands. Equal amount of fragments CTX1 and CTX2 were mixed, incubated at 90°C, and cooled to room temperature to form the paired double strands. Annealed product was digested with BamHI/EcoRI and cloned into plasmid pGEX-2T. Recombinant plasmid pGEX-2T-CTX was subjected to DNA sequencing, and DNA manipulation was performed.

2. *GST-CTX MVIIA's Expression and Purification*: The expression of plasmid pGEX-2T-CTX was induced with 0.1 mM isopropyl β-D-thiogalactopyranoside (IPTG) at 37°C for 3 h in *Escherichia coli* BL21, which were harvested and disrupted by sonication (Fig. 18.1). After centrifugation, the supernatants and pellets were run on 12% SDS-PAGE. The supernatant containing fusion protein was applied to a Glutathione-Sepharose 4B Fast Flow column (preequilibrated with 50 mM PBS) and eluted with 50 mM Tris-HCl buffer (pH 8.0) containing 10 mM reduced glutathione. After purification, fusion protein was analyzed on 12% SDS-PAGE. The activity and protein concentration were measured with hot plate and Bradford assay, respectively. In the preparation of fusion protein CTX MVIIA and thioredoxin (Trx), the synthetic DNA sequence encoding CTX MVIIA was cloned into the expression vector pET-32α (+). Fusion protein Trx-CTX MVIIA containing 6× His-tag was expressed in *Escherichia coli* BL21 and purified by one-step metal chelating affinity chromatography; its purity and activity were analyzed by HPLC and hot plate assay, respectively.

3. *Polyclonal Antibody's Preparation and Purification*: Two New Zealand adult male rabbits were simultaneously immunized with a mixture of 1 mg of GST-CTX MVIIA and complete Freund's adjuvant, 0.5 mg of which was given subcutaneously at 2-week intervals. Prior to each immunization, blood was collected via the marginal vein of the rabbit ear, and the sera were stored at −20°C. Polyclonal antibody was preliminarily purified via salting out (between 0.3 and 0.5 saturation with ammonium sulfate) and further purified through anion-exchange chromatography of a DE52 column.

4. *Fusion Proteins' Western Blotting*: Using a standard procedure, the purified fusion proteins were analyzed on 12% SDS-PAGE, the gel was immersed in the transfer buffer, the proteins were transferred to a polyvinyldifluoridine (PVDF)

membrane, electrophoresed at 100 V for 1.5 h, the membrane was at 37°C in blocking buffer incubated for 2 h, washed with Tris-buffered saline containing 0.05% Tween-20 (TBST), at 37°C with the purified polyclonal antibody incubated for 2 h, and washed with TBST three times to remove the primary antibody. The horseradish peroxidase-conjugated anti-rabbit secondary antibody was then added, and 3,3′-diaminobenzidine was used to show the specific protein bands.

5. *Anti-GST-CTX MVIIA Antibody's ELISA*: At 4°C, a microtiter plate was with 10 μL of GST-CTX MVIIA (10 μg/mL) coated overnight. Wells were at 37°C with 5% BSA saturated for 2 h and washed with 5 mM TBS three times. Into each well, the diluted purified antibody was added, and preimmunized rabbit serum was used as negative control (each concentration 5 wells). After washing, into each well 100 μL of horseradish peroxidase-conjugated anti-rabbit antibody was added, incubated for 45 min, and with TBS containing 0.1% Tween-20 washed at least four times. 3,3′,5,5′-Tetramethyl-benzidine was used for development. By adding 100 μL of H_2SO_4 (0.5 M), the reaction was terminated. OD value in each well was measured at 450 nm and at 630 nm on a microplate reader. To determine specific anti-CTX MVIIA antibody by ELISA, the fusion protein thioredoxin (Trx) and CTX MVIIA (10 μg/mL) was coated on microtiter wells at 4°C overnight.

18.18 ASSAYS FOR RITUXIMAB IN PATIENT PLASMAS[24]

1. *Raji cell-Based Flow Cytometry Assay*: Raji cells (or anti-mouse IgG beads) were incubated in 12 × 75 mm polystyrene tubes, washed twice with 4 mL of PBS, centrifuged at 1260 × *g* for 2 min, aspirated, and reconstituted in appropriate media. All assays with unknowns or standards were conducted in at least duplicate. In a typical assay, the rituximab (RTX) standards as well as the sera or EDTA-anticoagulated patient plasmas were in BSA/PBS (1% bovine serum albumin in PBS) diluted 400-fold. Raji cells having at least 90% viability were washed, resuspended in BSA/PBS to 3×10^6 cells/mL, 100 μL of suspension was mixed with 50 μL of diluted standards, samples were gently vortexed and at room temperature incubated for 30 min with mild shaking. The cells were washed twice, set to 200 μL, mixed with 50 μL of 10 mg/mL mouse IgG (to minimize nonspecific binding), and treated with 25 μL of 0.1 mg/mL Al488 or Al633 mAb HB43. The tubes were gently vortexed, covered, at room temperature incubated for 30 min with mild shaking, washed twice, set to 100 μL, and in 150 μL of solution of 1% paraformaldehyde in PBS fixed. Samples were given flow cytometry analysis using a BD FACScalibur flow cytometer for 10,000 events per sample. By use of fluorescent FACS calibration beads, mean values of fluorescence intensity were converted to values of molecules of equivalent soluble fluorochrome (MESF).

2. *Goat Anti-Mouse IgG Polystyrene Bead-Based Flow Cytometry Assay*: Spherotech goat anti-mouse IgG beads packaged as a slurry containing approximately 3×10^7 beads/mL were suspended in 10-fold BSA/PBS, 100 μL of suspension was mixed with 50 μL of 3000-fold dilution (in BSA/PBS) of rituximab standards

and test compounds, samples were gently vortexed and at room temperature incubated for 30 min with mild shaking. The cells were set to a volume of 200 μL, mixed with 10 μL of 20-fold dilution of Caltag FITC goat anti-human IgG to yield a final concentration of 2 μg/mL. After a 30-min incubation at room temperature, the samples were washed twice, set to a volume of 250 μL, and samples were given flow cytometry analysis using a BD FACScalibur flow cytometer for 10,000 events per sample.

3. *ELISA*: Rabbit antimouse IgG $F(ab')_2$-specific was diluted in 0.05 M carbonate-bicarbonate buffer (pH 9.6) to final 3 μg/mL, to each well of a 96-well Falcon Pro-bind flat-bottom polystyrene plate or a Corning plate, 100 μL of antibody solution was added to prepare ELISA plates, the plate was at 4°C in aluminum foil wrapped and held overnight, washed five times with PBST (PBS + 0.1% Tween-20) in a Titertek M96 96-well programmable plate washer, covered with adhesive plastic, wrapped in aluminum foil, and stored at 4°C until used. To each well of the ELISA plate, 100 μL of 4000-fold or 8000-fold dilution (in BSA/PBS) of rituximab and test compounds were added to perform the assay. The plate was sealed with adhesive plastic, at 37°C incubated for 60 min, and washed five times with PBST. To each well, 100 μL of HRP-conjugated goat anti-mouse IgG, $F(ab')_2$ specific dilution (3000-fold in BSA/PBST) was added, the plate was at 3°C incubated for 60 min, and washed five times with PBST. To each well, 100 μL of SigmaFast *o*-phenylenediamine developing reagent was added, and the plate was usually incubated 10–20 min in the dark at room temperature until the appropriate level of color had developed. By adding 50 μL of H_2SO_4 (1 M), the reaction was stopped, and the OD values of the wells were read in a Titertek Multiscan Plus MkII 96-well plate reader at 492 nm.

18.19 ASSAYS OF TNF-α, SUPEROXIDE AND THYMOCYTE FOR LIVER INJURY OF MICE[25–27]

A suspension of 2.5 mg of BCG (viable bacilli) in 0.2 mL of saline was injected via the tail vein into each mace; 10 days later a solution of 7.5 μg of LPS in 0.2 mL of saline was injected. The mice were anesthetized with ether, sacrificed by cervical dislocation 16 h after LPS injection, and trunk blood was collected into heparinized tubes (50 U/mL) and centrifuged (1500 r/min, 10 min, 4°C). Serum was aspirated and stored at −70°C until assayed. The liver was also removed and stored at −70°C until required.

For the *in vivo* experiment, Kunming mice (20 ± 2 g, Animal Department of Anhui Medical University, China) were equally divided into five groups randomly, including normal, model control, and test compound groups (three different doses). Mice in test compound groups received suitable doses using an 18-gauge stainless steel animal feeding needle for 10 days prior to LPS injection. Mice in normal and model control groups were fed the same volume of vehicle only. For *in vitro* experiment, the Kupffer cells isolated from normal and BCG priming rat were divided into seven groups randomly, including control cells, cells added with LPS (5 μg/mL) alone, cells added with LPS (5 μg/mL), and test compound.

The liver of normal rat was initially perfused through the portal vein with D-Hank's until blood free and finally by recirculation with Hank's containing 0.5 mg/mL collagenase IV until the vessels were digested (up to 20 min). The liver was scraped using a cell scraper, filtered by 100-μm filter, and stirred in Hank's containing 2.5 mg/mL pronase and 0.05 mg/mL Dnase for 20 min at 37°C. After centrifugation three times and washing at 300 × *g* for 10 min at 4°C in GBSS, the cells were centrifuged in an 180 mg/mL Nycodenz gradient at 2500 × *g* for 20 min. Kupffer cells were carefully sucked by cusp-straws at the pearl layer inderphase. The purified Kupffer cell fractions were finally collected by centrifugal elutriation. The Kupffer cells were washed with Hanks' and resuspended in RPMI 1640 medium containing antibiotics (100 IU/mL penicillin, 100 μg/mL streptomycin), 2 mM L-Gln, and 100 μl/mL FCS. One-milliliter aliquots containing 1×10^6 cells were added to 24-well culture plates. The cells were incubated for 60 min in a humidified atmosphere containing 50 μL/mL CO_2 at 37°C. Nonadherent cells were removed, and adherent cells were washed twice with PBS. To observe direct effect, the cells at a density of 1×10^6/mL were incubated with different concentrations of test compound. The cells (1×10^6/well) were cultured for 48 h with 5 μg/mL LPS, the supernatants were collected, and the concentration TNF-α and NO were measured.

According to *in vitro* liver injury model, BCG-induced Kupffer cells were isolated from the livers of the rats injected via the tail vein with 3 mg of BCG 10 days before, and hepatocytes were isolated from the normal rat. The hepatocytes (1×10^9/mL), different concentrations of test compound, and BCG-induced Kupffer cells (1×10^6/well) were co-cultured for 48 h with 5 μg/mL LPS, and the supernatants were collected. In a 96-well plate, 100 μL of cell culture supernatant and 100 μL of Griess reagent (10 mg/mL sulfanilamide and 1 mg/mL *N*-1-naphthyl-ethylenediamine dihydrochloride in 25 μl/mL phosphoric acid) were incubated at room temperature for 10 min. Absorbance was measured at 540 nm. The nitrite concentration was calculated by comparing samples with standard solutions of sodium nitrite produced in the culture medium.

Livers were thawed, weighed, homogenized with Tris-HCl (5 mM containing 2 mM EDTA, pH 7.4), centrifuged (1000 × *g*, 10 min, 4°C), and MDA and SOD in the supernatant were immediately analyzed. MDA in liver tissue was determined by the thiobarbituric acid method. The assay for total SOD was based on its ability to inhibit the oxidation of oxyamine by the xanthine-xanthine oxidase system. The absorbance of the red nitrite produced by the oxidation of oxyamine was determined at 550 nm.

Thymocytes (2×10^6/well) from mice were cultured for 48 h in 96-well plates containing RPMI 1640 medium supplemented with 5 μg/mL concanavalin A and 0.1 mL of collected supernatant. Three hours before the termination of culture, cells were pulsed with MTT (Sigma-Aldrich Chem., St. Louis, MO) stock (20 μL/well, 0.22 μM in 5 mg/mL sterile PBS, stored in the dark at 4°C for up to 1 week, before use was immediately filtered to remove any formazan precipitate), returned to 37°C, and incubated for another 3 h. The plates were centrifuged (1000 × *g*, 10 min) to form cell pellets and MTT formazan products. The supernatant was carefully aspirated without disturbing the pellets, and formazan was solubilized by adding isopropanol (100 μL of isopropanol/200 μL of supernatant). Insoluble material was then removed by centrifugation (1000 × *g*, 10 min). The solubilized formazan in isopropanol was collected and distributed into 12-well flat-bottom ELASA plates at a final volume of 100 μL/well. Plates were read at 570 nm within 1 h of adding isopropanol.

18.20 GENE TRANSFER OF KRINGLE 1-5 AND RELATED ASSAYS FOR MICE WITH HEPATOCELLULAR CARCINOMA[28]

1. *Plasmid Construction*: The human kringle 1-5 (K1-5) with hexahistidine (His)-tag complementary DNA (cDNA, 1756-base) was cloned into pcDNA3.1 mammalian expression vector to construct pcDNA3.1/K1-5, which were verified by polymerase chain reaction, restriction analysis, and sequencing. The pcDNA3.1 vector served as a control.

2. *Expression of Endogenous K1-5 by Hepatoma Cells and in vitro Transfection of K1-5*: According to the Qiagen protocol with Lipofectin, 10 μg of pcDNA3.1/K1-5 or pcDNA3.1 (control) was transferred to Cos-1 cells. Total protein (10 μg) from conditioned media of Cos-1 cells transfected with pcDNA3.1/K1-5 or pcDNA3.1 or 10 μg of protein from cell lysate of KYN-2, KIM-1, and HAK1-B run on 10% SDS-PAGE was blotted to a polyvinylidene difluoride membrane, and with rabbit anti-His prove antibody (1:200 dilution) or mouse anti-human K1-3 antibody, which also recognized human K1-5 (1:200 dilution), incubated at 4°C overnight. After a 1 h incubation with anti-rabbit horseradish peroxidase (HRP)-conjugated antibody (1:10,000 dilution) or anti-mouse HRP-conjugated antibody (1:5000 dilution), immunoreactive bands were stained in an enhanced chemiluminescence Western blot analysis system. The relative protein expression was determined by densitometry using a charge-coupled device camera-based image analyzer.

3. *In vitro Proliferation Assay of BCE and Hepatoma Cells*: K1-5 protein from Cos-1 cells transfected with pcDNA3.1/K1-5 was collected using His Trap HP kit. To each gelatinized well of 96-well plates, approximately 1000 BCE cells in 100 μL of DMEM containing 10% FBS and recombinant human fibroblast growth factor 2 (3 ng/mL) were added, incubated at 37°C for 24 h, the media were replaced with 100 μL media containing various amounts of purified K1-5 protein (0, 1, 10, 100 μM) and 5% FBS, incubated for 30 min, and fibroblast growth factor 2 (1 ng/mL) was added. To each well of 96-well plates, approximately 1000 hepatoma cells in 100 μL of Dulbecco's modified Eagle medium containing 10% FBS were added, incubated at 37°C for 24 h, the media were replaced with 100 μL of media containing various amounts of purified K1-5 protein (0, 1, 10, 100 μM) and 5% FBS, incubated for 72 h, and cell proliferation was measured by a tetrazolium-based assay. Conditioned medium from cells transfected with pcDNA3.1 served as control.

4. *Injection Schedule of Human Hepatoma Cells and Liposome-DNA Complexes into Mice*: After anesthesia, the livers of male 5-week-old nude mice were injected with 2×10^6 human hepatoma cells with a 27-gauge needle. The mice implanted with KYN-2 cells were randomly divided into treatment with PBS, treatment with liposome-pcDNA3.1 (control vector-treated), treatment with liposome 25 μg pcDNA3.1/K1-5 (25 μg K1-5 cDNA-treated), and treatment with liposome 50 μg pcDNA3.1/K1-5 (50 μg K1-5 cDNA-treated) groups. The mice implanted with KIM-1 cells or HAK-1B cells were divided into control vector–treated and 50 μg K1-5 cDNA-treated groups. Seven days after implantation of hepatoma cells, the mice via the tail vein received injections of 100 μL of liposome plasmid complexes at 3-day intervals. For investigating tumor growth, after 4 weeks of the implantation, the mice were killed, and the tumor size was measured with calipers in two dimensions. Tumor volume was calculated using the equation Length $\times$ width^2 $\times$ 0.52. For studying survival, the injections of 50 μg K1-5 cDNA or control vector were continued to mice death. For evaluating intrahepatic metastasis, the severe combined immunodeficient mice received implantation of 2×10^6 KYN-2 cells and were randomly divided into control vector and 50 μg K1-5 cDNA-treated groups. After 3 weeks of injecting liposome-plasmid complexes, the mice were killed, and the number of intrahepatic nodules was counted. The liver weight of each mouse was also evaluated.

5. *Immunohistochemistry*: At 4°C, the sections were incubated with goat anti-His prove antibody (1:100 dilution) overnight to identify the cells that produced K1-5 in the liver. Using the avidin-biotin procedure with a Vectastatin ABC kit, the sections were incubated, and reacted with 0.005% H_2O_2/3,3′-diaminobenzidine at room temperature for 2 min. Each incubation was followed by 3 washes with PBS.

6. *Assessment of Vascularity of Tumor Tissues*: Sections of tumor tissues were treated with liposome/pcDNA3.1/K1-5 and liposome/pcDNA3.1 and immunostained with rat anti-CD31 antibody (1:100 dilution). Using light microscopy, stained blood vessels in tumor tissues were counted in 50 blindly selected random fields at 200-fold magnification.

7. *Western Blot Analysis of K1-5, Vascular Endothelial Growth Factor, Angiopoietin-1, and Angiopoietin-2 in Tumor Tissue*: In protein lysate buffer, tumor tissues of KYN-2 cells treated with 50 μg K1-5 cDNA or control vector were homogenized. Protein samples (50 μg) were run on 10% SDS-PAGE and polyvinylidene difluoride membranes, the membranes were incubated overnight at 4°C with primary antibodies (200-fold dilution of rabbit anti-His prove antibody, 200-fold dilution of rabbit anti-vascular endothelial growth factor antibody, 200-fold dilution of goat anti-angiopoietin (Ang)-1 and Ang-2 antibodies, 500-fold dilution of goat anti-actin antibody. After 1 h incubation with secondary anti-rabbit HRP-conjugated antibody (1:10,000 dilution), anti-goat HRP antibody (1:2000 dilution), or anti-mouse HRP antibody (1:5000 dilution), immunoreactive bands were stained in an enhanced chemiluminescence Western blot analysis system.

8. *Serum α-fetoprotein and Alanine Aminotransferase Levels and Body Weight*: By radioimmunoassay, serum α-fetoprotein level in KYN-2-implanted mice was

measured. Using a Clinical Automatic Analyzer with a standard UV method 7600, serum alanine aminotransferase activity of mice was measured. The body and liver weights of mice were evaluated at the time the animals were killed.

18.21 BODY WEIGHT ASSAY FOR MICE WITH LIVER TUMOR INCIDENCE[29]

The B6C3F1 mice (at 9-week age point) from all assays were assign to 1 of 17 consecutive weight groups ranging from 20 to 57.5 g in 2.5-g intervals with extra groups for high and low outliers. Each weight group of mice developing a liver tumor were sorted for the calculation of the relative tumor risk. The mice with hepatocellular adenoma, carcinoma, or hepatoblastoma were designated as positive, otherwise designated as negative. The process was repeated for each age point to 68 weeks.

Male B6C3F1/Nctr BR mice were approximately 4 weeks old at receipt, 5 weeks old at the initiation of controlled feeding, and 6 weeks old on the first day of dosing. The health of the mice was monitored during the studies. Groups of 120 male mice received chloral hydrate in distilled water by gavage at doses of 0, 25, 50, or 100 mg/kg (all at 5 mL/kg), 5 days per week for 104 to 105 weeks, whereas vehicle controls received distilled water only. Each dose group was divided into two dietary groups of 60 mice. The *ad libitum*-fed mice had feed available *ad libitum*, and the dietary controlled mice received feed in measured daily amounts. All mice had water available *ad libitum*. Twelve mice from each diet/dose group were euthanized for pathologic and biochemical evaluation at 71-week age point. Survivors of the remaining 48 from each group were evaluated at 110-week age point. These mice were fasted overnight before necropsy. The mice with liver tumors were referred to mice bearing single or multiple hepatocellular adenomas, hepatocellular carcinomas, and/or hepatoblastomas.

From week 21, the weight data set of the first experimental group was imported with the preceding week's data set as a reference. According to the predefined sort criteria (including whether the weight value for the previous week was outside either the 5% or the 12% confidence limits of the idealized weight curve or whether the mouse gained or lost weight in the preceding week), each mouse weight from the data set was sorted into its appropriate weight group and assigned either the corresponding *ad libitum* or weight-reduced tumor risk value from the tumor risk tables. The resulting tumor risk data set in the appropriate week's column of a new table were stored, proceeding to the next week's data set. According to all the weekly data sets from 21 to 68 weeks, the mean tumor risk for each mouse in the data set was calculated, the survival time for each mouse was imported, and the Poly-3 weighting time at risk factor (a_{ij}) for each mouse was calculated. The mean (cumulative) tumor risk data set was sorted into 2% incremental percentage tumor risk groups, and the corresponding a_{ij} value for each mouse was assigned into the appropriate tumor risk group. This procedure can be adapted to sort individual data for each week rather than the means of all 48 weeks. The resulting tumor risk group table was used to calculate the mean tumor risk estimate and standard deviation of the experimental group to predict survival-adjusted liver tumor rate, overall liver tumor rate, and the number of animals bearing tumors. The procedure was repeated for the next experimental group.

18.22 P450 ACTIVITY AND INDUCIBILITY IN RAT HEPATOCYTES FOR CYTOTOXICITY ASSAY[30]

1. *Isolation of Rat Hepatocytes*: The rats were anesthetized with sodium pentobarbital at 80 mg/kg. The liver was at 37°C perfused *in situ* with GIBCO Liver Perfusion Medium supplemented with 200 IU/mL penicillin and 200 μg/mL streptomycin for approximately 10–15 min and with GIBCO Liver Digestion Medium for approximately 10–12 min. The hepatocytes were from the digested liver released, suspended in Leibovitz's L-15 medium containing 5% FBS, and filtered through a plastic mesh with 100-μm pores. Filtered cells were sequentially washed twice with GIBCO hepatocyte wash medium supplemented with 100 IU/mL penicillin and 100 μg/mL streptomycin, once with 42% Percoll, once with hepatocyte wash medium in the presence of penicillin/streptomycin supplements, and finally once with hepatocyte attachment medium (HAM, WME in the presence of 5% FBS, 100 IU/mL penicillin, 100 μg/mL streptomycin, pH 7.4) and counted in the presence of 0.04% Trypan blue. Cell viability was greater than 90%.

2. *Primary Culture of Rat Hepatocytes for Enzyme Activity and Inducibility Measurements*: Into each well, 2 μL of diluted and neutralized collagen solution was dispensed, and by placing the plates on a shaker at high speed, an even coating was achieved. Prior to seeding, all plates were hydrated for at least 30 min with 100 μL of supplemented FBS containing WME (WME in the presence of 5% FBS, 100 IU/mL penicillin, 100 μg/mL streptomycin, 2 mM L-Gln, 0.1 μM dexamethasone, 10 μg/mL insulin, 5.5 μg/mL transferrin, and 6.7 ng/mL selenium, pH 7.4). In each well, approximately 3×10^4 hepatocytes in 100 μL of hepatocyte attachment medium were seeded. Hepatocytes were according to the experimental design cultured, allowed to attach at 37°C in 92% air-humidified atmosphere and 5% CO_2 for 2 h, medium was replaced with supplemented FBS containing WME for the single-layer collagen and Matrigel plates, and with Matrigel overlay for the collagen/matrigel (C/M) sandwich plates, the plates were returned to the incubator for another 22 h, dosed with DMSO, 20 μM bNF, 200 μM PB, and 10 μM DEX in 100 μL of supplemented FBS-free WME (WME in the presence of 100 IU/mL penicillin, 100 μg/mL streptomycin, 2 mM L-Gln, 0.1 μM dexamethasone, 10 μg/mL insulin, 5.5 μg/mL transferrin, and 6.7 ng/mL selenium, pH 7.4) at 24, 48, and 72 h after seeding (dosing day 0, 1, and 2). Inducers' stocks were prepared in DMSO, and the final concentration of DMSO in the medium was 0.4%. At 24, 48, 72, and 96 h after seeding (dosing day 0, 1, 2 and 3), the enzyme activities were measured, which were averaged from at least 6 wells per time point (Fig. 18.2a).

3. *Enzyme Activities*: At the treating end, the control medium or medium containing the inducers were removed, and the cells were washed at 37°C for 5 min by adding 200 μL of HBSS (pH 7.4). The buVer was removed and per well replaced

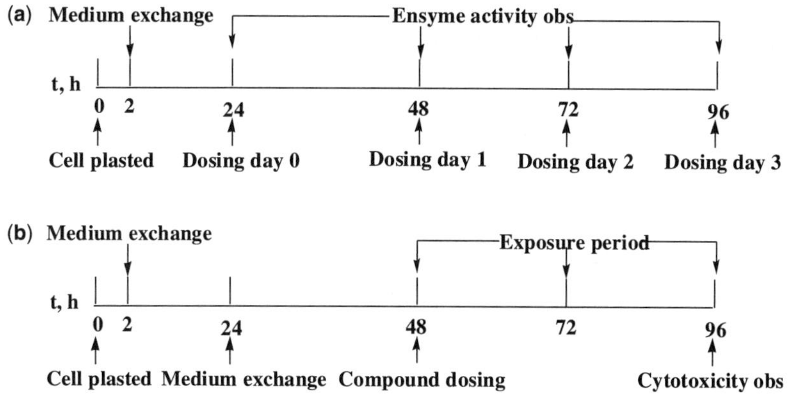

Figure 18.2 Schemes for the measurements of basal level enzyme activities: (a) enzyme inducibility or (b) cytotoxicity.

with 100 μL of HBSS containing 8 μM 7-ethoxyresorufin (7ER) and 10 μM dicoumarol, 15 μM 7-pentoxyresorufin (7PR) and 10 μM dicoumarol, 10 μM 7-methoxy-4-trifluoromethylcoumarin (MFC), or 100 μM 7-benzyloxy-4-trifluoromethylcoumarin (BFC). On a shaker, the plates were quickly mixed and returned to the incubator for 30, 30, 60, and 60 min for 7ER, 7PR, MFC, and BFC plates, respectively. At the incubation end, from each well a 75 μL aliquot of buffer was removed into a "V"-bottomed 96-well plate for storage at −80°C before analysis. The plates were thawed and to each well 10 μL of sodium acetate (0.5 M, pH 4.5) and 15 μL of β-glucuronidase/arylsulfatase (freshly diluted 1:100 with distilled water) were added, mixed on a shaker, and incubated at 37°C for 2 h. Enzymatic hydrolysis was terminated by adding 100 μL of methanol (for 7ER and 7PR plates) or 100 μL of Tris base (0.25 M) in 60% (v/v) acetonitrile (for MFC and BFC plates). After brief shaking, from each well a 100 μL aliquot was removed into a new 96-well plate, and 100 μL of glycine-NaOH (0.5 M, pH 10.5) was added to 7ER and 7PR plates. Fluorescence was measured with 530-nm excitation and 590-nm emission for 7ER and 7PR plates and 410-nm excitation and 510-nm emission for MFC and BFC plates on a SpectraMax Gemini EM plate reader. Standard curves were constructed with 0–60 pmol of resorufin or 0–160 pmol of HFC per well. The standards were added to the incubation buffer and taken through the enzymatic hydrolysis procedure.

4. *Cytotoxicity*: Using Promega CellTiter 96AQueous one cell proliferation assay, H4IIE cells were evaluated to determine the cell viability based on the reduction of 3-(4,5-dimethylthiazol-2-yl)-5-(3-carboxymethoxyphenyl)-2-(4-sulfophenyl)-2H-tetrazolium (MTS) by mitochondrial dehydrogenases in viable cells. Cells at passages 5–10 were seeded in 96-well plates (at 9000 cells/well) in 90 μL of FBS-containing DMEM. Before treatment with test compounds, cells were at 37°C in 5% CO_2 and 92% air-humidified atmosphere incubated overnight. Rat hepatocytes were at 3×10^4 cells/well seeded in 96-well C/M sandwich plates in 100 μL of supplemented FBS containing WME, allowed a 2 h attachment at 37°C in 5% CO_2

and 92% air-humidified atmosphere, and incubated for another 46 h (one medium change at 24 h), treated with test compounds, culture medium was replaced with 90 μL of supplemented FBS-free WME, and to each well 10 μL of dosing solutions were added. After a 48-h exposure to the reference compounds under the normal incubation conditions, to each well 20 μL of CellTiter 96AQueous One Solution was added, and the cells were further incubated for 2–4 h. By reading the absorbance at 490 nm on a SpectraMax Plus 384 plate reader, the amount of MTS metabolite formazan was quantified. Using the Neutral red uptake assay, cytotoxicity of reference compounds to 3T3 cells was evaluated. 3T3 cells at passages 5–10 were seeded in 96-well plates (9000 cells/well) in 90 μL of calf serum containing DMEM. Cells were at 37°C incubated for approximately 24 h in 5% CO_2 and 92% air-humidified atmosphere. At dosing, to each well 10 μL of dosing solutions were added, and the cells were returned to incubation for 48 h. Figure 18.2b is the scheme for hepatocyte culture. At the treating end, the medium was removed and the plates were rinsed once with PBS. To each well, 250 μL of Neutral red medium (25 μg/mL, in DMEM) was added, and incubation was continued for another 3 h. At incubation end, the Neutral red medium was removed, the cells were washed once with 250 μL of PBS, and extracted with 100 μL of water/ethanol/acetic acid (49/50/1) solution for approximately 30 min. Amount of intracellular Neutral red dye as a result of cell uptake was quantified by reading the absorbance at 540 nm on the SpectraMax Plus 384 plate reader.

18.23 LDH RELEASE IN RAT HEPATOCYTES FOR TOXICITY ASSAY[31]

1. *Culturing of Hepatocytes*: Isolated hepatocytes cultured on Matrigel were diluted with WME to 1 : 1. Hepatocytes were seeded at 6.4×10^5 cells/well (800 μL) in BD Falcon 12 well Multiwell plates for the LDH and the testosterone 6β-hydroxylation assays, at 2×10^6 cells/well (1.5 mL) in BD Falcon 6-well Multiwell plates for the [^{14}C]leucine incorporation assay, and at 3×10^6 cells/dish (3 mL) in 60-mm Permanox culture dishes for the CYP3A23 gene expression assay. Hepatocytes cultured in WME supplemented with FBS (10%), dexamethasone (100 nM), insulin (1 μM), penicillin (100 IU/mL), and streptomycin (100 μg/mL) were allowed to attach for 3 h in a humidified 37°C incubator with 5% CO_2, the medium was replaced with serum-free WME supplemented with dexamethasone (100 nM), insulin (1 μM), penicillin (100 IU/mL), and streptomycin (100 μg/mL), culture medium was changed daily and treated with test compounds. In the concentration response experiment, after the 3-h attachment period, cultured hepatocytes were treated with test compounds once every 24 h for a total of 72 h at selected concentrations or treated with culture medium as vehicle control.

2. *LDH Assay*: LDH release due to lysis or leakage from injured hepatocytes was considered as cytotoxicity indicator. At treatment end, 400 μL of supernatant from each well containing cultured hepatocytes was transferred to a 1.5-mL Eppendorf centrifuge tube placed on ice. The remaining supernatant was aspirated, mixed with 800 μL of Triton-X prepared in PBS (1%, pH 7.4), placed in a cold room (4°C)

for 2 h, cells were scraped into another 1.5-mL Eppendorf centrifuge tube using a cell scraper and centrifuged at 10,000 × *g* for 5 min. The pellets were sonicated (on ice for 30 s at 10 W) for obtaining cell lysates. To a well in a 96-well microplate containing 90 μL of PBS (pH 7.4) and 100 μL of reaction mixture (provided in the Cytotoxicity Detection Kit), a 10 μL aliquot of the supernatant or the lysate was added, the mixture was incubated at room temperature for 30 min, and the absorbance of each sample was determined at 492 nm. Amount of LDH released to the supernatant (LDH leakage) was expressed as a percentage of the total cellular LDH content (i.e., sum of the LDH in the supernatant and in the cell lysate).

18.24 RAINBOW TROUT HEPATOCYTES' EROD ACTIVITY ASSAY[32]

This assay was to validate and use a primary culture of rainbow trout hepatocytes as a multi-end-point *in vitro* assay for toxicity characterization of sediment extracts. The assay end points were chosen to include acute toxicity (cytotoxicity) as well as sub-lethal biomarker and effect end points such as metabolic inhibition, DNA damage, endocrine disruption (estrogenicity), and cytochrome P450 1A-catalyzed EROD activity. Within 72 h, the sampling sediments stored cool and dark were subjected to Accelerated Solvent Extraction. Resulting organic extracts, and the solutions of model compounds E2, 2,3,7,8-TCDD, 4-*n*-nonylphenol, and 4-nitroquinoline-1-oxide in DMSO were assayed in a primary culture of male rainbow trout hepatocytes. Cell media were assayed for the estrogenic biomarker vitellogenin, whereas the cells were analyzed for cytochrome P450 1A-catalyzed EROD activity, DNA damage by the Fast Micromethod, metabolic inhibition, and cytotoxicity (membrane integrity).

18.25 EROD ACTIVITY, ROS PRODUCTION AND CYTOTOXIC CONCENTRATION IN FISH HEPATOCYTES FOR DRUG TOXICITY ASSAYS[33]

1. Immature rainbow trout from a local hatchery were kept in tanks with aerated charcoal-filtered tap water at 15°C, fed with commercial fish food, and acclimatized to laboratory conditions for a minimum of 2 weeks before use. Rainbow trout was killed, in aseptic conditions the abdominal cavity was rapidly opened, the liver was perfused via the hepatic portal vein with about 30 mL of Dulbecco's PBS (D-PBS) without calcium and magnesium, and albumin 0.1% and sodium citrate 10 mM were adjusted to pH 7.5 as perfusion solution. The liver was dissected, cut in small pieces, transferred to a sterile dissociation D-PBS with calcium and magnesium, 0.5% albumin, 120 mM NaCl, and 15 mM sodium citrate, agitated at 4°C for 30 min, the mixture was passed through sterile 63-μm nylon gauze three times, and the cell suspension was washed three times by centrifugation at 2000 × *g* and 4°C for 4 min until the supernatant became clear. The supernatant was removed, the cells were resuspended in 5 mL of D-PBS with calcium and magnesium and using Trypan blue exclusion determined for viability (>90%). Freshly isolated hepatocytes

(15×10^5 cells per well) were seeded in 48-well untreated microplates, cultured at 15°C in M199 cell culture medium supplemented with 5% decomplemented fetal calf serum, penicillin and streptomycin (50 IU/mL), and 10 mM HEPES for 24 h before exposure to test chemicals. The PLHC-1 cell line was derived from the hepatocellular carcinoma of the top-minnow *Poeciliopsis lucida*. Cells were routinely grown at 30°C in EMEM salts supplemented with 10% (v/v) decomplemented FCS 1% (v/v) nonessential amino acids, and 50 IU/mL of penicillin and streptomycin in a 5% CO_2 humidified atmosphere. For chemical testing, 5×10^5 cells per well were subcultured, seeded in 96-well plate, and left to grow up to confluency before adding test chemicals. Test chemicals that dissolved either in DMSO or in ultrapure water were diluted with medium to final solvent concentration was always 0.5% or 1% (v/v), respectively. Solvent alone was used as control (carrier control). After a 24-h exposure to test chemicals and controls, both PLHC-1 cells and PRTH cells were subjected to MTT and EROD assays.

2. *MTT Assay*: Cytotoxic concentrations were determined by the MTT reduction test. After test chemical exposure, the medium was removed, cells were with 0.5 mg/mL MTT dissolved in RPMI medium incubated for 3 h, MTT was cleared out, the formazan salts were solubilized in 100 μL of isopropanol, and plates were read at 570 nm against a 660-nm reference wavelength on a microplate reader. A percentage of the corresponding control value was expressed as cell viability.

3. *EROD Assay*: Medium was removed and in 96- or 48-well microplates replaced by 100 or 300 μL of culture medium containing 2 μM 7-ethoxyresorufin. Kinetics of resorufin production were at 30°C and room temperature monitored in a microplate fluorimeter at 530-nm excitation and 590-nm emission during 15 min for PLHC-1 cells and PRTH, respectively. The positive control for EROD induction in each test was 1 μM BaP. The wells were washed with PBS and using BSA as a standard with the fluorescamine assay, total cellular proteins were determined. The EROD activity was expressed in pmol resorufin/min and pmol resorufin/mg proteins.

4. *ROS Generation*: PLHC-1 cells were washed three times with PBS, at 30°C incubated with 1 μL of dichlorofluorescein diacetate (DCFH-DA; 40 μM) diluted in PBS supplemented with 10 mM glucose (PBS-Gluc) for 30 min, the cell monolayers were washed twice with PBS and exposed to test chemicals in PBS-Gluc, the fluorescence of oxidized DCFH was measured immediately and at 1-h intervals during 5 h with a microplate fluorimeter at 490-nm excitation and 535-nm emission wavelengths. Positive control used 5 μM H_2O_2. The kinetics allowed determining the toxic exposure time, at which maximum of ROS production occurred. Results were expressed as a percentage of the basal fluorescence in the carrier control wells.

18.26 COHO SALMON'S IGF-I GENE SEQUENCING AND TAQMAN ASSAY[34]

1. *Animals*: Coho salmon (*Oncorhynchus kisutch*) raised in 1.3-m-diameter cylindrical tanks with recirculated fresh water at 11°C to 12°C under simulated natural

photoperiod at the Northwest Fisheries Science Center were fed BioOregon-grower1 at the rations specified. Two-year-old and 1-year-old coho salmon were used for tissue distribution and hepatocyte culture assays and comparison between TaqMan and RNase protection assays, respectively.

2. *Acidic Ribosomal Phosphoprotein P0 Sequencing*: Total RNA from coho salmon liver was reverse transcribed. Based on conserved regions' ARP sequences, forward TGGAAGCACTGCAAAGATGC and reverse ATGGTTCCTCTGGAGAT-CTTGG, of the zebrafish (*Dania rerio*) and catfish (*Ictalurus punctatus*) cloning primers were designed. Fifty-microliter PCR was performed using 5 μL of cDNA template, 100 nM forward and reverse primers, 2.5 U/μL Promega Taq polymerase, 200 μM dNTPs and 1.5 mM $MgCl_2$, and cycling conditions, Taq activation 95°C for 5 min followed by 36 cycles of 94°C for 45 s, 55°C for 45 s, and 72°C for 1 min 30 s, with a final 72°C extension for 5 min. With cloning primers, PCR products were sequenced in both directions. By filtration through a Nanosep membrane, unincorporated nucleotides and primers were from the cloning PCR removed, and the cleaned up PCR product was used as template for a BigDye Terminator cycle sequencing reaction. By chromatography with Sephadex G100 in Centrisep columns, unincorporated dyes were from the BigDye reaction removed, and on an Applied Biosystems (ABI) 3100 Genetic Analyzer, sequencing was performed.

3. *TaqMan Assays*: With ABI's Primer Express program, assays for salmon insulin-like growth factor-I (IGF-I) and acidic ribosomal phosphoprotein P0 (ARP) were designed. To avoid differences between salmonid species, primer and probe sequences were designed. To reduce signal from genomic DNA, one primer in each assay was placed across a cDNA intron/intron boundary.

After removing culture medium and adding Tri-Reagent (MRC, Cincinnati, OH), cell culture plates were at −80°C frozen and stored for at most 1 month. Plates were thawed, wells were scraped, and following the MRC protocol, RNA was isolated with bromochloropropane as the phase separation reagent and washed twice in 75% ethanol to yield 10–20 g RNA/well. By spectrophotometry, RNA was quantified and purity assessed, and 260 : 280 ratios were 1.8–2.0. Visualization on 1% agarose gels showed that RNA was not degraded and then diluted to 100 ng/μL.

First-strand cDNA was in 15 μL of RT reactions with 3 μL of RNA template, 2.5 U/μL SuperScript II reverse transcriptase, 5 μM random hexamer primers, 500 μM dNTPs, 0.4 U/μL RNase inhibitor, 10 mM dithiothreitol (DTT), and 1× RT buffer synthesized. RT reactions were set up on ice, loaded with a multichannel pipette to avoid loading order biases, and performed in 96-well plates with cycling conditions of 25°C for 10 min, 48°C for 60 min, and 95°C for 5 min.

On an ABI 7700 Sequence Detector using the standard cycling conditions recommended by the manufacturer (50°C for 2 min, 95°C for 10 min, followed by 40 cycles of 95°C for 15 s, 60°C for 1 min), TaqMan PCR was performed. Wells contained 25 μL of PCR mixture (0.01 U/μL AmpErase uracil *N*-glycosylase,

0.025 U/μL AmpliTaq Gold DNA polymerase, passive reference dye I, 200 μM dNTPs including dUTP, 3.5 mM $MgCl_2$) with 200 nM probe, 900 nM forward and reverse primers, and 2 μL (18S and ARP assays) or 4 μL of (IGF-I assays) cDNA template. To assess PCR efficiency, a fivefold serial dilution (12 wells) of sample cDNA was run for each assay on each plate. Assays were not multiplexed. TaqMan PCRs for target and reference genes were from the same RT reaction run.

18.27 DNA STRAND BREAK IN RAT HEPATOCYTE AND COMET ASSAY[35]

1. *Preparation and Cultivation of Hepatocytes*: Intact hepatocytes from male Wistar rats (250–300 g) by collagenase perfusion and Percoll gradient centrifugation sedimenting at the bottom of the tube were suspended in WME containing 10 mM *N*-(2-hydroxyethyl)piperazine-*N*′-(2-ethanesulfonic acid) (HEPES), 2 mM L-Gln, 100 IU/mL penicillin, 100 μg/mL streptomycin, and 0.5% ITS+ premix, plated onto collagen-coated 24 wells at a density of 2×10^5 cells/well or 6 wells at a density of 1×10^6 cells/well. For viability testing and comet assay, after 2 h incubation, the medium was changed and pretreated with antioxidants immediately. Five hours later, medium was again changed, and hepatocytes were exposed to acetaminophen with or without antioxidants for 12 h, then the cells were either immediately used or were incubated for an additional 24 h in the absence of acetaminophen and antioxidants. For assays without antioxidants aimed at malondialdehyde (MDA) measurement and RNA isolation, a 3 h preincubation phase was used. At 6 h after the start of acetaminophen treatment, RNA was recovered, and 20 h after the start of acetaminophen treatment, MDA measurement was performed.

2. *Determination of Cytotoxicity*: Cell viability was assessed by spectrophotometrically measuring the formation of formazan from MTT. At the end of the experiment, hepatocytes were at 37°C with 0.7 mg/mL MTT incubated for 30 min, washed with PBS, the blue formazan was with isopropanol/formic acid (95/5) from cells extracted, and cell viability was photometrically measured by the Neutral red test at 560 nm. For this, hepatocytes were incubated at 37°C with 2-amino-3-methyl-7-dimethylaminophenazonium chloride (50 mg/mL) for 2–4 h, washed with PBS, the red dye was extracted from the cells with a mixture of 50% ethanol and 1% acetic acid and photometrically determined at 550 nm.

3. *Comet Assay*: Using comet assay (single cell gel electrophoresis), DNA single strand break was measured. From the culture plate, by incubation with collagenase, cells were detached, dissolved in 0.5% low melting point (LMP) agarose, and spread on a microscopic slide precoated with 1.5% agarose. After solidification, cells were in a sodium laurylsarcosinate solution lysed overnight. Slides were washed, transferred to an electrophoresis chamber, and nuclei were with the alkaline electrophoresis buffer (pH 13) treated for 25 min and electrophoresed for another 25 min. After neutralization, nuclei were with ethidium bromide stained. In a fluorescence microscope, image length was measured.

18.28 DNA LADDERING IN RAT HEPATOCYTE AND TUNEL ASSAY[36]

1. *Hepatocyte isolation*: Female C57BL/6J mice (6 to 12 weeks old) were anesthetized, and a midline laparotomy was performed. The inferior vena cava was cannulated and at a rate of 2.5 mL/min perfused. The portal vein was sectioned, and the solution was allowed to flow through the liver. The liver was sequentially perfused with EGTA solution (EBSS, without Ca^{2+} and Mg^{2+}, with 0.5 mM EGTA and 10 mM HEPES, pH 7.5) and collagenase solution (EBSS, with Ca^{2+} and Mg^{2+}, 30 mg collagenase P/100 mL, 6 mg trypsin inhibitor/100 mL, and 10 mM HEPES, pH 7.5). By mechanical dissociation, hepatocytes were removed, through sterile 100-μm mesh nylon filtered, washed twice by centrifugation at 50 × *g* for 5 min, and resuspended in serum-containing culture medium (1/3 Waymouth's medium, 2/3 MEM, 10% FBS, 1% penicillin-streptomycin, and 1% HEPES). By Trypan blue exclusion (0.02% final concentration), cell count and viability were assessed.

2. *Cell Cultures*: Three million to 6 million isolated hepatocytes were in 8 mL of defined serum-containing medium cultured except in some assays where Hepatozyme serum-free medium was used, either attached to 100-mm Primaria plastic plates (nitrogen-containing hydrophilic surface improving attachment, spreading, and growth of primary cells) or on 100-mm polyhydroxyethylmethacrylate-coated glass Petri dishes under detached conditions to avoid cell anchorage. For some assays, additional attachment or survival factors were used. Plastic fibronectin plates were at 37°C with 50 μg/mL fibronectin in PBS coated for 2 h and washed twice with PBS before use. When specified, to the culture 20 ng/mL hepatocyte growth factor was added, at 37°C in 5% CO_2 incubated for 15 min to overnight, the cells were collected, counted, and their viability was determined by Trypan blue exclusion. After collection, detached cells were centrifuged at 50 × *g* for 5 min and tested. For the plates with attachment, the attached cells were harvested by trypsination at 37°C for 4–6 min, washed, and directly analyzed or mixed with the floating cells in the corresponding plate before testing. To explore the effect of temperature, 3×10^6 cells were plated on poly-HEMA–coated glass plates several times at either 37°C or 4°C. Cells were also at 4°C stored for different times in 15-mL tubes, then at 37°C on poly-HEMA–coated glass plates incubated for 1 h, their viability was assessed by Trypan blue exclusion, and DNA fragmentation was determined by the ELISA.

3. *DNA Laddering*: DNA laddering as the marker of the DNA fragmentation occurring in apoptotic cells was investigated in electrophoresis gel. Liver cell pellets were lysed with 0.4 mL of DNA extraction buffer (10 mM Tris-HCl, 0.1 M EDTA, 0.5% sodium dodecyl sulfate, pH 8.0), incubated overnight at 42°C with proteinase K (200 μg/mL), extracted twice with phenol, chloroform/isoamyl alcohol for DNA, and incubated with 50 μg/mL RNase A. Treating with 52 μL of sodium acetate (2.5 M, pH 7.3) and 2 μL of glycogen (20 mg/mL) at −70°C for at least 1 h, DNA was precipitated, centrifuged at 13500 rpm for 15 min, air dried, and dissolved in water to give DNA pellets. On 1.5% agarose gel mixed with 0.5 μg/mL ethidium bromide, 4 μg DNA was run and under ultraviolet light visualized.

4. *Cell Death ELISA*: Using a cell death ELISA kit, DNA fragmentation was also detected, which was based on a quantitative sandwich-enzyme-immunoassay principle using antibodies against DNA and histones and allowed the specific determination of mono- and oligonucleosomes in the cytoplasmic fraction of cell lysates. According to the manufacturer's instructions, assay was performed and 10^5 cells were used in duplicate. In negative and positive controls, no immunoreagent was provided whereas in positive sample immunoreagent was provided in the kit. In background control, the buffer contained no cells. OD value was at 405–490 nm measured and expressed by the enrichment factor in comparison with the basal status, which was defined as the apoptosis level in the plastic plate.

5. *TUNEL Assay*: The Oncor Apoptag Plus *in situ* detection kit was used to identify the 3′-OH DNA fragments produced by the action of endonuclease in the process of apoptosis as recommended by the manufacturer. Under attached or detached conditions, hepatocytes were cultured for 3 h, collected, fixed to glass slides with paraformaldehyde, digested with proteinase K, and incubated with terminal deoxynucleotidyl transferase (TdT) and digoxigenin nucleotides. To detect the presented signal and 4′-6′-diamino-2-phenylindole (DAPI) counter-stained nuclei anti-digoxigenin fluorescein labeled antibody was used. Under a confocal laser scanning microscope at 488 nm and 568 nm, cells from detached and attached cultures were scored for incidence of TUNEL(+) staining, and phase-contrast microscopy was used for apoptotic morphology. Using Olympus Fluoview 2.032 software, digital pictures were taken as record and to facilitate scoring. Two positive and one negative control were given in the kit and used in all assays. When green fluorescence of the nucleus (TUNEL) with condensation and disorganization of the chromatin (DAPI) demonstrated DNA fragmentation, cells were considered apoptotic.

18.29 COUNTING INCREASES IN VIABLE CELL NUMBER FOR HEPATOCYTE PROLIFERATION ASSAY[37,38]

Male SD rat (225–250 g) was anesthetized with sodium pentobarbital at 65 mg/kg ip. After the performance of tracheotomy, a catheter of PE 240 polyethylene tubing was inserted into the trachea to facilitate the respiration. The right femoral artery and vein were cannulated with PE 50 polyethylene tubing. To supplement fluid loss and maintain the anesthetic level, 0.5 mL saline/100 g body weight per hour, containing 1 mg/mL sodium pentobarbital, was infused through the vein by a Syringe Infusion Pump. To prevent blood clotting, the catheters were flushed with heparin (200 U/mL). A median-line incision (3 cm) was made on the abdomen posteriorly from the xiphoid process of the sternum. Left lateral and median liver lobes were ligated using surgical silk (size 0) and given 2/3 PH. The incision was sutured with two layers of the muscle and skin. At various time points after PH, the diaphragm was cut, and a blood sample (5–10 mL per rat) was drawn from the right ventricle of the heart within 2 min. In a tissue culture hood, the blood was transferred to a sterile centrifuge tube and centrifuged (2500 × *g*, 20 min) for collecting plasma. In the preparation of serum, the blood was allowed to clot on ice for 10 min then spun down. The serum from non-PH rats was used as 0 h control. For sham-operated rats, the liver was manipulated without PH.

In the liver perfusion, male SD rat (300 g) was anesthetized with sodium pentobarbital (ip, 65 mg/kg). An 18-gauge, 1.25-in. (32 mm) iv catheter was used as the portal vein catheter tied to the vein by surgical silk. The perfusion buffer was drained through a catheter inserted into the inferior vena cava via the right atrium of the heart. First, the liver was perfused with 400 mL of nonrecycled oxygenated Ca^{2+}-Mg^{2+} free Dulbecco's phosphate-buffered saline (DPBS, 2.68 mM KCl, 1.47 mM KH_2PO_4, 136.9 mM NaCl, 8.1 mM Na_2HPO_4, pH 7.4) containing 0.49 mM EDTA to reduce Ca^{2+} in the liver tissue. Second, the liver was perfused with 100 mL of DPBS without EDTA to wash out the EDTA in the system. Finally, the liver was perfused with 100 mL of recycled oxygenated collagenase/Swim's 77 (0.25 mg/mL) with 5 mM Ca^{2+} for 10–15 min. Upon finishing perfusion the liver was rinsed thoroughly with 10–20 mL of sterile DMEM. In the tissue culture hood, the liver was transferred into a sterile Petri dish containing fresh medium, the liver capsule membrane was slit using two forceps, and the cells were released into the medium by gently shaking the liver. Two Spectra/Mesh N filters (70 μM, 40 μM, spectrum) were used to filter the isolated liver cells to remove tissue chunks and cell debris. The filtered cell suspension was further purified by low-speed differential centrifugation (300 × g, 3 min, 4°C). The supernatant was discarded by aspiration. The cell pellet was gently mixed with fresh medium, and the centrifugation procedure was repeated three times. Viable cell concentration and percentage of nonhepatocytes was determined by the Trypan blue exclusion method. In the purification of hepatocyte, 20 μL of Trypan blue (0.4%) was added into 20 μL of cell suspension (dilution factor = 2), which was mixed and loaded onto the hemocytometer. All corner squares (64 squares/chamber) of the two chambers were counted (128 squares total; each square was 1 nl, 128 nl × 7.813 = 1 μL). Viable cells/μL = viable cells per 128 squares × 7.813 × 2. The final concentration was the average value of three separate countings. Viability (%) = viable cells/(viable + dead cells) × 100%. Nonhepatocyte (%) = nonhepatocyte/total cells × 100%.

Each well of 6-well plates was coated with 20 μL of solution of rat tail collagen (0.8 mg/mL double-distilled H_2O with 0.1% acetic acid) and allowed to dry in the hood. To the well, approximately 2 mL of DMEM/F-12 (supplemented with 25 mM sodium bicarbonate, 10 mM HEPES, 100 IU/mL penicillin, and 0.1 mg/mL streptomycin) was added. To each well, the suspension of viable cells (final cell concentration, 1×10^5 cells/mL) were seeded and constantly mixed by agitation. The plates were left in the humidified incubator at 37°C with 5% CO_2 about 18 h. When the attachment reached end, the medium was changed. The culture plates were mildly shaken, and the medium was aspirated. The wells were rinsed with 1 mL of medium and aspirated. From three randomly selected wells in the culture, the counts of the viable starting cell number were calculated. Serum or plasma samples in 2 mL of fresh medium (10%) were added to wells, to which heparin (35 U/mL final) was added to prevent the medium clotting during the culture. Cells grown in serum-free medium and in serum-free medium plus 100 ng/mL EGF and 20 mU/mL insulin were used as the negative and positive control, respectively.

The medium was changed by aspiration at 24 h with no rinse. To each well, fresh medium with serum or plasma and other additives were added. At the end of 48 h culture, attached hepatocytes were harvested and counted. The medium in each well was aspirated. Serum or plasma containing wells were rinsed once with saline and aspirated to wash off various plasma proteins. Trypsin-EDTA (400 μL) was added to each well. The plate was covered, left in the incubator for approximately 1 min, removed from the incubator, and checked under the microscope. The trypsin digestion was assessed. The plate was immediately placed on ice and serum (SD rats, 40 μL, 10% final concentration) was added into each well to inhibit further trypsin digestion of the cells (440 μL total volume per well). With the plate maintained on ice, the cells were detached using a Pasteur pipette to gently blow the cells from the well bottom. After all the cells were lifted, the cell suspension was transferred into a 1.5-mL Eppendorf tube and kept on ice for an immediate counting without centrifugation or trypsin removal. Then, 40 μL of cells and 10 μL of Trypan blue (0.4%) were well mixed in a plate (dilution factor = 1.25) and loaded onto the hemocytometer.

18.30 ASSAY OF HGF AND TGF-β IN CCL-64 CELLS[39]

CCL-64 cells (Mv-1-Lu) were grown in RPMI 1640 supplemented with 10% FCS, 2 mM L-Gln, and 40 μg/mL gentamicin (complete medium, CM). Serial dilutions of test samples were prepared on 96-well flat-bottomed microplates with 100 μL of CM per well. Prior to the assay, to activate latent TGF-B, the samples were transiently acidified by first adding HCl to pH 2 and subsequently neutralizing with NaOH. CCL-644 cells were seeded at 10^4 cells/well and grown for 224 h in a total volume of 0.2 mL of CM. After 20 h, the cells were pulsed for 4 h with 1 μCi/well of [methyl-^{3}H]thymidine and then harvested with a Micromate 196 B counter. The concentration of TGF-β in sample was determined by the growth inhibition caused by the sample, compared with a standard curve obtained by resting serial dilutions of porcine TGF-β. When the assay was used to measure HGF, the concentration was determined by this cytokine's ability to reverse the growth inhibitory effect of 350 pg/mL of TGF-β.

18.31 ASSAY FOR SERUM IFN-α OF PATIENTS WITH CHRONIC HEPATITIS C[40]

1. *Patient Sera*: Serum samples of various time points during different IFN-α treatment schedules were prepared within 2 h of blood sampling and stored at −70°C until assayed. All patients infected chronically with HCV were enrolled in investigator-initiated clinical trials after having given informed consent.

2. *Generation of a Selectable HCV Replicon with a Reporter Gene*: The basic replicon construct pFK-I389/NS3-3′/5.1 carried three cell culture adaptive mutations that enhance RNA replication cooperatively (E1202G, T1280I, and S2197P). By using standard recombinant DNA technologies, the reporter gene luciferase from

the firefly (*Photinus pyralis*) and the coding sequence for ubiquitin were inserted upstream of the neo gene. Huh-7 cells that carried persistently replicating HCV replicons-derived cell clones were generated.

3. *Cell Culture and Dose-Response Assays*: In DMEM supplemented with 2 mM L-Gln, nonessential amino acids, 100 IU penicillin, 100 μg streptomycin, and 10% FCS and 250 mg/mL G418, Huh-7 human hepatoma cells carrying the HCV replicon were grown, passaged twice per week at a dilution of 1:3 to 1:5, depending on cell growth by using PBS containing 0.05% trypsin and 0.05% EDTA. For the luciferase/RT-PCR assay, 2×10^5 cells were seeded in 9.6 cm^2 dishes in complete DMEM supplemented with 250 mg/mL G418, incubated for 24 h, and medium was replaced by complete DMEM without G418 and containing 100, 50, 25, 10, 7.5, 5, 2.5, 1, or 0 U/mL of interferon (IFN)α-2a in triplicate. In all other cases, cells (4×10^4 cells per well) were seeded in 24-well plates (2 cm^2 per well), incubated for 24 h, medium was replaced by 1 mL of complete DMEM per well containing 10% human serum and one of 1000, 500, 250, 100, 50, 25, 12.5, 6.25, 3.13, 1.56, 0.78, 0.39, 0.20, or 0 U/mL of IFN-α2a or IFN-α2b, or 180, 60, 30, 15, 7.5, 3.75, 1.88, 0.94, 0.47, 0.23, 0.12, 0.06, 0.02, or 0 ng/mL pegylated interferon (PEG-IFN)-α2a. All were assayed at least twice each in sextuple and on several plates, such that the positions of individual samples relative to the controls were varied to reduce positional effects. In cell proliferation assay, medium in triplicate wells was replaced by 1 mL of complete DMEM containing increasing concentrations of ribavirin. After staining the cells with the tetrazolium salt WST-1, by measuring the OD proliferation was determined. A positive (1000 U/mL IFN-α) and a negative control (0 U/mL IFN-α) were assayed on each plate such that the position was different from plate to plate.

4. *Incubation of Cells with Patient Sera*: Twenty-four hours after seeding, cells in each well of a 24-well plate were with 0.9 mL of complete DMEM and 0.1 mL of serum from patients incubated. All were assayed in triplicate. On each plate, a positive (1000 U/mL IFN-α) and a negative control (0 U/mL IFN-α) was included.

5. *Cell Harvesting and Measurement of Luciferase Activity*: After 48–50 h incubation (or after 72 h in case of luciferase/RT-PCR assays), cells were harvested. After removal of the medium, cells were washed twice with PBS, incubated with 180 μL or 350 μL of ice-cold luciferase lysis buffer (1% Triton X-100, 25 mM glycylglycine, 15 mM $MgSO_4$, 4 mM EGTA, and 1 mM DTT added just before usage) for 10 min, centrifuged at 13,000 × *g* and 4°C for 15 min to harvest the lysate, 100 μL of lysate was mixed with 360 μL of assay buffer (25 mM glycylglycine, 15 mM $MgSO_4$, 1 mM DTT, 2 mM ATP, 15 mM potassium phosphate buffer, pH 7.8) and 200 μL of a 200 mM luciferin stock solution in 25 mM glycylglycin was added. Luminescence was measured in a luminometer for 20 s and expressed as the number of relative light units (RLU) detected.

6. *RNA Preparation and Quantitative RT-PCR*: Cell lysate (150 μL) was mixed with 600 μL of GITC buffer (4 M guanidinium thiocyanate, 25 mM sodium citrate, 0.5% sarkosyl, 0.1 M β-mercaptoethanol), 15 μL of sodium acetate (2 M, pH 4.5), 150 μL of phenol, and 30 μL of chloroform, on ice incubated for 15 min, and samples were at 4°C and 13,000 × *g* centrifuged for 15 min. Aqueous phase nucleic acids

were transferred to a fresh tube, precipitated with 1 volume of isopropanol, washed once with 70% ethanol, and resuspended in 50 μL of doubly distilled water. Using an ABI PRISMTM 7700 Sequence Detector System, 3 μL of sample was used for quantitative RT-PCR analysis. With the One-step RT-PCR kit and using HCV 5′-6FAM (6-carboxyfluorescine)-TCCTGGAGGCTGCACGACACTCATTAMRA-(tetrachloro-6-carboxyfluorescine)-3′, HCV-S66 5′-ACGCAGAAAGCGTCTAG-CCAT-3′, HCV-A165 5′-TACTCACCGGTTCCGCAGA-3′, GAPDH 5′-TET (6-carboxy-4,7,2′,7′-tetrachlorofluorescein)-CAAGCTTCCCGTTCTCAGCCT-TAM-RA-3′ GAPDH-S 5′-GAAGGTGAAGGTCGGAGTC-3′, GAPDHA 5′-GAA-GATGGTGATGGGATTTC-3′, the HCV and GAPDH specific RT-PCRs were conducted in duplicate. Reactions were carried out at 50°C for 60 min (reverse transcription), at 95°C for 15 min (heat inactivation of RT and activation of Taq-polymerase), at 95°C for 15 s, and at 60°C for 1 min, total 40 cycles (amplification). The total volume of the reaction mixture was 15 μL and contained 2.66 mM 6-carboxy-X-rhodamine (Rox, passive reference), 4 mM $MgCl_2$, 0.66 mM dNTPs, 0.266 mM HCV probe, 0.3 mM GAPDH probe, 1 mM of each HCV and GAPDH primer, and 0.6 μL of enzyme. By comparison with serially diluted *in vitro* transcripts, the amount of HCV RNA was calculated, and by comparison with serially diluted HuH-7 total RNA, the amount of GAPDH RNA was calculated. Both RNA species were involved in the RT-PCR amplifications.

18.32 RAT MEDIUM-TERM LIVER, DNA MICROARRAY AND Cu^+-REDUCING ANTIOXIDATION ASSAYS FOR DEN-INDUCED HEPATOCARCINOGENESIS[41]

1. *Treatment of Animals*: As depicted in Fig. 18.3, in assay 1, male F344/DuCrlCrlj rats were housed in an air-conditioned room at 22 ± 1°C and a relative humidity of 60 ± 10%, with a 12-h light/dark cycle for a 2-week acclimation period, maintained on a basal pellet diet and tap water *ad libitum*, divided into five groups with 15 rats in groups 1, 2, 3, and 5 and 23 in group 4, administered a single intraperitoneal injection of DEN (200 mg/kg body weight) dissolved in 0.9% NaCl solution (saline) at 6 weeks of age to initiate hepatocarcinogenesis on the first day of the experiment. After 2 weeks on the basal diet, rats were administered the basal diet containing 1.00% (group 1), 0.10% (group 2), or 0.01% (group 3) or 0.00% (DEN alone control; group 4) D-allose and 1.00% (group 5) D-glucose as a control for the following 6 weeks. At the end of week 3, all rats were subjected to two-thirds PH. Body weight was recorded at 2-week intervals, and at the end of week 8 of the experiment, all surviving rats were killed by exsanguination under deep anesthesia, and the livers were excised. In assay 2, a total of 15 rats were divided into three groups (each five) and administered the basal diet containing 1.00% (group 1) or 0.00% (group 3) D-allose and 1.00% (group 2) D-glucose as a control for a week. At the end of the experiment, all rats were killed by exsanguination under deep anesthesia, and the livers were excised. In experiment 3, a total of 10 rats were divided into two groups (each five) and administered the basal diet containing 2.00% or 0.00% D-allose for a week.

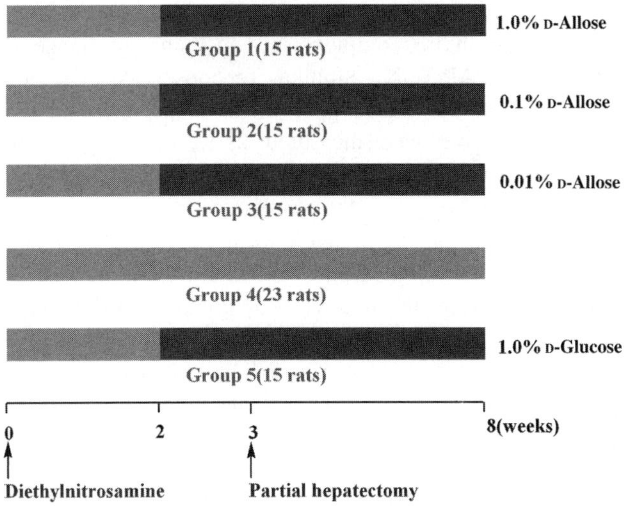

Figure 18.3 Design of assay 1.

At the end of the experiment, all rats were killed by exsanguination under deep anesthesia, and blood samples were collected. In assay 4, a total of 90 rats were divided into six groups (each 15) and administered a single intraperitoneal injection of DEN (200 mg/kg body weight) dissolved in 0.9% NaCl solution (saline) at 6 weeks of age to initiate hepatocarcinogenesis (groups 1–6) on the first day of the experiment. After 2 weeks on the basal diet, rats were administered the choline-deficient, L-amino acid-defined diet (CDAA) basal diet containing 2.00% (group 1), 0.00% (control; group 2), or 2.00% (group 3) D-glucose, the cholinesufficient, L-amino acid-defined diet (CSAA) control basal diet containing 2.00% (group 4) D-allose, 0.00% (control, group 5) or 2.00% (group 6) D-glucose for the following 6 weeks. Due to no modifying or toxic effects at the maximum dose of 1.00% in assay 1, the dose of 2.00% was selected. At the end of week 3, all rats were given two-thirds PH. Body weight was recorded at 2-week intervals, and at the end of week 8 of the experiment, all surviving rats were killed by exsanguination under deep anesthesia and their livers excised.

2. *Tissue Processing for Assays 1, 2, and 4*: In assays 1 and 4 analyzing GST-P positive foci, the livers, kidneys, and spleen were immediately excised at autopsy, weighed, and fixed in 10% neutral-buffered formalin for 48 h, 2- to 3-mm-thick sections of the liver were cut, one from the right and caudate lobes and two from the right anterior lobe, embedded in paraffin for hematoxylin and eosin staining for histopathologic examination and immunohistochemical staining for GST-P. Kidney and spleen sections were also embedded for histopathologic examination. For RNA analysis (assay 2), a liver was divided into three sections, one was fixed in 10% neutral-buffered formalin and embedded in paraffin for histopathologic examination and the other two were frozen immediately in liquid nitrogen for subsequent mRNA analysis.

3. *Immunohistochemistry (Assays 1 and 4)*: Using the VENTANA HX system, immunohistochemical staining was performed to determine the location of GST-P.

Sections were incubated with a rabbit anti-rat GST-P antibody (1:2000), secondary antibody, and the avidin-biotin complex (ABC). Using an Image Processor for Analytical Pathology (IPAP-WIN; Sumika Technoservice, Osaka, Japan), the number and area (foci/cm^2) of GST-P positive foci larger than 0.2 mm in diameter and the total area of the liver were measured, as well as the number and area of foci/cm^2 of a liver section were calculated and compared among the groups.

4. *RNA Isolation and PamChip Microarrays (assay 2)*: Using the RNAlater RNA Stabilization reagent and an RNeasy Midi kit for DNA microarrays, total RNA was isolated from liver tissues from five rats in each group. To obtain 100 μL of sample, isolated RNA from the five rats was pooled and the concentration was measured from absorbance at 260 nm. For customized PamChip microarray, a three-dimensional microarray system RatTox ver. 2.1 was chosen. With oligonucleotide DNA probes (60 mer), all genes were spotted in duplicate. For preparing fluorochrome-labeled cDNA, sample 20 μg aliquots of total RNAs were used, and according to the manufacturer's instructions, reverse transcriptase and FITC labeling reactions were performed. Samples were at 95°C denatured for 5 min, and 50 μL of hybridization solution containing 5 × SSPE was used to each test site of the PamChip microarray. Using the three-dimensional microarray system at 50°C and a rate of 5 μL/s with 150 cycles, hybridization was performed, and each test site was washed with 50 μL of 5×SSPE three times and using cooled CCD fluorescence images automatically captured. After each treatment, signals were converted to expression ratios relative to the controls using the gene expression analysis software integrated in FD10. Each gene expression level was normalized with respect to that of glyceraldehyde 3-phosphate dehydrogenase.

5. *Potential Antioxidant (PAO) Test (Assay 3)*: After treatment with 2.00% D-allose or the basal diet only and 50.00% D-allose saline solution, the antioxidation power of rat serum was evaluated by measuring Cu^+ reduction to Cu using a PAO test kit. The serum samples from two groups of five rats, each administered a basal diet containing 2.00% or 0.00% D-allose for a week, were corrected for PAO test. In 11 samples, 1 was 50.00% D-allose saline solution, and 5 samples each of 2 groups were rat serum. PAO assay provided a quantitative measurement for the antioxidation capability of biological fluids, such as plasma, attributable to non-enzymatic antioxidants such as fat-soluble and water-soluble agents. At 495 nm, the stable complexes formed between Cu^+ and bathocuproine can be assessed. Using three wells for solutions diluted in three steps of ×1, ×10, and ×100, a sample was measured. The sample data was selected from that with the highest reliability among three wells.

18.33 RAT MEDIUM-TERM LIVER ASSAY FOR DEN-INDUCED HEPATOCARCINOGENESIS ASSAY[42]

Male F344 rats at 5 weeks of age were housed in plastic cages on hardwood chip bedding in an air-conditioned room at 22 ± 2°C with a 12-h light/dark cycle, maintained on Oriental MF basal diet and tap water *ad libitum*, randomly allocated to 6 groups of 15 or 16 rats each, and given a single ip injection of DEN 200 mg/kg body weight. Two weeks thereafter, rats received Oriental MF (Oriental Yeast Co., Ltd., Tokyo, Japan) powdered basal diet containing 0.03% 2-amino-3,8-dimethylimidazo[4,5-f]quinoxaline either alone or in combination with 0.4% chitin, chitosan, chitin-oligo sugar, chitosan-oligo sugar, and chlorophyllin-chitosan for 6 weeks. Other groups received 0.1% chitin, chitosan, chitin-oligo sugar, chitosan-oligo sugar, and chlorophyllin-chitosan or basal diet alone. At the end of week 3, all rats were subjected to two-thirds PH. Body weights and food consumption were recorded at 2-week intervals. At week 8, under ether anesthesia, all surviving rats were killed. The livers were immediately excised and weighed for calculation of organ to body weight ratios, cut into 2- to 3-mm-thick slices, and fixed in ice-cold acetone for routine immunohistochemical examination of glutathione *S*-transferase placental form expression. With the aid of a video image processor, the numbers and areas of glutathione *S*-transferase placental form positive foci (>0.1 mm in diameter) and total areas of the liver sections were assessed.

18.34 ASSAY FOR RAT CARCINOGENESIS INITIATED BY THREE CARCINOGENS[42]

Female F344 rats at 5 weeks of age were housed in plastic cages on hardwood chip bedding in an air-conditioned room at 22 ± 2°C with a 12-h light/dark cycle, maintained on Oriental MF basal diet and tap water *ad libitum*, randomly allocated to 3 groups of 20 rats each and a further 3 groups of 10 rats each. For initiation during the initial 4-week period, rats in groups 1–3 received combined treatment with 5 sc doses of 40 mg/kg body weight DMH, a single ig administration of 40 mg/kg 7,12-dimethylbenz[a]anthracene (DMBA), and 0.1% BBN in the drinking water for 2 weeks. Throughout the remainder of the assay, rats then received diet containing 1.0% chitin, chitosan, or basal diet alone. The other groups of 10 rats each received diet containing 1.0% chitin, chitosan, or basal diet alone without carcinogen exposure. During the initiation period, body weight and food consumption were measured once a week, and then once every 4 weeks. From week 10 after starting assay, rats were carefully checked for the presence of mammary tumors once a week and data on the size, location, and number of tumors were recorded. To avoid cannibalism, rats with bleeding from tumors were isolated. Before the end of the assay, moribund rats were killed for autopsy. At week 39, under ether anesthesia all surviving rats were killed. Whole skin with mammary glands and tumors, as well as the lungs, liver, kidneys, spleen, stomach, small and large intestines, and gross lesions were removed, and before fixation, liver, kidneys, and spleen were weighed. The lung, stomach, and small and large intestines received injection of 10% buffered formalin solution; after 5 min the large intestine was from the cecum to the rectum opened and fixed flat between filter papers. Under a light microscope and after staining with 0.2% methylene blue for 5 min, aberrant crypts were counted. Using hematoxylin and eosin–stained sections, other tissues were processed routinely for histopathologic examination.

18.35 ASSAY FOR RAT CARCINOGENESIS INITIATED BY FIVE CARCINOGENS[43]

A total of 160 male F344/DuCrj rats were randomly divided into 8 groups (20 rats in each) and were allowed free access to a γ-ray-irradiated (6.0 kGy) powder diet MF (Oriental Yeast Co., Ltd., Tokyo, Japan) for a 9-day quarantine/acclimation period with body weight and health conditions monitored. Six-week-old rats at normal health status were entered into the assay. The rats were housed in a plastic cage on hardwood chip bedding in an environment-controlled room that had constant conditions of temperature 20°C to 24°C, humidity 51% to 66%, ventilation (more than 15 times/h), and artificial illumination for 12 h each day. For multi-organ initiation, rats were treated with *N*-nitrosodiethylamine (DEN, 100 mg/kg, in saline, i.p. single injection at the beginning), sequentially followed by *N*-methylnitrosourea (MNU, 20 mg/kg, in saline, i.p. 4 times during weeks 1 and 2), DMH (40 mg/kg, in saline, s.c. 4 times during weeks 3 and 4), BBN (0.05% in drinking water, during weeks 1 and 2), and DHPN (0.1%, in drinking water, during weeks 3 and 4), and then received test substances for an additional 24 weeks. The rats were observed daily for abnormalities, weighed weekly for the first 14 weeks, and then biweekly for the remaining weeks. The food and water consumption by cage was also measured when the body weight was measured. At the end of week 28, all surviving rats were killed under ether anesthesia by exsanguination via the abdominal aorta for examination of lesion development. At autopsy, all rats were grossly examined, the brain, liver, kidneys, spleen, heart, adrenals, testes, thymus were immediately excised and weighed, and organ to body weight ratios were calculated. After fixation, the pituitary and thyroids (including parathyroids) were also weighed and ratios similarly calculated. Using 10% buffered formalin solution, the tissues/organs were fixed. For each rat, spleen, lymph nodes (mandibular, mesenteric), bone marrow, thymus, thyroids (including parathyroids), nasal cavity, trachea, lungs (including bronchi), tongue, esophagus, stomach (forestomach and glandular stomach), small intestine (duodenum, jejunum, ileum), large intestine (cecum, colon, rectum), pancreas, liver, kidneys, urinary bladder, prostate, seminal vesicles, bone (femur, sternum), Zymbal's glands, brain (cerebrum, cerebellum), spinal cord (thoracic portion), along with any other tissues with an abnormal appearance were routinely processed for embedding in paraffin wax, sectioned, stained with hematoxylin and eosin solution, and examined histopathologically.

To measure glutathione S-transferase placental form (GST-P) positive foci, the livers were cut into slices about 3 mm thick with a razor blade and fixed in ice-chilled 10% buffered formalin solution for subsequent embedding and sectioning. Using the avidin-biotin-peroxidase complex method, sections were incubated overnight with rabbit anti-rat liver GST-P polyclonal antiserum (1 : 2000 dilution) for immunohistochemical staining of GST-P. Using a color video image processor, the numbers and areas of GST-P positive foci (>0.2 mm in diameter) were measured and evaluated quantitatively.

18.36 ASSAY FOR DEN-INITIATED RAT CARCINOGENESIS[44,45]

Five-week-old male F344/DuCrj rats were housed five to a plastic cage on hardwood chip bedding in an air-conditioned room at 20°C to 23°C with relative humidity of 52% to 63%, a 12-h light/dark cycle, and more than 15 air changes/h maintained positive air pressure.

From a total of 130 rats, 117 were allocated to 8 groups (groups 1–5 each 18 rats, groups 6–8 each 9 rats). To make each group have similar weight distribution and approximately equal initial mean body weights, a computerized stratified body weight technique was used. Rats of groups 1–5 received an ip injection of DEN at 200 mg/kg body weight, whereas rats of groups 6–8 were given saline vehicle instead of DEN. Two weeks after starting the assay, rats were administered Annatto C at dietary levels of 0% (as control), 0.03%, 0.1%, or 0.3% or phenobarbital sodium at a dietary concentration of 0.05% (as a positive control) for 6 weeks from week 3 to 8. Three weeks after assay beginning, all rats were given two-thirds PH. For all rats, mortality and clinical signs were checked twice a day, body weights, food and water consumption were recorded every week, and surviving rats were killed under ether anesthesia at week 8.

At autopsy, the organs in the thoracic and abdominal cavities were macroscopically examined and recorded, while the livers were immediately excised and weighed to allow calculation of the liver to body weight ratio. Four- to five-millimeter-thick sections from the three liver lobes (the cranial and caudal parts of the right lateral lobe and the caudal part of caudate lobe) of all surviving rats were fixed in 10% buffered formalin, embedded in paraffin wax, sectioned, and stained immunohistochemically for GST-P (ABC method). Using a color image processor, all GST-P positive hepatocytic foci (>0.2 mm in diameter, the lowest limit for reliable evaluation) were measured, and the numbers and areas of foci/square centimeter (cm^2) of liver section were calculated. By microscopic analysis, BrdU positive labeling indices (LIs) were quantified. By randomly counting the number of positive nuclei per 1000 hepatocytes or number per unit area (mm^2) in sections stained immunohistochemically for BrdU, the LI were determined.

18.37 ASSAY FOR NORMAL DOSE DEN-INITIATED AND LOW DOSE DEN-MAINTAINED RAT CARCINOGENESIS[44–46]

Male F344 rats were randomly divided into five groups, housed three per cage with wood chip bedding in an air-conditioned animal room at 24 ± 2°C and 55 ± 5% humidity, with 12-h light/dark cycle, at their age of 6 weeks given a single ip injection of DEN (99% purity, 200 mg/kg) in saline to initiate hepatocarcinogenesis, and 2 weeks after starting received 0, 1, 5, 10, or 20 ppm DEN in their drinking water.

Because melatonin secretion from the pineal gland has a clear circadian rhythm, with serum levels at night higher than in the daytime, treatment with melatonin was only performed during the dark period, in an attempt to maintain the natural melatonin circadian rhythm. At week 3, all rats were given a 2/3 PH. Body weights were monitored once a week. At the end of week 8, rats were anesthetized using ether, and from the aorta blood samples were collected for hormone analysis. In order to analyze hormonal circadian rhythm, at around midnight and the others at around 0900, three rats in each group were sacrificed. At autopsy, their livers were excised, weighed, and cut into 2 ± 3 mm thick slices. Four slices, one of which was from each of the liver lobes, were in ice-cold acetone fixed for immunohistochemical demonstration of glutathione *S*-transferase placental form positive foci. Liver, testes, and prostate were weighed. Prostate tissue was divided into the three ventral, dorsolateral, and anterior lobes, and after formalin fixation they were weighed separately. Organ weights relative to body weight were calculated. By radioimmunoassay, melatonin, adrenocorticotropic hormone (ACTH), corticosterone, testosterone, luteinizing hormone (LH), and follicle-stimulating hormone (FSH) were measured. Based on the Kennaway G280 anti-melatonin antibody, melatonin was measured by a double-antibody radioimmunoassay. Only in group 1 (control group) and group 5 (the highest dose melatonin group) were LH, FSH, and testosterone were measured. In order to calculate BrdU labeling indices of whole liver section hepatocytes and hepatocytes inside foci, immunohistochemical double staining was used. After BrdU-stained nuclei were visualized by peroxidase DAB using anti-BrdU (1 : 1000), with VECTOR red substrate GST-P (1 : 300) was stained by alkaline phosphatase. From data for positive cells per unit liver section or area of GST-P positive foci, the labeling indices were calculated. For analysis of preneoplastic lesion development, using a color video image processor, numbers and areas of GST-P positive foci ($>$0.2 mm in diameter) were measured, and data per cm^2 of liver section were calculated.

18.38 TGF-β-MEDIATED ANTIPROLIFERATION ASSAY[47]

1. *Cell Culture*: YAMC (young adult mouse colonocyte) and Smad3$^{-/-}$ cell lines were maintained under permissive conditions at 33°C in HAM's F12/RPMI 1640 supplemented with insulin, transferrin, selenium (ITS), 5 U/mL mouse IFN-β, and 5% FBS. Assays were performed with cells under nonpermissive conditions at 37°C without IFN-β in media and in 0.5% serum to inactivate tsTag. All assays were done with cell passages 8–25.

2. *[^{3}H]Thymidine Incorporation for Defining Growth Inhibition*: Cells were at permissive conditions and a density of 2.5×10^5 cells/well in 12-well plates seeded overnight, moved to nonpermissive conditions, and in the presence or absence of TGF-β (5 ng/mL) maintained for 48 h. For the final 2 h of the incubation period, [^{3}H]thymidine (5 μC) was added to each well, the cells were washed with ice-cold 10% TCA, solubilized with 0.2 N NaOH, and assayed in a β-counter. All assays were performed in triplicate.

3. *Western Blot for Defining Mediators of Cell Cycle Arrest*: Smad3$^{-/-}$ cells plated in 100-mm tissue culture grade plates and were allowed to grow to 70% confluence at permissive conditions, switched to nonpermissive conditions, TGF-β (5 ng/mL) was added to each well, and cells were lysed using CHAPS buffer with phosphatase and protease inhibitors. At 55 mA and 4°C, proteins were electrophoresed on 10% or 12% SDS-PAGE gels and transferred to nitrocellulose overnight. The blot was rinsed in PBS plus 0.2% Tween-20 (PBS-T), placed in blocking buffer (5% nonfat milk powder in PBS-T) at a dilution of 1 : 1000 and 4°C overnight, incubated in primary antibody, rinsed twice with blocking buffer, on a rocking platform washed in blocking buffer for one 15-min wash and three 5-min washes, and with appropriate species-specific HRP-conjugated secondary antibodies at room temperature incubated for 45 min. This was followed by the same washing regimen as before. The blot was subjected to luminal-based chemiluminescence and exposed to Kodak X-OMAT film.

18.39 ASSAY FOR DEN AND HCB-INITIATED RAT CARCINOGENESIS[48]

Six-week-old male rats (average body weight, 120 g) were divided into three groups (15 rats at the low dose, 16 rats at each of the other doses), given a single intraperitoneal injection of DEN at 200 mg/kg body weight dissolved in 0.9% NaCl to initiate hepatocarcinogenesis. After a 2-week recovery period on basal diet, rats received hexachlorobenzene (HCB, 99% purity) in the diet at 0.6, 3, 15, 75, or 150 ppm, and group 2 (15 rats, was given only DEN) and group 3 injected with saline alone and then HCB as group 1 (10 rats per dose). At week 3, a PH was given to all rats. At week 8, all surviving rats were killed and their livers were excised and prepared for immunohistochemical determination of liver cells positive for glutathione GST-P. The hepatocarcinogenic potential of HCB was assessed by comparing the number and area per square centimeter of GST-P$^+$ liver foci (>0.2 mm in diameter) induced by the compound with those in controls (group 2) given DEN alone.

REFERENCES AND NOTES

1. A. Hagiwara, N. Imai, Y. Doi, K. Nabae, T. Hirota, H. Yoshino, M. Kawabe, Y. Tsushima, H. Aoki, K. Yasuhara, T. Koda, M. Nakamura, T. Shirai. Absence of liver tumor promoting effects of annatto extract (norbixin), a natural carotenoid food color, in a medium-term liver carcinogenesis bioassay using male F344 rats. Cancer Lett 199 (**2003**) 9–17.
2. N. Ito, H. Tsuda, M. Tatematsu, T. Inoue, Y. Tagawa, T. Aoki, S. Uwagawa, M. Kagawa, T. Ogiso, T. Masui, K. Imaida, S. Fukushima, M. Asamoto. Enhancing effect of various hepato-carcinogens on induction of preneoplastic glutathione S-transferase placental form positive foci in rats – an approach for a new medium-term bioassay system. Carcinogenesis 9 (**1988**) 387–394.

3. R. Hasegawa, H. Tsuda, T. Shirai, Y. Kurata, A. Masuda, N. Ito. Effect of timing of partial hepatectomy on the induction of preneoplastic liver foci in rats given hepatocarcinogens. Cancer Lett 32 (**1986**) 15–23.
4. R. Hasegawa, N. Ito. Liver medium-term bioassay in rats for screening of carcinogens and modifying factors in hepatocarcinogenesis. Food Chem Toxicol 30 (**1992**) 979–992.
5. F. Pinheiro, R.R. Faria, J.L.V. de Camargo, A.L.T. Spinardi-Barbisan, A.F. da Eira, L.F. Barbisan. Chemoprevention of preneoplastic liver foci development by dietary mushroom Agaricus blazei Murrill in the rat. Food Chem Toxicol 41 (**2003**) 1543–1550.
6. D.L. Gustafson, A.L. Coulson, L. Feng, W.A. Pott, R.S. Thomas, L.S. Chubb, S.A. Saghir, S.A. Benjamin, R.S.H. Yang. Use of a medium-term liver focus bioassay to assess the hepatocarcinogenicity of 1,2,4,5-tetrachlorobenzene and 1,4-dichloro-benzene. Cancer Lett 129 (**1998**) 39–44.
7. J.M. Ward, A. Hagiwara, L.M. Anderson, K. Lindsey, B.A. Diwan. The chronic hepatic and renal toxicity of di(2-ethylhexyl)phthalate, acetaminophen, sodium barbital, and phenobarbital in male B6C3F1 mice: autoradiographic, immuno-histochemical, and biochemical evidence for levels of DNA synthesis not associated with carcinogenesis or tumor promotion. Toxicol Appl Pharmacol 96 (**1988**) 494–506.
8. T. Itoh, M. Moto, M. Takahashi, H. Sakai, K. Mitsumori. Liver initiation activity of norfloxacin but not nalidixic acid, pipemidic acid, and ciprofloxacin on in vivo short-term liver initiation assay in rats. Toxicology 222 (**2006**) 240–246.
9. C. Eipel, H. Schuett, C. Glawe, R. Bordel, M.D. Menger, B. Vollmar. Pifithrin-alpha induced p53 inhibition does not affect liver regeneration after partial hepatectomy in mice. J Hepatol 43 (**2005**) 829–835.
10. U.M. Gehling, M. Willems, M. Dandri, J. Petersen, M. Berna, M. Thill, T. Wulf, L. Müller, J.M. Pollok, K. Schlagner, C. Faltz, D.K. Hossfeld, X. Rogiers. Partial hepatectomy induces mobilization of a unique population of haematopoietic progenitor cells in human healthy liver donors. J Hepatol 43 (**2005**) 845–853.
11. K. Tsuji, A.H. Kwon, H. Yoshida, Z. Qiu, M. Kaibori, T. Okumura, Y. Kamiyama. Free radical scavenger (edaravone) prevents endotoxin-induced liver injury after partial hepatectomy in rats. J Hepatol 42 (**2005**) 94–101. *Notes*: (1) Blood and liver samples were collected with general method. Collected liver samples stored at −80°C were homogenized in four volumes of cell homogenizing buffer [50 mM Tris-HCl, pH 7.4, containing CPI (1X) and 1 mM phenylmethylsulfonyl-fluoride (PMSF)] and centrifuged at 16,500 × *g* for 20 min. Using commercial kits, serum levels of aspartate transaminase (AST), LDH, and TB were determined, while serum levels of MDA were measured by spectrophotometric assay using a commercially available kit (MDA-586). Using commercial kits TNF-α, IL-1β, IFN-g, IL-12, and CINC-1 were measured in serum and liver. (2) In Western blot analysis, frozen liver samples homogenized in 5 volumes of cell solubilizing buffer (10 mM Tris-HCl, pH 7.4, containing 1% Triton X-100, 0.5% NP-40, 1 mM EGTA, 1 mM EDTA, 150 mM NaCl, and 1 mM PMSF) were centrifuged at 16,500 × *g* for 15 min, the supernatant was subjected to SDS-PAGE (7.5% gel) and electroblotted onto a polyvinylidene-difluoride membrane. Using an ECL blotting detection agent and rabbit polyclonal antibody against mouse inducible nitric oxide synthase as the primary antibody, immunostaining was performed. (3) In Northern blot analysis, to isolate total RNA, 0.1 g of liver tissue frozen at −80°C was homogenized in 1.5 mL of TRIzol reagent according to the manufacturer's instructions, allowed to stand for 30 min, and centrifuged at 16,500 × *g* for 15 min, and with isopropanol RNA was precipitated. RNA pellets were with ice-cold

75% (v/v) ethanol washed twice, dried under vacuum, and dissolved in 500 μL of sterile H_2O. By agarose-formaldehyde gel electrophoresis, 10 μg of RNAs was fractionated, transferred to membranes, and immobilized by baking for hybridization with DNA probes. A cDNA probe of rat iNOS (830 bp) was provided. Using the random-primed method, cDNAs were radiolabeled with [α-^{32}P]-dCTP.

12. H. Tsuchiya, M. Kaibori, H. Yanagida, N. Yokoigawa, A.-H. Kwon, T. Okumura, Y. Kamiyama. Pirfenidone prevents endotoxin-induced liver injury after partial hepatectomy in rats. J Hepatol 40 (**2004**) 94–101.
13. H. Kojima, K.D.E. Kaita, Z. Xu, J.H. Ou, Y. Gong, M. Zhang, G.Y. Minuk. The absence of up-regulation of telomerase activity during regeneration after partial hepatectomy in hepatitis B virus X gene transgenic mice. J Hepatol 39 (**2003**) 262–268. *Notes*: (1) Prior to initiating the experiments, 8-week-old CD-1 male mice were acclimatized to the laboratory for 1 week. Mice were housed in an air-conditioned room at 24°C with a 12-h light/dark cycle. Under ether anesthesia, a 70% PH was performed to induce hepatic regeneration. For identifying peak telomerase activity following PH in CD-1 non-transgenic mice, a preliminary experiment was performed in which groups of four mice were killed at 0, 6, 12, 24, 36, or 48 h, their livers were removed and immediately frozen in liquid nitrogen. Four hepatitis B virus X (HBx) transgenic CD-1 mice were killed at 12 h post-PH when maximum telomerase activity was observed in the CD-1 non-transgenic mice. HBx mRNA and protein expression in these mice was limited to the liver, kidney, and testes. (2) Under the transient expression of HBx, CCL 13 and CRL 8024 cells were used for measuring telomerase activity. Cells were grown at 37°C in minimum essential medium supplemented with 10% cool calf serum, 110 mg/mL sodium pyruvate, 2.2 mg/mL sodium bicarbonate, and antibiotics (100 IU/mL penicillin, 100 mg/mL streptomycin, 250 mg/mL Fungizone) and 5% CO_2-supplemented atmosphere, plated 24 h before the transfection experiment to achieve 70% confluence in a 35-mm culture dish on the day of the experiment, were transfected with 1 mL of HBx expression plasmid or vector alone and 3 mL of FuGENE 6 Transfection Reagent according to the manufacturer's instructions, and collected at 0, 24, and 48 h after transfection. (3) In the construction of HBx expression plasmid, full-length HBx gene was amplified with forward primers 5′-ATGGCTGCTCGGGTGTGC-3′ and reverse primers 5′-AGATGATTAGGCAGAGGTG-3′. PCR product was cloned into a pcDNA3.1/V5-his A vector. By Wizard PureFection Plasmid DNA Purification System, the HBx expression plasmid/vector was prepared.
14. Y. Baruch, K. Neubauer, L. Shenkar, E. Sabo, A. Ritzel, T. Wilfling, G. Ramadori. Von Willebrand factor in plasma and in liver tissue after partial hepatectomy in the rat. J Hepatol 37 (**2002**) 471–477. *Notes*: (1) Eight-week-old male Wistar rats were kept in a 12-h light/dark cycle and had free access to food and water. PH of 70% was carried out. The left and middle lobes were excised under light ether anesthesia. Sham operation included a midline cut and liver exteriorization without cutting it. Healthy rats without any manipulation were used as controls. At time intervals ranging from 60 min to 5 days (two or three rats from each operated group at each time), rats were sacrificed. Before excising the liver, blood was withdrawn from the inferior vena cava, collected into a tube containing sodium citrate, and separated immediately. The liver was excised and snap-frozen in liquid nitrogen. (2) On stored plasma with ELISA kit, using rabbit anti-human vWF, assay was performed. Levels of control, sham-operated and hepatectomized rats' vWF levels were expressed as percentage of the controlled lyophilized pooled human plasma supplied within the ELISA kit. (3) By cryostat, from frozen livers thin sections (6 mm) were prepared, air dried, and at −20°C fixed in methanol (5 min) and acetone (10 s).

For the immunoperoxidase analysis, sections were with peroxidase-labeled immunoglobulin G (IgG) against murine or rabbit Ig incubated. The first antibody and second fluorescein isothiocyanate-conjugated antibody were against vWF/factor VIII (1:50) and murine or rabbit IgG, respectively. (4) With a complementary DNA (cDNA) specific to human vWF, Northern blot hybridization analysis was prepared. Total RNA was isolated, separated by agarose gel electrophoresis, transferred onto nylon membranes and hybridized with specific [^{32}P]dCTP-labeled cDNA probes labeled by random priming or nick translation. Using QuikHyb solution, hybridization was at 68°C carried out for 2 h, in 2× saline-sodium citrate containing 0.1% sodium dodecyl sulfate washed twice for 15 min at room temperature and once for 5–15 min at 50°C. At −80°C, nylon filters were exposed to x-ray films. By hybridization with an oligonucleotide complementary to 28S rRNA, Northern blot experiments were normalized. Hybridization procedures were repeated at least three times.

15. C. Picard, L. Lambotte, P. Starkel, C. Sempoux, A. Saliez, V. Van Den Berge, Y. Horsmans. Steatosis is not sufficient to cause an impaired regenerative response after partial hepatectomy in rats. J Hepatol 32 (**2002**) 645–652. *Notes*: (1) Under light ether anesthesia, a two-thirds PH was performed, part of the hepatectomized lobes were immersed in Bouin's solution and in 4% buffered formaldehyde for histologic and immunohistochemical analysis, and the rest was snap-frozen in liquid nitrogen and stored at −80°C to be used as controls before hepatectomy. After the operation, all rats were allowed free access to food and water, killed by puncture of the aorta and exsanguination under ether anaesthesia 24 h after liver resection. One hour before sacrifice, BrdU was injected intraperitoneally (50 mg/kg), pieces of the remaining liver were immersed in Bouin's solution and in 10% buffered formaldehyde for histologic and immunohistochemical analysis, and the rest was snap-frozen in liquid nitrogen and stored at −80°C for future analyses. (2) Immunoprecipitations and kinase assays were performed according to literature [Jaumot M, Estanyol JM, Serratosa J, Agell N, Bachs O. Activation of cdk4 and cdk2 during rat liver regeneration is associated with intranuclear rearrangements of cyclin-cdk complexes. Hepatology 29 (**1999**) 385–395], with slight modifications. Incubation of samples with anti-CDK-2 antibody was from 8 h shortened to 4 h, while incubation with Protein A/G Plus Agarose was carried out for 2 h instead of 1 h.
16. S. Sakuda, S. Tamura, A. Yamada, J.I. Miyagawa, K. Yamamoto, S. Kiso, N. Ito, S. Higashiyama, N. Taniguchi, S. Kawata, Y. Matsuzawa. NF-κB activation in non-parenchymal liver cells after partial hepatectomy in rats: possible involvement in expression of heparinbinding epidermal growth factor-like growth factor. J Hepatol 36 (**2002**) 527–533. *Notes*: (1) Male SD rats at 7 weeks of age (200–240 g) were used and received a two-thirds PH, after which three rats were sacrificed at each of the indicated times. Rats from the sham group had their livers exposed and manipulated but not removed. The control rats (time 0) received no surgery. (2) In morphometric analysis, each experimental group consisted of three rats. Three tissue blocks were taken from every rat and three sections were made from every block. Each group was given the mean ± standard deviation. To explore NF-κB activation distribution in the zones, the percentage of positively stained nuclei was determined by a person who was not aware of this purpose for morphometry in periportal, intermediate, and pericentral zones. In each zone, the Kruskal–Wallis test was used for evaluating the number of NF-κB-positive nuclei.
17. H. Sakai, T. Tsukamoto, M. Yamamoto, A. Hirata, A. Inagami, N. Shirai, T. Iidaka, T Yanai, T. Masegi, M. Tatematsu. Summation of initiation activities in the liver after partial hepatectomy. Cancer Lett 176 (**2002**) 1–5. *Note*: Male F344 rats were housed five per

plastic cage on wood chips for bedding, maintained under constant conditions (12:12 h light/dark cycle, 60% humidity at 22 $\pm$ 2°C) on Oriental NMF diet and tap water *ad libitum* and allowed to acclimatize for 1 week before the start of experiments, at which time they were 7 weeks old and weighed 140–150 g.

18. K.K. Anand, B. Singh, A.K. Saxena, B.K. Chandan, V.N. Gupta, V. Bhardwaj. 3,4,5-Tri-hydroxy bnzoic acid (gallic acid), the hepatoprotective principle in the fruits of terminal belerica-bioassay guided activity. Pharmacol Res 36 (**1997**) 315–321.

19. K.K. Anand, B. Singh, A.K. Saxena, B.K. Chandan, V.N. Gupta. Hepatoprotective studies of fraction from the fruits of Terminalia belerica Roxb. on experimental liver injury in rodents. Phytother Res 8 (**1994**) 287–292.

20. P.A. Bricarello, L.G. Zaros, L.L. Coutinho, R.A. Rocha, M.B. Silva, F.N.J. Kooyman, E. De Vries, A.P. Yatsuda, A.F.T. Amarante. Immunological responses and cytokine gene expression analysis to Cooperia punctata infections in resistant and susceptible Nelore cattle. Vet Parasitol 155 (**2008**) 95–103. *Note*: At the beginning of the experiment, 100 young Nelore bulls were 11–12 months old and their fecal and blood samples were collected fortnightly. At the end of the experimental period, according to mean fecal egg counts (FEC), cattle were classified as resistant or susceptible. One hundred five days after the beginning of the trial, cattle having the lowest or the highest FEC means were slaughtered for worm burden determination. The small intestine's tissue samples were collected for histologic evaluation and gene expression analysis, and mucus was collected for antibody levels assessment. Based on worm burden, those 20 cattle identified as resistant or susceptible were reselected, which resulted in seven animals in the resistant group and eight animals in the susceptible group, therefore results were obtained for these 15 bulls.

21. J. Kubler-Kielb, E. Vinogradov, G. Ben-Menachem, V. Pozsgay, J.B. Robbins, R. Schneerson. Saccharide/protein conjugate vaccines for Bordetella species: preparation of saccharide, development of new conjugation procedures, and physico-chemical and immunological characterization of the conjugates. Vaccine 26 (**2008**) 3587–3593.

22. T. Mouratidou, I. Kaniou-Grigoriadou, C. Samara, T. Kouimtzis. Detection of the marine toxin okadaic acid in mussels during a diarrhetic shellfish poisoning (DSP) episode in Thermaikos Gulf, Greece, using biological, chemical and immunological methods. Sci Total Environ 366 (**2006**) 894–904.

23. F. Wang, G. Chen, Y. Zhu, Z. Guo, S. Zheng, J. Zhan. Development of antibody-based assays for omega-conotoxin MVIIA. J Biochem Biophys Methods 67 (**2006**) 49–56.

24. P.V. Beum, A.D. Kennedy, R.P. Taylor. Three new assays for rituximab based on its immunological activity or antigenic properties: analyses of sera and plasmas of RTX-treated patients with chronic lymphocytic leukemia and other B cell lymphomas. J Immunol Methods 289 (**2004**) 97–109. *Notes*: (1) Tests for free and inhibitory CD20 in neat plasmas: Raji cells (5×10^6 cells/mL) were at room temperature with Al488 rituximab (10 μg/mL) in the presence of media or neat NHP or CLL patient plasmas incubated for 30 min. The samples were washed three times, fixed, and evaluated for the amount of bound Al488 rituximab using flow cytometry. (2) Tests for free and inhibitory CD20 in Raji cell lysates: Raji cells (1×10^8 cells) were pelleted, lysed by adding 0.5 mL of lysis buffer containing 0.5% Triton X-100 in 0.14 M NaCl, 0.01 M Tris-HCl, 0.025% sodium azide, and 50 μL of protease inhibitor cocktail, held at room temperature for 30 min, and centrifuged at 12,000 $\times$ *g* for 2 min, the supernatant was isolated and aliquots were spiked with rituximab (5 or 15 μg/mL final concentration) and tested in the three assays. Because the sample (400-fold) for Raji cell assay was highly diluted, lysis of the

indicator Raji cells by detergent was prevented. As a control, lysis buffer without Raji cells was spiked with rituximab and test compounds in the three assays under comparable conditions.

25. H. Wang, W. Wei, Y.X. Shen, C. Dong, L.L. Zhang, N.P. Wang, L. Yue, S.Y. Xu. Protective effect of melatonin against liver injury in mice induced by Bacillus Calmette-Guerin plus lipopolysaccharide. World J Gastroenterol 10 (**2004**), 2690–2696.
26. T. Hasegawa, K. Sakurai, Y. Kambayashi, A.R. Saniabadi, H. Nagamoto, K. Tsukada, A. Takahashi, H. Kuwano, M. Nakano. Effects of OPC-6535 on lipopolysaccharide-induced acute liver injury in the rat: involvement of superoxide and tumor necrosis factor-alpha from hepatic macrophages. Surgery 134 (**2003**) 818–826.
27. V.C. Gavino, J.S. Miller, S.O. Ikharebha, G.E. Milo, D.G. Cornwell. Effect of polyunsaturated fatty acids and antioxidants on lipid peroxidation in tissue cultures. J Lipid Res 22 (**1981**) 763–769.
28. T. Torimura, T. Ueno, M. Kin, E. Taniguchi, T. Nakamura, K. Inoue, R. Sakata, O. Hashimoto, M. Sakamoto, H. Ohira, R. Kumashiro, M. Sata, H. Yano, M. Kojiro, N. Vettonmaki, Y. Cao. Gene transfer of kringle 1-5 suppresses tumor development and improves prognosis of mice with hepatocellular carcinoma. Gastroenterology 130 (**2006**) 1301–1310.
29. J.E.A. Leakey, J.E. Seng, W.T. Allaben. Body weight considerations in the B6C3F1 mouse and the use of dietary control to standardize background tumor incidence in chronic bioassays. Toxicol Appl Pharmacol 193 (**2003**) 237–265.
30. R.T. Mingoia, D.L. Nabb, C.H. Yang, X. Han. Primary culture of rat hepatocytes in 96-well plates: Effects of extracellular matrix configuration on cytochrome P450 enzyme activity and inducibility, and its application in *in vitro* cytotoxicity screening. Toxicol In Vitro 21 (**2007**) 165–173.
31. G. Rajaraman, J. Chen, T.K.H. Chang. Ginkgolide A contributes to the potentiation of acetaminophen toxicity by Ginkgo biloba extract in primary cultures of rat hepatocytes. Toxicol Appl Pharmacol 217 (**2006**) 225–233. *Notes*: (1) Determination of total cellular protein synthesis: Using [^{14}C]leucine incorporation, total cellular protein synthesis in cultured hepatocytes was defined. A 0.5 mL aliquot of the solubilized cell pellet was treated with 5 mL of BD Scintiverse Liquid Cocktail. In a Beckman LS 6000 liquid scintillation counter, the radioactivity was measured. [^{14}C]Leucine incorporation was expressed as dpm/mg total cellular protein. (2) A spectrophotometric method was used to determine total cellular protein concentration. Cell lysate sample (12.5 μL) or various concentrations of the bovine serum albumin standard (12.5 μL, 10–100 μg/mL final concentrations) was added to 125 μL of Lowry stock in 96-well microplates. Lowry stock consisted of 2% sodium carbonate in 0.1 M NaOH, 1% copper sulfate, and 2% sodium potassium tartrate in a ratio of 98 : 1 : 1. After a 30-min incubation at room temperature, 12.5 μL of Folin-Ciocalteu's phenol reagent was added and the mixture was at room temperature incubated for additional 30 min and the absorbance of each sample was determined at 595 nm in a Multiskan Ascent multiplate reader.
32. K.E. Tollefsen, E. Bratsberg, O. BØyum, E.F. Finne, I.K. Gregersen, M. Hegseth, C. Sandberg, K. Hylland. Use of fish in vitro hepatocyte assays to detect multi-endpoint toxicity in Slovenian river sediments. Marine Environ Res 62 (**2006**) S356–S359.
33. N. Laville, S. Aït-Aïssa, E. Gomez, C. Casellas, J.M. Porcher. Effects of human pharmaceuticals on cytotoxicity, EROD activity and ROS production in fish hepatocytes. Toxicology 196 (**2004**) 41–55.

34. A.L. Pierce, J.T. Dickey, D.A. Larsen, H. Fukada, P. Swanson, W.W. Dickhoffa. A quantitative real-time RT-PCR assay for salmon IGF-I mRNA, and its application in the study of GH regulation of IGF-I gene expression in primary culture of salmon hepatocytes. General Comparat Endocrinol 135 (**2004**) 401–411.

35. V. Lewerenz, S. Hanelt, C. Nastevska, C. El-Bahay, E. Röhrdanz, R. Kahl. Antioxidants protect primary rat hepatocyte cultures against acetaminophen-induced DNA strand breaks but not against acetaminophen-induced cytotoxicity. Toxicology 191 (**2003**) 179–187.

36. F.N. Smets, Y. Chen, L.-J. Wang, H.E. Soriano. Loss of cell anchorage triggers apoptosis (anoikis) in primary mouse hepatocytes. Mol Genet Metab 75 (**2002**) 344–352.

37. H.H. Wang, W.W. Lautt. Hepatocyte primary culture bioassay: a simple tool to assess the initiation of the liver regeneration cascade. J Pharmacol Toxicol Methods 38 (**1997**) 141–150.

38. G. Michalopoulos, H.D. Cian, A.R. Novotny, A.D. Kligerman, S.C. Strom, R.L. Jirtle. Liver regeneration studies with rat hepatocytes in primary culture. Cancer Res 42 (**1982**) 4673–4682.

39. M. Borset, A. Waage, A. Sundan. Hepatocyte growth factor reverses the TDF-β-induced growth inhibition of CCL-64 cells: a novel bioassay for HGF and implication for the TDF-β bioassay. J Immunol Methods 189 (**1996**), 59–64.

40. J.M. Vrolijk, A. Kaul, B.E. Hansen, V. Lohmann, B.L. Haagmans, S.W. Schalm, R. Bartenschlager. A replicon-based bioassay for the measurement of interferons in patients with chronic hepatitis C. J Virol Methods 110 (**2003**) 201–209.

41. M. Yokohira, K. Hosokawa, K. Yamakawa, K. Saoo, Y. Matsuda, Y. Zeng, T. Kuno, K. Imaida. Potential inhibitory effects of D-allose, a rare sugar, on liver preneoplastic lesion development in F344 rat medium-term bioassay. J Biosci Bioeng 105 (**2008**) 545–553.

42. M. Kawabe, M. Futakuchi, S. Tamano, T. Shirai, M. Hirose. Modifying effects of chitin, chitosan and their related compounds on 2-amino-3,8-dimethylimidazo-[4,5-f]quinoxaline (MeIQx) in a rat medium-term hepatocarcinogenesis model, and their post-initiation effects in a female rat 2-stage multi-organ carcinogenesis model. Food Chem Toxicol 46 (**2008**) 2758–2763.

43. T. Ichihara, H. Yoshino, Y. Doi, K. Nabae, N. Imai, A. Hagiwara, S. Tamano, O. Morita, Y. Tamaki, H. Suzuki. No enhancing effects of diacylglycerol oil on tumor development in a medium-term multi-organ carcinogenesis bioassay using male F344 rats. Food Chem Toxicol 46 (**2008**) 157–167.

44. A. Hagiwara, N. Imai, Y. Doi, K. Nabae, T. Hirota, H. Yoshino, M. Kawabe, Y. Tsushima, H. Aoki, K. Yasuhara, T. Koda, M. Nakamura, T. Shirai. Absence of liver tumor promoting effects of annatto extract (norbixin), a natural carotenoid food color, in a medium-term liver carcinogenesis bioassay using male F344 rats. Cancer Lett 199 (**2003**) 9–17.

45. M. Kawabe, A. Hagiwara, H. Yoshino, K. Miyashita, T. Nakanishi, M. Sano, S. Tamano, K. Okamoto, K. Miura, H. Tokado. Epoprostenol sodium, a prostaglandin I2, lacks tumor promoting effects in a medium-term liver carcinogenesis bioassay in rats. Cancer Lett 162 (**2001**) 141–147.

46. K. Imaida, A. Hagiwara, H. Yoshino, S. Tamano, M. Sano, M. Futakuchi, K. Ogawa, M. Asamoto, T. Shirai. Inhibitory effects of low doses of melatonin on induction of preneoplastic liver lesions in a medium-term liver bioassay in F344 rats: relation to the influence of electromagnetic near field exposure. Cancer Lett 155 (**2000**) 105–114.

47. S.K. Mithani, G.C. Balch, S.-R. Shiou, R.H. Whitehead, P.K. Datta, R.D. Beauchamp. Smad3 has a critical role in TGF-β-mediated growth inhibition and apoptosis in colonic epithelial cells. J Surg Res 117 (**2004**) 296–305.
48. R. Cabral, T. Hoshiya, K. Hakoi, R. Hasegawa, N. Ito. Medium-term bioassay for the hepatocarcinogenicity of hexachlorobenzene. Cancer Lett 100 (**1996**) 223–226.

19

METHODS AND APPLICATIONS OF ESTROGEN ASSAYS

Shiqi Peng

During the reproductive life of women, the prevention of cardiovascular disease has been attributed to the modulating actions of endogenous estrogens, and the difference of estrogenic status between prepubertal girls and boys has been considered to account for the earlier onset of puberty of the former. It is well established that the exogenous hormone replacement therapy and oral contraceptives used by millions of women worldwide are beneficial during menopause based on the observed decreased risk of arteriosclerosis and myocardial infarction, which have been mainly explained because of their influence on lipoproteins metabolism, prevention of arteriosclerosis, vasodilator effects, and decreases in blood pressure with a subsequent improvement of blood flow. The diagnosis and treatment of various pediatric endocrine diseases such as prepubertal gynecomastia, premature thelarche, and gonadal dysgenesis greatly benefit from estrogen evaluation. On the other hand, xenoestrogens, one type of endocrine disrupting chemical (EDC), can mimic the action of physiologic estrogens through interaction with ERs, possibly resulting in adverse health effects in an intact organism, or its progeny, consequent to changes in endocrine functions in humans and wildlife. Important efforts have been made toward the development of screening assays to evaluate and understand the actions of both endogenous estrogens and exogenous estrogen-like compounds in humans and wildlife. In this chapter, 21 assays are presented: yeast estrogen assay,[1–5] *in vivo*, *ex vivo*, and *in vitro* assays for estrogen-like effect,[3] ameliorative yeast assay,[6] in yeast two-hybrid assay,[7]

Pharmaceutical Bioassays: Methods and Applications. By Shiqi Peng and Ming Zhao

estrogen-sensitive yeast strain RMY/ER-ERE assay,[8] bioluminescent yeast assays,[9] LYES assay,[10] ERα and ERβ stable transactivation assay,[11] HELNα and HELNβ transfected cell assay,[12] luciferase reporter gene assay,[13] target AKT pathway assay,[14] luciferase assay,[15] androgen receptor transactivation assay,[16] GFP expression assay,[17] luteinizing hormone releasing hormone (LHRH) assay,[18] Ishikawa endometrial cancer cell assay for estrogen receptor expression,[19] Ishikawa cell assay for 11β-HSD2 activity,[20] Ishikawa cell assay for estrogen activity,[21] Ishikawa cell assay for plasminogen activator inhibitor-1 (PAI-1),[22] testosterone induction assay with Sertoli cell,[23] and electrophoretic mobility and antibody shift assays.[24]

19.1 YEAST ESTROGEN ASSAY[1–5]

Yeast, of which the steroid-binding domain was fused by GAL4-VP16, was incubated with SC-medium without histidine at 30°C and 130 rpm overnight. By adding DMSO up to a final concentration of 15% (v/v), stock cultures were prepared from exponentially growing cultures and stored in 0.5 mL aliquots at −80°C. Exponentially growing overnight cultures were diluted with SC-medium to an $OD_{600\ nm}$ of 0.75. To 10 mL aliquots, 100 μL of DMSO (negative controls), 100 μL of E2 in DMSO (positive controls), or 100 μL of test compound in DMSO were added, incubated at 30°C and 130 rpm for 2 h, diluted to fivefold volume, and determined to get $OD_{600\ nm}$. To 200 μL of test culture, 600 μL of Z-buffer (60 mM Na_2HPO_4-$7H_2O$, 40 mM NaH_2PO_4-H_2O, 10 mM KCl, 1 mM $MgSO_4$-$7H_2O$, 35 mM β-mercaptoethanol), 20 μL of SDS solution (3.5 mM) and 50 μL of chloroform were added, carefully mixed and preincubated at 28°C for 5 min, and then 200 μL of *o*-nitrophenyl-3-D-galactopyranoside in Z-buffer (13.3 mM) was added to initiate the enzyme reaction. The cultures were incubated at 28°C until a significant yellow color developed. For 17β-estradiol-induced positive controls, yellow color occurred within 20 min. For test compound-induced assays, yellow color occurred after 120 min. To the cultures, 500 μL of Na_2CO_3 (1 M) was added to stop the reaction, the cell debris was pelleted by centrifugation (25,500 × *g*, 15 min), and the $Ex_{420\ nm}$ of the supernatants was determined. β-Galactosidase activity of the test cultures was calculated according to the equation $U\ [\mu mol/min] = Cs/t \cdot V \cdot Ods$, where t is incubation time (min) of the enzyme reaction, V is volume (0.2 cm^3) of the used test culture aliquot, Ods is $OD_{600\ nm}$ of test culture, and Cs is concentration of *o*-nitrophenyl-3-D-galactopyranoside (μM) in the reaction supernatant calculated according to $Cs\ (\mu M) = 10^6 \times [Exs - Ex_B]/\varepsilon_N \times d$, where Exs is $Ex_{420\ nm}$ of the enzyme reaction supernatant of test compound, ExB is $Ex_{420\ nm}$ of the enzyme reaction supernatant of the blank control, ε_N is ε for *o*-nitrophenyl-3-D-galactopyranoside in the enzyme assay reaction mixture ($4666 \times 10^3\ cm^2$/mole), and d is diameter of the cuvette (1 cm). EC_{50} were calculated from dose-response curves obtained by fitting the data by $Y = D + [A - D]/[1 + (C/X)^B]$, where X is estrogen concentration in the test, Y is β-galactosidase activity, B is relative slope of the middle region of the curve as estimated from a linear/log regression of the linear part of the dose response curve,

C is estrogen concentration at half maximal response, and D is minimum β-galactosidase activity. Test compound screening results, the relative β-galactosidase induction factor, was defined as the ratio of β-galactosidase activity in a sample and in the corresponding negative control.

19.2 *IN VIVO, EX VIVO,* AND *IN VITRO* ASSAYS FOR ESTROGEN-LIKE EFFECT[3]

1. In Vivo *Uterotropic Assay*: Impubertal female Sprague-Dawley rats (21 days old, 45–50 g) were housed four per cage with a multiple rat rack, 21 ± 2°C, 55 ± 15% humidity, a 12-h light/dark cycle, and given water and food *ad libitum*. All rats had been acclimatized for 3 days in the animal room prior to the first assay. Estradiol valerate and test compounds were dissolved in the minimum ethyl alcohol and diluted in tap water, and the same amount of ethyl alcohol was added to tap water (as control) and to the solution of test compounds in water. Rats were administered the tested samples or 1.0 mg/10 mL/kg estradiol valerate for 3 days by oral gavage, body weight of each rat was recorded daily, and detailed clinical observation carried out simultaneously. Twenty-four hours after the final dose, rats were sacrificed by inhalation of CO_2, their uteri were excised, trimmed free of any fat and adhering non-uterine tissue, pierced and blotted to remove excess fluid. Just above its junction with the cervix and at the junction of the uterine horns with the ovaries, the body of the uterus was cut. Wet weights of uteri were recorded and absolute uterine weight was used as an index of uterine growth and reported as mean ± SEM% variation (versus control group) of uterine weight.

2. Ex Vivo *Assay*: Rat uterine horns were used for the uterotropic assay. From each horn a 1.5 ± 0.2 cm piece was excised and the two pieces were placed together in 10 mL of isolated organ baths containing Tyrode solution consisted of 80.0 mg/mL NaCl, 2.0 mg/mL KCl, 2.0 mg/mL $CaCl_2 \cdot 2H_2O$, 1.0 mg/mL $MgCl_2 \cdot 6H_2O$, and NaH_2PO_4, 0.5 mg/mL. The solution was maintained at 32°C and continuously bubbled with a 95/5 mixture of oxygen/CO_2. The tissues were connected to a Basile High Sensitivity transducer, stretched to a passive tension of 2 g, allowed to equilibrate for 1 h, and tested for the response to oxytocin (0.2–10 nM), prostaglandin $F_{2\alpha}$ (14–28 nM), and bradykinin (1–5 nM). Tension changes were recorded on a Basile 7050 Unirecord. By carefully cutting and weighing the paper response traces, the under the curve areas (UCAs) were defined. Data were reported as mean ± SEM of the UCAs.

3. In Vitro *Assay*: Prepubertal female SD rats (28 days old, 90–105 g) were housed four per cage with a multiple rat rack, 21 ± 2°C, 55 ± 15% humidity, a 12-h light/dark cycle, and given water and food *ad libitum*. Rats were sacrificed by inhalation of CO_2, their uterine horns were excised as formerly described, and weighed. From each horn, a 1.5 ± 0.2 cm piece was excised and placed in Tyrode solution in 10 mL of isolated organ baths, connected to a Basile High Sensitivity transducer, stretched to a passive tension of 2 g, allowed to equilibrate for 1 h, and tested for the response to oxytocin (0.2 nM), prostaglandin $F_{2\alpha}$ (14 nM), and bradykinin (0.5 nM) without (control) or with 17β-estradiol (0.05–2.72 μg, corresponding with 0.02–1.00 μM), and test

compounds. Tension changes were recorded on a Basile 7050 Unirecord. Carefully cutting and weighing the paper response traces, the UCAs were obtained, and data were reported as mean ± SEM of percentage variation of the UCAs.

19.3 AMELIORATIVE YEAST ASSAY[6]

1. *Fish Treatment*: Adult female Taiwan cyprinid fishes (about 7–8 inches of body length) in a semirecirculating tank at room temperature were fed a commercial fish diet and treated with 0.04 μg/mL E2 for 2 weeks, sacrificed, livers were removed, immediately frozen in liquid nitrogen, and total RNA was extracted using Trizol reagent following the manufacturer's instructions.

2. *Cloning the Full-Length ER Genes*: To isolate the estrogen receptor cDNA of *Varicorhinus barbatulus*, six different fish species including gilthead seabream, medaka, rainbow trout, channel catfish, Atlantic croaker, and goldfish were used. To identify conserved nucleotide sequences, clustalW software was used for multiple sequence alignment of these gene sequences. Based on the region overlapping the DNA and hormone binding domains, two degenerate primers 5′-CA(G/A)GG(T/A)-CACAATGA(T/C)TA(C/T)AT-3′ and 5′-TG(C/G)TCCATGCCTTTGTT(A/G)-CT-3′ were designed. Using the Superscript II one-step reverse transcriptase PCR kit and random primers, total RNA from fish liver was reverse-transcribed. PCR amplification was carried out with the initial denaturation at 94°C for 3 min, and each sample had 30 cycles consisting of denaturation for 30 s at 94°C, annealing at 55°C for 30 s, and extension at 72°C for 1 min. Generated PCR fragment was purified and cloned into the pGEM-T Easy vector for sequencing, and using BLAST on NCBI the sequence was confirmed as part of *Varicorhinus barbatulus* (vb)ERα or *Varicorhinus barbatulus* (vb)ERβ2). To obtain the full-length cDNA, the 5′ end of vbERα or vbERβ2 was extended using gene specific primers 5′-CCAACACCTGCCTGCTGAGA-3′ (for vbERα) and 5′-ACCATCACCATCCAGTTGCTG-3′ (for vbERβ2) for 5′-RACE with SMART RACE cDNA Amplification Kit according to the manufacture's instructions. A 3′-RACE procedure using gene-specific primers 5′-TGTACTCTGGATC-AAGAGCCG-3′, 5′-GTCAGTGCTTTATGTATGCCTC-3′ and 5′-TCATTCTGC-TCCAGTCCAGT-3′ (for vbERα) and 5′-GAAACTCATGTTCTCACCTGACC-3′, 5′-CAGCAACAGTCCATCCGGCT-3′ and 5′-CATCGAGTGGACATGGACACAG-3′ (for vbERβ2) was also practiced to obtain the full 3′ extensions of both genes. Various length 5′- and 3′-RACE products were cloned and sequenced. By overlapping sequences, the full-length cDNAs of vbERα and vbERβ2 were assembled. Using the package of DNAMAN software, amino acid composition, protein molecular weights, and domain signatures were assessed.

3. *Constructing the Recombinant Yeast Assays*: Using the yeast Match-maker One-Hybrid System, the ER-dependent transcriptional effect of xenoestrogens was monitored. By PCR with primer-αf containing an EcoR1 cleavage site (5′-AGGAATTCATGTACCCTAAGGAGGAGCA-3′) and primer-αr containing a BamH1 cleavage site (5′-CAGGATCCTCAGGGGTCTGGACTCTGGT-3′), the

full-length cDNA of vbERα was reconstructed. Using primer-βf (5′-CAGGAATTC-ATGAGCTCTTCCCTTGGTCC-3′) and primer-βr (5′-CAGGATCCTCAATCTG-TGTCCATGTCC-3′) with EcoR1 and BamH1 sites, respectively, the full-length cDNA of vbERβ2 was reconstructed. To confirm sequence, the PCR products were cloned into the Topo TA cloning vectors. With EcoR1 and BamH1, the full-length cDNA of vbERα or vbERβ2 was excised and directionally cloned into the multiple cloning sites of the pGAD424 expression vector (named as pGAD424-vbERα and pGAD424-vbERβ2, respectively) to produce ER protein subtypes in yeast. To drive its production, β-galactosidase reporter (ERE-pLacZi) construct was prepared by cloning three tandem repeats of the double-stranded cis-element (ER responsive element 5′-GGTCACAGTGACC-3′) for transcription activity of vbERα and vbERβ2 into the minimal target promoter (PCYC1). The individual subtype (pGAD424-vbERα or pGAD424-vbERβ2) was transformed together with the reporter ERE-pLacZi and via double crossover integrated into genomic DNA of the host yeast (YM4271). After mating 17–24 h, the yeast cells were harvested, washed with 1x TE buffer (pH 8.0) containing 25 μg/mL kanamycin, resuspended in 5 mL of same buffer, and spread on plates with selective media (Leu, His) containing 15 mM 3-amino-1,2,4-triazole (3-AT) to reduce possible leaky expression of the HIS3 gene. Yeast containing the individual pGAD424-vbERα or pGAD424-vbERβ2 subtype together with the reporter ERE-pLacZi were used to assay estrogenic activity of compounds simply by observing the color reaction produced by the β-galactosidase reporter activated through vbERα or vbERβ2.

4. *Transcriptional Activity Between vbERα and vbERβ2-Mediated Assay*: Using *o*-nitrophenyl-D-galactopyranoside (ONPG) as a substrate, the liquid culture was assayed for β-galactosidase activity (μU). First, the recombinant yeast was incubated overnight on SD medium containing 3-AT without uracil and leucine at 30°C and 180 rpm, with SD medium diluted to 0.5 OD_{600}, and individual test compound or DMSO solvent alone were added and incubated for an additional 4 h. In the final screening test, the concentration of DMSO was controlled to not exceed 2%. After incubation, the culture (V1) was collected by centrifugation and resuspended in 0.1 mL of Z buffer (V2, 0.1 M sodium phosphate, 10 mM KCl, 1 mM $MgSO_4$, pH 7) and the V1/V2 ratio was defined as a concentration factor. Via three freeze/thaw cycles, the cell extracts were obtained. After adding 0.7 mL of Z buffer containing β-mercaptoethanol by mixing with 160 μL of ONPG (4 mg/mL), the activity of β-galactosidase was immediately measured. When the yellow color developed, 0.4 mL of Na_2CO_3 (1 M) was added to stop the reaction. By centrifugation at 14,000 rpm for 5 min, the cell debris was removed, and OD was recorded at 420 nm. β-Galactosidase activity was calculated by the equation $1000 \times OD_{420}/OD_{600} \times$ incubation time (min) $\times 0.1 \times$ concentration factor.

19.4 IN YEAST TWO-HYBRID ASSAY[7]

By lithium acetate method according to the small-scale yeast transformation procedure, two expression plasmids pGBT9-hERα (human estrogen receptor alpha) and pGAAD424-TIF2 were introduced into yeast cells (*Saccharomyces cerevisiae* Y190). Yeast cells were plated with synthetic dextrose (SD) medium free from Trp and Leu to select transformants containing the introduced plasmids. To assay estrogen activity of test compounds, yeast cultures were incubated overnight, diluted with SD medium without Trp and Leu to bring the OD_{600} up to 0.2–0.3. The fresh cultures (250 μL) were mixed with 2.5 μL of test compounds and 17β-estradiol as positive control, at 30°C incubated for 4 h, 150 μL of yeast suspension was taken to measure the OD_{600}, the remaining cells were washed in Z-buffer (0.1 M sodium phosphate, pH 7.0, 10 mM KCl, 1 mM $MgSO_4$ and 0.27% β-mercaptoethanol), disintegrated with liquid nitrogen with the triple freeze-thaw procedure, the cell lysate was mixed with 40 μL of ONPG (4 mg/mL), allowed to react about 30 min until a yellow color developed, and 100 μL of Na_2CO_3 (1 M) was added to stop the reaction. An aliquot (150 μL) was placed into each of 96 wells of a microplate, and absorbance was read on a microplate reader μQuant at 420 and 520 nm to estimate estrogenic activity. β-Galactosidase activity was calculated with $U = 1000 \times (OD_{415} - 1.75 \times OD_{570}/OD_{600}Vt)$, where t is the reaction time (min), V is the yeast culture volume (mL), OD_{600} is the yeast suspension density at the beginning of experiment, OD_{415} is the *o*-nitrophenol adsorption in the suspension at the reaction end, and OD_{570} is the light dispersion of suspension at the reaction end.

19.5 ESTROGEN-SENSITIVE YEAST STRAIN RMY/ER-ERE ASSAY[8]

The budding *Saccharomyces cerevisiae* yeast construct RMY/ER-ERE expressed the wild-type human estrogen receptor α, contained an estrogen response element (ERE) and a reporter gene upstream of the lacZ gene encoding β-galactosidase and was selectively cultured using an 0.2-μm pore filter-sterilized selective medium SM without His, Met, and Trp (20 mg/mL dextrose, 6.7 mg/mL yeast nitrogen base without amino acids, 0.032 mg/mL adenine sulfate, 5 mg/mL casamino acids, 2% bacto-agar for agar plate) in nanopure water, with a sterile pipette tip collected from an SM agar plate streak and mixed in 5 mL of liquid SM medium to provide a homogeneous cell suspension. The suspension was poured into a polystyrene Petri dish (60 mm × 15 mm), the cells were at 30°C grown to confluence in humid air, and medium was refreshed at 4- to 5-day intervals or as required. When confluence was attained, the cells were pelleted by centrifugation at 4°C and 2000 × *g* for 15 min, the spent medium was discarded, cells were re-suspended in 5 mL fresh cold yeast-peptone-dextrose (YPD) medium (10 mg/mL yeast extract, 20 mg/mL peptone, 20 mg/mL dextrose) and placed on ice until used.

To 750 μL of SM with 50 μM $CuSO_4$ and 200 μL of yeast cell suspension, 250 μL of each standard in aqueous medium was added to give a final volume of 1.2 mL (0.2% EtOH), the mixture was incubated at 30°C for 20 h, cells were collected by centrifugation at 2140 × *g* for 10 min, the supernatant was discarded, the cells were re-suspended in Z-buffer (60 mM $Na_2HPO_4 \cdot 7H_2O$, 40 mM $NaH_2PO_4 \cdot H_2O$, 10 mM KCl, 50 mM β-mercaptoethanol, and 1 mM $MgSO_4 \cdot 7H_2O$). To 100 μL of ONPG in Z-buffer, 40 μL of chloroform and 40 μL of SDS in a 30°C water bath, 100 μL of cell suspension was added and incubated for 6 min. The reaction was stopped, by centrifugation cell debris were removed, and the absorbance of the supernatant was measured at 420 nm. All standards were assayed in triplicate.

To study the time-lapsed response of live yeast cells to estrogen following non-transcriptional fluorescent cellular sensor, 0.1 mM fluorescein di-β-D-galactopyranoside (FDG) was used as a substrate. In a reaction catalyzed by β-galactosidase, FDG was hydrolyzed and gave a fluorescent product that can be monitored with fluorescence microscopy. The cells were treated with 10^{-6} M, a concentration capable of having maximal response to estrogen without toxicity to the yeast cells, 17β-estradiol. An aliquot of 2800 μL of cells was washed and re-suspended in PBS, incubated in a 30°C water bath for 10 min, 100 μL of cell suspension was immediately added to 100 μL 30°C prewarmed 0.1 mM FDG substrate in a 2-mL amber glass vial to give the T-1 sample. To the remaining cells, 300 μL of medium with or without E2 (sham control) was added, from which a sample was collected immediately (0 h), and incubation was continued to remove sample aliquots at the desired time points (0, 0.5, 1, 2, 4, and 24 h), thus gave E2-exposed cells a 10^{-6} M final E2 medium concentration. With 1800 μL of ice-cold stop medium (PBS containing 10 mM HEPES, 4% FBS and 1 pg/mL propidium iodide identifying dead cells having fluoresce red), the FDG loading was stopped and the cells were kept at 4°C in the dark until viewed. For imaging, 5 μL of stained cell suspension was added to a microscope slide, a coverslip was added, and cells were viewed at ×1000 magnification using a Zeiss microscope with a ×10 eyepiece and ×100 oil-immersion objective. Images were captured with an in-built Leica digital camera and PictureFrame software. A GFP/FITC filter set was used for fluorescent images (excitation 488 nm/blue, emission 512 nm/green). To avoid photobleaching, all the work was carried out in dim oblique light. Using Adobe Photoshop 7.0 and IrfanView 3.90 software, brightness and contrast were corrected.

19.6 BIOLUMINESCENT YEAST ASSAYS[9]

Reporter plasmid YipAREluc with androgen responsive promoter was constructed by inserting a sequence containing two copies of an androgen responsive element (ARE) into the XhoI site of Yipluc. To construct a constitutively luminescent reporter plasmid, plasmid pGEMluc-skl was cut with EcoRI and SalI, and the 1670-bp band containing the firefly luciferase gene without peroxisomal targeting signal was isolated from the gel. The luciferase coding sequence was inserted into the vector pRS316/GPD-PGK between the GPD promoter and PGK terminator yielding pRS316luc. Using standard techniques, all DNA manipulations were performed. By restriction enzyme digestions and sequencing, the plasmid constructions were verified. Using the bacterial hosts MC1061 or XL-1 Blue, the plasmids were constructed. *Saccharomyces cerevisiae* strain BMA64-1A was the parent strain of all the new strains in this assay. Using lithium acetate method, the luciferase reporter and receptor expression plasmids were transformed into yeast cells. By linearizing the plasmid with EcoRV, the chromosomal integrations of YipEREluc and YipAREluc were directed into the URA3-locus.

At 30°C with vigorous shaking in SD medium supplemented with required amino acids, 4 mL of yeast culture was overnight grown for 24 h, the culture was diluted to OD_{600} 0.6, grown at 30°C until the OD_{600} reached 0.8, 90 μL aliquots of this culture were pipetted on to a 96-well plate, and 10 μL of test compounds or moisturizer (diluted in 10% ethanol or 10% dimethylsulfoxide) were added. After pipetting, the plate was shaken for 20 s, incubated at 30°C for 2.5 h, incubated, shaken for another 20 s, and 100 μL of 1 mM D-luciferin (in 0.1 M Na-citrate buffer pH 5) was pipetted into the wells containing induced cultures. The plate was briefly shaken and immediately measured with a Victor multilabel counter in the luminescence mode using 1-s counting time.

19.7 LYES ASSAY[10]

In yeast estrogen screen (YES) assay, to the microtiter plate wells, 10 μL of stock solutions were added at each concentration level. Immediately, 190 μL of growth medium containing chlorophenol red-β-D-galactopyranoside (CPRG) and yeast cells were added, shaken vigorously on a titer plate shaker (2 min), and the plates were in a ventilated heating cabinet incubated for 3 days. All test compounds were tested in quadruplicate in three independent assays with the exception of E2 (24 independent measurements). After 3 days, the absorbances were measured at wavelengths of 630 nm and 550 nm. The corrected absorbances were plotted versus the concentrations to calculate the EC_{50} values. In each assay, E2 was used as positive control and double-distilled water was used as blank.

In LYES assay, a modified version of the YES-assay including a digestion step with the enzyme lyticase, the conventional procedure was followed. The recombinant *Saccharomyces cerevisiae* cells were at 30°C on an orbital shaker incubated for 48 h, 50 mL of yeast suspension was centrifuged at 3500 rpm for 10 min, and the supernatant was replaced by 10 mL of fresh growth medium. To the wells, 10 μL of test compounds at the concentration levels and 90 μL of yeast suspension were added, the microtiter plate was sealed with autoclave tape, vigorously shaken for 2 min, and at 36°C incubated for 4 h. To prepare 1 mg/mL lyticase stock solution, 10 mg of lyticase, 1 mL of potassium phosphate buffer (1 M, pH 7.5), 0.2 mL of NaCl (5 M), and 5 mL of glycerol were combined and diluted with double-distilled water to 10 mL. To each well, 4 μL of this stock solution and 36 μL of Z-buffer (60 mM $Na_2HPO_4 \cdot 7H_2O$, 40 mM $NaH_2PO_4 \cdot H_2O$, 10 mM KCl, 1 mM $MgSO_4 \cdot 7H_2O$, and 50 mM β-mercaptoethanol) were added, at room temperature incubated for 45 min, to each well 35 μL of Triton X-100 (0.1%) was added, the plates were incubated again for 20 min, to wells 25 μL of CPRG solution (1 g/L) was added, and the absorbance was at 550 nm immediately measured. While after a 2-h incubation the absorbances were at 630 and 550 nm determined again. Except for E2 (7 fold) and diethylstilbestrol (2 fold) all substances were tested in duplicate and with three independent experiments. Double distilled water was used as blank.

In E-screen assay, the estrogenic activity of test compounds was determined by a proliferation test with human estrogen receptor positive MCF-7 breast cancer cells. The medium containing charcoal dextran stripped human serum inhibited MCF-7 breast cancer cells' proliferation. By the release of this inhibition, estrogens or estrogen-like substances induced the cell proliferation. With 96-well microtiter plates, E-screen assay was performed, and in one single assay each test compound was tested eightfold.

19.8 ERα AND ERβ STABLE TRANSACTIVATION ASSAY[11]

To generate HELN cells, HeLa cells were transfected with the ERE-βGlobin-Luc-SVNeo plasmid. HELN-ERα and HELN-ERβ cell lines were obtained by a second transfection of the corresponding pSG5-puro plasmids pSG5-ERα-puro and pSG5-ERβ-puro, respectively. Selection by geneticin and puromycin was performed at 1 mg/mL and 0.5 mg/mL, respectively. Luminescent and inducible clones were identified with a photon counting camera to isolate the most responsive clones. HELN cell line was grown in phenol red containing DMEM (1 mg/mL glucose) supplemented with 5% FCS and 1% antibiotic (penicillin/streptomycin) and a 5% CO_2 humidified atmosphere at 37°C. Due to FCS and phenol red's estrogenic activity, HELN-ERα and HELN-ERβ cell lines were grown in phenol red-free medium supplemented with 6% dextran-coated charcoal (DCC)-treated FCS and 1% penicillin/streptomycin (6% DCC-FCS). HELN-ERα and HELN-ERβ cells were seeded in 96-well white opaque tissue culture plates at a density of 4×10^4 cells/well, maintained in 6% DCC-FCS for 8 h, test compounds were added, and cells were incubated with compounds for 16 h. Assays were performed in quadruplicate and repeated three times. At the incubation end, effector containing medium was replaced by 0.3 mM luciferin containing 6% DCC-FCS. The 96-well plate was placed in a microplate luminometer to measure intact living cell's luminescence for 2 s. Using Graph-Pad Prism statistics software, EC_{50} values were evaluated.

HELN-ERα or HELN-ERβ cells were in 24-well tissue culture plates at a density of 10^5 cells/well seeded and in 6% DCC-FCS grown, at 37°C with [^{3}H]-E2 (41.3 Ci/mmol specific activity) labeled for 16 h in the absence or presence of nonradioactive E2 or test compounds. The incubation medium was finally diluted to 400 μL, and each dilution was assayed in duplicate. After incubation, unbound material was aspirated, and cells were washed three times with 400 μL of cold PBS, 250 μL of lysis buffer (400 mM NaCl, 25 mM Tris phosphate pH 7.8, 2 mM DTT, 2 mM EDTA, 10% glycerol, 1% Triton X-100) was added, plates were shaken for 5 min, 200 μL of total cell lysate was mixed with 4 mL of LSC-cocktail, and [^{3}H] bound radioactivity was as liquid scintillation counted. By Bio-Rad protein assay, protein concentrations were measured and used to normalize bound radioactivity values expressed in dpm. For saturation ligand binding analysis experiments, cells were with 0.01–3 nM [^{3}H]-E2 and 100 nM of nonradioactive E2 labeled. By subtracting nonspecific binding from total binding, specific binding was determined. By subtracting bound ligand from total ligand, the free ligand concentration was calculated. By fitting data to the Hill equation and by linear Scatchard transformation, the dissociation constant (Kd) was calculated as the free concentration of radioligand at half-maximal binding. In ligand competition assays, each test compound was tested at least three times. Cells were labeled with 0.1 nM [^{3}H]-E2 (in the absence or presence of increasing concentrations of nonradioactive competitive compounds). Without competitor, specific bound radioactivity was 750–1000 dpm. Results were plotted as dpm versus concentration. Using the relative binding affinity (RBA) to E2 compound, selectivity toward ERα and ERβ was evaluated. RBA for each competitor was calculated as the ratio of E2 to competitor concentration required reducing specific radioligand binding by 50% (ratio of IC_{50} values). The RBA value for E2 was arbitrarily set at 100.

19.9 HELNα AND HELNβ TRANSFECTED CELL ASSAY[12]

After incubation with 30 mg of C18 Oasis HLB batch for 10 min, 800 mL of every patient's serum was desteroided and centrifuged (13,000 × *g*, 10 min) to separate the supernatant. To this stripped serum, a known amount of estradiol was added to obtain a set of estradiol standards with concentrations from 10^{-9} to 10^{-4} mM and at 37°C incubated for 4 h to permit an equilibrium between free estrogens and estrogens bound to sex hormone-binding globulin. HeLa cells transfected with ERE-β Glob-Luc-SVNeo and pSG5-Puro-hERα or β, so-called HELNα and HELNβ, were selected by geneticin and puromycin at 1 mg/mL and 0.5 μg/mL, respectively. Luminescent and inducible clones were identified using photon-counting cameras. HELN Erα was cultured in 150 cm^2 plastic flasks in DMEM without phenol red, supplemented with 5% dextran-coated and charcoal-treated FCS, 0.25 μg/mL puromycin, and 0.5 mg/mL geneticin. To the culture medium, aromatase inhibitor aminogluthetimide (AG) was added at a concentration of 50 μM. The cells were seeded in 96-well plates (3×10^4 cells per well) in presence of DMEM without phenol red, supplemented with 3% dextran-coated and charcoal-treated FCS, 50 μm AG, and at 37°C incubated for 8 h. Culture medium was replaced by 100 μL of DMEM without phenol red, supplemented with 50 μM AG, to which 20 μL of human serum in triplicate was added.

19.10 LUCIFERASE REPORTER GENE ASSAY[13]

Sera were selected from the serum bank on two criteria, (a) treated with a single IFNβ product, (b) availability of antibody resulting from the routine BAb testing by the sandwich ELISA, and when positive were further tested with the A549/encephalomyocarditis virus (EMCV) CPE method. Sera were collected more than 12 h after the last interferon injection, a time when circulating levels of injected interferon are below threshold detectable by bioassay. For the presence of neutralizing antibodies (NAbs) using the luciferase reporter gene assay as well as the sandwich ELISA and A549/EMCV CPE assays, a total of 163 sera were tested, of which 82 samples were from patients treated with IFNβ-1b and 81 from patients treated with IFNβ-1a. Samples represented a range of NAb titers from negative to high positive and were blinded before being assayed.

Cell line HL116 was in DMEM supplemented with 10% FBS, 2 mM L-Gln, 0.1 mM sodium hypoxanthine, 0.4 μM aminopterin, 0.016 mM thymidine, 100 IU/mL penicillin, 100 μg/mL streptomycin sulfate, and 0.25 μg/mL amphotericin B at 37°C and 5% CO_2 maintained. Cells were passed twice weekly at a 1:8 dilution with a limit of 25 passages and grown in culture medium without antibiotics and antimycotics during the assay. In this assay, different parameters such as the effect of plate color, plating cell density, plate incubation time, amount of substrate, duration of plate shaking, and methods of drawing dilution curves were monitored to optimize the assay procedure.

Sera of patients were at 56°C inactivated for 30 min and stored at 4°C for up to 2 weeks until use. Cultured HL116 cells were visually inspected for viability, in 96-well black plates with a transparent bottom at 4×10^4 cells/well seeded, and at 37°C and 5% CO_2 incubated for 16 h. Serum samples were serially diluted, with interferon preincubated for 1 h, to the microtiter plates at 100 μL/well added, incubated overnight, and the culture media were discarded. After a 37°C and 5 h incubation, plates were cooled to room temperature (10 min), the plate bottom was sealed with backing tape to prevent light contamination, and 50 μL of Steadylite substrate was added. After 1 min shaking, luminescence signal was immediately read.

For NAb screening assay, 1:10 and 1:500 diluted sera were added to an equal volume of IFNβ, IFNβ-1b, or IFNβ-1a at a concentration of 20 LU/mL. Positive samples were reassayed in the NAb titer assay, where sera were diluted 1:2 fold serially with 1:10, 1:100, or 1:500 as an initial dilution and added to an equal volume of IFNβ at 20 LU/mL; for each test sera six dilutions were used. For quality control, at least two points above and below 50% must be cut off for the serum dilution to be valid. In parallel, a homologous IFN dilution curve was involved on each plate to determine the antigen activity. One Laboratory Unit (LU) was defined as the end point of this assay at 50% luminescence signal for the interferon dilution curve on a log-log scale. Only assays in which the challenge dose of IFNβ was between 7 and 15 LU/mL were accepted. Positives were determined by signals >20 TRU/mL, and positive samples were serially diluted and reassayed. With the Kawade-Grossberg Formula $t = f^*(n - 1)/9$, titers were calculated, where t is titer in Ten Fold Reduction Units (TRU)/mL, and f is reciprocal of serum dilution at end point, n is actual amount of IFN antigen in test as LU/mL.

19.11 TARGET AKT PATHWAY ASSAY[14]

A DNA fragment encoding the Myc-His-tagged active form of mouse AKT1 from the retroviral plasmid pCX4bleo-myrAKT was subcloned into pCX4pur vector. DNA fragments for firefly luciferase and Renilla luciferase from the pGL3-enhancer and pRL-TK vectors, respectively, were subcloned into the pCX4bleo vector. Full sequence information for the resultant recombinants was available upon request. Retroviral-mediated gene transfer was carried out, and infected cell populations were selected in culture media containing puromycin (1 μg/mL) and zeocine (500 μg/mL) for 2 weeks. In all cases, cultures arose from the polyclonal expansion of infected cells.

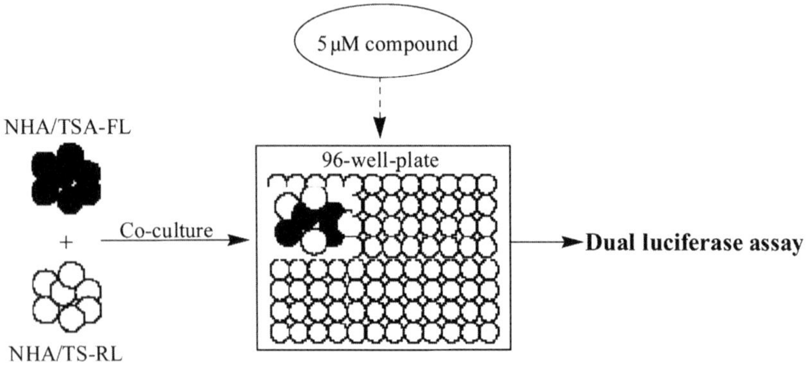

Figure 19.1 Schematic representation of screening strategy.

In a 96-well plate (100 μL final volume per well), a mixture of equal amounts (1500 cells each) of NHA/TS cells (normal human astrocytes immortalized by hTERT and SV40ER) infected with Renilla-luciferase (NHA/TSRL) and myrAKT-expressing NHA/TS cells infected with firefly luciferase (NHA/TSA-FL) was cultured. Twenty-four hours later, to wells each test compound of SCADS inhibitor kit was added individually to a final concentration of 5 μM, and the cells were incubated for 48 h. Using the Dual-Luciferase Reporter Assay System, luciferase assays were performed, and the ratio of firefly- to Renilla-luciferase activities was calculated. A corresponding schematic representation of screening strategy is demonstrated in Fig. 19.1.

19.12 LUCIFERASE ASSAY[15]

The p403-GPD and p405-GPD yeast expression vectors and the p406-CYC1 yeast expression vector were used to express human estrogen receptor α and β and to construct the reporter plasmid, respectively. On the isolated mRNA of T47D human breast cancer cells and intestinal Caco-2 cells, cDNA was synthesized. Using T47D cDNA, marathon uterus cDNA, and human intestine cDNA, by PCR full-length human estrogen receptor β (ERβ) cDNA was obtained. To perform the first PCR, 34.2 μL ultrapure water, 5 μL of $MgCl_2$ (25 mM), 5 μL of Expand HF 10x concentrated buffer (without $MgCl_2$), 0.8 μL of dNTP mix (25 mM), 1 μL of enzyme mix, 2 μL of different cDNAs, and 2 μL of primer mix containing 10 μM each primer were pipetted into a thin-walled PCR tube and (1) denature template 3 min at 95°C, (2) denature template 30 s at 94°C, (3) anneal primers 1 min at 60°C, (4) elongation 2 min at 72°C, (5) go to step 2 and repeat 35 times, (6) elongation 7 min at 72°C and (7) for over 10°C. The second PCR was performed with the same conditions but 2 μL of the first PCR mixture was used instead of 2 μL of different cDNAs. The 5′- and 3′-primer 5′-CGTCTAGAGCTGTTATCTCAAGACATGGATATAA-3′ and 5′-TAGGATCCGTCACTGAGACTGTGGGTTCTG-3′ contained a restriction site for XbaI just before the ATG started codon and for BamHI just after the TGA stopped codon, respectively. Full-length ERβ PCR product was isolated and ligated into a pGEM-T Easy Vector. The uterus/intestine human Erβ cDNA clone that fully corresponded with the ERβ sequence was cut out of the pGEM-T Easy plasmid with XbaI and BamHI and cloned into the corresponding XbaI–BamHI site of the p405-GPD-ERβ vector, which was used to transform Epicurian Coli XL-2 Blue Cells.

Yeast K20 was transfected with p406-ERE2s2-CYC1-yEGFP reporter vector and integrated at the chromosomal location of the Uracil gene via homologous recombination to construct yeast hERα and hERβ cytosensors. The transformants were grown on MM/LH plates using PCR and Southern blot hybridization to select clones, in which the integration had occurred at the desired URA3 site with only a single copy of p406-ERE2s2-CYC1-yEGFP. This strain was transformed with p403-GPD-Erα, p405-GPD-ERβ, or both expression vectors, and transformants were grown on MM/L, MM/H, or MM plates, respectively.

The plate with selective MM/L, MM/H, or MM medium was inoculated with −80°C stock yeast ERα, ERβ, or ERα/β cytosensor (20% glycerol, v/v), respectively, incubated at 30°C for 24–48 h and stored at 4°C. The day before assay, a single colony of the yeast cytosensor was at 30°C with vigorous orbital shaking at 225 × *g* in 10 mL of the corresponding selective medium inoculated overnight. At the late log phase, the yeast ERα, Erβ and ERα/β cytosensor culture was diluted (1 : 10) in MM/L, (1 : 20) in MM/H, and (1 : 20) in MM, respectively. This minimal medium consisted of yeast nitrogen base without amino acids or ammonium sulfate (1.7 mg/mL), dextrose (20 mg/mL), and ammonium sulfate (5 mg/mL). The MM/L and MM/H medium were supplemented with L-leucine (60 μg/ml) or L-histidine (2 μg/mL), respectively.

Aliquots of 200 μL of yeast culture were pipetted into each well of 96-well plates, and 1 μL of ethanol or DMSO stock solution of test compound was added to result in 0.5% final concentration. The plates were included for 4 h and 24 h, fluorescence was at 485 nm and 530 nm directly measured, and the OD of the yeast culture at 630 nm was determined to check the toxicity of test compound for yeast.

19.13 ANDROGEN RECEPTOR TRANSACTIVATION ASSAY[16]

Yeast expression plasmid YepBUbi-FLAG1 carrying the CUP promoter and a tryptophan selection marker was used as a backbone for cloning human androgen receptor (AR) expression plasmid YEpBUbiFLAG-AR. By PCR, the sequence of human AR was amplified with primers creating sticky SalI/NotI ends and ligated into the respective cloning sites on plasmid YepBUbi-FLAG1. Modifying the reporter plasmid YRpE2 carrying two copies of the vitellogenin ERE, the iso-1-cytochrome c (CYC1) promoter in fusion to the lacZ gene, and a uracil auxotrophy marker, the reporter plasmid YRpE2-ARE was created. With XhoI, ERE was cut from the plasmid YRpE2. One of the two XhoI sites surrounding the ERE was not complete and needed to be reconstituted by insertion of one base by means of site-directed mutagenesis. Androgen response element (ARE) 5′-TCGAGCTAGAACAGTATGTTCTCACC-AGAGAACAGTATGTTCTCTGCAGC-3′ containing a consensus AR binding element composed of two 6-bp asymmetrical elements separated by a 3-bp spacer was with sticky XhoI overhangs synthesized and using T4 DNA ligase ligated into YRpE2. For creating reporter plasmid YRpE2-GFP-ARE using GFP as a reporter gene, vector YRpE2-GFP was used as a backbone. Into this vector ARE was cloned. By automated sequencing, the sequence of the cloned vectors was verified. For amplification, by electroporation all plasmids were transformed into electrocompetent *Escherichia coli*. Using a standard lithium acetate method, transformation of plasmids into hyperpermeable yeast cells (*Saccharomyces cerevisiae*, strain 188R1) was performed. By growth on dropout medium lacking Trp and uracil, transformed yeast cells were selected. Detecting expressed AR expression of AR was shown with Western blotting. Overnight cultures of yeast were grown on selective medium without Trp and uracil, diluted into the expression medium YPHSM, again grown overnight, adjusted to $OD_{600} = 0.5$ or 0.3, and induced with 100 μM $CuSO_4$. After 4 or 6 h of incubation, yeast cells were harvested, washed once, resuspended in the same volume of disintegration buffer (20 mM Tris, pH 7.4, 150 mM NaCl, 2 mM dithiothreitol, 10% glycerol, 1% Tween-20, 1 mL/20 g protease inhibitor cocktail), glass beads were added, and by vortexing three times for 30 s with a rest on ice for 15 s between the intervals, disintegration was performed, and using the MES buffer system according to the supplier, yeast cell extracts were immediately loaded onto a NuPage 4% to 12% Bis-Tris gel and SDS-PAGE. By electroblotting, samples were transferred to a Protran nitrocellulose transfer membrane. With an anti-AR antibody as a primary antibody and an anti-mouse IgG AP conjugate as a secondary antibody, the blot was incubated. As a substrate for alkaline phosphatase, Lumi-Phos WB Chemiluminescent Substrate was used, and the blot was viewed in a Lumi-Imager.

To a yeast culture standard substance 5α-DHT or a test compound and 10 μM $CuSO_4$, the inducer, were added to form a mixture having OD_{600} value of 0.4. DMSO alone was used as a blank. A calibration curve with 5α-DHT was performed within each test run, and each determination was performed in duplicate. After 4 h incubation, cells were harvested, washed in lacZ buffer [60 mM Na_2HPO_4, 40 mM NaH_2PO_4 (pH 7.0), 10 mM KCl, 1 mM $MgSO_4$, 1 mM dithiothreitol] and disintegrated with glass beads. Of the clear cell lysates, β-galactosidase activity and protein concentration were determined. The specific enzyme activity was expressed in Miller units, which take the amount of total protein into account.

For each assay run, a new overnight culture was diluted to $OD_{600} = 0.4$ before use. To 5 mL of diluted yeast culture, 5 μL of different standard concentrations and 45 μL of DMSO were added. By the adding 100 μM $CuSO_4$, expression of AR was induced. DMSO alone was used as a blank. Within each assay run, a calibration curve with 5α-dihydrotestosterone was performed. Assays were carried out in duplicate. After a 30°C overnight incubation, cells were centrifuged at 3500 rpm for 5 min, washed in 1 mL of washing buffer (20 mM Tris, pH 7.4, 150 mM NaCl, 10% glycerol), resuspended in 250 μL of washing buffer, and 2 × 100 μL of this suspension of each vial was transferred into two black microplates. Fluorescence of yeast cells was on a Spectramax Gemini XS fluorimeter at 473 nm excitation and 509 nm emission measured (Fig. 19.2).

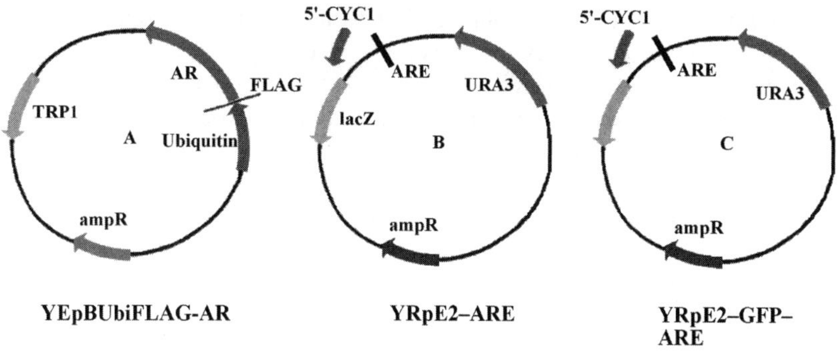

Figure 19.2 Maps of the plasmids.

19.14 GREEN FLUORESCENT PROTEIN EXPRESSION ASSAY[17]

Aliquots of 2 mL of blank calf urine and E2 (1 ng/mL), diethylstilbestrol (DES; 1 ng/mL) and 17α-ethynylestradiol (EE2; 1 ng/mL), α-zearalanol (30 or 50 ng/mL) and mestranol (10 ng/mL) spiked calf urine samples were adjusted to pH 4.8, and 20 μL of β-glucuronidase/arylsulfatase (3 U/mL) was added. The samples were deconjugated by enzyme at 37°C overnight, treated with 2 mL sodium acetate buffer (0.25 M, pH 4.8), and subjected to solid phase extraction (SPE) on a C18 column conditioned previously with 2.5 mL of methanol and 2.5 mL of sodium acetate buffer. The column was successively washed with 1.5 mL of 10% (w/v) sodium carbonate solution, 3.0 mL of water, 1.5 mL of sodium acetate buffer (pH 4.8), 3.0 mL of water, and 2 mL of methanol/water (50/50 v/v). The column was air dried and eluted with 4 mL of acetonitrile, and the eluate was applied to an NH_2-column conditioned previously with 3.0 mL of acetonitrile. Acetonitrile eluate was evaporated to 2 mL by nitrogen gas stream, of which 100 μL part (equivalent to 100 μL urine) was transferred to a 96-well plate and mixed with 50 μL of water and 2 μL of DMSO. The plate was dried overnight in a fume cupboard to remove the acetonitrile and screened on estrogenic activities with the yeast estrogen bioassay. In the same way, a reagent blank was prepared, using 2 mL of sodium acetate buffer (0.25 M, pH 4.8) instead of urine.

The yeast cytosensor (20% glycerol v/v) expressing the human estrogen receptor α (hERα) and yeast EGFP (yEGFP) in response to estrogens was incubated in an agar plate containing the selective MM/L medium at 30°C for 24–48 h and then stored at 4°C. The day before assay, a single colony of the yeast cytosensor was inoculated in 10 mL of selective MM/L medium overnight at 30°C with vigorous orbital shaking at 225 × *g*. At the late log phase, the yeast ERα cytosensor was diluted in MM/L, giving an OD at 604 nm in the range 0.07–0.13. For exposure in 96-well plates, aliquots of 200 μL of this diluted yeast culture were pipetted into each well already containing the extracts of the urine samples. E2 dose-response curve was included in each exposure experiment. Aliquots of 200 μL of diluted yeast culture were pipetted into each well of a 96-well plate, and exposure to different doses of E2 performed through the addition of 2 μL of E2 stock solutions in DMSO. Each urine sample extract and each E2 stock was assayed. Exposure was performed for 0 h and 24 h. Fluorescence at these time intervals was measured directly using excitation at 485 nm and measuring emission at 530 nm. The densities of the yeast culture at these time intervals were also determined by measuring the OD at 630 nm to check whether a urine sample was toxic for yeast.

In the determination of the decision limit (CCα) and detection capability (CCβ) of the yeast estrogen assay, extracts of 20 blank calf urine, 20 spiked calf urine samples (E2, DES, and EE2 at 1 ng/mL, α-zearalanol at 30 ng/mL or 50 ng/mL, and mestranol at 10 ng/mL) and a reagent blank were analyzed. After 24 h exposure, the obtained fluorescence signals of the 20 blank urine samples, 20 spiked urine samples, and a reagent blank sample were corrected for the signals obtained at 0 h ($t_{24}-t_0$). On three different days, the extracts of the blank urines and their corresponding spikes were prepared and analyzed in the yeast estrogen assay in three separate exposures. Within a time period of 10 days, these three sample treatments and exposures were performed. In another experiment, extracts of 20 blank urines and spikes of 50 ng of zearalanol per mL urine were prepared in 1 day and were analyzed in the yeast estrogen assay in one exposure. In the context of EC Decision 2002/657, the mean signal of 20 blank calf urine samples plus 2.33 times the corresponding standard deviation was defined as the decision limit CCα ($\alpha = 1\%$), and the decision limit CCα plus 1.64 times the standard deviation of the signal of the spiked calf urine sample was defined as the detection capability CCβ ($\beta = 5\%$).

In the determination of the specificity of the yeast estrogen assay, three blank calf urine samples were spiked with a high dose of testosterone or progesterone (1000 ng/mL), and extracts were analyzed. In the determination of interference, these blank calf urine samples were spiked with high dose of either testosterone or progesterone in combination with low dose of estrogens (E2, DES, and EE2 at 1 ng/mL, α-zearalanol at 50 ng/mL, and mestranol at 10 ng/mL).

19.15 LUTEINIZING HORMONE RELEASING HORMONE (LHRH) ASSAY[18]

1. *Animals and Treatments*: Fifty-four male broiler chickens (Ross 508, 26 days old) were randomly divided into three homogenous experimental groups, allocated in separate cages and housed in a light- and temperature-controlled room, and food and water were provided *ad libitum*. Broiler chickens were fed with an isoenergetic and isonitrogenous pelleted finisher diet consisted of 705 g maize, 245 g soya bean meal, 20 g of calcium biphosphate, 10 g of soya bean oil, 8 g of mineral and vitamin premix, and 5 g of calcium carbonate per kg of dry matter (chemical composition per kg dry matter: metabolizable energy 12.9 MJ, crude protein 190 g, ether extract 43 g, ash 44 g, water 120 g) for 10 days. Mineral and vitamin premix per kg consisted of retinol 30,000 UI, cholecalciferol 7,500 UI, alpha-tocopherol acetate 100 mg, Cu 40 mg, and Na 250 mg. After acclimation, experimental broiler chickens were with the standard diet containing placebo (group 1, control), 1 ppm of clenbuterol (group 2),

and 1 ppm of cimaterol (group 3), respectively, fed for 21 days. At the age of 57 days, broiler chickens were fasted overnight, after anesthesia sacrificed by cervical dislocation, after slaughtering immediately sampled for tissue of heart, lung, brain, testicle, spleen, thymus, and bursa of Fabricius, tissue samples were chilled and washed in ice-cold saline solution (154 nM NaCl), blotted dry, frozen in liquid nitrogen, and stored at −80°C until assayed.

2. *Preparation of Cytosol and Membrane Fractions*: In ice-cold buffer (50 mM Tris-HCl, 1 mM EDTA, 12 mM thioglycerol, 10 mM sodium molybdate, 10% glycerol, pH 7.4), frozen tissues were homogenized, the homogenates were filtered through a double layer of cheese cloth, at 4°C and 3000 × *g* centrifuged for 20 min, and the supernatants were at 4°C and 105,000 × *g* centrifuged for 45 min. The resulting supernatants (cytosolic fractions) from testicle, spleen, thymus, and bursa of Fabricius were used for measuring AnR (testicle) and GR (spleen, thymus, and bursa of Fabricius) at 6 mg/mL of protein and 2.5 mg/mL of protein, respectively. At 4°C in the incubation buffer (75 mM Tris-HCl, 25 mM $MgCl_2$, pH 7.4), using a sonicator, the resulting pellets (membrane fractions) were suspended, diluted to a final concentration of 2 mg/mL protein, and used for determining β-AR.

3. *Androgen Receptor Assay*: Cytosol (200 μL) aliquots were at 4°C with 50 μL of increasing concentrations of [^{3}H]mibolerone (0.13–2.6 nM) incubated overnight. To estimate nonspecific binding, a parallel set of tubes containing 500-fold molar excess of nonlabeled testosterone was prepared as an androgen competitor. In order to avoid possible binding of mibolerone to GR, to each incubation tube 100 nM triamcinolone acetonide was added. After absorption of free hormones on dextran-coated charcoal and precipitation, 200 μL of supernatant was added to 4 mL of scintillation liquid, and the radioactivity was measured for 2 min on a β-counter at a 60% efficiency.

4. *Glucocorticoid Receptor Assay*: To measure spleen, thymus, and bursa of Fabricius GR density, aliquots of cytosol were incubated at the same conditions as for AnR with increasing concentrations (0.3–10 nM) of [^{3}H]dexamethasone. The aspecific binding was assessed in presence of 5 μM unlabeled corticosterone, and tubes were then processed as for AnR.

5. *β_2-Adrenergic Receptor (β_2-AR) Assay*: Measurement of and subtypes was performed. Aliquots (200 μg) of heart, lung, brain, spleen, thymus, bursa of Fabricius membrane proteins were at 37°C with increasing concentrations of the nonselective β-AR antagonist (−)[^{3}H]CGP 12177 (0.06–4 nM) in a total volume of 200 μL of incubation buffer incubated for 1 h. Nonspecific binding was calculated in the presence of 100 μM isoproterenol. To the incubation mixture by adding 1 μM unlabeled ICI 118551, a β2-selective antagonist, the β1-AR density was measured, whereas β2-AR concentrations were measured by adding 3 μM CGP 20712 A, a β1-selective antagonist. By adding 2 mL of ice-cold buffered saline solution (154 mM NaCl, 50 mM Tris-HCl, pH 7.4) incubation was stopped, the incubation mixtures were under vacuum over presoaked glass microfiber filters immediately filtered, the filters were washed with buffered saline, solubilized with 4 mL of scintillation fluid, and the radioactivity was measured by use of the β-counter.

19.16 ISHIKAWA ENDOMETRIAL CANCER CELL ASSAY FOR ESTROGEN RECEPTOR EXPRESSION[19]

1. *Electrophoretic Mobility Shift Assays*: Ishikawa endometrial cancer cells were treated with 10 nM environmental toxicant 2,3,7,8-TCDD or DMSO (vehicle) for 30 min, harvested by trypsinization, pelleted at 1000 × *g*, and nuclear extracts were obtained. Nuclear protein (10 μg) and poly(dI-dC) (1 μg) were at 25°C incubated for 15 min, 1 nM ^{32}P-labeled specific DRE probe was added and the mixture was at 25°C incubated for 15 min. For competition with specific unlabeled or mutant DRE probe prior to the addition of ^{32}P-labeled DRE, a 100-fold excess was incubated for 15 min. The reaction was carried out in 25 mM HEPES, 1.5 mM EDTA, 10% glycerol, and 1 mM dithiothreitol in 30 mL of final volume. Reaction mixtures were loaded onto a 5% polyacrylamide gel and run at 120 V in 90 mM Tris, 90 mM borate, and 2 mM EDTA (pH 8.0), gels were dried, and protein-DNA binding was visualized by autoradiography.

2. *Cell Proliferation Assay*: In DME/F12 maintenance media at a density of 5 × 10^4 cells/well, Ishikawa cells were seeded into 35-mm 6-well tissue culture plates, on the following day the maintenance media was with phenol-free and serum-free DME/F12 media supplemented with sodium bicarbonate (2.2 mg/mL), 10 μL/mL antibiotic-antimycotic solution, insulin (6.25 μg/mL), transferrin (6.25 μg/mL), selenium (6.25 ng/mL), BSA (1.25 mg/mL) and linoleic acid (5.35 μg/mL) replaced, cells were with test compounds treated for 8–12 days, the media was changed and cells were re-dosed with compounds at 2-day intervals, harvested by trypsinization, and counted using a Coulter cell counter.

19.17 ISHIKAWA CELL ASSAY FOR 11β-HSD2 ACTIVITY[20]

1. *Cell Culture and Treatments*: Ishikawa cells were routinely grown in DMEM/F-12 (1:1) supplemented with 10% FBS, penicillin-streptomycin, and sodium pyruvate, maintained in a humidified incubator at 37°C and 95% air–5% CO_2 in T-25 Corning flasks, the medium was changed every day, and the cells were passed as required. For 11β-HSD2 activity assay, cells were passed onto 12-well Corning plates, cultured to 60% to 70% confluence. Prior to treatment, the cells were cultured in serum-free medium for 24 h, and under serum-free conditions all the treatments were performed in triplicate wells. Controls were similarly incubated in triplicates without treatment.

2. *Radiometric Conversion Assay*: 11β-HSD2 activity in intact cells was determined by measuring the rate of cortisol to cortisone conversion. In an attempt to exclude them having a possible competitive inhibition at the end of the treatment, the cells were washed three times in serum-free medium to remove the test compounds, and the cells were incubated at 37°C in serum-free medium containing about 1 × 10^5 cpm [^{3}H]cortisol and 10 nM unlabeled cortisol for 4 h. At incubation end, the medium was collected, steroids were extracted, the extracts were dried, and

the residues were resuspended. A fraction of the resuspension was spotted on a TLC plate and developed with chloroform/methanol (9/1, v/v). By UV light of the cold carriers, the bands containing the labeled cortisol and cortisone were identified, cut out into scintillation vials, and counted in ScintisafeTM Econol 1. From the radioactivity of cortisol and cortisone, the blank values (defined as the rate of conversion without cells) were subtracted to gain the percentage of cortisol to cortisone conversion. The data were presented as mean $\pm$ SEM of 3–6 independent experiments.

19.18 ISHIKAWA CELL ASSAY FOR ESTROGEN ACTIVITY[21]

The AlkPhos assay used as an indicator of ER activation was conducted on quiesced Ishikawa endometrial adenocarcinoma cells. In 7% FBS stripped of endogenous estrogens with dextran-coated charcoal in phenol red-free MEM, cells were cultured for 3 days, in the same medium passed into microtiter cell plates, after 24 h the medium was changed to serum-free MEM, and prior to addition of test steroids, cells were allowed to quiesce for 24 h. To ensure the final concentration of ethanol in the medium was consistently less than 0.1%, the test steroids were added as ethanol stocks. Control cells received the same concentration of ethanol alone. After 4 days of exposure, to test steroids the AlkPhos assay was performed in microtiter plates. After adding chromogenic substrate until maximally stimulated, control Ishikawa endometrial adenocarcinoma cells had an absorbance at 405 nm of about 1.2 during the 1- to 3-h incubation period, and plates were monitored periodically at 405 nm in an ELISA plate reader. This development ensured a linear enzymatic analysis, and all test conditions were assayed in quadruplicate.

19.19 ISHIKAWA CELL ASSAY FOR PLASMINOGEN ACTIVATOR INHIBITOR-1 (PAI-1)[22]

1. *Cell Culture*: Ishikawa cells were cultured in 90% Eagle's MEM with 10% FBS, afterwards the culture was proceeded in Eagle's MEM without FBS or phenol red, 48 h later various sex steroids were settled in culture dishes, and the steroid concentration was selected according to the need of individual assay.

2. *Enzyme Immunoassay for Determination of Human PAI-1 Antigen*: All steps were performed at 4°C, except where indicated. Ishikawa cells (10–20 mg) were in HG buffer [5 mM Tris-HC1, pH 7.4, 5 mM NaC1, 1 mM $CaC1_2$, 2 mM ethyleneglycol-bis(β-aminoethylether)-*N,N,N′,N′*-tetraacetic acid, 1 nM $MgCl_2$, 2 mM dithiothreitol, 25 μg/mL aprotinin, and 25 μg/mL leupeptin] with a Polytron homogenizer homogenized to a suspension and in a Microfuge at 12,000 rpm centrifuged for 3 min to remove the nuclear pellet. Protein concentration of samples was measured with the Bradford method to standardize PAI-1 antigen levels. Using a TintElize PAI-1 kit, PAI-1 antigen level in Ishikawa cells treated with sex steroids was determined by ELISA. The levels of PAI-1 expressed in Ishikawa cells were standardized with their corresponding cellular protein concentration.

19.20 TESTOSTERONE INDUCTION ASSAY WITH SERTOLI CELL[23]

1. *Sertoli Cell Culture*: Sertoli cells were isolated from testis of 19-day-old SD rats and cultured for 2–3 days. In the isolation, testes were collected from four rats, in Hank's balanced salts solution containing 0.01 mg/mL DNaseI decapsulated and chopped into small pieces, the chopped tissues were digested in an Erlenmeyer flask with HBSS containing 1 mg/mL collagenase, 1.4 mg/mL trypsin, 1 mg/mL hyaluronidase, and 0.01 mg/mL DNase I, and at 32°C and 180 rpm in an oscillating incubator for 30 min, digested cell suspension was washed twice with HBSS at 800 × *g* for 2 min and finally resuspended in DMEM/F-12 (1:1) medium containing 10% charcoal dextran stripped serum, 5 μg/mL insulin, 5 μg/mL transferrin, and 5 ng/mL selenite, plated ($5 \times 10^4/cm^2$) in matrix gel coated 6-well plates, and maintained at 34°C with 5% CO_2. To obtain Sertoli cells with >95% purity, cultures were at 22°C with 20 mM Tris (pH 7.4) hypotonically treated for 2.5 min to lyse contaminating germ cells, which was followed by two successive washes with DMEM/F-12, and media were replaced at 24-h intervals thereafter. Day 0 of Sertoli cell cultures was set as 24 h after the hypotonic treatment. Using 3β-hydroxysteroid dehydrogenase (3β-HSD), AP staining, and testosterone induction assays, the possible contamination of Leydig cells and peritubular cells was verified.

2. *Histochemical Staining 3β-HSD, AP Activity, and Testosterone Induction Assay*: For 3β-HSD staining, cells were at 34°C in 50 mM PBS (pH 7.4) containing 0.2 mg/mL nitroblue tetrazolium, 1 mg/mL NAD, and 0.12 mg/mL dehydroepisoandrosterone incubated for 90 min with the positively stained cells having a dark blue color. For localizing alkaline phosphatase activity, the culture was incubated in dark with an alkaline solution containing 0.5 mg/mL Fast Blue RR and 40 μg/mL β-napthol phosphate for 30 min. For testosterone induction assay, cells were exposed to 10 ng/mL hCG for 24 h, and using a testosterone ELISA kit according to the manufacturer's instructions, the conditioned media were assayed for the testosterone content. Briefly, 25 μL of sample/standard was mixed with 100 μL of testosterone-HRP conjugate reagent and 50 μL of rabbit anti-testosterone reagent sequentially in wells and at 37°C incubated for 90 min, the wells were rinsed five times with distilled water, mixed with 100 μL of TMB solution, at 22°C incubated for 20 min, the reaction was stopped, and absorbance was read at 450 nm within 15 min.

19.21 ELECTROPHORETIC MOBILITY AND ANTIBODY SHIFT ASSAYS[24]

1. *Cell Culture and Transient Transfections*: Porcine ovarian granulosa cells (JC-410 cells, 37°C) or mouse Sertoli cells (15P1 cells, 32°C) were routinely maintained in a humidified incubator with 5% CO_2 and cultured in F-12/DMEM containing 10% FBS, 2 mM L-Gln, 100 IU/mL penicillin, and 100 μg/mL streptomycin. Using electroporation, transient transfections were performed. With 5 μg of luciferase reporter and 0.5 μg of pCMVlacZ with or without 5 μg of

expression vector of the Krupple transcription factor, the cells were transfected. After 10 min incubation at room temperature, using a Gene Pulser the cells were electroporated in a 0.4-cm gap width electroporation cuvette with 960 μF at 220 V. The cells were at 37°C incubated for 10 min, cultured in 60-mm Petri dishes for 48 h, and lysed in a lysis buffer containing 100 mM Tris-HCl, 0.5% NP-40, and 1 mM DTT. Appropriate cell extract aliquots were used for luciferase and β-galactosidase assays.

2. *Electrophoretic Mobility and Antibody Shift Assays*: Double-stranded oligonucleotides (TCCCACCCCACCCCCACCAAAG) synthesized by BioCorp were labeled at the 5′ ends using T4 polynucleotide kinase and [γ-^{32}P]ATP. Nuclear extract (10 μg) from either 15P1 or JC-410 cells was incubated at room temperature in a binding buffer containing 50 μg/mL poly(dI-dC) and 20 fmol of labeled DNA probe for 20 min. Using a 5% nondenaturing polyacrylamide gel in 1x TBE buffer (50 mM Tris-borate-EDTA, pH 8.0), the reaction mixture was analyzed. For the antibody supershift assay, into the reaction mixture, 1 μg of antibody against ZNF202 or preimmune control IgG was added and at room temperature incubated for an additional 15 min, and gels were dried and visualized by autoradiography.

REFERENCES AND NOTES

1. K. Rehmann, K.W. Schramm, A.A. Kettrup. Applicability of a yeast oestrogen screen for the detection of oestrogen-linke activity in environmental samples. Chemosphere 38 (**1999**) 3303–3312.
2. X. Denier, J. Couteau, M. Baudrimont, E.M. Hill, J. Rotchell, C. Minier. In vitro study of the effects of cadmium on the activation of the estrogen response element using the YES screen. Marine Environ Res 66 (**2008**) 108–110.
3. P. Bolle, S. Mastrangelo, F. Perrone, M.G. Evandri. Estrogen-like effect of a Cimicifuga racemosa extract sub-fraction as assessed by in vivo, ex vivo and in vitro assays. J Steroid Biochem Mol Biol 107 (**2007**) 262–269.
4. L. Salste, P. Leskinen, M. Virta, L. Kronberg. Determination of estrogens and estrogenic activity in wastewater effluent by chemical analysis and the bioluminescent yeast assay. Sci Total Environ 378 (**2007**) 343–351.
5. I.C. Beck, R. Bruhn, J. Gandrass. Analysis of estrogenic activity in coastal surface waters of the Baltic Sea using the yeast estrogen screen. Chemosphere 63 (**2006**) 1870–1878.
6. K.Y. Fu, C.Y. Chen, W.M. Chang. Application of a yeast estrogen screen in non-biomarker species Varicorhinus barbatulus fish with two estrogen receptor subtypes to assess xenoestrogens. Toxicol In Vitro 21 (**2007**) 604–612.
7. S.N. Kovalchuk, V.B. Kozhemyako, L.N. Atopkina, A.S. Silchenko, S.A. Avilov, V.I. Kalinin, V.A. Rasskazov, D.L. Aminin. Estrogenic activity of triterpene glycosides in yeast two-hybrid assay. J Steroid Biochem Mol Biol 101 (**2006**) 226–231.
8. E. Wozei, S.W. Hermanowicz, H.Y.N. Holman. Developing a biosensor for estrogens in water samples: Study of the real-time response of live cells of the estrogen-sensitive yeast strain RMY/ER-ERE using fluorescence microscopy. Biosensors Bioelectronics 21 (**2006**) 1654–1658.

9. P. Leskinen, E. Michelini, D. Picard, M. Karp, M. Virta. Bioluminescent yeast assays for detecting estrogenic and androgenic activity in different matrices. Chemosphere 61 (**2005**) 259–266.
10. T. Schultis, J.W. Metzger. Determination of estrogenic activity by LYES-assay (yeast estrogen screen-assay assisted by enzymatic digestion with lyticase). Chemosphere 57 (**2004**) 1649–1655.
11. A. Escande, A. Pillon, N. Servant, J.-P. Cravedi, F. Larrea, P. Muhn, J.-C. Nicolas, V. Cavaillés, P. Balaguer. Evaluation of ligand selectivity using reporter cell lines stably expressing estrogen receptor alpha or beta. Biochem Pharmacol 71 (**2006**) 1459–1469.
12. F. Paris, N. Servant, B. Térouanne, P. Balaguer, J.C. Nicolas, C. Sultan. A new recombinant cell bioassay for ultrasensitive determination of serum estrogenic bioactivity in children. J Clin Endocrinol Metab 87 (**2002**) 791–797.
13. R. Lam, R. Farrell, T. Aziz, E. Gibbs, G. Giovannoni, S. Grossberg, J. Oger. Validating parameters of a luciferase reporter gene assay to measure neutralizing antibodies to IFNβ in multiple sclerosis patients. J Immunol Methods 336 (**2008**) 113–118.
14. L. Wang, K. Sasai, T. Akagi, S. Tanaka. Establishment of a luciferase assay-based screening system: Fumitremorgin C selectively inhibits cellular proliferation of immortalized astrocytes expressing an active form of AKT. Biochem Biophys Res Commun 373 (**2008**) 392–396.
15. T.F.H. Bovee, R.J.R. Helsdingen, I.M.C.M. Rietjens, J. Keijer, R.L.A.P. Hoogenboom. Rapid yeast estrogen bioassays stably expressing human estrogen receptors α and green fluorescent protein: A comparison of different compounds with both receptor types. J Steroid Biochem Mol Biol 91 (**2004**) 99–109.
16. V. Beck, E. Reiter, A. Jungbauer. Androgen receptor transactivation assay using green fluorescent protein as a reporter. Anal Biochem 373 (**2008**) 263–271.
17. T.F.H. Bovee, H.H. Heskamp, A.R.M Hamers, R.L.A.P. Hoogenboom, M.W.F. Nielen. Validation of a rapid yeast estrogen bioassay, based on the expression of green fluorescent protein, for the screening of estrogenic activity in calf urine. Anal Chim Acta 529 (**2005**) 57–64.
18. P. Badino, R. Odore, S. Pagliasso, A. Schiavone, C. Girardi, G. Re. Steroid and β-adrenergic receptor modifications in target organs of broiler chickens fed with a diet containing β_2-adrenergic agents. Food Chem Toxicol 46 (**2008**) 2239–2243.
19. M. Wormke, E. Castro-Rivera, I. Chen, S. Safe. Estrogen and aryl hydrocarbon receptor expression and crosstalk in human Ishikawa endometrial cancer cells. J Steroid Biochem Mol Biol 72 (**2000**) 197–207. *Note*: In the preparation of nuclear extracts, cells were harvested by trypsinization, resuspended in HED hypotonic buffer (25 mM HEPES, 1.5 mM EDTA, 1 mM dithiothreitol, pH 7.6), allowed to swell on ice for 10 min (all subsequent steps were performed on ice), pelleted at 1000 × *g* for 5 min, homogenized in 0.5 mL of HEGD buffer (HED plus 10% glycerol, pH 7.6) using a pestle drill apparatus resulting in >90% membrane disruption determined by Trypan blue staining, the homogenate was centrifuged at 4000 × *g* for 10 min, the supernatant fraction was immediately centrifuged at 40,000 × *g* for 30 min to obtain cytosolic extract, the pellet containing nuclei and cellular debris was resuspended in an equal volume of HEGD buffer containing 0.5 M potassium chloride, on ice incubated for 1 h, centrifuged at 40,000 × *g* for 30 min to obtain high salt nuclear extract.

20. A.D. Darnel, T.K. Archer, K. Yang. Regulation of 11β-hydroxysteroid dehydrogenase type 2 by steroid hormones and epidermal growth factor in the Ishikawa human endometrial cell line. J Steroid Biochem Mol Biol 70 (**1999**) 203–210.

21. A. Kotov, J.L. Falany, J. Wang, C.N. Falany. Regulation of estrogen activity by sulfation in human Ishikawa endometrial adenocarcinoma cells. J Steroid Biochem Mol Biol 68 (**1999**) 137–144.

22. J. Fujimoto, M. Hori, S. Ichigo, T. Tamaya. Sex steroids regulate the expression of plasminogen activator inhibitor-1 (PAI-1) and its mRNA in uterine endometrial cancer cell line Ishikawa. J Steroid Biochem Mol Biol 59 (**1996**) 1–8.

23. K.P. Lai, M.H. Wong, C.K.C. Wong. Effects of TCDD in modulating the expression of Sertoli cell secretory products and markers for cell-cell interaction. Toxicology 206 (**2005**) 111–123.

24. W. Xing, M.R. Sairam. Cross talk of two Krupple transcription factors regulates expression of the ovine FSH receptor geneq. Biochem Biophys Res Commun 295 (**2002**) 1096–1101.

20

METHODS AND APPLICATIONS OF ANTIMALARIAL ASSAYS

Shiqi Peng

The malaria situation is continuing to worsen in many parts of the world. Chemotherapy, indoor residual spraying, and insecticide-treated nets will remain the three most useful tools for controlling this deadly disease, which is believed to kill at least 1 million people each year, a death every 30 seconds. As a result of the resurgence of malaria and the inevitable phenomenon of drug resistance, there is an urgent need for discovering or designing new antimalarial drugs with various mechanisms of action for chemotherapy and indoor residual spraying, for achieving early diagnosis, and for treatment with an artemisinin-based combination. The tremendous progress in study of the biology and the biochemistry of malarial parasites has led to the identification of drug targets that are both parasite specific and essential for parasite growth and survival, and some of them are now exploited in the mechanisms-based assays of various screening programs and in determination of concentration of antimalarials. In this chapter, 22 assays are presented: *Plasmodium falciparum* and murine P388 leukemia cell assay,[1,2] antiplasmodial activity assay,[3,4] *Plasmodium falciparum* growth *in vitro* assay,[5] antibody assay for red blood cell polymorphisms,[6] *Leishmania* macrophage assay,[7] lactate dehydrogenase-based antiplasmodial assay,[8] histidine-rich protein II assay,[9,10] histidine-rich protein II assay,[11,12] antimalarial activity assay,[13–15] *Plasmodium yoelii* liver stage parasites inhibition assay,[16] *in vivo* antimalarial assay,[17] β-hematin inhibition assay,[18] non-radiolabeled ferriprotoporphyrin IX biomineralization inhibition

Pharmaceutical Bioassays: Methods and Applications. By Shiqi Peng and Ming Zhao

assay,[19] survival of *Anopheles gambiae* assay,[20,21] chloroquine assay,[22,23] MS assay for quantification of chloroquine in dog plasma,[24] sensitive fluorescence HPLC assay for AQ-13,[25] HPLC and HPTLC assays for chloroquine, primaquine, and bulaquine,[26] real-time PCR-based chloroquine sensitivity assay,[27] rat embryos *in vitro* assay,[28] HPLC-MS assay for β-DHA in rat plasma,[29,30] and *Plasmodium falciparum* clone and DHA assay.[2]

20.1 *Plasmodium falciparum* AND MURINE P388 LEUKEMIA CELL ASSAY[1,2]

The required quantity of *Plasmodium falciparum* parasites maintained in a complete medium (RPMI-1640, 25 mM HEPES, 25 mM $NaHCO_3$, and 10% pooled human serum, with uninfected human red blood cells at 2.5% hematocrit) was introduced into flat-bottomed 96-well plates. The cell suspension (1% parasitemia) was distributed at 0.2 mL/well containing the test compound in triplicate alongside untreated controls, and the plates were shaken vigorously using a microculture plate shaker. The culture was at 37°C incubated for 18 h in microaerophilic conditions. To each well, tritiating hypoxanthine with a specific activity of 14.1 μCi/mmol (DuPont NEN, Boston, MA, USA) was added (0.5 μCi/well) and at 37°C incubated for another 24 h. To lyse the cells, the contents of the well were then frozen at −30°C and unfrozen at 50°C, harvested by filtration, and washed several times with water. Thereafter, the disks were dried and added to toluene scintillator in vials, and the radioactivity incorporated into parasites was estimated.

Test compounds inhibiting 75% or more of the parasite growth were systematically submitted to cytotoxicity assays. Murine P388 leukemia cells were grown in RPMI 1640 medium containing 0.01 nM β-mercaptoethanol, 10 mM L-Gln, 100 IU/mL penicillin, 100 μg/mL streptomycin, 50 μg/mL gentamicin, and 50 μg/mL nystatin, supplemented with 10% FCS. Cells were maintained at 37°C in humidified atmospheric air containing 5% CO_2. The inoculum seed at 10^4 cells/mL (0.1 mL/well) was introduced into flat-bottomed 96-well plates containing serial concentrations of test compound. Culture was at 37°C incubated for 72 h in the required atmosphere. Thereafter, cells were at 37°C with 0.02% Neutral red dissolved in 1/9 methanol/water (0.1 mL/well), incubated for 1 h, and washed with 1 N PBS and finally lysed with 1% SDS. After a brief agitation on a microculture plate shaker, the plates were transferred to a Titertek Twinreader equipped with a 540-nm filter to measure absorbance of the extracted dye. Cell viability was expressed as the percentage of cells incorporating dye relative to the untreated controls, and IC_{50} values were determined by linear regression method.

20.2 ANTIPLASMODIAL ACTIVITY ASSAY[3,4]

1. In vitro *Assay for* Plasmodium falciparum: Using the K1 strain of *Plasmodium falciparum* (resistant to chloroquine and pyrimethamine), antiplasmodial activity was

assayed. Infected human red blood cells in RPMI 1640 medium with 5% Albumax were exposed to serial test compound dilutions in microtiter plates, at 37°C in a reduced oxygen atmosphere incubated for 48 h, to each well 0.5 μCi [^{3}H]hypoxanthine was added, cultures were incubated for a further 24 h, harvested onto glass fiber filters, and washed with distilled water. Using a BetaplateTM liquid scintillation counter, the radioactivity was counted, recorded as counts per minute (cpm) per well at each test compound concentration, and expressed as a percentage of the untreated controls. From the sigmoidal inhibition curves, IC_{50} values were calculated with artemisinin as reference.

2. *In vitro Assays for Trypanosoma brucei rhodesiense, Trypanosoma cruzi, Leishmania donovani, and L6 cell cytotoxicity*: For *Trypanosoma brucei rhodesiense* and *Trypanosoma cruzi*, antiparasitic assay was used, for L6 cell cytotoxicity rat skeletal myoblasts assay was used, and for *Leishmania donovani* Alamar blue assay was used. Axenic amastigotes were grown in SM medium (pH 5.4) supplemented with 10% FBS, 100 μL of culture medium with 10^5 amastigotes from axenic culture with or without a serial test compound dilution was seeded in 96-well microtiter plates, incubated for 72 h, to each well 10 μL of Alamar blue was added, the plates were incubated for another 2 h and read with a microplate fluorometer. Benznidazole, melarsoprol, miltefosine, and podophyllotoxin were used as reference.

3. *PfFabI Inhibition Assay*: Test compounds were dissolved in DMSO and tested at 10–100 μg/mL in the presence of 1 μg of *Plasmodium falciparum* enoyl-ACP reductase (*Pf*FabI, 22 nM) and 200 μM NADH in 1 mL of HEPES (20 mM, pH 7.4) and 150 mM NaCl. The reaction was started by adding 50 μM crotonoyl-CoA (substrate). At 340 nm, by monitoring the oxidation of NADH to NAD^+ for 1 min, the enoyl-reductase activity of PfFabI was assayed. From graphically plotted dose-response curves, the initial velocities were determined, and IC_{50} values were estimated. Triclosan was used as positive control (IC_{50}: 50 nM = 14 ng/mL).

4. *MtFabI and EcFabI Inhibition Assays*: MtFabI (50 nM) and *Escherichia coli* enoyl-ACP reductase (*Ec*FabI, 10 nM) were assayed in 30 mM PIPES (pH 6.8) and 150 mM NaCl buffer (pH 8.0), using 25 μM 2-dodecenoyl-CoA and 250 μM NADH. By following the oxidation of NADH at 340 nm, enzyme activity was monitored.

20.3 *Plasmodium falciparum* GROWTH *IN VITRO* ASSAY[5]

Plasmodium falciparum parasites were cultured in human erythrocytes at 3% to 5% hematocrit in RPMI medium supplemented with 25 mM HEPES (pH 7.5), 0.225% sodium bicarbonate, 40 μg/mL gentamicin sulfate, 11 mM glucose, 200 μM hypoxanthine, and 0.5% Albumax II (Invitrogen), in a modular chamber and a gas mixture of 1% O_2 and 5% CO_2 at 37°C maintained, and synchronized at the ring stage by sorbitol lysis. By counting the number of infected red blood cells per 2000 red blood cells from Giemsa-stained thin smears, parasitemia was determined. Cultures containing low concentrations of hypoxanthine (10 μM) were diluted to containing 1% parasitemia and 2% hematocrit, adozelesin or DMSO was added, and 0.2 mL of cultures were plated in triplicate in 96-well flat-bottomed plates with uninfected erythrocytes as controls. Plates were at 37°C incubated for 24 h, to each well 1 μCi [^{3}H(G)]hypoxanthine (Amersham) was added, the plates were incubated for a further 24 h, and using a cell harvester, cells were harvested onto glass fiber filter mats. From the experimental values, the background cpm values from the uninfected samples were subtracted. [^{3}H]Hypoxanthine incorporation of test compound-treated samples was expressed as the percentage uptake relative to the DMSO control. Using a nonlinear regression sigmoidal dose response analysis (Prism v. 4.0a), IC_{50} values were calculated.

20.4 ANTIBODY ASSAY FOR RED BLOOD CELL POLYMORPHISMS[6]

In June, before the annual malaria transmission season, venous blood was collected into heparinized Vacutainer tubes. By serology, ABO blood groups were determined, and by using genomic DNA, the other red blood cell polymorphisms were detected. Plasma was collected for 413 children. The other cases corresponded either with children who refused to have blood drawn or to samples that lacked blood.

To determine plasma antibodies directed to the three recombinant proteins MSP2/3D7, MSP2/FC27, and RESA from the asexual blood stages of *Plasmodium falciparum*, ELISA was used. Three recombinant proteins were expressed in *Escherichia coli* and stocked at 1 mg/mL in sterile distilled water and used in a carbonate buffered (pH 9.6) at a final concentration of 0.5 mg/mL. Plasma samples were diluted in PBS containing 5% BSA, 1% Tween-20, and 0.02% sodium azide (IgG2 and IgG4, 1/10; IgG1 and IgG3, 1/50; total IgG, 1/200) and tested in duplicate. These plasma dilutions were chosen in a sufficient range of OD values from the negative to the positive reference pooled control plasmas included in each plate. Monoclonal antibodies mouse anti-IgG1 (clone HP6069, 1 mg/mL), anti-IgG3 (clone HP 6047, 0.5 mg/mL), anti-IgG2 (clone HP 6014, 0.25 mg/mL), and anti-IgG4 (clone HP 6023, 0.25 mg/mL) were used for determining the immunoglobulin isotypes. Two polyclonal antibodies conjugated to alkaline phosphatase, an anti-human IgG (Fc specific, Sigma, Saint-Quentin-Fallavier, France) diluted 1/2000 for IgG as well as a goat anti-mouse IgG (H + L) (Tebu) diluted 1/1000, 1/1000, 1/2000, and 1/4000 for IgG2, IgG4, IgG1, and IgG3, respectively, were used. With *p*-nitrophenylphosphate, bound enzyme was detected, and the OD was read at 405 nm (reference filter 620 nm). Reference positive and negative pooled control plasmas were included in each plate, and results were expressed in arbitrary units (AU) calculated. The thresholds for positivity were determined from the mean reactivities +2SD of 30 plasma samples from non-immune subjects.

20.5 *Leishmania* MACROPHAGE ASSAY [7]

Mouse peritoneal macrophages were seeded in RPMI 1640 medium with 10% FBS into Lab-tek 16 chamber slides. After 24 h, *Leishmania donovani* amastigotes were added at a ratio of 3 : 1 (amastigotes to macrophages). The next day, the medium was replaced by fresh medium containing different test compound concentrations and incubated at 37°C for 96 h. The medium was removed, the slides were fixed with methanol, and stained with Giemsa. The ratio of infected to non-infected macrophages was determined microscopically and expressed as percentage of the control, and the IC_{50} value was calculated by linear regression.

20.6 LACTATE DEHYDROGENASE-BASED ANTIPLASMODIAL ASSAY[8]

Chloroquine-resistant Gombak A and chloroquine-sensitive strain D10 were continuous *in vitro* subcultured from cryopreservation, maintained in human red blood cells, with RPMI 1640 medium supplemented with 10% human serum, 25 mM HEPES, 0.2% sodium bicarbonate, and gentamicin diluted to 7% hematocrit. Following the lactate dehydrogenase method with slight modifications, test compounds were assessed for antiplasmodial activity *in vitro* in human blood. Test compounds were dissolved in DMSO to produce a stock solution of 20 mg/mL, whereas the concentrations of chloroquine diphosphate and artemisinin were 1000 μg/mL and 1 μg/mL, respectively. Before being transferred into two 96-well microtiter plates in triplicate of 10 μL each at 22 concentrations of twofold dilutions, these stock solutions were diluted with 10% human serum, and to each well 190 μL of parasitized red blood cell suspensions (1% to 2% parasitemia) was added. For the positive control wells, parasitized red blood cells were devoid of test compounds, whereas non-parasitized red blood cells were prepared for the negative control wells. The plates were incubated at 37°C in a candle jar for 72 h, cooled at −20°C to lyse the red blood cells, allowed to reach room temperature, and 20 μL of supernatant blood suspension was dispensed into a new microtiter plate containing 100 μL of Malstat reagent and 25 μL of mixture of nitroblue tetrazolium and phenazine ethosulfate. With an ELISA plate reader, absorbance was at 630 nm measured. The percentage inhibition at each concentration was determined, and the mean of at least three IC_{50} values of parasite viability was calculated using probit analysis.

20.7 HISTIDINE-RICH PROTEIN II ASSAY[9,10]

The complement-inactivated plasma samples, from uncomplicated falciparum malaria patients and a healthy volunteer treated with oral sodium artesunate (100 mg followed by 50 mg orally every 12 h for 5 days), were serially diluted and applied in two columns to 96-well microculture plates at 50 μL/well. In addition, two columns with serial dilutions of spiked plasma were added to each plate as controls. On each 96-well plate, samples plus the controls were therefore tested. In addition to the plates with unknown samples, one plate was dosed with six serial dilutions in duplicate of known test compound concentrations covering the whole test range.

The culture and ELISA procedures largely followed the literature. Synchronized parasitized (clone W_2) blood samples from continuous culture were diluted to a 0.1% parasite density and 1.7% hematocrit. This cell medium mixture (175 μL) was added to each well (resulting in total 225 μL/well), and 1.05 mL of the remaining cell medium mixture per plate was mixed with 0.3 mL of test compound-free plasma and frozen as negative controls. The plates were incubated at 37°C for 72 h in a candle jar or gas mixture (5% CO_2, 5% O_2, 90% N_2) before being frozen-thawed.

20.8 HISTIDINE-RICH PROTEIN II ASSAY[11,12]

Heparinized blood samples from clinical falciparum malaria patients were cryopreserved in liquid nitrogen until used. The samples were thawed, washed twice, mixed with RPMI 1640 medium (with 10% human serum) at 5% hematocrit, and transferred into cell culture flasks. The cell medium mixture was incubated at 37°C in a 5% CO_2, 5% O_2, and 90% N_2 gas mixture until a minimum parasite density of 2% was achieved. The samples were synchronized with 5% sorbitol to ensure synchronous parasite cultures. For assay, the samples were with RPMI 1640 medium supplemented with 10% serum diluted to 0.05% to 0.1% parasitemia and 1.5% hematocrit. The cell medium mixture was dispensed in aliquots of 200 μL per well into the wells of the microculture plates.

Five assay series were performed with each strain on test compound-free plates, plates predosed with artesunate and mefloquine (mefloquine hydrochloride: 601.3 nM for MM97-09 and 37.6 nM for LS97-57; artesunate: 160 nM for MM97-09 and 2.75 nM for LS97-57), and initially test compound-free plates with addition of artesunate and mefloquine after 24 h. The plates were incubated at 37°C in a gas mixture (5% CO_2, 5% O_2, and 90% N_2) for 72 h. At 12-h intervals, 100 μL of the upper layer of the supernatant was carefully pipetted out of two wells of the culture plates. The remaining cell medium mixture in the wells (at 3% hematocrit) was as a separate sample frozen. For microscopic evaluation of parasite density and developmental stage, each of the other two wells was used to make thick and thin films.

To determine 50% and 90% inhibitory concentrations, the samples were exposed to ascending concentrations of artesunate and mefloquine (mefloquine hydrochloride 9.4–601.3 nM, artesunate 0.7–44.0 nM). After serial twofold dilutions (7 concentrations and one test compound-free control well) of test compound (25 μL/well) were dispensed into standard 96-well microculture plates, to each well 200 μL of cell medium mixture was added, the plates were at 37°C in a gas mixture (5% CO_2, 5% O_2, and 90% N_2) incubated for 72 h, freeze-thawed twice to obtain complete hemolysis, and assayed in the histidine-rich protein II (HRP2)-ELISA to obtain growth estimates for each test compound concentration.

HRP2 in the supernatant and cell samples was quantified using a commercial ELISA test kit (Malaria Ag CELISA, Cellabs, Brookvale, NSW). One hundred microliters of the hemolyzed (freeze-thawed) culture samples was diluted, transferred into the ELISA plates, precoated with monoclonal antibodies against HRP2, incubated at room temperature for 1 h in a humid chamber, washed four times with the washing solution, to each well 100 μL of diluted Ab-conjugate were added, incubated for 1 h, washed four times, to each well 100 μL of diluted TMB chromogen (1 : 20) was added, in the dark incubated for another 15 min, and 50 μL of stopping solution was added. With an ELISA microplate absorbance reader (SpectraMAX 340 Microplate Spectrophotometer, Molecular Devices, Sunnyvale, CA) at 450 nm, colorimetric analyses was performed.

20.9 ANTIMALARIAL ACTIVITY ASSAY[13–15]

The chloroquine-resistant strain FcB1 of *Plasmodium falciparum* (IC_{50} of chloroquine = 62 ng/mL) was maintained continuously in culture on human erythrocytes and used for the evaluation of antimalarial activities of the test compounds. *In vitro* antiplasmodial activity was determined by a modification of the semiautomated microdilution technique. Prior to the addition of 0.5 μCi of [^{3}H]hypoxanthine (Amersham, Les Ulis, France) per well for 24 h, the stock solutions of test compounds in DMSO (10 mg/mL) were serially diluted with culture medium and introduced to asynchronous parasite cultures (0.5% parasitemia and 1% final hematocrit) on 96-well plates at 37°C for 24 h. The growth inhibition for each test compound concentration was determined by comparison of the radioactivity incorporated into the treated culture with that in the control culture maintained on the same plate. IC_{50} was obtained from test compound concentration response curve, and the results were expressed as the mean determined from three independent assays. The DMSO concentration never exceeded 0.1% and did not inhibit the parasite growth.

20.10 *Plasmodium yoelii* LIVER STAGE PARASITES INHIBITION ASSAY[16]

Plasmodium yoelii yoelii (265 BY strain) sporozoites were obtained by the dissection of infected *Anopheles stephensi* salivary glands.

Primary mouse hepatocytes isolated from mice liver biopsies were seeded in 8-well Lab-Tek plastic chamber slides previously coated with rat tail collagen I (BD Biosciences, Le Pont de Claix, France) at a density of 10^5 cells per well. Before inoculation of *Plasmodium yoelii* sporozoites (10^5 per well), mouse hepatocytes were cultured for 24 h at 37°C in 5% CO_2 in WME supplemented with 10% FCS, 1% L-Gln, 1% sodium pyruvate, 1% insulin-transferrin-selenium, 1% nonessential amino acids, and 1% penicillin-streptomycin. Test compounds were solubilized in DMSO, diluted in culture medium and added to hepatocyte cultures at the time of sporozoite inoculation, and medium was changed daily until 48 h. On the last day of incubation, the cultures were fixed with cold methanol and parasite-specific stained by a mouse polyclonal serum raised against the *Plasmodium falciparum* heat shock protein 70, which also cross-reacts with the hsp70 of *Plasmodium yoelii*, and parasites revealed with FITC-conjugated goat anti-mouse immunoglobulin were quantified by immunofluorescence analysis. Under a fluorescence microscope with a ×25 light microscope, objective parasite numbers were counted. IC_{50} values of test compounds were calculated by linear regression using Excel software and derived from three independent experiments, where each concentration was assayed in triplicate.

Primary mouse hepatocytes were seeded in 96-well plates coated with rat tail collagen I at a density of 5×10^4 cells per well. Before inoculation of *Plasmodium yoelii* sporozoites (10^5 per well), mouse hepatocytes were cultured for 24 h at 37°C in 5% CO_2 in WME supplemented with 10% fetal calf serum, 1% L-Gln, 1% sodium pyruvate, 1% insulin-transferrin-selenium, 1% non-essential amino acids, 1% penicillin-streptomycin (Invitrogen, Cergy-Pontoise, France), test compound was added to each well, plates were incubated for 48 h, 50 μL of Neutral red in PBS (0.02%) was added per well, incubated for an additional 24 h, washed with PBS, the Neutral red was extracted with 1% SDS from viable cells and quantified by measuring OD_{540} on a microplate autoreader photometer. TC_{50} values were calculated by linear regression using Excel software and derived from three independent assays, where each concentration was tested in quadruplicate.

20.11 *IN VIVO* ANTIMALARIAL ASSAY[17]

A chloroquine-sensitive *Plasmodium berghei berghei* was used to assess the *in vivo* intrinsic antimalarial activity. The parasite strain was maintained live by continuous reinfestation in NMRI mice. Male NMRI mice (28–32 g, 8 weeks old) were maintained in an air-conditioned room, fed standard mouse cubes and clean drinking water *ad libitum*, caged in groups of five, and housed in the animal house. Parasitized erythrocytes from a donor infected mouse by cardiac puncture in heparin were diluted with sterile blood from same age male mice. Mice were inoculated intraperitoneally with infected blood suspension (0.2 mL) containing 10^6 parasitized erythrocytes lethal inoculum on day zero (D0), and infected mice with 5% to 7% parasitemia were allocated to several groups of five mice each.

The blood schizonticidal activity of test compound doses was assessed by the classic 4-day suppressive assay, with chloroquine diphosphate salt (Sigma, Steinheim, Germany) as standard drug and normal saline as control. According to their relative toxicity, different doses of each test compound were screened. On D0 *Plasmodium berghei*–infected mice, test compound aqueous solutions were administered at 12-h intervals for each dose from D0 to the fourth day (D3). On day 4, tail blood smear were taken, stained with Grunwald-Giemsa, parasitized red blood cells were counted, and parasitemia was determined on at least 9000 red blood cells. Percentage growth inhibition was defined as Growth inhibition (%) = [(parasitemia in control − parasitemia with test compound)/parasitemia in control] × 100. Qualitative parasitemia was used to find the degree of activity at the screening doses. All doses were assayed in triplicate by using three sets of mice for each group. Statistical analyses of the data were done using one-way ANOVA, and the differences between treated group and control were identified with two-way ANOVA test. The data were considered significant at $p < 0.05$.

20.12 β-HEMATIN INHIBITION ASSAY[18]

For x-ray powder diffraction determination, 180 mg of hemin was used, and volumes were proportionately scaled up. At 60°C for 60 min, 4.5 M β-hematin acetate was from the solution of hemin in 0.1 M NaOH prepared. To 4.5 M β-hematin acetate, 0.1 M NaOH was added, the mixture was neutralized with one equivalent HCl, and the formed hematin precipitates were immediately filtered. Infrared spectra were recorded using KBr disks between 2000 and 1000 cm^{-1} using a Perkin-Elmer PARAGON 1000 FT-IR spectrometer. Using a Phillips PW 1050/80 vertical goniometer with Cu Kα radiation ($\lambda = 1.541$ Å) and in 5–40° of 2θ the powder diffraction pattern of β-hematin was collected.

Serial dilutions of test compound solutions were performed in triplicate in a 96-well plate using a multichannel pipette. In the final mixture, concentrations of test compound solution corresponded to 0–10 equivalents relative to hematin, namely each well contained 10.12 μL of test compound solution. To each well, 101.2 μL of hematin stock solution (1.680 mM in 0.1 M NaOH) was added. After mixing the solutions, 58.7 μL of acetate solution (12.9 M, pH 5.0) was preincubated at 60°C, added to the plate, at 60°C incubated for 60 min, at room temperature 80 μL of 30% (v/v) pyridine solution in 20 mM HEPES (pH 7.5) was added, solids were resuspended and allowed at ambient temperature to settle for 15 min. Then 38 μL of supernatant was transferred to another plate and with 30% (v/v) pyridine solution (pH 7.5, 20 mM HEPES) diluted to 250 μL. Data were collected using a microplate reader and analyzed.

20.13 NON-RADIOLABELED FERRIPROTOPORPHYRIN IX BIOMINERALIZATION INHIBITION ASSAY[19]

The procedure for ferriprotoporphyrin IX (FP) biomineralization assay consisted of incubating at 37°C in a normal nonsterile flat-bottom 96-well plate a mixture of 50 μL of test compound solution or 50 μL of solvent (for control), 50 μL of 0.5 mg/mL hemin chloride freshly dissolved in DMSO and 100 μL of 0.5 M sodium acetate buffer (pH 4.4) for 18–24 h, and the final pH of this mixture was 5–5.2, for which the addition order, namely first the hemin chloride solution, second the buffer, and finally the solvent or the solution of test compounds, was adhered to. The samples were incubated and with the help of a multichannel pipette (sucking up and forcing back 10 times to ensure an adequate transfer of all material) transferred into a 96-well Multi-Screen plate size. Under vacuum, the solvent was through the filter aspirated, and the wells were washed once with 200 μL of DMSO per well. β-Hematin adhering to the filters was dissolved in 200 μL of NaOH (0.1 N) per well at the same time the vacuum was turned off to avoid leaking the solubilized FP through the fiber filters. Finally from each well, 150 μL of solution was transferred into a parallel well in a flat-bottom 96-well plate and the absorbance was read at 405 nm with a micro-ELISA reader. The data were expressed as the percentage of inhibition of FP biomineralization and calculated by the equation $\%\ \text{inhibition} = 100 \times [(OD_{control} - OD_{test\ compound})/OD_{control}]$.

In FP biomineralization inhibition assay, DMSO, water, methanol, ethanol/water (70/30), and Triton X-100 were used as controls. Chloroquine phosphate (serial dilution in water from 1 mg/mL stock solution to 0.001 mg/mL), quinine (1 mg/mL to 0.125 mg/mL in ethanol/water 70/30, Sigma-Aldrich, St. Louis, MO, USA), methylene blue, phenosafranin, thionine (idem chloroquine, Sigma-Aldrich, St. Louis, MO, USA), Na_2HPO_4, NaH_2PO_4, Na_2SO_4, NaCl and $CaCl_2$ (10–0.312 mg/mL in water) were assayed as the test compounds.

FP biomineralization inhibition was also assayed in normal nonsterile flat-bottom 96-well plates by incubating the mixture in the same conditions and centrifuging the plate at $1600 \times g$ for 5 min. By vigorously flipping the plate upside down twice, the supernatant was discarded. To remove unreacted FP, the remaining pellet was resuspended in 200 μL of DMSO, the plate was centrifuged once again, and the supernatant discarded similarly. The pellet consisted of precipitate of β-hematin and was dissolved in 150 μL of NaOH (0.1M) for direct *in situ* spectroscopic quantification.

20.14 SURVIVAL OF *Anopheles gambiae* ASSAY[20,21]

Adults of laboratory-reared *Anopheles gambiae* mosquitoes (from a colony established from wild gravid females, all mosquito life stages were maintained under semi-field conditions) were maintained on a 6% glucose solution on filter-paper wicks, water, and routine human blood meals three times per week. Three days later, Oviposition cups were placed inside the cages 3 days after each blood meal, which were kept in screen houses under ambient conditions, to collect eggs of the following day. In semi-field conditions, the larvae for both the colony and the experiments were maintained on fish food, and the plastic pans ($25 \times 20 \times 14$ cm) filled with fresh water.

For direct observation, approximately 100 pupae/cage and the test compound were placed in 30-cm cubic metal frame cages covered with mesh netting. *Anopheles gambiae* mosquitoes were allowed to emerge, and evening observations of test compound feeding were conducted from 20:00 to 22:00 h, for which at intervals of 30 min a flashlight was used. On a second night of observation, a mosquito net was suspended and placed over *Ricinus communis* in a screenhouse. A cup of water containing approximately 500 pupae was placed under the net, and emerging adults allowed to feed on the test compound for 24 h. On the following night, mosquitoes were observed from 20:00 to 22:00 h in intervals of 30 min until they were observed probing or feeding on the test compound.

For survival assay, mosquitoes were divided into regime groups, and approximately 100 pupae per regime were harvested from the main mosquito colony, transferred to plastic cups, and placed in cages and held in a screenhouse under ambient conditions. Each cage of mosquitoes had access to test compound sources, one group of mosquitoes had access to a cotton pad moistened with distilled water, representing negative controls, and another group of mosquitoes with neither test compound source nor water was used as an additional negative control. Dead mosquitoes were removed at 4-h intervals from 07:00 to 23:00 h until all died.

20.15 CHLOROQUINE ASSAY[22,23]

Blood samples (10 mL) from healthy volunteers who were on chloroquine (2 × 150 mg base weekly, Fisons, Australia) and doxycycline (50 mg or 100 mg daily, Doryx, Faulding, Australia) malaria prophylaxis were centrifuged at 1200 × *g* for 10 min and the separated plasma was stored at −20°C and shipped on dry ice. The chloroquine-sensitive FC27 strain of *Plasmodium falciparum* was used for the determination of antimalarial activity of each plasma sample. Plasma samples were diluted twofold on microculture plates in drug-free serum and then inoculated with a suspension of parasitized erythrocytes in culture medium. The drug susceptibility and minimum concentration of spiked drug that inhibited parasites, relative to control, from developing to schizonts was recorded. The assumption was made that the maximum inhibitory dilution (MID) of the volunteers' samples contained a chloroquine concentration equivalent to that the minimum inhibitory concentration (MIC) observed in the sensitivity test. The chloroquine equivalent concentration in the prediluted volunteers' specimen was estimated by multiplying the MID by the MIC.

20.16 MS ASSAY FOR QUANTIFICATION OF CHLOROQUINE IN DOG PLASMA[24]

1. *Instrumentation*: Perkin-Elmer 200 Series (Perkin Elemer Instruments LLC, Shelton, CT, USA) HPLC was equipped with a quarternary pump, a degasser, and autosampler with a thermostatted column compartment. On a Chromolith SpeedROD RP-18e reversed phase chromatographic column (50 mm × 4.6 mm i.d.), test compounds were analyzed using a mobile phase containing a mixture of 10 mM ammonium acetate buffer/methanol (25/75, v/v, pH adjusted to 4.6 with formic acid), which was pumped at 0.8 mL/min flow rate. Mass detection was performed on an API 4000 triple quadrupole instrument (Applied Biosystems MDS SCIEX, Toronto, Canada) using MRM, and a turbo electrospray interface in positive ionization mode was used.

2. *Sample Preparation*: With addition of 300 μL of working solution of internal standard (prepared in methanol), 50 μL aliquots of plasma were subjected to protein precipitation. After a 20 s vortex, the samples were at 12,000 rpm centrifuged for 10 min, 200 μL of the upper layer was transferred into autosampler vials, and a 10 μL aliquot was injected into the chromatographic system.

3. *Bioanalytical Method Validation*: For assaying dog plasma, a full validation according to U.S. FDA guidelines was followed. For assaying human plasma, partial validation evaluating precision and accuracy, specificity and recovery was followed. Standard stock solutions of chloroquine (1 mg/mL) and the internal standard (IS, 1 mg/mL) were in 10-mL volumetric flasks with diluent (1% formic acid/methanol, 50/50, v/v) prepared separately. From the stock solution, by adequate dilution using diluent (water/methanol, 50/50, v/v), working solutions for calibration and quality controls were prepared. By diluting its stock solution with methanol, IS working solution (50 ng/mL) was prepared. To obtain chloroquine concentration levels of

2.0, 4.9, 9.8, 19.6, 48.9, 97.8, 244.6, and 489.1 ng/mL, 50 μL of working solutions were added to 950 μL of drug-free plasma. Based on an independent weighing of standard drug at concentrations of 2.0 ng/mL (LLOQ), 5.8 ng/mL (low), 194.1 ng/mL (medium), and 388.2 ng/mL (high) as a single batch at each concentration, bulk QC samples were prepared. These samples were divided into aliquots in microcentrifuge tubes (Tarson, 1.5 mL) and stored in the freezer below −50°C until analyzed. A calibration curve consisted of a blank dog plasma sample (plasma without the IS), a zero sample (plasma with the IS), and eight nonzero samples covering the concentration range 2.0–489.1 ng/mL including LLOQ. Using the analyte to IS peak area ratios by weighted ($1/x^2$) least squares linear regression on 5 consecutive days, the calibration curves were generated. The acceptance criteria for calibration curve standards and quality control samples were as per U.S. FDA guidelines.

By comparing the peak area of chloroquine in spiked plasma samples (five low, medium, and high quality controls) with the peak area of chloroquine in aqueous samples having the same amounts of chloroquine, which was prepared at the step immediately prior to chromatography, the recovery of chloroquine from the extraction procedure was determined. Similarly, by comparing the mean peak areas of extracted quality control samples ($n = 15$) with mean peak areas of IS in aqueous samples having the same amounts of IS, which was prepared at the step immediately prior to chromatography, the recovery of IS was determined.

The stability of chloroquine and IS stock solutions was assayed at room temperature and under refrigeration (ca. 4°C). Prior to analysis, the stock solutions of chloroquine and IS were with diluent (water/methanol, 50/50, v/v) diluted to 50 ng/mL. After three freeze-thaw cycles, the stability of chloroquine in dog plasma was investigated. Over the concentration range equivalent to that used for the calculation of precision and accuracy, these samples were quantified against the freshly spiked calibration curve standards. Under processing (ambient temperatures) and storage conditions (−50°C), the stability of chloroquine in dog plasma was also evaluated. Finally, stability in the final extract for dog plasma was determined in the autosampler.

20.17 SENSITIVE FLUORESCENCE HPLC ASSAY FOR AQ-13[25]

1. *Instrumentation and Chromatographic Conditions*: A Waters fluorescence HPLC system (Waters, Milford, MA) consisted of a 600E solvent delivery pump connected to a refrigerated 717 autosampler, and UV and fluorescence detectors. The 486 UV detector was set at 340-nm wavelength, while the 474 fluorescence detector was set at 320-nm excitation wavelength and 380-nm emission wavelength. Using the Millenium32 software, Version 4.0, the detectors' responses were recorded. At 25°C, using an Xterra RP18 analytical column (5-μm particle size, 250 mm × 4.6 mm i.d.) and an Xterra RP18 guard column (3.9 mm × 4.6 mm i.d.) between the injector and the analytical column, chromatography was performed. At 25°C, using a mobile phase containing 60% borate buffer (20 mM, pH 9.0) and

40% acetonitrile in the isocratic mode at 1.0 mL/min flow rate, chromatographic separations were accomplished. To prepare the mobile phase, a 20 mM solution of boric acid was adjusted to pH 9.0 using NaOH (10 N). Using Oasis MCX 3.0 mL solid phase extraction cartridges with a vacuum manifold (Waters), solid phase extraction of blood samples was performed. After elution from the Oasis MCX cartridges, using a Thermo Savant System with an SPD Speed Vac, an RVT4104 refrigerated Vapor Trap, and an OFP-400 vacuum pump, the AQs and their metabolites were concentrated.

2. *Preparation of Stock Solutions, Standard Samples, and IS*: Stock solutions were prepared in 10 mL of MeOH at a concentration of 1.0 mM (10 μmol of AQ in 10 mL of MeOH) diluted with MeOH to 1 : 10 preparing 100 μM working solutions. Stock and working solutions were stored at −15°C for up to 7 days. Standard samples of AQs and their metabolites were prepared similarly in 1.0 mL of MeOH at 250, 500, 1000, and 2000 nM. A 10 μL aliquot of the 100 μM IS working solution was added to 990 μL of each standard sample for internal standardization, and the final concentration of the IS in all samples was 1.0 μM.

3. *Extraction of AQ-13, CQ, and Their Metabolites from Whole Blood*: From blood to extract AQ-13 (a candidate 4-aminoquinoline antimalarial), CQ (chloroquine, Sigma-Aldrich, St. Louis, MO), and their metabolites, 1.0 mL (990 μL) aliquots of blood in buffered sodium citrate (3.2% sodium citrate tubes with a 5 mL draw) were mixed with 10 μL of IS working solution (100 μM) and 9.0 mL of KH_2PO_4 (0.1 M) in 15-mL screw-crapped tubes producing 1.0 μM final concentrations of the IS. After 15 min shaking, the samples were at room temperature and 3000 × *g* centrifuged for 15 min, the supernatants (5.0 mL) were aspirated with a Pasteur pipette and loaded onto Oasis MCX 3.0 mL cartridges, which had been washed with 2.0 mL of methanol and 2.0 mL of water. The matrix of the Oasis MCX solid phase extraction cartridges was poly(divinylbenzene-*co*-*N*-polyvinyl-pyrrolidone) used for the HLB sorbent with a 30-μm particle size to which HSO_3 groups had been bonded to produce a cation exchange resin that facilitated the retention of basic compounds such as AQs (aminoquinolines). After loading with supernatant, the cartridges were washed with 0.1 N HCl (4 mL × 1 mL), eluted with 5% NH_4OH in MeOH (4 mL × 1 mL), using an SPD Speed Vac evaporated to dryness, the residues were dissolved in 200 μL of MeOH, vortex-mixed, and filtered with PVDF Durapore 0.22-μm filters, and 2 aliquots of each filtered sample (20 μL of apiece) were injected onto the column and run in duplicate.

20.18 HPLC AND HPTLC ASSAYS FOR CHLOROQUINE, PRIMAQUINE, AND BULAQUINE[26]

1. *Apparatus and Chromatographic Conditions*: The HPLC system consisted of 250 binary gradient pump (Perkin-Elmer), a 7125 injector with a 20 μL loop, and 235 diode array detector. HPLC separation was based on a RP select-B C8 lichrospher (Merck) analytical column (250 mm × 4 mm i.d., 5-μm particle size). Column effluent was at 265 nm monitored. Using an IndTech HPLC interface and software,

the data were acquired and processed. The mobile phase consisted of 0.01 M sodium acetate buffer (pH 5.6) and acetonitrile (45/55 v/v), both of which were filtered and degassed before use. Chromatography was performed at 27 $\pm$ 3°C and 1.0 mL/min flow rate. The HPTLC workstation included Automatic TLC Sampler-III equipped with a 50-μL Hamilton syringe, TLC Scanner-3 equipped with mercury, tungsten, and deuterium lamp for scanning of TLC plate. The separation was based on thin layer plates of precoated Si-gel 60 F_{254} (20 × 20 cm, 0.2 mm, Merck). The plates were scanned at 254 nm. Using a CATS4-HPTLC, software data were acquired and processed. TLC was performed at 27 $\pm$ 3°C in a Camag TLC twin trough glass chamber, and hexane/diethyl ether/methanol/diethylamine in the ratio of (37.5/37.5/25/0.5, v/v) was used as solvent system.

2. *Preparation of Stock and Working Standard Solutions and Sample Solutions*: By individually weighing 25 mg of compound and 50 mg of potassium carbonate into 1 mL of water, with methanol diluting to 25 mL, stock standard solutions of primaquine diphosphate and chloroquine phosphate were prepared. By dissolving 25 mg of compound into methanol containing 0.5% dimethyl octyl amine, the stock standard solution of bulaquine was prepared. By serial dilution containing bulaquine in the range 0.5–35 mg/mL, chloroquine (free base) in the range 1.5–50 mg/mL, and primaquine (free base) in the range 0.5–35 mg/mL, working standard solutions (calibration concentrations) containing three compounds were prepared in methanol.

Ten tablets of primaquine diphosphate or chloroquine phosphate were weighed and crushed evenly to powder. The powder equivalent to 5 mg of each of the compounds was taken, mixed with 25 mg of potassium carbonate in 1 mL of water, and with methanol diluted to10 mL. Capsule material equivalent to 5 mg of bulaquine was taken, extracted with 3 mL of methanol solution of 1% dimethyloctylamine three times and with the same solvent diluted to 10 mL. To remove any insoluble matter before analysis, all the sample solutions were filtered, of which 0.5 mL was with methanol further diluted to 25 mL. The compound solution in methanol was injected into the HPLC and as well spotted onto a TLC plate and analyzed for the compound content.

20.19 REAL-TIME PCR-BASED CHLOROQUINE SENSITIVITY ASSAY[27]

Plasmodium falciparum multi-drug resistance (*pfmdr1*) and *Plasmodium falciparum* chloroquine-resistance transporter (*pfcrt*) mutation assay: Mutation assay was carried out with LightCycler using hybridization probes. For *pfmdr1* mutations, hybridization probes consisting of two oligonucleotides bind to an internal sequence amplified by forward and reverse primers. Sensor probe labeled at the 3′ end with fluorescein was designed to the mutation sites. Anchor probe labeling at the 5′ end with LightCycler Red 640 and phosphorylating at the 3′ end to prevent extension by Taq polymerase was designed for conserved sequences adjacent to the mutation sites. Both probes localized on the same DNA strand could hybridize in a head to tail arrangement and thus bring the two fluorescent dyes into close proximity. During fluorescence energy resonance transfer (FRET), fluorescein was excited by the LightCycler instrument's light, and the excitation energy was transferred to LightCycler red 640. During the melting phase (period added after the PCR), the emitted fluorescence was measured by the photohybrids. With the temperature slowly increasing the fluorescence was decreased, while one of the probes melted off moving away the two fluorescent dyes. A specific melting temperature was obtained for each genotype: A sensor probe spanning one mismatch could still hybridize with the target sequence but at lower temperature melted off than a sensor probe with a perfect match. For *pfcrt* mutations, an amplification primer iLC labeling with LightCycler red 640 on the third base from the 3′ end was used. During amplification, iLC extension occurred allowing hybridization with 3′-FITC labeled sensor probe. FRET occurred between the sensor probe and the whole PCR product working as anchor probe. To detect five mutations of *pfmdr1* gene and three mutations of *pfcrt* gene, primers and probes (for *pfmdr1* gene: 5′-TGTATTATCAGGAGGAACATTACC-3′, 5′-TCTTTAATATTACACCAAACACAGATAT-3′, 5′-ATTAATATCATCACCTAAATACATG-3′, 5′-TAAAAAATGCACGTTTGACTTTATGTATTA-3′, 5′-CCTTTTTAGGTTTATTTATTTGGTCAT-3′, 5′-ACCACAAACATAAATTAACGGA-3′, 5′-TTCTGTAATTTGATAGAAAAAGCTA3′, 5′-CCCCATAAAGCTGCATTTACAATAATTCTTCT-3′, 5′-AACTATCAATAAATAATTGAGCGCTTTGACAGAA-3′, 5′-TTTTCAATAGTTAGTCAAGAACCCATGT-3′, 5′-TGTGATTATAACTTAAGATATCTTAGAAAC-3′, and 5′-CCAAATTTGATATTTTCATATATGGAC-3′; for *pfcrt* gene: 5′-CTTGTCTTGGTAAATGTGCTCA-3′, 5′-TGTGTAATTGAAACAATTTTTGCTAA-3′, and 5′-GTTACCAATTTTGTTTAAAGTTCT-3′) were designed and synthesized by TIBR MOLBIOL.

PCR amplification: 2 μL of LightCycler DNA Master Hybridization Probes buffer, $MgCl_2$ (3 mM for pfmdr mutations and 4 mM for *pfcrt* mutations), 5 μL DNA template, primers and probes were by sterile water completed to 20 μL. The final concentrations of primers F1/R1 and F2/R2 each were 0.5 and 0.6 μM, respectively; both F/iLC were 0.5 μM. Final concentration of probes was 0.2 μM. For *pfmdr1* codons 86 and 184, the PCR program was 9 min at 95°C, 35 cycles of 95°C for 10 s, 54°C for 10 s, 65°C for 40 s and followed by a melting curve analysis from 35°C to 75°C at 0.2°C/s. For *pfmdr1* codons 1034–1042 and 1246, the conditions for cycling were 9 min at 95°C, 40 cycles of 95°C for 10 s, 54°C for 10 s, 65°C for 55 s and followed by a melting curve analysis from 35°C to 80°C at 0.2 °C/s. For *pfcrt* codons, the PCR program was 10 min at 95°C, 40 cycles of 95°C for 10 s, 45°C for 10 s, and 65°C for 15 s. Temperature transition rates were 20°C/s for denaturation and annealing and 5°C/s for extension. Melting curve program included one cycle of 95°C for 2 s, 32°C for 20 s, and heating at 75°C. Temperature change rates were 20°C/s except for the final step, which had a 0.2°C/s temperature transition rate.

20.20 RAT EMBRYOS *IN VITRO* ASSAY[28]

CD(SD)BR male and primiparous female rats (11–12 weeks old) were maintained under standard conditions (21.5 ± 1.5°C, 55 ± 5% humidity, and 6:00 a.m. to

6:00 p.m. artificial lights) with food and tap water *ad libitum*. Female rats were with male breeder rats mated overnight and the paired rats were left in cohabitation from 4:00 p.m. to 9:00 a.m the next day. By vaginal smear, copulation was ascertained. The day that spermatozoa were found in the vaginal smear was defined as day 0 of pregnancy.

In the afternoon of gestation day 9, pregnant female rats were randomly distributed to experimental groups, euthanized to extract embryos (9.5 days of age, 1–3 somites), and cultured in a 25-mL glass bottle (5 embryos/bottle) containing 5 mL of heat-inactivated sterile rat serum and antibiotics (100 IU/mL penicillin and 100 μg/mL streptomycin). At the start of the culture, bottles were with a mixture of 5% O_2, 5% CO_2, and 90% N_2 flushed for 1 min and placed in a thermostatic (37.8 ± 0.1°C) roller (ca. 20 rpm) apparatus. In the next morning and afternoon, the bottles were with a mixture of 10% O_2, 5%CO_2, and 85% N_2 and a mixture of 20% O_2, 5% CO_2, and 75% N_2 re-gassed, respectively. The latter gas mixture was also used for re-gassing the culture in the morning of the third day. The total culture time was 48 h. To obtain a range of final concentrations of 0.01–2 μg/mL, dihydroartemisinin (DHA, Holleykin, Guanzhou, China) was dissolved in DMSO and added to the culture medium (5 μL/bottle). Group 1 (control) DMSO 0.1% from gestation days (GD) 9.5 to 11.5; group 2 DHA 0.01, 0.05, 0.1, 0.5, 1, and 2 μg/mL from GD 9.5 to 11.5; group 3 DHA 0.05, 0.1, 0.5, 1, and 2 μg/mL at GD 9.5 for 1.5 h, and then the medium was changed and embryos were cultured in normal serum until GD 11.5; group 4 DHA 1 and 2 μg/mL at GD 11 1.5 h before the end of the culture. At the culture end, the embryos (11.5 days old) were transferred into Tyrode's salt solution and under the stereomicroscope examined, heart beat and yolk sac circulation were observed; embryos without heart beat were considered dead. Yolk sac diameter, crown rump length, and head length were measured, the somites were counted, and live embryos were scored.

For each embryo, observed morphologic abnormalities were divided into major (marked deviations from the species' normal morphogenetic/organogenetic process) and minor (minor divergences from the normal morphogenetic/organogenetic process) abnormalities. Embryos showing abnormalities were divided into embryos with major abnormalities (with or without minor abnormalities) and embryos with minor abnormalities (never associated with major abnormalities). After measuring yolk sac diameters, the first three of five embryos in each bottle were used to collect embryonic blood cells. Embryos with intact yolk sac were immersed in PBS (pH 7.4). Yolk sac vessels were cut at the vitelline artery and vein level allowing blood cells to pool in the PBS. By centrifugation at 700 rpm for 3 min, collected cells were concentrated and using a Cytospin 3 (Shandon, China) transferred onto slides. The slides were in methanol immediately fixed and stained with Wright-Giemsa method. Under a light microscope, embryonic red blood cells were analyzed to detect possible alterations. After blood collection, embryos were scored. After scoring, the remaining two embryos of each bottle were in 4% formaldehyde fixed overnight, embedded in paraffin, sectioned, stained with hematoxylin and eosin, and under a light microscope histologically examined to detect possible tissue alterations.

20.21 HPLC-MS ASSAY FOR β-DHA IN RAT PLASMA[29,30]

1. *Instrumentation*: HPLC based on an Agilent 1100 system (Palo Alto, CA, USA) equipped with a G1313A autosampler, a vacuum de-gasser unit, and a G1312A binary pump was via a TurboIonSpray ionization (ESI) interface for mass analysis and detection coupled to an API 4000 triple-quadrupole mass spectrometer (Applied Biosystems/MDS Sciex, Concord, ON, Canada). A 10-port switching valve directed HPLC eluate to waste in the first 2.5 min of the chromatographic run and afterwards to the ionization source. With the Analyst 1.3 data acquisition and processing software (Applied Biosystems/MDS Sciex), data were collected and analyzed.

2. *Chromatographic and Mass Spectrometric Conditions*: Chromatographic separation was achieved on a Luna ODS C18 column (150 mm × 4.6 mm i.d., 5 μm, Phenomenex, Torrance, CA, USA) and a Security Guard C18 guard column (4.0 mm × 3.0 mm i.d., 5 μm, Phenomenex, Torrance, CA, USA). The chromatography was performed at 20°C, with acetonitrile/10 mM aqueous ammonium acetate containing 0.1% (v/v) formic acid (85/15, v/v) as mobile phase and 0.8 mL/min flow rate. Mass spectrometer was operated in positive ion mode. By infusing an acetonitrile solution containing 100 ng/mL analytes at 30 μL/min flow rate into the mobile phase (0.8 mL/min) using a T-connection, the tuning parameters were optimized for β-dihydroartemisinin (β-DHA) and artemisinin (IS). Following optimization of the settings, the instrument was operated with a 5.5 kV ion spray voltage, 3 psi collision gas, 12 psi curtain gas, 50 psi nebulizer gas, and 50 psi heater gas, and the heater gas temperature was set at 550°C. Nebulizer, heater, curtain, and collision activated dissociation gas was ultrapure nitrogen. For β-DHA and IS, the fragmentation transitions for the multiple reaction monitoring were m/z 267.4–163.4 and m/z 300.4–209.4, respectively, with a 200 ms dwell time per transition.

3. *Preparation of Calibration Standards and Quality Control Samples*: By dissolving the accurately weighed reference (β-DHA or IS) in acetonitrile, the stock solution of β-DHA or IS was obtained. β-DHA stock solution (400 μg/mL) was with acetonitrile diluted to 0.2, 0.5, 2, 5, 20, 50, and 100 ng/mL working solutions. Three stock solutions of β-DHA for quality control-method validation were prepared from a separately weighed β-DHA. Via dilution, 0.5, 5, and 80 ng/mL QC working solutions were prepared. By diluting 100 μg/mL IS stock solution with acetonitrile, 200 ng/mL IS solution was prepared. All stock and working solutions were stored at 4°C for use. Calibration standards and quality control samples ranging from 0.2 to 100 ng/mL were prepared for calibration, accuracy and precision, quality control, and stability assessment. By spiking 100 μL of working solutions and 20 μL of IS (200 ng/mL) with 100 μL of drug-free plasma and 200 μL of phosphate buffer (pH 7.4), calibration standards and QC samples were prepared. By shaking with 3 mL of ether for 10 min, this mixture was extracted. By centrifugation at 3000 × *g* for 10 min, the upper ether and lower water were separated. The ether was transferred to another tube and at 40°C under a gentle stream of nitrogen evaporated to dryness. The residue was dissolved in 100 μL of mobile phase, vortex-mixed for 1 min, and a 10 μL aliquot of the solution was injected into the LC-MS/MS system for analysis.

Matrix matched calibration standards were obtained with 0.2, 0.5, 2, 5, 20, 50, and 100 ng/mL β-DHA in plasma. QC samples were obtained with 0.5, 5, and 80 ng/mL β-DHA in plasma.

4. *Sample Preparation*: To 100 μL of rat plasma, 20 μL of IS (200 ng/mL), 100 μL of acetonitrile, and 200 μL of phosphate buffer (pH 7.4) were added, mixed, the mixture was extracted with 3 mL of ether, and treated. Plasma samples were diluted with blank plasma and re-analyzed when the concentration of β-DHA was higher than the upper limit of quantification (ULOQ, 100 ng/mL).

5. *Application to Pharmacokinetic Assay*: Male Wistar rats (230–250 g) were maintained at 22 ± 2°C and 55 ± 5% relative humidity on a 12-h light/12-h dark cycle for at least 5 days before being used. Before drug administration, rats were fasted for 12 h and further for 3 h after dosing. During experiments, water was freely available for rats. β-DHA in sesame oil was given orally to rats ($n = 6$, 10 mg/kg), 250 μL of blood samples were withdrawn from one jugular vein before dosing and at 0.25, 0.5, 0.75, 1.0, 1.25, 1.5, 1.75, 2.0, 2.5, 3.0, 4.0, and 5.0 h postdosing. After each sampling, 250 μL of physiologic saline was replenished to the blood from another jugular vein. All heparinized blood samples were at 3000 × *g* centrifuged for 15 min to obtain plasma which were at −20°C stored and assayed within 1 month.

20.22 *Plasmodium falciparum* CLONE AND DHA ASSAY[2]

By centrifugation for 5 s, 50 μL of Affigel protein A (binding capacity of 20 mg of purified human immunoglobulin G[IgG]/mg of gel) was washed with PBS (pH 7.2, 50 μL × 2). The RPMI 1640 medium (50 μL) supplemented with 5.94 mg/mL HEPES and 2.1 mg/mL sodium bicarbonate was used in a third washing step, and after centrifugation the supernatant was discarded, and Affigel protein A gel was incubated with the plasma and serum (250 μL) spiked with DHA or plasma from patients at room temperature for 30 min. The stock solution of DHA in 70% ethanol (1 mg/mL) was kept at −10°C for up to 1 month before use, which was further diluted in fresh plasma or serum to give a working concentration range of 1.25 to 100 ng/mL.

Affigel protein A-treated plasma (100 μL) and heat-inactivated serum (50 μL/well) were transferred to row A and B of a flat-bottom plate, respectively. Using the heat-inactivated serum added through G into row B as the diluent for plasma or serum containing DHA, twofold serial dilutions were prepared. Fifty microliters of heat-inactivated control plasma or serum was added to row H for parasitizing and nonparasitizing erythrocyte controls. To all wells for rows A to G and eight wells for row H, the suspension of 175 μL of malarial parasite infected erythrocytes (W2 clone; 0.5% parasitemia with 80% young rings at a 1.7% hematocrit) was added. To the remaining four wells of row H, for nonparasitized controls the similar suspension of uninfected erythrocytes was added. The microtiter plates were incubated at 37°C for 24 h, and pulsed with 1 Ci/mmol [^{3}H]hypoxanthine by adding 25 μL of 0.5 μCi isotope solution to each well. The microtiter plates were incubated for another 18 to 20 h. The contents of the plates were harvested, and the particulate material of the water lysed cell suspension was filtered through glass fiber filter paper. The filter paper was dried in an oven and placed in a plastic bag, and 10 mL of scintillation fluid was added and the plastic bag was sealed. The level of incorporation of [^{3}H]hypoxanthine by the malarial parasites in each well was determined by counting in a liquid scintillation counter.

Fresh whole blood was collected and placed in heparinized tubes, the plasma and buffy coat were removed, the erythrocytes were lysed and serially diluted (2- to 64-fold) with sterile water, of which 50 μL aliquots were diluted in 250 μL of plasma or medium spiked with DHA to give 50 ng/mL final concentration and a range of hemoglobin concentrations from 1.5 to 0.05 g%. To determine the possible interference of hemolysate in the detection of DHA activity by the assay, the effect of hemolysate in both plasma and culture medium (RPMI 1640 medium with 15% heat-inactivated human serum) was examined. The final concentration of plasma and serum in each well was 30% and 15%, respectively, which was similar to that in the normal drug susceptibility test. In addition to the testing of hemolysates of normal erythrocytes, plasma samples collected from two patients with severe falciparum malaria were spiked with 50 ng/mL DHA and subjected to the same assay. The plasma samples were obtained from both patients prior to treatment with any antimalarial drugs.

REFERENCES

1. P. Rasoanaivo, D. Ramanitrahasimbola, H. Rafatro, D. Rakotondramanana, B. Robijaona, A. Rakotozafy, S. Ratsimamanga-Urverg, M. Labaied, P. Grellier, L. Allorge, L. Mambu, F. Frappier. Screening extracts of Madagascan plants in search of antiplasmodial compounds. Phytother Res 18 (**2004**) 742–747.
2. T. Isavadharm, J.O. Peggins, T.G. Brewer, N.J. White, H.K. Webster, D.E. Kyle. Plasmodium falciparum-based bioassay for measurement of artemisinin derivatives in plasma or serum paktiya. Antimicrob Agents Chemother 48 (**2004**) 954–960.
3. D. Tasdemir, B. Topaloglu, R. Perozzo, R. Brun, R. O'Neill, N.M. Carballeira, X. Zhang, P.J. Tonge, A. Linden, P. Rüedi. Marine natural products from the Turkish sponge agelas oroides that inhibit the enoyl reductases from plasmodium falciparum, mycobacterium tuberculosis and Escherichia coli. Bioorg Med Chem 15 (**2007**) 6834–6845.
4. S.E. Lee, M.R. Kim, J.H. Kim, G.R. Takeoka, T.W. Kim, B.S. Park. Antimalarial activity of anthothecol derived from Khaya anthotheca (Meliaceae). Phytomedicine 15 (**2008**) 533–535.
5. S.K. Yanow, L.A. Purcell, T.W. Spithill. The A/T-specific DNA alkylating agent adozelesin inhibits Plasmodium falciparum growth in vitro and protects mice against Plasmodium chabaudi adami infection. Mol Biochem Parasitol 148 (**2006**) 52–59.
6. J.B. Sarr, S. Pelleau, C. Toly, J. Guitard, L. Konaté, P. Deloron, A. Garcia, F. Migot-Nabias. Impact of red blood cell polymorphisms on the antibody response to Plasmodium falciparum in Senegal. Microbes Infect 8 (**2006**) 1260–1268.
7. S. Van Miert, S. Van Dyck, T.J. Schmidt, R. Brun, A. Vlietinck, G. Lemiére, L. Pieters. Antileishmanial activity, cytotoxicity and QSAR analysis of synthetic dihydrobenzofuran lignans and related benzofurans. Bioorg Med Chem 13 (**2005**) 661–669.

8. K.L. Chan, C.Y. Choo, N.R. Abdullah, Z. Ismail. Antiplasmodial studies of Eurycoma longifolia Jack using the lactate dehydrogenase assay of Plasmodium falciparum. J Ethnopharmacol 92 (**2004**) 223–227.
9. H. Noedl, P. Teja-Isavadharm, R.S. Miller. Nonisotopic, semiautomated plasmodium falciparum bioassay for measurement of antimalarial drug levels in serum or plasma. Antimicrob Agents Chemother 48 (**2004**) 4485–4487.
10. W. Trager, J.B. Jensen. Plasmodium falciparum in culture: use of outdated erythrocytes and description of the candle jar method. J Parasitol 63 (**1977**) 883–887.
11. H. Noedl, C. Wongsrichanalai, R.S. Miller, K.S.A. Myint, S. Looareesuwan, Y. Sukthana, V. Wongchotigul, H. Kollaritsch, G. Wiedermann, W.H. Wernsdorfer. Plasmodium falciparum: effect of anti-malarial drugs on the production and secretion characteristics of histidine-rich protein II. Exp Parasitol 102 (**2002**) 157–163.
12. R.E. Hayward, D.J. Sullivan, K.P. Day. Plasmodium falciparum: histidine-rich protein II Is expressed during gametocyte development. Exp Parasitol 96 (**2000**) 139–146.
13. S. Krief, M.T. Martin, P. Grellier, J. Kasenene, T. Sevenet. Novel antimalarial compounds isolated in a survey of self-medicative behavior of wild chimpanzees in Uganda. Antimicrob Agents Chemother 48 (**2004**) 3196–3199.
14. H. Noedl, W.H. Wernsdorfer, R.S. Miller, C. Wongsrichanalai. Histidine-rich protein II, a novel approach to antimalarial drug susceptibility testing. Antimicrob Agents Chemother 46 (**2002**) 1658–1664.
15. W. Trager, J.B. Jensen. Human malarial parasites in continuous culture. Science 193 (**1976**) 673–675.
16. M. Carraz, A. Jossang, P. Rasoanaivo, D. Mazier, F. Frappier. Isolation and antimalarial activity of new morphinan alkaloids on Plasmodium yoelii liver stage. Bioorg Med Chem 16 (**2008**) 6186–6192.
17. A. Hilou, O.G. Nacoulma, T.R. Guiguemde. In vivo antimalarial activities of extracts from Amaranthus spinosus L. and Boerhaavia erecta L. in mice. J Ethnopharmacol 103 (**2006**) 236–240.
18. K.K. Ncokazi, T.J. Egan. A colorimetric high-throughput β-hematin inhibition screening assay for use in the search for antimalarial compounds. Anal Biochem 338 (**2005**) 306–319.
19. E. Deharo, R.N. García, P. Oporto, A. Gimenez, M. Sauvain, V. Jullian, H. Ginsburg. A non-radiolabelled ferriprotoporphyrin IX biomineralisation inhibition test for the high throughput screening of antimalarial compounds. Exp Parasitol 100 (**2002**) 252–256.
20. D.E. Impoinvil, J.O. Kongere, W.A. Foster, B.N. Njiru, G.F. Killeen, J.I. Githure, J.C. Beier, A. Hassanali, B.G. Knols. Feeding and survival of the malaria vector Anopheles gambiae on plants growing in Kenya. Med Vet Entomol 18 (**2004**) 108–115.
21. R.E. Desjardin, C.J. Canfield, J.D. Haynes, J.D. Chulay. Quantitative assessment of anti-malarial activity in vitro by a semiautomated microdilution technique. Antimicrob Agents Chemother 16 (**1979**) 710–718.
22. B. Kotecka, M.D. Edstein, K.H. Rieckmann. Chloroquine bioassay of plasma specimens obtained from soldiers on choloroquine plus doxycycline for malaria prophylaxis. Int J Parasitol 26 (**1996**) 1325–1329.
23. B.G.J. Knols, B.N. Njiru, E.M. Mathenge, W.R. Mukabana, J.C. Beier, G.F. Killeen. MalariaSphere: a greenhouse enclosed simulation of a natural Anopheles gambiae (Diptera: Culicidae) ecosystem in western Kenya. Malaria J 1 (**2002**) 19.

24. P. Singhal, A. Gaur, V. Behl, A. Gautam, B. Varshney, J. Paliwal, V. Batra. Sensitive and rapid liquid chromatography/tandem mass spectrometric assay for the quantification of chloroquine in dog plasma. J Chromatogr B 852 (**2007**) 293–299.

25. H. Deng, H. Liu, F.M. Krogstad, D.J. Krogstad. Sensitive fluorescence HPLC assay for AQ-13, a candidate aminoquinoline antimalarial, that also detects chloroquine and *N*-dealkylated metabolites. J Chromatogr B 833 (**2006**) 122–128.

26. A.K. Dwivedi, D. Saxena, S. Singh. HPLC and HPTLC assays for the antimalarial agents chloroquine, primaquine and bulaquine. J Pharm Biomed Anal 33 (**2003**) 851–858.

27. F. de Monbrison, D. Raynaud, C. Latour-Fondanaiche, A. Staal, S. Favre, K. Kaiser, F. Peyron, S. Picot. Real-time PCR for chloroquine sensitivity assay and for pfmdr1-pfcrt single nucleotide polymorphisms in Plasmodium falciparum. J Microbiol Methods 54 (**2003**) 391–401.

28. M. Longo, S. Zanoncelli, D. Manera, M. Brughera, P. Colombo, J. Lansen, G. Mazué, M. Gomes, W.R.J. Taylor, P. Olliaro. Effects of the antimalarial drug dihydroartemisinin (DHA) on rat embryos in vitro. Reprod Toxicol 21 (**2006**) 83–93.

29. R.S. Dungan, U. Kukier, B. Lee. Blending foundry sands with soil: effect on dehydrogenase activity. Sci Total Environ 357 (**2006**) 221–230.

30. J. Xing, H.-X. Yan, R.-L. Wang, L.-F. Zhang, S.-Q. Zhang. Liquid chromatography-tandem mass spectrometry assay for the quantitation of β-dihydroartemisinin in rat plasma. J Chromatogr B 852 (**2007**) 202–207.

21

METHODS AND APPLICATIONS OF CYTOGENETIC RECEPTOR AND ENZYME ASSAYS

Shiqi Peng

In past decades, the spread and subsequent effects of ecotoxic chemicals in the environment, either from anthropogenic sources or from natural sources, has been a major concern. In some cases, the contamination exhibits toxicity to animals and humans via modifying the cytogenetic receptor and enzyme system. To monitor the contamination of different chemicals and their mixtures as well as that of ionizing and non-ionizing radiation, reliable and easy-to-use detector systems are particularly needed in environmental care. For biosensor designs, especially when responses to DNA-effecting agents are considered, the enormous diversity of microbes with respect to their genetic responses to environmental changes is an attractive resource. Because DNA-damaging agents induce a variety of specific lesions having different implications in cells (for instance the repaired lesion has no further consequences for the cell, or the unrepaired lesion leads to cell death, or the lesion induces error-prone repair pathways leading to mutagenesis and cancer induction in higher organisms), apart from assays for enzyme detection, a number of specifically cytogenetic receptor assays have been established. In this chapter, 18 assays are described: UT-7/EPO cell (an immortalized cell line, obtained from the New York Blood Center, New York, NY) proliferation and neutralizing anti-erythropoietin (anti-EPO) antibody assays,[1] assay system for biotin protein ligase (BPL) from *Escherichia coli*,[2] β-galactosidase

Pharmaceutical Bioassays: Methods and Applications. By Shiqi Peng and Ming Zhao

and α-amylase biosynthesis assays,[3,4] penicillin-binding protein (PBP 5) and vancomycin activity assays,[5] recombinant bacteria assay evaluating androgen biosynthesis,[6] UDP-Glc (uridine diphosphate-glucose) 4-epimerase assay,[7] ATPs exchange assays,[8] lux-fluoro assay for combined genotoxicity and cytotoxicity,[9] reporter gene assay for aryl hydrocarbon receptor,[10] CYP1A enzyme assay,[11] CYP1A activity (EROD) assay,[12] deubiquitinating enzyme (DUB) activity assay,[13] flow cytometric assay,[14] serum neutralization assay based on rHMPV-GFP,[15] yellow fluorescent protein-based assay,[16] red fluorescent cytotoxic T lymphocyte assay,[17] green fluorescent protein-based assay,[18,19] and DNA break assay in HepG2 cells.[20]

21.1 UT-7/EPO CELL PROLIFERATION AND NEUTRALIZING ANTI-EPO ANTIBODY ASSAYS[1]

After at least three passages, UT-7/EPO cells starved for recombinant human EPO (rHuEPO) were stored in liquid nitrogen, thawed in vials containing approximately 1.0×10^7 cells in 10% dimethyl sulfoxide (DMSO) and 90% fetal bovine serum (FBS), incubated in a basal medium containing Iscove's modified DMEM, FBS (HyClone, Logan, UT), penicillin-streptomycin sulfate (50 U/mL penicillin G and 50 μg/mL streptomycin sulfate in 0.85% saline, GIBCOTM BRL, Invitrogen Corp., Carlsbad, CA), L-Gln 2 mM stock (29.2 mg/mL in 8.5 mg/mL NaCl, Mediatech Inc., Herndon, VA), and rHuEPO of 1 U/mL final concentration at 37°C in a 5% CO_2 atmosphere for 16–28 h, washed in medium containing Iscove's modified Dulbecco's medium plus 10% FBS, and resuspended at a density of 8×10^5 cells/mL in basal medium. Suspensions containing ≥80% viable cells were used.

1. *Proliferation Assay*: To wells of a round-bottom 96-well tissue culture plate, 24 μL rHuEPO standards at 2.5, 5, 8, 10, 12, 20 and 50 mU/mL final concentrations or 20 μL of the quality controls (QCs) at 5 and 10 mU/mL final concentrations (all prepared from the 10 U/mL rHuEPO reference standard), 50 μL of UT-7/EPO cell suspension (4×10^4 cells/well) in 130 μL of basal medium culture, and 50 μL of [^{3}H]thymidine at 1 μCi/mL final concentration were added in triplicate, the 96-well plate was at 37°C in a 5% CO_2 atmosphere incubated for 6 h, and using a Filtermate cell harvester onto a UniFilter microfilter plate with GF/C filters (UniFilter-96 GF/C white 96-well Barex microplates with 1.2 μm pore size, PerkinElmer Life and Analytical Sciences, Inc., Boston, MA) UT-7/EPO cells were harvested. The plate was at room temperature dried overnight, BackSeal bottom-sealing tape (PerkinElmer Life and Analytical Sciences, Inc., Boston, MA) was applied to the bottom of the plate, of which each well was filled with 50 μL of MicroScint 20 liquid scintillant solution (PerkinElmer Life and Analytical Sciences, Inc., Boston, MA), the top of the plate was sealed with TopSeal-A self-adhesive sealing film (PerkinElmer Life and Analytical Sciences, Inc., Boston, MA), radioactivity in each well was counted in a TopCount scintillation counter (1 min/well), and using a 4-parameter logistic model (SoftMax® Pro v. 4.0, Molecular Devices Corporation, Sunnyvale, CA), the results of rHuEPO standards were fit to a standard curve. When the mean back-calculated

concentrations of each standard and the QCs were within $\pm 25\%$ of their nominal concentrations and the percentage coefficient of variation (%CV) between replicates was $\leq 25\%$, the data were deemed acceptable. At least one set of duplicates of one QC was required to meet the criteria for acceptability. The stimulation of UT-7/EPO cell proliferation was defined as the fold stimulation by dividing the mean cpm in the test sample wells by the mean cpm in the control wells.

2. *Neutralization Assay*: To detect neutralizing anti-EPO antibodies, EPO-responsive UT-7/EPO cells were used. In assay, 20 μL of rHuEPO standards at 2.5–50 mU/mL final concentration or 20 μL of QCs at 5 and 10 mU/mL final concentration were added in triplicate to the wells of a round-bottom 96-well tissue culture plate, while all remaining wells of the plate received 20 μL of rHuEPO at 10 mU/mL final concentration. As the positive and negative controls, to the appropriate wells 20, 10, or 5 μL pool of sera positive and negative for anti-EPO antibodies were added, respectively, in the same volume as the test sample. Depending on EPO levels in the sample determined by ELISA and levels of anti-EPO antibodies determined by the RIP assay, test samples were diluted. For borderline positive samples for anti-EPO antibodies (approximately 0.7% to 0.9% cpm bound in the RIP assay) with EPO levels $\geq$10 mU/mL, 20 μL of 1:2, 1:4, 1:8, and 1:16 dilutions in human serum were assayed (a final dilution of 1:10 in each well), while for highly positive samples for anti-EPO antibodies determined by the RIP assay ($\geq$30% cpm bound) and samples of EPO levels $\leq$15 mU/mL, 5 μl of 1:2, 1:4, 1:10, and 1:50 dilutions in human serum were assayed (a final dilution of 1:40). For samples of RIP results between low and high values, 10 μL of sample was assayed as 1:2, 1:4, 1:8, and 1:16 dilutions in serum (final dilution of 1:20 in each well). To all wells of the 96-well plate, the suspension of starved UT-7/EPO cells (4×10^4 cells/50 μL), 125, 120, or 110 μL of basal medium were added, depending on the amount of test sample or positive and negative control in the wells, at 37°C the plate was incubated in a 5% CO_2 atmosphere for 42–48 h, to all wells 50 μL of 20 μCi/mL [^{3}H]thymidine in basal medium (1 μCi/well) was added, similar to the proliferation assay, the plate was incubated for 6 h, the cells were harvested onto a UniFilter microfilter plate with GF/C filters, the plate was dried, and the radioactivity in each well was measured. By interpolation from the rHuEPO standard curve that was fitted using a 4-parameter logistic model, measurable rHuEPO levels in the test samples, QCs, positive control, and negative control were estimated. If the mean back-calculated concentrations of each rHuEPO standard were $\pm 25\%$ of the nominal concentration of each standard and the %coefficient of variation (CV) between replicates was $\leq 25\%$, the data were acceptable. As an additional acceptability parameter, QCs mean back-calculated concentrations had to be $\pm 25\%$ of their nominal concentration and %CV had to be $\leq 25\%$. At each concentration, at least one QC had to meet these criteria. If the positive control had $\geq 25\%$ neutralizing anti-EPO antibodies and the negative control had $<25\%$ neutralizing anti-EPO antibodies, the test samples' results were acceptable. The assay had a 25% cutoff, therefore any value $<25\%$ was negative and $\geq 25\%$ was positive. Percentage neutralization for each sample was calculated by determining EPO amount recovered from the original 10 mU/mL added using the equation [(10 – concentration recovered)/10] $\times$ 100%.

21.2 ASSAY SYSTEM FOR BIOTIN PROTEIN LIGASE (BPL) FROM *Escherichia coli*[2]

Black plates (96-well, BMG Technologies) were at 4°C with 100 μL of 1% casein dissolved in TrisbuVered saline blocked overnight and stored at 4°C for up to 2 months. For BirA assay, a master mixture of 50 mM Tris (pH 8.0), 3 mM ATP, 5.5 mM $MgCl_2$, 8 μM biotin, 0.1 mM dithiothreitol, 100 mM KCl, 7 μM peptide 85-11 (MAGGLNDIFEAQKIEWHE, the residue K is the biotinylation target) and 3 μM fluorescein-labeled peptide fl-85-11 (FITC-β alanine-MAGGLNDIFEAQKI-EWHE) was prepared, of which a 47.5 μL aliquot was added into each well of the 96-well plates and preequilibrated at 37°C for 5 min. By adding 2.5 μL of BirA of 11 nM final concentration, BirA reaction was initiated. After 37°C and 45 min incubation, by adding 5 μL of stop solution containing 50 mM EDTA and 13.6 μg of avidin (specific binding activity 14 U/mg) per well, the reaction was terminated. Before using a BMG Laboratories Polar-Star Galaxy Plate Reader measuring the plate for fluorescence polarization, the reaction was at 37°C incubated for a further 10 min. The plate reader was set in polarization mode with 485-nm excitation filter and 520-nm emission filter, and millipolarization units (mP) were calculated, where P = (Int|| – Int^)/(Int|| + Int^), Int|| = intensity of emission in the plane parallel to excitation (channel 1) and Int^ = intensity of emission in plane perpendicular to excitation (channel 2). Using 2.3 μM fluorescein in 50 mM Tris (55 μL, pH 8.0), the gain was adjusted for channels 1 and 2 such that an mP value of 35 was obtained.

21.3 β-GALACTOSIDASE AND α-AMYLASE BIOSYNTHESIS ASSAYS[3,4]

Epicurian coli K12 strain was grown by inoculating 50 mL of Luria-Bertani (LB) media in a 125-mL Erlenmeyer flask, at 37°C in a shaking water bath incubated for 12–18 h, the cells were transferred to fresh media to get an optical density of 0.2 (420 nm), incubated to harvest the bacteria in the log growth phase with 0.5–0.7 optical density, 5 mL aliquots were centrifuged at 5000 rpm for 10 min, and the pellet was used for the subsequent toxicant exposures.

Bacillus subtilis (Microbiology Department, Nottingham University, UK) was grown in LB media, at 37°C in a shaking water bath preincubated overnight. In a 125-mL Erlenmeyer flask, 50 mL of fresh media was inoculated with 50 mL of overnight cultures and at 37°C in a shaking water bath incubated for 8 h.

1. *β-Galactosidase Activity Assay*: To the bacterial pellet, 0.9 mL of toxicants (heavy metals and pesticides) were added, in a shaking water bath at 37°C incubated for 30 min, 0.1 mL of IPTG, 0.4 mL Z buffer ($Na_2HPO_4 \cdot 2H_2O$, 6.20 g/L; $NaH_2PO_4 \cdot 2H_2O$, 10.69 g/L; KCl, 0.75 g/L; and $MgSO_4 \cdot 7H_2O$, 0.25 g/L), and 0.5 mL of fresh media were added, at 37°C incubated for another 30 min, 0.8 mL of Z buffer, 100 μL of SDS (0.1% w/v), and 0.1 mL of ONPG were added, incubated for another 15 min until color development, and the reaction was stopped by adding

1 mL of Na_2CO_3 (1 M). β-Galactosidase activity was measured spectrophotometrically at 420 nm.

2. *α-Amylase Activity Assay*: To the bacterial pellet, 0.9 mL of toxicants (heavy metals and pesticides) were added, the samples were centrifuged at 8000 rpm for 5 min, at 25°C 1 mL of supernatant was with 1 mL of starch (0.5%, as substrate) in 50 mM Tris-HCl (pH 7.0) incubated for 10 min, 2 mL of 3,5-dinitrosalicylic acid (DNSA) was added to stop the reaction, and the solution was placed in boiled water for 5 min to initiate the reaction of the released maltose units with DNSA to give maximum absorbance at 489 nm. Protein content in the supernatant was determined in a UV spectrophotometer at 570 nm using BSA as the standard. Before SDS-PAGE, α-amylase was partially purified, 100 μL of samples were loaded on each well, run on 10% polyacrylamide gels using mini electrophoresis equipment, the gels were stained in Coomassie Brilliant Blue (CBB-250) and destained. SDS-PAGE molecular weight markers were commercial purified α-amylase (58 kDa) and BSA (66 kDa).

21.4 PENICILLIN-BINDING PROTEIN (PBP 5) AND VANCOMYCIN ACTIVITY ASSAYS[5]

1. *Purification of Epicurian coli PBP 5*: Using overnight cultures of MC1061/pRN10.3 (Pharmacia), by ampicillin affinity chromatography a truncated, soluble form of PBP 5 from *E. coli* was purified, and by passage over a Sephacryl S-200 column (1 × 20 cm) equilibrated in the same buffer exchanged into 0.1 M Tris/150 mM NaCl/pH 8.0, to which glycerol was added to 20%; using micro Bradford assay, PBP 5 concentration was determined, and PBP 5 was stored at −80°C.

2. *OPD-Based D-Ala Assay* [Tris, D-Ala, HRP, ampicillin, FAD, *o*-phenylenediamine (OPD), and vancomycin were from Sigma Chemical Company (St. Louis, MO)]: D-Amino acid oxidase (DAO), HRP, and OPD concentrations were assayed for detecting D-Ala by measuring the observed signal at various D-Ala concentrations as a function of reagent concentration. As DAO's coenzyme, FAD was required to stabilize the activity of DAO. Commercially available HRP was sufficiently used to completely suppress any apparent kinetic lag in the assays. OPD concentration dependence demonstrated a maximum at around 0.4 mM and was used at 0.4 mM in all assays. Formation rate of H_2O_2 (and chromophore) was proportional to DAO concentration. The amount of DAO was set to provide an easily detectible signal without using excess DAO. Either by increasing DAO concentration or by allowing the reaction to proceed for a longer period of time, the sensitivity of this assay can be increased marginally (about 20%).

3. *OPD-Based [D-Ala] Standardization Reactions in Microtiter Plates*: A series of samples in quadruplicate with variable (0–100 nmol) D-Ala in 50 μL of assay buffer in microtiter plates were prepared and 150 μL of detection reagent containing all the components necessary to convert D-Ala to a detectible signal was added. Tris buffer and pH 8.5 were used to increase assay sensitivity. In the assay, the first step, where D-Ala was produced by the action of a PBP (or some other enzyme), can be generally performed in nearly any buffer and at any pH. If there was a pH

change between assay and detection steps, in detection step at least a fivefold higher buffer concentration should be used. In the first, the buffer used in the PBP assays was 0.1 M Tris (pH 8.5). By adding 150 μL of detection reagent, which consisted of 0.53 mM OPD (0.4 mM in final assay volume), 0.5 units of HRP, 1.67 μg/mL FAD (1.25 μg/mL in final assay volume), and 0.015 units DAO in 0.1 M Tris (pH 8.5), the detection reaction was performed. In addition, to stop PBP-catalyzed reactions, 66.7 μg/mL ampicillin (50 μg/mL in final assay volume) was involved in the detection reagent. After 1 h by adding 100 μL of 1 N HCl, OPD-based detection reactions were stopped, and the absorbance of plates was read on Tecan SpectraFluor Plus microtiter plate reader (Research Triangle Park, NC) at 490 nm. In OPD-based D-Ala detection, plots of *A*490 versus [D-Ala] were approximately linear up to 60 nmol.

21.5 RECOMBINANT BACTERIA ASSAY EVALUATING ANDROGEN BIOSYNTHESIS[6]

1. *Expression of pJL17/OR* (University of Kentucky, Lexington, KY, USA) *and pCWH17mod(His)4* (Vanderbilt University, Nashville, TN, USA) *in* Escherichia coli (La Jolla, CA, USA): As depicted in Fig. 21.1, *Escherichia coli* XL1 was transformed with pJL17/OR or pCWH17mod(His)4. On a fresh Luria–Bertani (LB) agar plate (per mL containing 100 μg of ampicillin), ampicillin-resistant colonies were streaked. A single colony was selected, inoculated into 10 mL of LB (per mL containing 100 μg of ampicillin), grown overnight at 37°C with vigorous shaking, 2 mL aliquot was transferred into 200 mL of terrific broth (TB, per mL containing 50 μg of ampicillin), at 37°C with vigorous shaking bacteria were grown up to 0.4–0.5 OD_{578}, isopropyl-β-D-thiogalactopyranoside (IPTG) was added to 1 mM final concentration to induce enzyme expression, 5-aminolevulinic acid (ALA) and riboflavin were added to 1 mM final concentration, the temperature was decreased to 29°C,

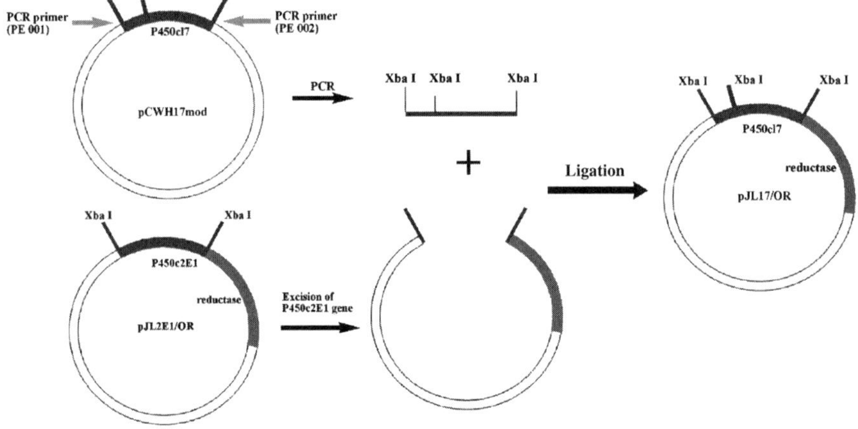

Figure 21.1 Construction of the human $P450_{C17}$-reductase coexpression plasmid pJL17:OR.

the shaking rate was decreased to 120 rpm, and expression time was extended to 40 h. By sonication (sonicator Bandelin Sonopuls HD 60, Bandelin electronic; Berlin, Germany) of this recombinant bacteria, the protein was extracted for measurement.

2. *Storage of Recombinant Bacteria*: At 4°C by centrifugation at 5000 × *g* for 15 min, the recombinant bacteria was harvested, washed with 0.1 M sodium phosphate (pH 7.4), resuspended in 0.1 M sodium phosphate (pH 7.4, containing 20% glycerol) to result in 15.0 OD_{578}, which was adjusted to 5.0 OD_{578} for testing in 96-well plates, frozen in an ethanol/dry ice bath, and stored at −70°C.

3. *Assay with Intact Bacterial Cells Coexpressing Human P 450c 17 and Rat NADPH-P 450-Reductase*: To measure the inhibitory effect of test compounds on $P450_{C17}$, 10 μL of progesterone (1.25 mM, containing 600 nCi [^{3}H]progesterone) and 10 μL of appropriate dilution of inhibitor (dissolved in 96% ethanol) were transferred to 1.5-mL tubes and evaporated to dryness, 400 μL of sodium phosphate (0.1 M, pH 7.4, containing 12.5 mM glucose) was added, mixed for 5 min, and at 37°C incubated for 10 min, by adding 100 μL of thawed recombinant bacterial suspension the assay was started, at 37°C the tubes were horizontally shaken (200 rpm) for 45 min, by heating the tubes at 95°C for 5 min the reaction was stopped, steroids were extracted with 600 μL of cold ethyl acetate for 5 min, at 4°C and 3000 × *g* centrifuged for 5 min, 400 μL of supernatant was removed, evaporated to dryness, dissolved in 50 μL of methanol, and analyzed by HPLC. As a control, using the same conditions the incubation was performed without inhibitor.

4. *Assay for Inhibitors of $P450_{C17}$ in 96-Well Plates* (Corning Costar, Bodenheim, Germany): For screening high numbers of compounds, an assay was established using 96-well polypropylene plates with polystyrol mats. Ten microliters of progesterone with 600 nCi [^{3}H]progesterone (0.125 mM, dissolved in 96% ethanol) was added to each well, evaporated to dryness, 15 mL of sodium phosphate (0.13 M, pH 7.4, containing 33.3 mM glucose) and 5 μL of appropriate solution of inhibitor (dissolved in DMSO) were added, at 37°C with shaking incubated for 10 min, the assay was started by adding 30 μL of thawed recombinant bacterial suspension, the plate covered by a polystyrol mat was incubated at 37°C with vigorous shaking (250 rpm) for 45 min, by adding 5 μL of stop-mixture (containing 5 mg of ethacridine lactate and 6 mg of mercury dichloride per mL water), the reaction was stopped, steroids were extracted with 95 μL of cold ethyl acetate in the covered 96-well plate for 15 min, 40 μL of supernatant was removed, transferred into a 1.5-mL tube, evaporated to dryness, dissolved in 50 μL of methanol, and analyzed by HPLC.

5. *HPLC Analysis:* By use of the autosampler system 851-AS (Jasco, Tokyo, Japan), high pressure solvent pump M 6000 A (Waters, Milford, USA), radioactive flow detector LB 506C (Berthold, Wildbad, Germany), and HLABE 1.6.5 software (Berthold, Wildbad, Germany), HPLC was performed. Using methanol/water (1/1) with 0.4 mL/min flow rate, the steroids were separated on a Nukleosil RP 8 (125 × 2 mm) column (Macherey-Nagel, Düren, Germany). Retention time of 16α-hydroxyprogesterone (formed as a by-product by $P450_{C17}$) was 8 min, of 17α-hydroxyprogesterone was 14 min, and of progesterone was 29 min.

21.6 UDP-Glc 4-EPIMERASE ASSAY[7]

Based on detecting hexoses and (N-acetyl) hexosamines obtained after hydrolysis of reaction products formed from appropriate uridine diphosphate (UDP)-sugars, in the presence of the *Bacillus subtilis* Gne (GneA)-(His)6 protein the assay was established. The reaction mixture containing 1.5 μg of purified GneA-(His)6 protein and 0.5 mM of either of UDP-*N*-acetylglucosamine (UDP-GlcNAc), UDP-*N*-acetylgalactosamine (UDP-GalNAc), UDP-glucose (UDP-Glc), and UDP-galactose (UDP-Gal) was in 100 μL of 100 mM phosphate buffer (pH 7) at 37°C incubated for 2 h, mixed with 100 μL of 1 M TFA, under vacuum in a sealed tube, and heated at 100°C for 30 min to hydrolyze. Onto a Dionex Series DX500 HPLC system, 10 μL of hydrolyzed sample or 100 μM standards were injected, using a Dionex CarboPAc PA1 anion exchange column equilibrated in 8 mM NaOH, the components were separated isocratically at 1 mL/min flow rate using 8 mM NaOH, at 26 min a linear gradient from 0 to 450 mM of sodium acetate in 100 mM NaOH for 40 min was applied, the eluate was monitored with a pulse-electrochemical detector (Dionex), and chromatograms were analyzed with the Igor Pro program (WaveMetrix, Lake Oswego, OR).

21.7 ATPs EXCHANGE ASSAYS[8]

1. *ATP-PPi Exchange Assay*: In the presence of 2 mM 2′-deoxy ATP, 0.1 μCi tetrasodium (^{32}P) pyrophosphate (NEN Life Science Products), 0.1 mM cold pyrophosphate, 0.1 mM cognate amino acid, 50 mM HEPES buffer (pH 8.0), 10 mM $MgCl_2$, 100 mM NaCl, 2 mM dithioerythrol (DTE), 1 mM EDTA, and 10 mM KF in a total volume of 100 μL, 0.057 mg of enzyme fraction prepared from *Bacillus subtilis* PY79 was at 37°C incubated for 15 min, the reaction was treated with 800 μL of cold solution of 1.2% activated charcoal (Sigma, Steinheim, Germany) in 560 mM perchloric acid and 100 mM tetrasodium pyrophosphate for 15 min to terminate the reaction, the reaction mixture was incubated on ice, and the suspension was pelleted and washed two times with 1 mL of water. Using a 1900CA Tri-Carb liquid scintillation counter (LSC), the radioactivity of adsorbed materials was counted.

2. *ATP-Pi Exchange Assay*: In the presence of 2 mM ATP, (0.1 μCi) H_3PO_4 (Amersham/Buchler), 0.1 mM cold orthophosphate, 0.2 mM cognate amino acid, 50 mM HEPES buffer (pH 8.0), 10 mM $MgCl_2$, 100 mM NaCl, 2 mM dithioerythrol (DTE), 1 mM EDTA, and 10 mM KF in a total volume of 100 μL, 0.057 mg of enzyme fraction prepared from *Bacillus subtilis* PY79 was at 37°C incubated for 15 min, and enzyme fraction was incubated. After 15 min incubation at 37°C, the reaction was treated with 800 μL of cold solution of 1.2% activated charcoal in 560 mM perchloric acid and 100 mM Na_2HPO_4 for 15 min to terminate the reaction, the reaction mixture was incubated on ice, and the suspension was pelleted and washed three times with 1 mL water. Using a 1900CA Tri-Carb liquid scintillation counter (LSC), the radioactivity of adsorbed materials was counted.

21.8 LUX-FLUORO ASSAY FOR COMBINED GENOTOXICITY AND CYTOTOXICITY[9]

Recombinant *Salmonella typhimurium* TA1535 (pPLS-1) or S. choleraesuis TA1535 (pGFPuv, Clontech Laboratories Inc., CA, USA, 6079-1) culture was initiated from a single colony from an agar plate within 10 mL of NB-medium with 50 μg/mL ampicillin, the bacteria were at 37°C grown for 16 h, on a rotary shaker diluted (1:50) in fresh warm NB-medium containing 50 μg/mL ampicillin, at 37°C incubated until A_{600} reached 0.2–0.3, from this culture 90 μL aliquots were added per well of the prepared microplate, which was covered with a gas-permeable self-adhesive seal (Advanced Biotechnologies Ltd., Surrey, UK) and placed into the temperature-controlled microplate reader (Multilabel Counter 1420 Victor2 from EG&G Wallac, Perkin Elmer, USA). The microplate reader capable of sequential reading of absorbance, luminescence, and fluorescence was programmed for repeating the measurement cycle at 30°C for up to 8 h of continuous incubation with a duration of about 10 min per measurement cycle (total of 50 cycles). Each measurement cycle was started with orbital shaking for 120 s, luminescence was measured without filter for 0.2 s per well, absorbance was measured at 490 nm (20 nm band width) for 0.1 s per well, and fluorescence was measured (excitation at 405 nm and emission at 510 nm) for 0.1 s per well. The raw data were transferred into an EXCEL macro sheet, induction factors (Fi) as well as deduction factors (Fd) were calculated for genotoxic and cytotoxic potential of the applied agents, based on $Fi = Luxi \times A_0/Lux_0 \times A_x$ and $Fd = Flu_x/Flu_0$, respectively, where Lux_0 is light emission data of the untreated culture, Luxi is light emission data of the samples treated with the toxin, A_0 is cell growth absorbance of the untreated culture, Ax is cell growth absorbance of the treated sample, Flu_0 is the autofluorescence of the untreated culture, and Flu_x is the autofluorescence of the samples treated with the toxin.

21.9 REPORTER GENE ASSAY FOR ARYL HYDROCARBON RECEPTOR[10]

From lobectomy segments resected from adult patients, hepatocytes were prepared, isolated, plated for 24 h, and treated for 16 h with 10 nM 2,3,7,8-TCDD, BCP instrument Lyon, France) or 100 μM omeprazole (OM, Astra Hassle Molndal, Sweden) in the absence or presence of 3–50 μM omeprazole-sulfide (OMS, Astra Hassle Molndal, Sweden), while control cultures received DMSO (Sigma, St. Louis, MO) alone. In some assays, before treatment with TCDD and OMS, hepatocytes were with 10 μM rifampicin (RIF, Sigma, St. Louis, MO), 1 μM ketoconazole (KT, Sigma, St. Louis, MO), or both (RIF + KT) pretreated for 3 days.

The reporter gene construct pTXINV containing two inverted repeats of XRE in front of thymidine kinase promoter and luciferase reporter gene and pSV-β-galactosidase were transiently transfected for 16 h, fresh medium containing DMSO, 2,3,7,8-tetrachlorodibenzo-p-dioxin(TCDD), 3-MC, or OM without or with OMS was added, incubated for 16 h, the cells were harvested in reporter lysis buffer, and luciferase activity was normalized with respect to β-galactosidase activity.

21.10 CYP1A ENZYME ASSAY[11]

The core sequence of AhR binding motif, dioxin response element (DRE, 5′-TGNGCGTGAACGCAA-3′), identified in the regulatory region of many 2,3,7,8-tetrachlorodibenzo-*p*-dioxin (TCDD)-inducible genes including CYP1A2, CYP1B1, UDP glucuronosyltransferase 1A1, and glutathione *S*-transferase Ya C, was used to construct an oligonucleotide. A pair of complimentary oligonucleotides 5′-TTGCGTGCGATTGCGTGCGATTGCGTGCGA-3′ and 5′-TCGATCGCACGC-AATCGCACGCAATCGCACGCAAGTAC-3, containing three copies of the DRE in tandem with the restriction enzymes KpnI and XhoI as linkers, were constructed. After phosphorylation and annealing, the synthesized oligonucleotides were ligated into analogous sites of a modified luciferase vector that contained an antibiotic selection cassette and the SV-40 viral promoter.

HepG2 cells (human hepatoma cell line) were at 37°C in 95% air and 5% CO_2 maintained in DMEM (Invitrogen Life Technologies, Gaithersburg, MD) supplemented with 10% FBS, trypsinized, seeded at 4×10^6 cells in a P150 dish, recovered for 24 h, by lipofectamine 2000 according to the manufacturer's instructions with 20 μg of DRE-luciferase construct transfected, while the control cells received the same construct without DRE oligonucleotide. After 16 h exposure to the DNA, the medium was removed, replaced with fresh culture medium, incubated for 48 h, and the medium was replaced with fresh medium containing selection antibiotic. Medium was changed at 2-day intervals for 3 weeks until small colonies were visible, the colonies were tested for luciferase activity, and positive colonies were selected. Purified colonies in confluent T75 flasks were trypsinized and seeded onto 96-well plates to measure TCDD-mediated luciferase response of individual clones testing for the presence of recombinants. From several identified colonies, one exhibiting the greatest luciferase activity after exposure to TCDD (DRE12.6, NCI Chemical Carcinogen Reference Standard Repository, Kansas City, MO) was selected, purified by cloning, and DRE12.6 cells were used for all subsequent assays.

1. In T75 flasks, DRE12-6 and the control cells were grown as monolayers with DMEM containing 10% FBS and the selection antibiotic, after 30 passages cells were grown to confluency, confluent cells were removed from flasks by trypsinization, diluted to 1×10^5 cells/mL, seeded on 96-well plates with 150 μL/well DMEM containing 10% FBS for 18–20 h, medium was changed in plates containing both the DRE12.6 and control cells to DMEM containing 10% FBS and the test agents including 0.1, 1, 2.5, and 10 nM TCDD, 100 μM OM or natural products including garlic (10 μg/mL), chrysin (25 μM), quercetin (10 μM), resveratrol (20 μM), kaempferol (10 μM), apigenin (10 μM), curcumin (25 μM), green tea extract (GTE, 0.1 mg/mL), grapeseed extract (0.6 μg/mL), kava (50 μg/mL) and ginseng (50 μg/mL). Both cell lines were also exposed to environmental chemicals such as 20 μM

DMBA, 25 and 100 μM 2-AAF (Sigma, St. Louis, MO), 0.1, 0.5, and 2 μM 3-methylcholanthrene (3-MC), 20 μM dibenzofuran, 100 μM benzanthracene, 20 μM dieldrin, 50 μM methoxychlor, and 20 μM of each polychlorinated biphenyl (PCBs, Ultra Scientific, North Kingstown, RI) including 2,3,3′,4,5-pentachlorobiphenyl, 2,3,4,5-tetrachlorobiphenyl, Aroclor 1254, and 2,3,4,40-tetrachlorobiphenyl for 24 h. Selected concentrations of agents for analysis produced optimal responses without cell toxicity. To determine chemicals' AhR antagonist properties, DRE12.6 and control cells were preexposed to the same concentrations of quercetin, curcumin, chrysin, resveratrol, kaempferol, apigenin, GTE, garlic, grapeseed extract, ginseng, kava, DMBA, PCBs, dieldrin, Aroclor 1254, dibenzofuran, benzanthracene, 2-AAF, or methoxychlor for 3 h, treated with 2.5 nM TCDD, and incubated for 28 h. All inducers were dissolved in DMSO. DMSO (0.1%) was added to wells containing DRE12.6 or control cells and represented control. After treatment, media was aspirated, and to each wall 100 μL of PBS was added. Luciferase activities from the cell line harboring the control vector was compared with those generated in the DRE12.6 cells. Activities from treated DRE12.6 cells were expressed as the fold increase above DMSO-exposed DRE12.6 cells and normalized to the fold increase in treated control cells above DMSO-exposed control cells, which accounted for any chemical-mediated alterations in luciferase expression produced by the SV40 promoter.

2. In T75 flasks, HepG2 cells were grown in DMEM supplemented with 10% FBS, confluent HepG2 cells were trypsinized, plated in 6-well plates for 24 h, treated with 0.1, 1, 2.5, and 10 nM TCDD or natural products including resveratrol (20 μM), chrysin (25 μM), kava (50 μg/mL), kaempferol (10 μM), quercetin (10 μM), or GTE (0.1 mg/mL) for 24 h, also treated with 100 μM omeprazole, 20 μM DMBA, 25 and 100 μM 2-AAF, 100 μM benzanthracene, 20 μM Aroclor 1254, 20 μM dibenzofuran, or 0.5 and 2 μM 3-MC for 24 h, as well as exposed to inducers for 24 h. In assays where TCDD was incubated with an AhR antagonist, cells were preexposed to concentrations identical to those described above for resveratrol, curcumin, chrysin, kava, GTE, kaempferol, 2-AAF, or dibenzofuran for 3 h, treated with 2.5 nM TCDD, and incubated for 24 h. All inducers were dissolved in DMSO. DMSO (0.1%) was added to wells containing DRE12.6 or control cells and represented control. After treatment, HepG2 cells were harvested, with RNeasy filters RNA was isolated and at 260 nm spectrally quantified. From the 260/280 nm absorbance ratios and by integrity of the 28s and 18s bands on agarose gels, RNA purity was assessed. By electrophoresis of total RNA (10 μg) through a 1.0% agarose gel containing 2.2 M formaldehyde followed by blotting onto a nylon membrane, Northern blot analyses were performed. RNA was cross-linked to the membranes, the membranes were hybridized to random-primed cDNA probe encoding human CYP1A1, visualized autoradiographically, the blots were washed, rehybridized to the rRNA 18S probe, and used for normalization. By scanning with a Scan-Maker II and digitizing with Un-Scan-It software, autoradiographs of Northern blots were quantified. Exposure times were in the linear range for Kodak XAR-5 film.

3. DRE12-6 cells were maintained in DMEM supplemented with 10% FBS, confluent DRE12-6 cells were trypsinized, seeded on 96-well plates at 2×10^4 cells per

well, exposed to TCDD for 24 h to overexpress CYP1As, rinsed with PBS, 50 μM luciferin 6′ chloroethyl ether (luciferin-CEE) and luciferin 6′ methyl ether (luciferin-ME) were added for CYP1A1 and CYP1A2, respectively, plates were incubated at 37°C for 3 h, by adding 50 μL of luciferase detection reagent, the reaction was terminated, and luciferase activity was assayed. To determine the ability of ketoconazole and resveratrol to inhibit CYP1A1, 0, 0.01, 0.03, 0.1, 1.0, 3.0, 10 μM ketoconazole and 0, 0.001, 0.1, 1.0, 10, 30, 100 μM resveratrol were co-incubated with luciferin-CEE. The plates containing TCDD preinduced DRE12.6 cells, substrate, and inhibitor were incubated at 37°C in a CO_2 incubator for 3 h, and luciferase activity was assayed. To determine inhibition of CYP1A2, furafylline was added to media containing 0.001, 0.01, 0.1, 1, 3, 10, 30, 100 μM luciferin-ME, and theophylline was added to media containing 0, 0.001, 0.1, 1.0, 10, 30, 100, 250 μM luciferin-ME, incubated for 3 h, and luciferase activity was determined.

21.11 CYP1A ACTIVITY (EROD) ASSAY[12]

Primary rat hepatocytes were isolated by two-step collagenase perfusion, plated on collagen-coated culture dishes at 2×10^5 cells/cm^2 in WME supplemented with 2 mM L-Gln, 10 μM streptomycin, 100 IU/mL penicillin, 350 nM insulin, and 1 μM dexamethasone, enriched for plating with fetal calf serum (5%, v/v) for 3 h, the medium was exchanged for a serum-free one, the culture was stabilized for additional 24 h and maintained at 37°C in 5% CO_2 (air/CO_2, 95/5) humidified incubator.

Both primary rat hepatocytes and HepG2 cells were seeded on 24-well dishes at 2.5×10^4 cells/cm^2 using EMEM with nonessential amino acids and further supplemented with 1 mM pyruvate, 100 IU/mL penicillin, 100 μg/mL streptomycin, 2 mM L-Gln, and FCS (10%, v/v), stabilized for 24 h, treated with TCDD (1 nM final) and β-naphthoflavone (BNF; 25 μM final) for 72 h as well as with TCDD (1 nM final) and 3-MC (5 μM final, Sigma Chemicals St. Louis, MO) for 20 h, and culture medium was changed at 24-h intervals regardless of treatment. After inducer treatment, serum-free medium was added in case of HepG2 cells, and cells were treated for 3 h with test compounds or DMSO as vehicle control. Lactate dehydrogenase (LDH) leakage and MTT were measured to assess cytosolic membrane damage and metabolic capability of cells, respectively. Triton X-100 (1.5%, v/v, Sigma Chemicals, St. Louis, MO) treated cells had 100% LDH leakage, while DMSO (vehicle control) treated cells had 100% metabolic activity in the MTT assay.

From multiorgan donors who met an accidental death, liver samples were obtained, small pieces of tissue were well homogenized, by differential centrifugation microsomal fraction was prepared, microsomes were stored at −80°C in 25 mg protein/mL aliquots, and CYP (Sigma Chemicals, St. Louis, MO) was typically estimated as 500 pmol/mg protein.

Human liver microsomes (0.5 mg protein/mL) were suspended in 2 mL of potassium phosphate buffer (50 mM, pH 7.4), divided in half, pipetted into two quartz microcuvettes, a baseline was recorded, into the sample cuvette 7-ethoxyresorufin (4 μM final concentration) was added, on Shimadzu UV-1601 spectrophotometer absorbance spectrum was recorded in the range 360–460 nm, sequential additions of test compounds were made into the sample cuvette, and after each addition absorbance spectrum was recorded. To circumvent interference of the ligands with CYP absorbance spectrum, using the same sequence of substance additions, additional absorbance spectra were also recorded in the absence of microsomes. Using the spectrophotometer software, the corresponding spectra obtained with or without microsomes were subtracted. In the microsomal suspension, recombinant CYP enzymes (10 pmol) were used; other additions and set up were the same as for liver microsomes.

All preincubations of microsomes with test compounds in the presence or absence of other effectors, including lipoic acid, DTT, GSH, L-cysteine, and NADPH (Sigma Chemicals, St. Louis, MO) were in 200 μL of total volume in a separate Eppendorf at 37°C carried out for 2, 4, and 6 min, preincubation medium was added to 1.8 mL of assay buffer HEPES (0.1 M, pH 7.8) in a thermostatted fluorescence cuvette for a 6 s measurement, 7-ethoxyresorufin was added to initiate the reaction, and 5 s prior to 7-ethoxyresorufin addition to the assay cuvette, NADPH, if omitted from the preincubation mixture, was added. Inhibition assay used the same protocol only was adjusted to fit 96-well plate format where 0.5 pmol of recombinant CYP1A1 or CYP1A2 or 15 μg of total protein from human liver microsomes were used. Except methanol, which was used to terminate the reaction after 5 min, all other parameters were the same. Luminometer LS50B equipped with either thermostatted cuvette holder with continuous mixing (time course measurements) or well plate reader accessory was used as detection apparatus, and 530-nm excitation wavelength and 585-nm emission wavelength were used to construct a calibration curve.

21.12 DEUBIQUITINATING ENZYME (DUB) ACTIVITY ASSAY[13]

Using standard methods from *Escherichia coli*, an expression plasmid encoding a his-tagged YFP-ubiquitin fusion with a C-terminal alanine-cysteine (AC) addition (YFP-Ub-AC) was expressed, over Ni-nitrotriacetic acid purified, with Tris-HCl (25 mM, pH 7.5) containing 100 mM NaCl and 5% glycerol (v/v) diluted to 5 mg/mL (based upon $\varepsilon_{280} = 33{,}350$/M cm), DTT was added to 10 mM final concentration, and this YFP-Ub-AC intermediate was stored at −80°C until required for terbium labeling.

To prepare the terbium-labeled substrate, 1 mL of YFP-Ub-AC intermediate was thawed, using a NAP-10 desalting column desalted to remove DTT, the eluted protein was immediately combined with 200 μg of thiol-reactive Tb chelate, allowed to react at room temperature for 2 h, the reaction mixture was dialyzed against HBS to remove unreacted Tb chelate, the dialyzed Topaz-ubiquitin-Tb substrate was quantified by absorbance ($\varepsilon_{280} = 42{,}150$/M cm), with HBS diluted to 20 μM final concentration and stored at −80°C until used.

To assay relative activity of each DUB toward the YFP-ubiquitin-Tb substrate, serial dilutions of enzymes (UCH-L1, UCH-L3, USP-5, and USP-14, Boston

Biochem, Cambridge, MA) in 10 μL of assay buffer (20 mM Tris, pH 7.4, 0.01% Nonidet-P40, 1 mM DTT) were prepared in a black 384-well low-volume plate, each well was treated with 10 μL of YFP-ubiquitin-Tb substrate (20 nM) in the same buffer for 50 min, and using the LanthaScreen filter module, the plate was read on a BMG Labtech Pherastar plate reader, when measured using a 200-μs signal integration window following a 100-μs delay, the emission ratio was calculated as the raw acceptor intensity divided by the raw donor intensity, and no background subtraction or cross-talk correction was required. Kinetic reads were carried out similarly against varying concentrations of UCH-L3 at 1-min intervals for 90 min. In a 1-h reaction against a dilution series of ubiquitin aldehyde or ubiquitin using 15 pM UCH-L3 and 10 nM YFP-Ubiquitin-Tb, inhibitor titrations were performed. Using different concentration of UCH-L3 at various percentage conversions of substrate, Z' values were determined, 24 positive control wells and 24 negative control wells were measured, and Z' was calculated according to $Z' = 1 - [(3\sigma_{c+} + 3\sigma_{c-})/|\mu_{c+} - \mu_{c-}|]$, where σ_{c+} is the standard deviation of the positive control wells, σ_{c-} is the standard deviations of the negative control wells, μ_{c+} is average values for the positive control wells, and μ_{c-} is average values for the negative control wells. Normalized emission ratios were calculated relative to wells containing maximal and minimal FRET signal and multiplied by 100.

Using a Tecan Safire2 plate reader, time-resolved spectra of 10 nM YFP-ubiquitin-Tb incubated with or without excess UCH-L3 were measured. In a white 384-well low-volume plate, 20 μL of samples were read. Excitation was set to 332 nm (20 nm bandwidth), using a 200-μs signal integration window following a 100-μs delay and averaged over 100 measurements (flashes) per wavelength, emission measurements were collected from 475 to 650 nm in 1-nm increments. Using a BMG Labtech Pherastar plate reader, emission signal decays were measured. A simple mechanism of this assay is summarized in Fig. 21.2.

21.13 FLOW CYTOMETRIC ASSAY[14]

Escherichia coli MG1655/pANO1::cda′ was used in this assay. At 37°C, biosensors were grown with moderate shaking (200 rpm) in LB medium containing 0.2%

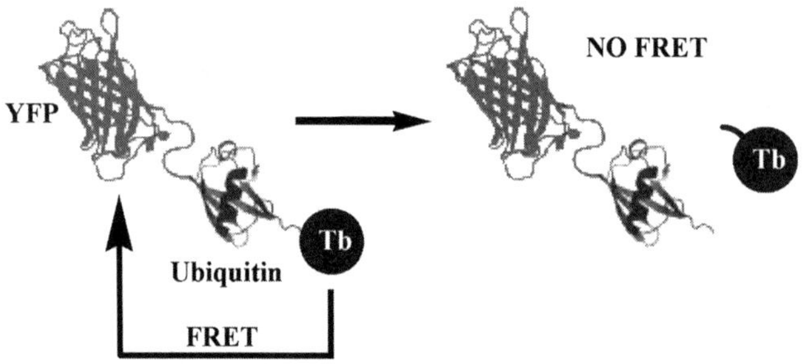

Figure 21.2 Mechanism of DUB activity assay.

glucose. To avoid SOS-induction, overnight cultures were grown at 30°C in LB medium with 0.4% glucose and supplemented with 100 μg/mL ampicillin. For induction of the SOS-response, *N*-methyl-*N*-nitro-*N*-nitrosoguanidine (MNNG) was used as the alkylating agent and added to the growth medium from 1 mg/mL stock solution.

In 300 mL triple baffled shake flasks, mycelium of *Streptomyces caespitosus* [also known as *Streptomyces lavendulae* (*Streptoverticillum caespitosus*)] for producing mitomycin C (MMC) was grown at 30°C with vigorous shaking (300 rpm) in 100 mL of Nishikori medium (1/100 with frozen mycelium) to ensure adequate aeration and prevent mycelium from forming clumps.

For the preincubation, an overnight culture of *Escherichia coli* MG1655/pANO1::cda′ was grown in 100 mL of fresh medium diluted to 10^2, 10^3, 10^4, or 10^6 until an optical density giving about 0.180 of OD_{600} and MNNG was added. In the dose-response assay, the preincubation used a 10^5 dilution of overnight culture and was grown for 150 min. Into separate 25-mL glass tubes containing 50 μL aqueous MNNG of 0.0015–10 μM final concentrations, 2.95 mL aliquots were transferred from each culture. Control sample tubes contained 50 μL of double-distilled water (ddH_2O). Cultures were grown for 90 min and 20 μL aliquots were in PBS diluted 50-fold for direct flow cytometry sampling.

On a FACScalibur™ (BD Biosciences, San Jose, CA, USA) flow cytometer equipped with a 488-nm argon laser, flow cytometric assays were performed. In the liquid culture assays, depending on cell density, samples were run at high flow rate (54 μL/min) for 8–120 s until 5000 events had been counted in the R1-gate, and samples containing bacteria from soil Nycodenz extractions were run at low flow rate (18 μL/min) for 60 s. Before assay each sample was mixed by vortexing

During assay, forward scatter (FSC) was set to E01, side scatter (SSC) was set to 350 V, and the fluorescence detectors (values in parenthesis are wavelength and band-pass, respectively) FL1 (530/30 nm) and FL2 (585/42 nm) were both set to 566 V. Either 80 (liquid culture) or 115 (soil samples) threshold was set on the FL1 detector, and no compensation was used. With CellQuest software package, data analysis was performed.

To distinguish biosensors in liquid culture, apolygonal-gate R1 was defined in bivariate FSC versus SSC dot-plots. Gate R2 was defined in FL2 versus FL1 dot-plots for the soil microcosm assay. To determine the mean fluorescence values per cell R1 (liquid) or R2 (soil) gated FL1 histogram-plots were used. Induction factors (Fi) were defined as MF_x/MF_0 where MF*x* indicated the relative FL1 mean fluorescence of a given sample and MF_0 indicated the mean fluorescence of the control sample. MF_{tot} and MF_{gate} values referred to MF of all events in the plots and of either R1- or R2-gated events, respectively.

A 6-day-old *Streptomyces caespitosum* mycelium culture was spun down (4000 × *g*, 5 min) and the supernatant was poured into a separate container. To measure the response to a theoretical *in situ* production, *Streptomyces caespitosum* mycelium was chosen over commercially purified MMC, which might include additional genotoxic substances related to the synthesis or breakdown of MMC. With equal volumes of ethyl acetate, crude broth was extracted twice and in a 15-cm glass Petri dish under a laminar flow hood evaporated to remove ethyl acetate, the formed residue was completely redissolved in 5 mL of ddH_2O and kept at $-18°C$ until needed. By correlating an SOS-green fluorescent protein (GFP) dose-response curve from a dilution-series of the crude broth against an MMC standard, the MMC concentration in this crude broth concentrate was estimated to be 0.5 mg/mL.

The soil was collected from an organic fallow field (sandy loam), at room temperature air dried overnight, sieved (0.2-cm mesh size), and 4.5 g of dried soil was transferred to sterile 50-mL plastic tubes with screw caps. By adding 250 μL of serial dilutions of 10^0, 10^1, 10^2, 10^3, and 10^4, *Streptomyces caespitosus* crude broth extract microcosms were spiked and left at 30°C overnight, while the control microcosm was made by adding 250 μL of ddH_2O. By centrifuging ($5000 \times g$) a culture grown in LB medium with 0.4% glucose to 1.0 OD_{600} *Escherichia coli* MG1655/pANO1::cda′ was harvested, with 100 mL of PBS washed twice, and resuspended in 10 mL of PBS. Microcosms were each inoculated with 250 μL of *Escherichia coli* MG1655/- pANO1::cda′ (1×10^8 *Escherichia coli* MG1655/- pANO1::cda′/g soil) and at 30°C incubated for 3 days. Final liquid amount in microcosms corresponded with 74% water holding capacity (WHC). After 1- and 3-day incubation, samples were taken for flow cytometry assay.

21.14 SERUM NEUTRALIZATION ASSAY BASED ON rHMPV-GFP[15]

Polyclonal rabbit serum against sucrose gradient purified human metapneumovirus virus (HMPV) and exhibiting high HMPV-neutralizing titer (R+), African green monkey (AGM, *Cercopithecus aethiops*) serum having high HMPV-neutralizing antibody titer (AGM+), and AGM serum lacking HMPV-neutralizing activity (AGM−), as well as four AGM sera from individual monkeys infected with HMPV (A-D), were used as control sera. By seeding 96-well plates with approximately 4×10^4 Vero cells per well in OptiPRO Serum Free Medium or with approximately 5×10^4 LLC-MK2 cells per well in OptiMEM containing 5% FBS, monolayer cultures of the LLC-MK2 rhesus kidney cell line (ATCC CCL7.1) or the Vero AGM kidney cell line (ATCC CCL-81) lacking the structural genes for interferon α and β were prepared.

On the next day, the cell monolayers were approximately 95% confluent, the sera were in Dulbecco's phosphate-buffered saline prediluted to 1:20, at 56°C heat-inactivated for 30 min to remove any residual complement activity, and in 96-well round-bottom plates with MEM medium further diluted.

Among 12 samples (150 μL each) arranged across the first-row wells, the first two samples were duplicates of the R+ control serum, followed by duplicates of the four test AGM sera, followed by two control wells containing medium without a control or test serum (S−). The remaining row of wells each contained 75 μL of MEM with 2 mM L-glutamine and 0.1 mg/mL gentamicin. Using a 12-channel micropipette, by transferring 75 μL per well from each row to the next, the samples were serially twofold diluted yielding 20- to 2560-fold final dilution, followed by mixing and changing tips. From the final row, the 75 μL of transfer was discarded. To every well 75 μL of medium containing approximately 240 50%-tissue-culture-infectious-dose ($TCID_{50}$) of recombinant HMPV expressing the enhanced green fluorescent protein (rHMPV-GFP) stock solution was added and mixed.

The titer of this stock solution was confirmed by back-titration on separate 96-well plates. The mixtures were incubated at 32°C for 1 h, from each well containing approximately 80 $TCID_{50}$ of virus, 50 μL was transferred to the corresponding position of 96-well plate of LLC-MK2 or Vero cells, the plates were at 32°C incubated for 1 h, by aspiration the virus-antibody mixtures were removed, the monolayers were with OptiMEM medium washed twice, covered with 180 μL of OptiMEM medium containing 2 mM L-glutamine, 0.1 mg/mL gentamicin, and 5 μg/mL trypsin, incubated at 32°C, and to each well 20 μL of OptiMEM medium containing 50 μg/mL fresh trypsin was added (to 5 μg/mL final fresh trypsin concentration) at 3-day intervals to promote infectivity.

Using an inverted UV light microscope, the plates were read daily, which can be done directly on living cells without preparing or disturbing the culture. On day 3 post-infection, on Vero cells HMPV foci were easily detectable, and by scoring wells as HMPV positive (GFP expression in one or more cells detectable by fluorescent microscopy) or negative (no GFP expression) and calculating 50% end point, a preliminary titer could be determined on that day. By visual inspection on day 4 to 5 post-infection, foci were detectable on LLC-MK2 cells, indicating that replication and gene expression was somewhat less efficient in this cell line. By scanning with a variable mode imager (excitation wavelength 485 nm, emission wavelength 538 nm, resolution of 100 μm), GFP expression was also monitored.

21.15 YELLOW FLUORESCENT PROTEIN-BASED ASSAY[16]

HEK293 cells were grown in either DMEM or MEM, which was supplemented with 10% fetal calf serum and 1% penicillin/streptomycin, maintained in a humidified atmosphere of 95% oxygen, 5% carbon dioxide at 37°C, and passaged weekly. Human GlyR α1 subunit in the pCIS2 vector, rat GABA α1 subunit in pCDNA3 vector, rat GABA β1 subunit in pCDNA3 vector, rat GABA ρ1 subunit in pCDNA1 vector, rat GABA ρ2 subunit in pCDNA1 vector, YFP-I152L in the pCDNA3 vector, and YFP-V163S in the pCDNA3 vector were used for assay.

T-25 tissue culture flasks were seeded with 3 mL of cells at 2×10^6 cells/mL, upon 40% to 80% confluent, cells were transfected with Effectene according to the manufacturer's instructions, each flask with 1 μg of plasmid cDNA encoding either YFP-I152L or YFP-V163S. In addition, ion channel CDNAs such as 1 μg of α1 glycine receptor (GlyR) cDNA, 0.5 μg of both α1 and β1 GABA type-A receptor ($GABA_AR$) cDNAs, 1 μg of ρ1 subunit cDNA, and 0.5 μg of ρ1cDNA as well as 2 μg of ρ2 cDNA subunit cDNA were added to form homomeric α1 GlyRs, heteromeric α1β1 $GABA_ARs$, homomeric ρ1 GABA type-C receptors ($GABA_CRs$), and heteromeric ρ1ρ2 $GABA_CRs$, respectively. The 24-h transfection culture media were exchanged to terminate transfection, and the cells were immediately plated into 96-well plate or 384-well plate at approximately 3×10^5 cells/well or 2×10^3 cells/well, respectively, 24–72 h after plating cells were used in assays. Expression efficiency was judged by the proportion of green fluorescent cells and varied from 20% to 80%. Approximately 1 h prior to assay commencement, the culture media in each well were replaced with the standard NaCl control solution containing 140 nM NaCl, 5 nM KCl, 2 nM $CaCl_2$, 1 nM $MgCl_2$, 10 nM HEPES, 10 nM glucose, and setting pH 7.3 with NaOH or NaI glutamine control solution containing 140 nM NaI, 5 nM KCl, 2 nM $CaCl_2$, 1 nM $MgCl_2$, 10 nM HEPES, 10 nM glucose, and setting pH 7.3 with NaOH or Na glutamine control solution containing 140 nM Na glutamine, 5 nM KCl, 2 nM $CaCl_2$, 1 nM $MgCl_2$, 10 nM HEPES, 10 nM glucose, and setting pH 7.3 with NaOH. Glycine, GABA, and strychnine were directly dissolved in the appropriate control solution. Picrotoxin was dissolved in DMSO and at −20°C stored as 20 mM aliquots in a mixture of 20% DMSO and 80% water. After placing the 384-well plates onto a motorized stage on a Nikon Diaphot 300 inverted microscope, cells were imaged with a ×10 objective.

Illumination from a 150-W halogen lamp passed through an yellow fluorescent protein (YFP) dichroic mirror, and YFP fluorescence was excited. With a CCD camera (SVGA Sensicam, Applied Scientific Instrumentation, Eugene, OR, USA), emitted fluorescence was imaged and digitized to disk on an IBM PC. To automate the imaging experiments, a suite of LabView VI software (National Instruments) routines purpose-written in the laboratory was used. These programs coordinated (1) the sequential positioning of the wells into the microscope field of view, (2) the automated focus of cells in each well, (3) the image acquisition via the CCD, and (4) the addition and removal of solutions from another 384-well plate via a 100 μL pipetting syringe under control of an LC PAL autosampler (CTC Analytics, Zwingen, Switzerland). The primary resolution for camera was 1024 × 1280 pixels, while the cell images were binned (2 × 2) resulting in a 512 × 640 pixels of resolution. The maximum image acquisition rate was 0.2 Hz. Typically, each image contained 10–150 cells with sufficient fluorescence. CCD image acquisition time was 1.6 s. Image assays were conducted at room temperature (18°C to 28°C).

Fluorescence assays were performed using a Photomultiplier-based FLUOstar Optima microplate reader equipped with an automated dual injection system, 500 ± 10 nM excitation filter and 540 ± 10 nM emission filter. In assay, the YFP emission signal was recorded for 5 s to determine the baseline fluorescence, NaI test solution was injected, and the net fluorescence response was recorded at 1-s intervals for 30 s. From the top of the plate, the machine was configured to both excite and detect YFP emission. Assays were performed at 37°C. Glycine concentration-response curves were fitted with Hill equation $F = F_{init}/1 + (EC_{50}/[\text{glycine}])^{nH}$, where F is the fluorescence corresponding with a glycine concentration, [glycine], F_{init} is the initial (or control) fluorescence value, EC_{50} is the glycine concentration eliciting a half-maximal response, and nH is Hill coefficient. Antagonist dose-responses were fitted with $F = F_{init} - \{F_{init}/[1 + (EC_{50}/[\text{glycine}])^{nH}]\}$. Using a least-squares fitting routine, curve fits were performed.

21.16 RED FLUORESCENT CYTOTOXIC T LYMPHOCYTE ASSAY[17]

By inserting the HIV central polypurine tract to previously modified EF.DsRed2, gene cppt.EF.DsRed was constructed. Lentiviruses were produced by co-transfection of cppt.EF.DsRed with the packaging construct pCMVdl-8.4 and using conventional calcium phosphate precipitation method enveloped pMD.G to 293T cells. P815 cells were transduced with cppt.EF.DsRed at 5 MOI, 5 days posttransduction using a FACScalibur (Becton Dickinson) sorted to >98% purity. P815 cells and 293T cells were cultured as stable lines in the absence of additional selection.

DsRed expressing P815 cell lines was added to each well in triplicates for 2–3 h and used as target cells (T) with the other cells as reference cells (R) at 1/1T/R ratio, at 37°C in V-bottomed 96-well plate with various number of activated HA-specific T cells (E, effectors) incubated for 4–24 h. A serial dilution of the target cells was also plated out in triplicate in the same plate for standard curve construction. At the end of 4 h incubation, T/R and effector mixture was harvested, washed three times with PBS, transferred to a 96-well flat, clear-bottom, white polystyrene framed microplate, and the fluorescent level of remaining target cells was determined using a Bio-Tek (Winooski, Vermont) FL600 Fluorescence microplate reader. All the viable P815 cells, which were larger in size than effectors, were gated for assaying relative percentage of GFP^+ versus $DsRed^+$ cells via FACScalibur. GFP and DsRed signals were determined with a filter set of 485/20 nm excitation wavelength and 530/20 nm emission wavelength and with a filter set of 550/20 nm excitation wavelength and 620/40 nm emission wavelength, respectively. The remaining number of GFP^+ or $DsRed^+$ targets in each well was calculated based on the standard curve constructed for each cell type, and the antigen specific cytolysis was determined as [(1 – cell number in experimental well)/(1 – cell number in control well)] × 100%.

21.17 GREEN FLUORESCENT PROTEIN-BASED ASSAY[18,19]

To construct GFP reporter plasmid pCIneo expression vector backbone (Promega, Ingelheim, Germany) was used, in which the CMV promoter was replaced by a series of six cres from diVerent cAMP inducible genes followed by a minimal bovine globin promotor. The GFP coding sequence of an enhanced green fluorescent protein (pEGFP; GFP carrying two mutations, a Phe-64 to Leu and a Ser-65 to Thr) gene was PCR amplified, equipped with a Kozak-consensus sequence upstream of the start codon, and cloned into the multiple cloning site of the modified vector (Fig. 21.3). The coding sequences of the G protein–coupled receptors were cloned into eukaryotic expression vectors pcDNA3.1zeo or pEAK-10, equipped with Kozak-consensus sequences, and in some cases vectors were engineered with a Kozak-consensus sequence followed by the signal sequence of the human 5HT3 receptor (the C-terminal 23 peptide) and the coding region of the receptor gene of interest in frame with the signal sequence, and the start codon was omitted. After cloning and transformation of *Escherichia coli*, the plasmid DNA DH5 or TOP 10F was isolated with a commercial purification system (Qiagen, Hilden, Germany) or preparative electrophoresis.

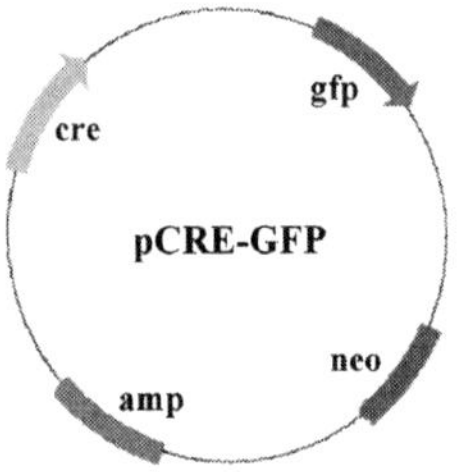

Figure 21.3 pGFP-CRE expression plasmid.

According to the manufacturer's instructions, using the electroporator II (Invitrogen, DeSchelp, The Netherlands), transfection of cos-7 cells was achieved. With calcium/phosphate precipitation, HEK 293 cells were usually transfected. Cells were seeded either in 24-well plate or in 96-well plate. In the presence of 1 mg/mL G418 or other appropriate antibiotics, by culturing for at least 2 weeks, stable cell lines were established. In the morning following transfection medium (DMEM supplemented with 10% FCS, the FCS was dialyzed against PBS using a 5-kDa cutoff membrane) was replaced, and using the indicated concentrations of the substances to be assayed, cells were activated for about 10 min. To achieve this, test compounds or their combinations were added to fresh medium, incubated for about 10 min (longer incubation leaded no higher GFP expression), the cells were washed twice with PBS, new medium was added to the cells, incubated for 12–24 h, the quantification of intrinsic GFP expression of the cells was evaluated with an automated photographic approach, and a direct measurement of the GFP fluorescence using a microplate fluorometer was used for controls.

In photographic assay, cells were plated at a density of less than 1×10^4 cells/well. At the indicated time, using an inverted fluorescence microscope equipped with a GFP filter set (excitation at 488 nm and emission at 505–550 nm) and a $\times 10$ objective (500–1000 cells were within view field), two representative areas per well were analyzed and the photos were taken with a conventional digital camera. Cells transfected with a plasmid containing an eGFP-coding sequence (cloned into the multiple cloning site of the pcDNA3.1zeo plasmid) and nontransfected cells were used at identical cell densities as controls and standards for downstream assay. After photography, pictures were converted to grayscale pictures and using the anlySIS software package of Soft Imaging Systems (Olympus, Hamburg, Germany) analyzed. All pictures were binarized, zero was set as the background (nontransfected cells), and all pixels with >10% intensities of the maximal brightness (eGFP-expressing HEK cells transfected with the pcDNA3.1-egfp) were set as foreground. Automated object recognition was tuned to objects whose area was between 30% and 300% of a typical EGFP-expressing HEK cell. For all downstream applications, only the recognized objects' number was used and defined as maximal number percentage of the cells activated with 10 μM forskolin or the cells maximally stimulated with the natural ligand of the receptor. Using a microplate fluorometer excited at 420 nm and analyzed at 530 nm, GFP expression was quantified. About 5×10^4 to 8×10^4 cells/well were plated, and GFP expression was monitored.

21.18 DNA BREAK ASSAY IN HepG2 CELLS[20]

HepG2 monolayer cell cultures were at 37°C maintained in a humidified 5% CO_2 in MEM supplemented with 5% FBS, 10^{-4} IU/mL penicillin, 0.1 mg/mL streptomycin, and 0.25 μg/mL amphotericin B, at 37°C subcultured weekly by digestion with 0.25% and 1 μmol/mL Na-EDTA for 7 min, passed 8 times through a 20-gauge needle to separate cells, and plates were inoculated with an appropriate number of cells to reach confluency within a given period. Medium was replaced with fresh medium at 2- to 3-day intervals.

In 24-well plates (16-mm well diameter) containing complete media supplemented with 50 nCi/mL [^{3}H]thymidine, HepG2 cells were grown, the medium was replaced with fresh MEM, cells were treated with various concentrations of test compound delivered in acetone for 24 h, 1% final concentration of solvent alone was used as negative control, medium was replaced with 0.5 mL of ice-cold PBS containing 20 μmol/mL Na-EDTA, culture medium was immediately alkalinized with 0.5 mL of NaOH (0.1 N) in darkness for 30 min then neutralized with 0.5 mL of HCl (0.1 N), culture medium was immediately treated with 0.25 mL of sodium lauryl sarcosinate (2%, v/v) and 20 μmol/mL EDTA in water, plates were sonicated for 10 s, and cell lysates were stored at 4°C prior to hydroxylapatite chromatography.

Using a circulating heated water bath fitted with 24 disposable columns (11 × 1 cm) and thermostatted to 60°C, hydroxylapatite chromatography was performed. Hydroxylapatite gel was prepared as a slurry containing 2.5 g of gel, suspended in 25 mL of KPO_4 buffer (12 μmol/mL, pH 7.0), refluxed, and allowed to cool. To each sample well containing cell lysate, 1 mL of gel slurry was added, at 60°C incubated for 10 min, poured into individual columns, columns were washed with 2 × 5 mL of KPO_4 (12 μmol/mL), which was previously equilibrated at 60°C. Single- and double-stranded DNA were eluted with 3 mL of KPO_4 (100 μmol/mL, pH 7.0) and 3 mL of KPO_4 (500 μmol/mL, pH 7.0), respectively. For DNA quantification, to 1 mL of each fraction, 0.1 mL of concentrated HCl was added, samples were at 80°C digested for 1 h, combined with 10 mL of scintillation cocktail (0.5% w/v PPO, 0.015% w/v POPOP in 2/1 toluene/Triton X-100), analyzed with liquid scintillation counting, and data were expressed as F-values corresponding with radioactivity (disintegrations/min).

REFERENCES AND NOTES

1. M. Kelleya, C. Cooper, A. Matticoli, A. Greway. The detection of anti-erythropoietin antibodies in human serum and plasma Part II. Validation of a semi-quantitative 3H-thymidine uptake assay for neutralizing antibodies. J Immunol Methods 300 (**2005**) 179–191.
2. B. Ng, S.W. Polyak, D. Bird, L. Bailey, J.C. Wallace, G.W. Booker. Escherichia coli biotin protein ligase: characterization and development of a high-throughput assay. Anal Biochem 376 (**2008**) 131–136.
3. K. Guven, S. Togrul, F. Uyar, S. Ozant, D.I. De Pomerai. A comparative study of bioassays based on enzyme biosynthesis in Escherichia coli and Bacillus subtilis exposed to heavy metals and organic pesticides. Enzyme Microbial Technol 32 (**2003**) 658–664.

4. K.H. Engler, R. Coker, I.H. Evans. A novel colorimetric yeast bioassay for detecting trichothecene mycotoxins. J Microbiol Methods 35 (**1999**) 207–218. *Note*: By adding 5 μL of SDS (0.1%, w/v) and 3 μL of chloroform to each well of the microtiter plate, cells were first permeabilized. Either 1 μL of X-gal in DMF (100 mg/mL, *in vivo* experiments and experiments examining the effects of different carbon sources), 5 μL of X-gal in DMF (20 mg/mL, effects of altering glucose concentration and inoculum cell density), or 8 μL of X-gal in aqueous DMF (20 mg/mL, methanol and ethanol toxicity experiments and the standardized assay) were added, the contents of the wells were mixed, the plates were at 35°C in a plate-shaker incubated for a maximum of 30 min and read on the Titertek plate-reader using a test filter at 666 nm and a reference filter at 560 nm. Activity of β-galactosidase was defined as product formation ($A_{666} - A_{560}$) as a function of cell density (A_{560}) if required.

5. W.G. Gutheil, M.E. Stefanova, R.A. Nicholas. Fluorescent coupled enzyme assays for D-alanine: application to penicillin-binding protein and vancomycin activity assays. Anal Biochem 287 (**2000**) 196–202.

6. P.B. Ehmer, J. Jose, R.W. Hartmann. Development of a simple and rapid assay for the evaluation of inhibitors of human 17a-hydroxylase-C17,20-lyase (P450cl7) by coexpression of P450cl7 with NADPH-cytochrome-P450-reductase in Escherichia coli. J Steroid Biochem Mol Biol 75 (**2000**) 57–63.

7. B. Soldo, C. Scotti, D. Karamata, V. Lazarevic. The *Bacillus subtilis* Gne (GneA, GalE) protein can catalyse UDP-glucose as well as UDP-N-acetylglucosamine 4-epimerisation. Gene 319 (**2003**) 65–69. *Note*: (1) At 37°C, *Escherichia coli* TOP10 and *Bacillus subtilis* 168 trpC2 were grown in Luria-Bertani (LB) medium or on LB agar plates, unless otherwise specified. For *Escherichia coli* transformants, 50 μg/mL ampicillin was routinely used. Using oligonucleotides VL444 (TTAAATTAGTTGATGCTGATGGTA)/BS448 (TGCGTGATTTGCATACAGCCGACA) from strains LAI123 (gneA3) and LAI121 (gneA1), the 0.8-kb DNA segment containing the distal moiety of galE (starting with residue 476) and the 5′ portion of the downstream gene, yxkC was amplified by PCR on genomic DNA. The PCR products were cloned into pBAD-TOPO to yield plasmids p649 (gneA3) and p652 (gneA1b). With oligonucleotides VL442 (CGATCCAAAGGTGATT-CACCAGTA)/VL-443 (CGGAACGCCGTATACTGTCGCAGA), the 5′ segment of galE (ending at position 393) and its upstream region were amplified by PCR on strain LAI121 (gneA1) DNA. This 0.7-kb segment ligation to pBAD-TOPO gave rise to plasmid p651 (gneA1a). (2) By adding phage U11 at a multiplicity of infection of 0.1 to a culture of strain CU1050 growing at 37°C and 0.3 OD_{600} in LB medium supplemented with 0.1% glucose, 5 mM $CaCl_2$, 5 mM $MgCl_2$, and 0.1 mM $MnCl_2$, phage stock was prepared. One hour after beginning lysis, cell debris was pelleted (10,000 × *g* for 10 min), the supernatant was filtered through an 0.45-μm membrane filter and stored at 4°C. Phage stock was spotted onto fresh streaks of *Bacillus subtilis* strains on LA plates, which were incubated at 30°C for 16 h. Sensitivity to the phage was revealed by growth inhibition and lysis, whereas resistance was indicated by the absence of a clearing zone. (3) With the QIAGEN Genomic-tip System, *Bacillus subtilis* genomic DNA was isolated. Using the alkaline lysis method, plasmid DNA was prepared. By primer-walking, DNA sequences were determined. By Microsynth, sequencing was carried out. From PCR products generated with oligonucleotides BS433 (TAGATGCTGAGCTGAATCTTCAGC) and BS440 (ATTC-AGTAATGATGTTATTGTGAA), the sequence of the gneA region of different strains was determined. With the Taq PCR Master Mix Kit, using 1 ng of *Bacillus subtilis* genomic DNA and 20 pmol of each primer, PCRs were carried out. Amplified PCR involved 30

cycles of denaturation at 95°C for 30 s, annealing at 48°C for 30 s, and extension at 72°C for 1 min/kb. (4) To construct $(His)_6$-GneA fusion protein, using oligonucleotides BS453 and BS454, gneA was amplified by PCR. By positioning gneA in frame with the C-terminal $(His)_6$ fusion tag of the vector, ligation of the amplification product into pBADTOPO yielded plasmid pBS653. At 37°C, *Escherichia coli* TOP10 containing plasmid pBS653 inoculated in LB medium supplemented with 50 μg/mL ampicillin was incubated with continuous shaking at 200 rpm. At 0.5 OD_{600}, protein expression was induced with 0.02% arabinose and incubated for 5 h. Cells of a 25-mL culture were at 4°C and 3000 × *g* centrifuged for 10 min, pellets were stored at −80°C, when required, thawed for 15 min at room temperature, resuspended in 720 μL of lysis buffer (50 mM NaH_2PO_4, pH 8, 300 mM NaCl, 10 mM imidazole) containing 1 mg/mL lysozyme, and sonicated with six 10-s bursts separated by 30 s of cooling in ice water. The lysate was cleared by centrifugation (10,000 × *g*, 20 min, 4°C), applied onto an Ni-NTA Spin Column, preequilibrated in lysis buffer, the column was washed (700 × *g*, 2 min, 4°C) three times with 600 μL of wash buffer (50 mM NaH_2PO_4, pH 8, 300 mM NaCl, 20 mM imidazole), the protein was eluted (700 × *g*, 2 min, 4°C) with 150 μL of elution buffer (50 mM NaH_2PO_4 pH 8, 300 mM NaCl, 250 mM imidazole), and the eluate was aliquoted and kept at −80°C.

8. A. Yazgan, G. Özcengiz, E. Özcengiz, K. Kılınc, M.A. Marahiel, N.G. Alaeddinoğlu. Bacilysin biosynthesis by a partially-purified enzyme fraction from Bacillus subtilis. Enzyme Microb Technol 29 (**2001**) 400–406.
9. C. Baumstark-Khan, A. Rode, P. Rettberg, G. Horneck. Application of the Lux-Fluoro test as bioassay for combined genotoxicity and cytotoxicitymeasurements by means of recombinant Salmonella typhimurium TA1535 cells. Anal Chim Acta 437 (**2001**) 23–30.
10. S. Gerbal-Chaloin, L. Pichard-Garcia, J.M. Fabre, A. Sa-Cunha, L. Poellinger, P. Maurel, M. Daujat-Chavanieu. Role of CYP3A4 in the regulation of the aryl hydrocarbon receptor by omeprazole sulphide. Cell Signalling 18 (**2006**) 740–750.
11. M. Yueh, M. Kawahara, J. Raucy. Cell-based high-throughput bioassays to assess induction and inhibition of CYP1A enzymes. Toxicol In Vitro 19 (**2005**) 275–287.
12. J. Vrba1, P. Kosina, J. Ulrichová, M. Modrianský. Involvement of cytochrome P450 1A in sanguinarine detoxication. Toxicol Lett 151 (**2004**) 375–387.
13. R.A. Horton, E.A. Strachan, K.W. Vogel, S.M. Riddle. A substrate for deubiquitinating enzymes based on time-resolved fluorescence resonance energy transfer between terbium and yellow fluorescent protein. Anal Biochem 360 (**2007**) 138–143.
14. A. Norman, L.H. Hansen, S.J. Sørensen. A flow cytometry-optimized assay using an SOS-green fluorescent protein (SOS-GFP) whole-cell biosensor for the detection of genotoxins in complex environments. Mutat Res 603 (**2006**) 164–172.
15. S. Biacchesi, M.H. Skiadopoulos, L. Yang, B.R. Murphy, P.L. Collins, U.J. Buchholz. Rapid human metapneumovirus microneutralization assay based on green fluorescent protein expression. J Virol Methods 128 (**2005**) 192–197.
16. W. Kruger, D. Gilbert, R. Hawthorne, D.H. Hryciw, S. Frings, P. Poronnik, J.W. Lynch. A yellow fluorescent protein-based assay for high-throughput screening of glycine and GABAA receptor chloride channels. Neurosci Lett 380 (**2005**) 340–345.
17. K. Chen, L. Chen, P. Zhao, L. Marrero, E. Keoshkerian, A. Ramsay, Y. Cui. FL-CTL assay: Fluorolysometric determination of cell-mediated cytotoxicity using green fluorescent protein and red fluorescent protein expressing target cells. J Immunol Methods 300 (**2005**) 100–114.

18. T. Roeder, Derk Göich, D. Heyden, M. Gewecke. A green-xuorescent- protein-based assay for the characterization of G-protein-coupled receptors. Anal Biochem 332 (**2004**) 38–45.

19. Z. Wang, C. Mo, G. Kemble, G. Duke. Development of an efficient fluorescence-based microneutralization assay using recombinant human cytomegalovirus strains expressing green fluorescent protein. J Virol Methods 120 (**2004**) 207–215. *Note*: MRC-5 cells (ATCC) and low-passage (<15 passages) human embryonic lung fibroblasts (HEL) were cultured in DMEM supplemented with 10% FBS. Human cytomegalovirus (HCMV) strains TownevarRIT and Toledo were obtained from E. Mocarski. HCMV TownevarRIT was derived originally from a vaccine preparation of Towne. HCMV Toledo was provided to E. Mocarski by Stuart Starr after eight passages in human fibroblast cells. Prior to use, human serum samples, goat and sheep antisera samples were heated at 56°C for 30 min to inactivate any complement activity and tested for CMV reactivity by ELISA. (2) By overlapping cosmids/plasmids, the technique used for rescue of recombinant HCMV was performed. A DNA fragment containing $B_{2.7}$ (TRL4) sequences (NcoI to BamHI, 2325 bp) including the transcription start site was subcloned from a Toledo cosmid encompassing the TRL region of the Toledo genome into pCITE2A (between NcoI and BglII sites) for generating pCITE TRL4 to construct Toledo $B_{2.7}$ EGFP. To generate pRC2.7 EGFP, the TRL4 sequences in this construct between EagI and SmaI sites were replaced with EGFP gene from pEGFP-N2 (EagI to blunt-end repaired NotI). EGFP gene was inserted in pRC2.7 EGFP and located between the transcription start site and prior to the first open reading frame in Toledo TRL4 RNA transcript. (3) To construct Toledo $B_{2.7}$ EGFP recombinant virus, eight Toledo cosmids were digested with PacI, and to separate the viral insert from the vector DNA, pRC2.7 EGFP was digested with PpuMI and XbaI. With the calcium phosphate method, these DNAs were combined and used to transfect HEL cells. To generate Toledo $B_{2.7}$ EGFP, infectious EGFP virus was plaque purified three times. To facilitate Towne $B_{2.7}$ EGFP construction by overlapping cosmid/plasmid DNAs, a plasmid was constructed in yeast by inserting a unique PmeI site into Towne TRL4 DNA. From the cosmid library, the TRL4 region of Towne was removed and cloned into a vector composed of pACYC184 and yeast ARS CEN HIS sequences to generate pTn50y. A kanamycin gene with flanking PmeI sites and TRL4 sequences to direct homologous recombination in yeast was with pTn50y co-transformed into *S. cerevisiae* CGY2570, and the transformants were selected for kanamycin resistance. To confirm the insertion of the kanamycin gene with flanking PmeI sites by homologous recombination into the TRL4 gene, restriction enzyme digestion and sequence analysis was used. As one transformant with the correct structure, pTn50ykan was digested with *Pac*I to release the Towne insert and with PmeI to release the kanamycin insert within the TRL4 gene. Combination of *Pac*I and *Pme*I digested pTn50ykan with seven Towne cosmids, each digested with PacI, released their Towne inserts, while PpuMI XbaI digested pRC2.7 EGFP resulted in an overlapping set of clones representing the Towne genome with an EGFP insertion in TRL4. By the calcium phosphate method, this mixture was used to transfect HEL cells, and resulting EGFP positive virus was plaque purified seven times to isolate Towne B2.7 EGFP, which contained an EGFP insertion in both TRL4 and IRL4 genes. After plaque purification, the genome structure of both Toledo B2.7 EGFP and Towne B2.7 EGFP was confirmed by Southern blot analysis of restriction enzyme digestions of viral DNA. (4) In 24-well dishes, MRC-5 cell monolayers were inoculated in triplicate at a multiplicity of infection (moi) 0.3 with each virus, which was allowed at 37°C to attach for 1 h, monolayers were washed once, overlayed with 1.5 mL of DMEM containing 10% heat-inactivated FBS, and incubated at 37°C. Supernatants were collected from each

well and replaced with fresh media at 24-h intervals. Prior to DNA extraction and use in quantitative PCR reactions, collected supernatants were stored at −80°C. Using real-time quantitative PCR, virus replication was measured for the accumulation of HCMV genomes in the cell supernatant. With a MasterPure DNA Purification Kit, from cell supernatants viral DNAs for PCR reactions were extracted. UL54 gene (viral DNA polymerase) was selected for real-time PCR amplification due to its high degree of sequence conservation in HCMV. Using Primer Express 1.5 Software, forward primer (50 nM) 5′CCGAGG-TGGGTTACTACAACG3′, reverse primer (300 nM) 5′GGAAGGGTAGAGGCTGGC-A3′, and fluorogenic probe (75 nM) 5′FAM-CCCCGTGGCCGTGTTCGACT-TAMRA3′ were designed. Using TaqMan Universal PCR Master Mix, reactions (50 μL) were carried out at 50°C for 2 min, at 95°C for 10 min, at 95°C for 40 cycles of 15 s, and at 60°C for 1 min. An ABI PRISM 7700 sequence detection system was used for performing reactions, data acquisition, and data analysis. Using known quantities of purified HCMV genomic DNA in real-time quantitative PCR reactions, standard curves were constructed and used to determine viral genome copy number for unpurified samples. (5) Twofold serial dilutions of each serum sample were prepared, mixed with equal volumes of virus (100–200 pfu), incubated at 37°C for 1 h, plated onto MRC-5 cell monolayers in 6-well dishes, plates were incubated at 37°C for 1 h, the inoculum was removed, and cell monolayers were washed with media followed by a media overlay of DMEM containing 10% heat-inactivated FBS. After incubating the plates for 7 days at 37°C, with methanol the cells were fixed and by Giemsa staining plaques were visualized. Similarly, samples consisting of virus alone and virus plus HCMV negative serum were treated and assayed as a control. (6) Serum samples were diluted twofold in wells of 96-well plates containing DMEM/10% heat-inactivated FBS in triplicate. To each well, an equal volume of virus (Toledo B2.7 EGFP or Towne B2.7 EGFP, about 8500 pfu) was added, plates were incubated at 37°C for 1 h, the antibody-virus mixtures were transferred onto MRC-5 cell monolayers in 96-well plates, at 37°C incubated for 1 h, the antibody-virus mixtures were replaced with DMEM/10% heat-inactivated FBS, and at 37°C incubated. On day 4, 5, and 6 after inoculation with the antibody-virus mixtures, the fluorescence in each well was measured. Each assay included control wells of mock infected, virus alone, and virus plus HCMV negative serum. For microscopic microneutralization assay, plates were examined on day 2 or 3 after inoculation with the antibody-virus mixtures, and the wells containing distinct fluorescent foci in a countable range were selected for assay, of which the pictures were captured automatically with ORCA-100 digital camera using Compix Automated Image Capture and Dynamic Intensity Analysis Software. Each image size was 3.82 mm × 3.06 mm, which corresponded with 41.4% whole well. With Compix Customized Image Processing and Analysis Software, captured image data were integrated and analyzed.

20. B. Hasspiler, F.N. Ali, M. Alipour, G.D. Haffner, K. Adeli. Human bioassays to assess environmental genotoxicity: development of a DNA break bioassay in HepG2 cells. Clin Biochem 28 (**1995**) 113–116.

INDEX

Pharmaceutical Bioassays: Methods and Applications. By Shiqi Peng and Ming Zhao